国家卫生健康委员会"十三五"规划教材配套教材

全 国 高 等 学 校 配 套 教 材

供基础、临床、预防、口腔医学类专业用

Basic Chemistry for Higher Medical Education

基础化学

第3版

主　编	傅迎　王兴坡
副主编	钮因尧　刘娜　胡新
编　委	（以姓氏笔画为序）

于　昆	大连医科大学	陈志琼	重庆医科大学
马汝海	中国医科大学	武世奎	内蒙古医科大学
王兴坡	山东大学	林　毅	武汉大学
尹计秋	大连医科大学	胡　新	北京大学
申小爱	中国医科大学	钮因尧	上海交通大学
刘　娜	大连医科大学	章小丽	昆明医科大学
刘国杰	中国医科大学	喻　芳	昆明医科大学
乔　洁	山西医科大学	傅　迎	大连医科大学
李雪华	广西医科大学	赖泽锋	广西医科大学
杨　静	南京医科大学		

人民卫生出版社

图书在版编目（CIP）数据

基础化学 = Basic Chemistry for Higher Medical Education：英文 / 傅迎，王兴坡主编. —3 版. —北京：人民卫生出版社，2019

全国高等学校五年制本科临床医学专业第九轮规划教材配套教材

ISBN 978-7-117-28803-3

Ⅰ.①基⋯　Ⅱ.①傅⋯②王⋯　Ⅲ.①化学－医学院校－教材－英文　Ⅳ.①O6

中国版本图书馆 CIP 数据核字（2019）第 171291 号

人卫智网　www.ipmph.com　医学教育、学术、考试、健康，
　　　　　　　　　　　　购书智慧智能综合服务平台
人卫官网　www.pmph.com　人卫官方资讯发布平台

版权所有，侵权必究！

Basic Chemistry for Higher Medical Education
基础化学
第 3 版

主　　编：傅　迎　王兴坡
出版发行：人民卫生出版社（中继线 010-59780011）
地　　址：北京市朝阳区潘家园南里 19 号
邮　　编：100021
E - mail：pmph @ pmph.com
购书热线：010-59787592　010-59787584　010-65264830
印　　刷：北京市艺辉印刷有限公司
经　　销：新华书店
开　　本：787×1092　1/16　印张：29　插页：1
字　　数：761 千字
版　　次：2014 年 10 月第 1 版　2019 年 12 月第 3 版
　　　　　2019 年 12 月第 3 版第 1 次印刷（总第 3 次印刷）
标准书号：ISBN 978-7-117-28803-3
定　　价：65.00 元

打击盗版举报电话：010-59787491　　E-mail：WQ @ pmph.com
质量问题联系电话：010-59787234　　E-mail：zhiliang @ pmph.com

PREFACE

The 3rd edition English textbook of *Basic Chemistry for Higher Medical Education* corresponds to the 9th edition of Chinese edition textbook *Basic Chemistry* published by the People's Medical Publishing House in August 2018.This textbook can be used for chemistry course in higher education.

The first and the second editions of this textbook have been used widely by freshmen in the past nine years.

Some changes have been made in the 3rd edition. We carefully chose the more appropriate contents, broke down the long chapters into the short texts, adjusted the order of some chapters, added some new examples and exercises, put the key terms and some answers of exercises at the end of each chapter, so as to make it be read more smoothly, better understood and easier learnt.

As we all know, for the students in most majors, Chemistry is always an important course in the higher education. The students will study basic chemistry lesson in the first semester generally. This textbook introduces the basic chemical concepts, the chemical knowledge, the chemical principles and their practical applications, even some living examples related to the medicine. The main contents of basic chemistry include the basic properties of various solutions, the essential and important chemical theories or principles as well as their applications, the changing regularity of elements and normal chemical reactions, the structure of atoms and molecules, the relationship between molecular structure and its properties, qualitative and quantitative analysis to substances, etc.

We still set "Question and thinking" part in each chapter to help students to understand the text better, and still keep the detail solutions of the exercises after each chapter, put them at the end of this textbook to make the students study the main contents conveniently.

Learning chemistry well will be conducive to the subsequent courses, such as organic chemistry, biochemistry, physiology, etc. We expect that the students can master the basic chemical theory and experimental skills by experimental training, and we also hope that chemistry curriculum can improve students' independent thinking ability, provide more ideas and methods, and inspire innovative spirit in their career.

Finally, we would like to sincerely thank professor Wei Zuqi, the editor-in-chief of the first and the second editions, for his contribution to this textbook. We hope that professor Wei Zuqi will continue to pay attention to this book and make valuable comments for us.

Editors-in-Chief
Fu Ying(傅迎)　Wang Xingpo(王兴坡)
May 2019

CONTENTS

Chapter 1　Colligative Properties of Dilute Solution ································ 1
 1.1　The Composition Scales of Mixture ··· 1
  1.1.1　Mass Fraction, Volume Fraction and Mass Concentration ············· 1
  1.1.2　The Mole Fraction ··· 2
  1.1.3　The Amount-of-Substance Concentration and Molality ················ 3
 1.2　Vapor Pressure Lowering ··· 4
  1.2.1　The Vapor Pressure of Liquid ······································· 4
  1.2.2　Vapor Pressure Lowering-Raoult's Law ······························· 6
 1.3　Boiling Point Elevation and Freezing Point Depression ························· 8
  1.3.1　Boiling Point Elevation and Freezing Point Depression of Solution ······ 8
  1.3.2　Colligative Properties of Electrolyte Dilute Solution ···················· 12
 1.4　Osmotic Pressure of Solution ·· 13
  1.4.1　Osmosis and Osmotic Pressure ····································· 13
  1.4.2　The Relationship between Osmotic Pressure of Solution and Concentration, Temperature of Solution ··· 14
  1.4.3　Application of Osmotic Pressure in Medicine ························ 15
 Summary ·· 20
 Exercises ·· 21
 Supplementary Exercises ·· 21

Chapter 2　The Basis of Chemical Thermodynamics ································ 23
 2.1　Thermodynamic System and State Function ··································· 23
  2.1.1　System and Its Surroundings ······································ 23
  2.1.2　State Functions and Process ······································· 24
  2.1.3　Heat and Work ·· 26
 2.2　Energy Conservation and Heat of Chemical Reaction ························· 28
  2.2.1　Thermodynamic Energy and the First Law of Thermodynamic ········· 29

2.2.2　Enthalpy Change and Isobar Heat ……………………………………………………… 30
  2.2.3　Extent of Reaction, Equation of Thermo-Chemistry and Standard State ………………… 30
  2.2.4　Hess's Law and Calculation of Reaction Heat ……………………………………………… 32
 2.3　Entropy and Gibbs's Free Energy ……………………………………………………………… 35
  2.3.1　Spontaneous Process and Its Characteristics ………………………………………………… 35
  2.3.2　Entropy of A System …………………………………………………………………………… 36
  2.3.3　Gibbs's Free Energy of System ……………………………………………………………… 38
 Summary ……………………………………………………………………………………………… 42
 Exercises ……………………………………………………………………………………………… 43
 Supplementary Exercises …………………………………………………………………………… 44

Chapter 3　Chemical Equilibrium ………………………………………………………………… 46
 3.1　The Extent of Chemical Reaction and Standard Equilibrium Constant ………………… 46
  3.1.1　The Standard Equilibrium Constant ………………………………………………………… 46
  3.1.2　Predicting Reaction Direction by Standard Equilibrium Constant $K^{\ominus}$ ……………… 49
 3.2　Experimental Equilibrium Constant ………………………………………………………… 50
 3.3　The Shift of Chemical Equilibrium …………………………………………………………… 51
  3.3.1　Effect of A Change in Concentration ………………………………………………………… 51
  3.3.2　Effect of A Change in Pressure ……………………………………………………………… 52
  3.3.3　Effect of A Change in Temperature ………………………………………………………… 53
 Summary ……………………………………………………………………………………………… 55
 Exercises ……………………………………………………………………………………………… 56
 Supplementary Exercises …………………………………………………………………………… 57

Chapter 4　Rate of Chemical Reaction …………………………………………………………… 60
 4.1　Chemical Reaction Rate and Its Expression ………………………………………………… 60
  4.1.1　The Rate of Chemical Reaction ……………………………………………………………… 60
  4.1.2　Average Rate and Instantaneous Rate of A Chemical Reaction …………………………… 61
 4.2　The Effect of Concentration on Reaction Rate ……………………………………………… 63
  4.2.1　Rate Law of Chemical Reaction ……………………………………………………………… 63
  4.2.2　Characteristics with the Simple Reaction Order …………………………………………… 64
 4.3　Brief Introduction for Rate Theory of Chemical Reaction ………………………………… 68
  4.3.1　Collision Theory and Activation Energy …………………………………………………… 68
  4.3.2　Brief Introduction of Transition State Theory ……………………………………………… 71
 4.4　The Effect of Temperature on Chemical Reaction …………………………………………… 72
  4.4.1　The Relationship between Temperature and Rate Constant—Arrhenius Equation …… 72

| 4.4.2 | The Effect of Temperature on Rate of Chemical Reaction | 73 |

4.5 Brief Introduction of Catalysis ⋯ 74
    4.5.1 Catalyst and Catalysis ⋯ 74
    4.5.2 Brief Introduction of Catalysis Theory ⋯ 75
    4.5.3 Biological Catalyst — Enzyme ⋯ 76

4.6 Brief Introduction of Chemical Reaction Mechanism ⋯ 77
    4.6.1 Elementary Reaction ⋯ 77
    4.6.2 Complex Reaction ⋯ 78

Summary ⋯ 79
Exercises ⋯ 80
Supplementary Exercises ⋯ 82

Chapter 5　Electrolyte Solutions ⋯ 84

5.1 Strong Electrolytic Solutions ⋯ 84
    5.1.1 Electrolytes and Degree of Dissociation ⋯ 84
    5.1.2 Debye-Hückel's Ion Interaction Theory ⋯ 85
    5.1.3 Activity and Ionic Strength ⋯ 86

5.2 Acid-Base Theory ⋯ 87
    5.2.1 Brønsted-Lowry Theory of Acid and Base ⋯ 88
    5.2.2 Autoionization of Water ⋯ 90
    5.2.3 The Lewis Acid-Base ⋯ 92

5.3 Dissociation Equilibria of Weak Electrolytes ⋯ 93
    5.3.1 Ionization Constants for Weak Acids and Weak Bases ⋯ 93
    5.3.2 The Relationship between K_a and K_b of A Conjugate Acid-Base Pair ⋯ 94
    5.3.3 Shifting An Acid-Base Equilibrium ⋯ 95

5.4 Calculating pH of Acid-Base Solutions ⋯ 98
    5.4.1 Strong Acid or Strong Base Solutions ⋯ 98
    5.4.2 Monoprotic Weak Acid or Monoacid Weak Base Solutions ⋯ 98
    5.4.3 Polyprotic Acid and Polyacid Base Solutions ⋯ 100
    5.4.4 The pH of Amphoteric Substances Solutions ⋯ 102

Summary ⋯ 104
Exercises ⋯ 105
Supplementary Exercises ⋯ 106

Chapter 6　Buffer Solution ⋯ 108

6.1 Buffer Process and Composition of Buffer Solutions ⋯ 108

  6.1.1 Buffer Process ··· 108
  6.1.2 Composition of Buffer Solutions ·· 109
 6.2 Calculating the pH of Buffer Solutions ·· 110
  6.2.1 Henderson-Hasselbalch Equation ·· 110
  6.2.2 Correction Formula to Calculate the Accurate pH ························ 111
 6.3 Buffer Capacity and Buffer Range ··· 111
  6.3.1 Buffer Capacity ·· 111
  6.3.2 Buffer Range ··· 113
 6.4 Preparation of Buffer Solutions ·· 113
  6.4.1 Main Steps to Prepare A Buffer Solution ·································· 113
  6.4.2 Standard Buffer Solutions ·· 116
 6.5 Main Buffer Systems in Blood ··· 117
 Summary ·· 118
 Exercises ·· 119
 Supplementary Exercises ·· 120

Chapter 7 Equilibria of Slightly Soluble Ionic Compounds ·························· 122
 7.1 The Solubility Product Constant and Solubility Product Rule ················ 122
  7.1.1 Solubility Product Constant ·· 122
  7.1.2 Solubility and Solubility Product Constant ································ 123
  7.1.3 Solubility Product Rule ··· 124
 7.2 Shifting the Precipitation Equilibrium ·· 125
  7.2.1 Influence Factors of Precipitation Dissolution Equilibrium Shift ······· 125
  7.2.2 Formation of Precipitation ·· 126
  7.2.3 Dissolving Precipitation ··· 127
 7.3 Applications of Precipitation Dissolution Equilibrium ························· 129
  7.3.1 Tooth Decay and Fluoridation ··· 129
  7.3.2 Urinary Calculus ··· 130
 Summary ·· 130
 Exercises ·· 131
 Supplementary Exercises ·· 132

Chapter 8 Oxidation-Reduction Reaction and Electrode Potential ··················· 134
 8.1 Primary Cell and Electrode Potential ··· 134
  8.1.1 Oxidation Number ··· 134
  8.1.2 Redox Half-reaction and Primary Cell ···································· 135

    8.1.3   Formation of Electrode Potential ········· 139
    8.1.4   Standard Electrode Potential ········· 140
8.2   Electromotive Force and Gibbs Free Energy ········· 144
    8.2.1   Electric Work and Gibbs Free Energy ········· 144
    8.2.2   Electromotive Force and the Direction of Spontaneous Redox Reaction ········· 144
    8.2.3   Standard Electromotive Force and Equilibrium Constant ········· 145
8.3   Nernst Equation and Some Factors Affecting the Electrode Potential ········· 147
    8.3.1   The Nernst Equation of Electrode Potential ········· 147
    8.3.2   Some Factors Affecting the Electrode Potential ········· 148
8.4   Measurement of the pH of A Solution by Potentiometry ········· 150
    8.4.1   Reference Electrode and Indicator Electrode ········· 150
    8.4.2   Potentiometric Determination of pH ········· 153
8.5   Electrochemistry and Biosensors ········· 153
Summary ········· 156
Exercises ········· 157
Supplementary Exercises ········· 159

Chapter 9   Atomic Structure and Periodic Law ········· 161

9.1   Foundations of Quantum Mechanics and Characteristics of Electronic Motion ········· 161
    9.1.1   Hydrogen Spectrum and Bohr's Model of Hydrogen Atom ········· 161
    9.1.2   Wave-Particle Duality of Electron ········· 164
    9.1.3   Heisenberg's Uncertainty Principle ········· 166
9.2   Quantum Mechanical Explanations of Hydrogen Atom ········· 166
    9.2.1   Wave Function and Physical Meanings of Three Quantum Numbers ········· 167
    9.2.2   Angular Distribution Graphs of Atomic Orbital and Electron Cloud ········· 170
    9.2.3   Radial Distribution Function Graph ········· 173
9.3   Structure of Many-Electron Atom ········· 175
    9.3.1   Energy Levels of Many-Electron Atom ········· 175
    9.3.2   Electron Spin ········· 176
    9.3.3   Electron Configuration of Atom ········· 177
9.4   Periodic Table and Trends in the Properties of Element ········· 179
    9.4.1   Electron Configuration of Atoms and Periodic Table ········· 179
    9.4.2   Periodic Trends in the Properties of Elements ········· 183
9.5   Elements and Health of Human Body ········· 185
    9.5.1   Essential Elements of Human Body and Brief Introduction on Their Biofunctions ······ 185

    9.5.2   Harmful Elements to Human in Environment ································ 187

  Summary ··· 190

  Exercises ·· 191

  Supplementary Exercises ·· 192

Chapter 10 Covalent Bond and Intermolecular Forces ················ 195

  10.1 Valence Bond Theory ··· 195

    10.1.1   Formation of Hydrogen Molecule ··· 196

    10.1.2   Main Points of Valence Bond Theory ·· 197

    10.1.3   Types of Covalent Bonds ·· 198

    10.1.4   Covalent Bond Parameters ·· 199

  10.2 The Polarity of Bond ·· 201

  10.3 Valence Shell Electron Pair Repulsion Model ···································· 201

  10.4 Hybridization Theory of Atomic Orbitals ··· 203

    10.4.1   Main Points of Hybridization Theory ··· 204

    10.4.2   Types of Hybrid Orbitals ·· 204

  10.5 Molecular Orbital Theory ·· 209

    10.5.1   The Central Themes of Molecular Orbital Theory ·························· 209

    10.5.2   Application of Molecular Orbital Theory ······································· 213

  10.6 Delocalized π Bond and Free Radical ··· 216

    10.6.1   Delocalized π Bond ·· 216

    10.6.2   A Brief Introduction of Free Radical ·· 217

  10.7 Intermolecular Forces ·· 218

    10.7.1   Polarity and Molecular Polarizability ·· 218

    10.7.2   Van der Waals' Forces ·· 220

    10.7.3   Hydrogen Bond ··· 222

  Summary ·· 224

  Exercises ··· 225

  Supplementary Exercises ·· 227

Chapter 11 Coordination Compounds ······································· 230

  11.1 Basic Concept of Coordination Compound ······································ 230

    11.1.1   Formation of Coordination Compound ·· 230

    11.1.2   Constitutes of Coordination Compound ······································ 231

    11.1.3   Formulas and Nomenclature of Coordination Compounds ·············· 233

11.2 Chemical Bond Theory of Coordination Compound ······ 234
 11.2.1 Valence Bond Theory of Coordination Compound ······ 234
 11.2.2 Crystal Field Theory ······ 237
11.3 Coordination Reaction Equilibrium ······ 244
 11.3.1 Coordination Equilibrium Constant ······ 244
 11.3.2 Equilibrium Shift of Coordination Reaction ······ 245
11.4 Chelate and Biological Ligands ······ 249
 11.4.1 Chelating Effect ······ 249
 11.4.2 Impact Factors on the Stability of Chelate ······ 251
 11.4.3 Biological Ligands ······ 252
Summary ······ 254
Exercises ······ 256
Supplementary Exercises ······ 258

Chapter 12 Colloids ······ 261

12.1 Colloidal Dispersions ······ 261
 12.1.1 Introduction ······ 261
 12.1.2 Classification and Characteristics of Colloidal Dispersions ······ 262
12.2 Sols ······ 264
 12.2.1 Properties of Sols ······ 264
 12.2.2 Micellar Structures and Stabilization of Sols ······ 266
 12.2.3 Aerosols ······ 270
12.3 Macromolecular Solutions ······ 271
 12.3.1 Introduction ······ 272
 12.3.2 Properties of Protein Solutions ······ 273
 12.3.3 Destabilization of Protein Solutions ······ 275
 12.3.4 Osmotic Pressure and Membrane Equilibrium of Macromolecular Solutions ······ 275
 12.3.5 Gels ······ 277
12.4 Surfactants and Emulsions ······ 279
 12.4.1 Surfactants ······ 279
 12.4.2 Association colloids ······ 280
 12.4.3 Emulsions ······ 281
Summary ······ 283
Exercises ······ 284
Supplementary Exercises ······ 285

Chapter 13　Titrimetric Analysis ... 287

13.1　Error and Deviation ... 287
13.1.1　Causes and Classification of Error ... 287
13.1.2　Evaluation of Analytical Result ... 288
13.1.3　Methods to Improve the Accuracy of Measurement ... 290

13.2　Significant Figures ... 291
13.2.1　Concepts of Significant Figures ... 291
13.2.2　Arithmetic Rules for Significant Figures ... 292

13.3　Principle of Titrimetric Analysis ... 293
13.3.1　Introduction to Titrimetric Analysis ... 293
13.3.2　Classification of Titrimetric Analysis ... 294
13.3.3　Preparation of Standard Solution ... 294

13.4　Acid-Base Titration ... 295
13.4.1　Acid-Base Indicator ... 295
13.4.2　Titration Curves and Choices of Indicator ... 297
13.4.3　Preparation and Standardization of Acid and Base Standard Solutions ... 304
13.4.4　Applications of Acid-Base Titrations ... 305

13.5　Oxidation-Reduction Titrations ... 306
13.5.1　Potassium Permanganate Method ... 306
13.5.2　Iodometric Titration ... 307

13.6　Complexometric Titration ... 309
13.6.1　Basic Principle of Complexometric Titration ... 309
13.6.2　Preparation and Standardization of EDTA Standard Solution ... 310
13.6.3　Applications of Complexometric Titration ... 310

13.7　Precipitation Titration ... 311

Summary ... 313
Exercises ... 313
Supplementary Exercises ... 316

Chapter 14　Ultraviolet-Visible Spectrophotometry ... 318

14.1　Absorption Spectrum of Substance ... 318
14.1.1　Selective Absorption of Light by Substance ... 318
14.1.2　Absorption Spectrum of Substance ... 319

14.2　Fundamental Principle of Spectrophotometry ... 320
14.2.1　Transmittance and Absorbance ... 320
14.2.2　Beer-Lambert Law ... 321

14.3 Improving Sensitivity and Accuracy in Determination ······ 323
   14.3.1 Main Sources in Spectrophotometric Errors ······ 323
   14.3.2 The Way to Improve Sensitivity and Accuracy ······ 325
14.4 Visible Spectrophotometry ······ 328
   14.4.1 Spectrophotometer ······ 328
   14.4.2 Quantitative Analysis Methods ······ 329
   14.4.3 Application in the Clinical Examination ······ 330
14.5 Ultraviolet Spectrophotometry ······ 331
   14.5.1 Main Classifications of Ultraviolet Spectrophotometer ······ 331
   14.5.2 Application of Ultraviolet Spectrophotometry ······ 332
   14.5.3 Analysis of Organic Compound Structures ······ 333
Summary ······ 335
Exercises ······ 335
Supplementary Exercises ······ 336

Chapter 15 Brief Introduction to the Modern Instrumental Analysis ······ 338

15.1 Atomic Absorption Spectrometry ······ 338
   15.1.1 Fundamental Principles of Atomic Absorption Spectrometry ······ 339
   15.1.2 Atomic Absorption Spectrophotometers ······ 339
   15.1.3 Brief Introduction to the Experimental Techniques ······ 340
   15.1.4 Application of Atomic Absorption Spectrometry ······ 343
15.2 Molecular Fluorescence Spectroscopy ······ 343
   15.2.1 Fundamental Principles of Molecular Fluorescence Spectroscopy ······ 344
   15.2.2 Quantitative Analysis of Fluorescence Spectroscopy ······ 346
   15.2.3 Fluorescence Spectrophotometer ······ 348
   15.2.4 Applications of Fluorescence Spectroscopy ······ 348
15.3 Chromatography ······ 349
   15.3.1 Brief Introduction to Chromatography ······ 349
   15.3.2 General Principle of Chromatography for Separation ······ 350
   15.3.3 Chromatographic Instruments ······ 352
   15.3.4 Qualitative and Quantitative Analysis of Chromatography ······ 354
   15.3.5 Application of Chromatography ······ 355
15.4 Mass Spectrometry and GC-MS or LC-MS ······ 356
   15.4.1 Brief Introduction of Mass Spectrometry ······ 356
   15.4.2 Mass Spectrogram ······ 356
   15.4.3 Chromatography-Mass Spectrometry ······ 357

15.5　Inductively Coupled Plasma Atomic Emission Spectrometry ······ 358
Summary ······ 360
Exercises ······ 361
Supplementary Exercises ······ 363

Chapter 16　Nuclear Chemistry and Its Applications ······ 365

16.1　Basic Concept of Nuclear Chemistry ······ 365
　16.1.1　Nuclear Chemistry Evolution and Its Applications ······ 365
　16.1.2　Nucleon, Nuclide and Isomers ······ 366
　16.1.3　Radioelement and Radioactive Series ······ 366
　16.1.4　Mass Defect and Nuclear Binding Energy ······ 367
16.2　Radioactive Decay and Nuclear Equation ······ 367
　16.2.1　Radioactive Decay ······ 367
　16.2.2　Nuclear Equations ······ 368
　16.2.3　Half-Life and Radioactivity ······ 369
　16.2.4　Carbon-14 Dating Method ······ 369
16.3　The Introduce of Radioactive Tracer Technique — PET-CT ······ 370
　16.3.1　Principles of Radioactive Tracer Technique ······ 370
　16.3.2　PET-CT Imaging Process ······ 371
　16.3.3　The Features of PET-CT ······ 371
　16.3.4　Radionuclide for Clinically Therapy and Diagnosis ······ 371
16.4　Nuclear Reactions and Radioactive Emissions ······ 372
　16.4.1　Nuclear Reaction ······ 372
　16.4.2　Nuclear Fission Reaction and Nuclear Fusion Reactions ······ 372
　16.4.3　Nuclear Chain Reaction ······ 372
　16.4.4　Nuclear Radiation and Nuclear Radiation Protection ······ 373
Summary ······ 375
Exercises ······ 376
Supplementary Exercises ······ 376

Appendix ······ 377

Appendix 1　Legal Units of Measurement ······ 377
Appendix 2　Physical and Chemical Constants ······ 379
Appendix 3　Equilibrium Constants ······ 380
Appendix 4　Thermodynamic Data ······ 385

Appendix 5　Standard Potentials in Aqueous Solution at 298.15K ·············· 388

Appendix 6　Greek Letters ·············· 389

Detailed Solutions to Exercises ·············· 390

References ·············· 445

Name, Atomic Number and Relative Atomic Weight of Elements ·············· 446

Chapter 1
Colligative Properties of Dilute Solution

A **solution** is a homogeneous and stable dispersion system being composed of two or more substances. The solution can exist in three states: gaseous, liquid or solid. The solution usually refers to aqueous solution.

The properties of a solution are neither the same as those of solute nor those of solvent. The properties of solution can be divided into two categories: the first category depends on the nature of solute, such as the color, the volume change and the conductivity of the solution; the second is called **colligative property**, which depends on the number of dissolved solute particles (molecules or ions) instead of the involved solute. These properties include mainly: (1) the lowering of the vapor pressure of the solution relative to that of the pure solvent; (2) the boiling point elevation; (3) the freezing point depression; (4) the phenomenon of osmotic pressure. For example, both of glucose ($C_6H_{12}O_6$) and sucrose ($C_{12}H_{22}O_{11}$) solutions have the same vapor pressure depression, boiling point elevation, freezing point depression and osmotic pressure, as long as they have the same molality.

The colligative property applies only to the change in the properties of dilute solutions, otherwise the results discussed will deviate the facts. So the colligative properties are also called colligative properties of dilute solution.

The colligative properties of solutions on the exchange and transport of substances inside and outside cells, clinical infusion, water and electrolyte metabolism, have some theoretical guiding significance. This chapter mainly introduces the colligative properties of the solution that containing nonvolatile, non-electrolyte of solute in the dilute solution.

1.1 The Composition Scales of Mixture

The **mixture** is a system being composed of two or more substances. Changes of the proportion for each component in the mixture system may lead to changes in the properties of the mixture. For a mixture, the **composition scale** should be specified while determining its composition. The composition scale is the relative content of each component in the mixture. We will learn some terms about the composition scale in this section.

1.1.1 Mass Fraction, Volume Fraction and Mass Concentration

1. Mass fraction

ω_B, the **mass fraction** of solute B is defined as the mass of solute B divided by the mass of the mixture. For solutions, the mass fraction is defined as the mass of solute divided by the mass of solution.

$$\omega_B \stackrel{\text{def}}{=\!=} \frac{m_B}{m} \tag{1.1}$$

Where, m_B is the mass of the solute B, m is the mass of the solution.

It is simple and convenient to express the composition scale by using the mass fraction, especially in the manufacture, such as commercial sulfuric acid, hydrochloric acid, nitric acid and ammonia.

2. Volume fraction

The **volume fraction** φ_B is defined as the volume of B divided by the sum of volume of each component in the mixture at the same temperature and the same pressure, i.e.

$$\varphi_B \stackrel{\text{def}}{=\!=} \frac{V_B}{\sum V_i} \tag{1.2}$$

Where, V_B is the volume of solute B, $\sum V_i$ is the sum of volume for each of component i in the mixture. For example, in the arterial blood of body, the volume fraction of O_2 $\varphi_B=0.196$ (or 19.6%) and volume fraction of sterilizing alcohol $\varphi_B=0.75$ (or 75%).

3. Mass concentration

The **mass concentration** ρ_B is defined as the mass of a solute B divided by the volume of the solution.

$$\rho_B \stackrel{\text{def}}{=\!=} \frac{m_B}{V} \tag{1.3}$$

Where, m_B is the mass of a solute B, V is the volume of the solution.

The SI unit of ρ_B is $kg \cdot m^{-3}$. The units $g \cdot L^{-1}$ and $g \cdot mL^{-1}$ are also used usually, such as dilute hydrochloric acid, dilute sulfuric acid, dilute nitric acid are all the $0.10 g \cdot mL^{-1}$ of solution in the Chinese Pharmacopoeia.

1.1.2 The Mole Fraction

The **mole fraction** also is called the **amount of substance fraction** or the ratio of the amount of substance. x_B, the mole fraction of B is defined as the ratio of the amount of substance B and the sum of amount of substance of the all components in the mixture, i.e.

$$x_B \stackrel{\text{def}}{=\!=} \frac{n_B}{\sum n_i} \tag{1.4}$$

Where, n_B is the amount of substance of B, $\sum n_i$ is the sum of the amount of substance for each of components i in the mixture.

Supposes the solution is composed of solute B and solvent A, then the mole fraction of solute B is

$$x_B = \frac{n_B}{n_A + n_B}$$

Likewise, the mole fraction of solvent A is

$$x_A = \frac{n_A}{n_A + n_B}$$

Obviously, $x_A + x_B = 1$. The sum of mole fraction of all components in the mixture(or solution) is equal to one. Mole fraction is applied widely in the researching on the properties of dilute solution because it is independent of the temperature.

1.1.3 The Amount-of-Substance Concentration and Molality

1. The amount-of-substance concentration (molarity)

The **amount-of-substance concentration** or **molarity** is defined as the amount of solute B divided by the volume of mixture, i.e.

$$c_B \stackrel{\text{def}}{=\!=} \frac{n_B}{V} \tag{1.5}$$

In Equation 1.5, c_B is molarity of B, n_B is the amount of substance of B, V is the volume of mixture. For solutions, the molarity is defined as the amount of substance of solute B divided by the solution volume. The molarity may be referred to as "concentration" for short generally.

The SI unit of molarity is $mol \cdot m^{-3}$. Some units are generally used instead, such as $mol \cdot L^{-1}$, $mmol \cdot L^{-1}$ and $\mu mol \cdot L^{-1}$.

When the molarity is used, **elementary entity** must be specified. For example, $c(H_2SO_4) = 1 mol \cdot L^{-1}$, $c(\frac{1}{2}Ca^{2+}) = 4 mmol \cdot L^{-1}$, and so on. The chemical formulas in parenthesis represent the elementary entities.

In medicine field, the World Health Organization proposed that the content of all substances known relative molecular mass in body fluids should be expressed as the molarity. For example, for the normal human body blood glucose content, the expression was 70mg%~100mg% in the past custom, which means that there are 70~100mg of glucose per 100mL of blood, but now it should be expressed as $c(C_6H_{12}O_6) = 3.9 \sim 5.6 mmol \cdot L^{-1}$. For the substance B unknown relative molecular mass, its content can be expressed as mass concentration.

Sample Problem 1-1 100mL of normal human plasma contains Na^+ 326mg, HCO_3^- 164.7mg and Ca^{2+} 10mg, what are their molarity in $mmol \cdot L^{-1}$ respectively?

Solution $c(Na^+) = \dfrac{326mg}{23.0g \cdot mol^{-1}} \times \dfrac{1}{100mL} \times \dfrac{1g}{1000mg} \times \dfrac{1000mL}{1L} \times \dfrac{1000mmol}{1mol} = 142 mmol \cdot L^{-1}$

$c(HCO_3^-) = \dfrac{164.7mg}{61.0g \cdot mol^{-1}} \times \dfrac{1}{100mL} \times \dfrac{1g}{1000mg} \times \dfrac{1000mL}{1L} \times \dfrac{1000mmol}{1mol} = 27.0 mmol \cdot L^{-1}$

$c(Ca^{2+}) = \dfrac{10mg}{40g \cdot mol^{-1}} \times \dfrac{1}{100mL} \times \dfrac{1g}{1000mg} \times \dfrac{1000mL}{1L} \times \dfrac{1000mmol}{1mol} = 2.5 mmol \cdot L^{-1}$

Sample Problem 1-2 The density of commercial concentrated sulfuric acid is $1.84 kg \cdot L^{-1}$, in which the mass fraction of H_2SO_4 is 96%, calculate the molarity $c(H_2SO_4)$ and $c(\frac{1}{2}H_2SO_4)$.

Solution The molar mass of H_2SO_4 is $98 g \cdot mol^{-1}$, that of $\frac{1}{2}H_2SO_4$ is $49 g \cdot mol^{-1}$.

$c(H_2SO_4) = \dfrac{96}{100} \times \dfrac{1}{98g \cdot mol^{-1}} \times 1.84 kg \cdot L^{-1} \times \dfrac{1000g}{1kg} = 18 mol \cdot L^{-1}$

$c(\frac{1}{2}H_2SO_4) = \dfrac{96}{100} \times \dfrac{1}{49g \cdot mol^{-1}} \times 1.84 kg \cdot L^{-1} \times \dfrac{1000g}{1kg} = 36 mol \cdot L^{-1}$

In a known chemical reaction, it is simple to express the relationship of amount of the substance and the mole fraction between the components in a balanced reaction directly.

2. The molality of solute B

b_B, the **molality** of solute B is defined as the amount of B divided by the mass of solvent, i.e.

$$b_B \stackrel{\text{def}}{=\!=} \frac{n_B}{m_A} \qquad (1.6)$$

Where, n_B is the amount of solute B, m_A is the mass of solvent A. The unit of b_B is $mol \cdot kg^{-1}$.

The molality is applied widely in the physical chemistry because it is independent of the temperature.

Sample Problem 1-3 Dissolved 7.00g of crystallized oxalic acid ($H_2C_2O_4 \cdot 2H_2O$) in 93.0g of water, calculate the molality b ($H_2C_2O_4$) and the mole fraction x ($H_2C_2O_4$).

Solution The molar mass of crystallized oxalic acid $MM.(H_2C_2O_4 \cdot 2H_2O) = 126 g \cdot mol^{-1}$, $MM.(H_2C_2O_4) = 90.0 g \cdot mol^{-1}$, then the mass of oxalic acid in 7.00g crystallized oxalic acid is

$$m(H_2C_2O_4) = \frac{7.00g \times 90.0g \cdot mol^{-1}}{126g \cdot mol^{-1}} = 5.00g$$

the mass of water in solution is

$$m(H_2O) = 93.0g + 7.00g - 5.00g = 95.0g$$

then

$$b(H_2C_2O_4) = \frac{5.00g}{90.0g \cdot mol^{-1} \times 95.0g} \times \frac{1\,000g}{1kg} = 0.585 mol \cdot kg^{-1}$$

$$x(H_2C_2O_4) = \frac{\dfrac{5.00g}{90.0g \cdot mol^{-1}}}{(\dfrac{5.00g}{90.0g \cdot mol^{-1}}) + (\dfrac{95.0g}{18.0g \cdot mol^{-1}})} = 0.010\,4$$

1.2 Vapor Pressure Lowering

1.2.1 The Vapor Pressure of Liquid

Consider some liquid in a closed container at the given temperature. Due to the thermal motion of the liquid molecule, a part of the molecules with enough kinetic energy overcoming the liquid's intermolecular forces escapes from the liquid surface and diffuses to form the gas phase molecule. The process of molecules leaving liquid state and going into a gaseous state is called **evaporation**. The evaporation process is reversible. With the increase of the gaseous molecules above the liquid phase, some gas molecules coming into contact with the surface of liquid can be trapped by intermolecular forces in the liquid. The process of vapor changing to liquid is called **condensation**. The evaporation and condensation process of water can be expressed as follows:

$$H_2O(l) \rightleftharpoons H_2O(g)$$

If evaporation takes place in a closed vessel, the evaporation process dominates at the beginning, but with the increase of vapor density, the rate of condensation also increases. When the evaporation rate is equal to the condensation rate, vapor and liquid is just at equilibrium, and the density of the vapor is no longer changed. At that time the vapor is "saturated" and the vapor pressure is called the **saturated vapor pressure of liquid**, or **vapor pressure of liquid**, p, the unit is Pa or kPa.

The vapor pressure of liquid depends on the nature of liquid. Different substance has different vapor pressure at a given temperature. For example, the vapor pressure of water and diethyl ether is 2.34kPa and 57.6kPa, respectively at 20℃.

The vapor pressure of liquid varies with the temperature. As the temperature rises, the equilibrium between the liquid phase and the gas phase will shift to right and the vapor pressure of gases will increase. Table 1-1 shows the relationship between vapor pressure and temperature for water.

Table 1-1 The Vapor Pressure of Water at Different Temperature

$t/℃$	p/kPa	$t/℃$	p/kPa
0	0.611 15	50	12.352
5	0.872 58	60	19.946
10	1.228 2	70	31.201
15	1.705 8	80	47.414
20	2.339 3	90	70.182
25	3.169 9	100	101.42
30	4.247 0	110	143.38
35	5.629 0	120	198.67
40	7.384 9	130	270.28

Figure 1-1 shows the changes of vapor pressure along with the changes of temperature for diethyl ether, ethanol, water and polyglycol.

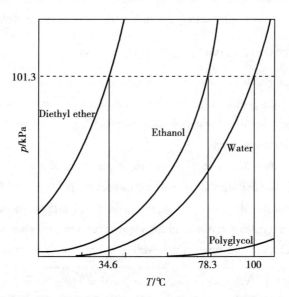

Figure 1-1 The Relationship of Vapor Pressure and Temperature

Just as the molecules in liquid can pass into the gaseous state by evaporation, the molecules in solid can do the same thing. Solids substance can also lose particles from their surface to form vapor, the process in this case is called the **sublimation** rather than evaporation. Sublimation is the direct change from solid to vapor (or vice versa) without going through the liquefying step.

In most cases, at ordinary temperatures, the saturated vapor pressures of solids range from low to extreme low. Attraction forces between molecules in many solids are too strong to allow much loss of particles from the solid surface.

However, there are some solids which do easily form vapors. For example, naphthalene used in old-fashioned "moth balls" to deter clothes moths usually has quite a strong smell, which is because the molecules must sublimate easily otherwise you wouldn't smell it.

Another example is solid carbon dioxide or "dry ice" which never forms liquid at atmospheric pressure and always converts directly from solid to vapor. That's why it is known as dry ice.

The vapor pressure of solid is influenced by temperature also as Table 1-2 shows for the ice.

Table 1-2   The Vapor Pressure of Ice at Different Temperature

$t/°C$	p/kPa	$t/°C$	p/kPa
0	0.611 15	−15	0.165 27
−1	0.562 66	−20	0.103 24
−2	0.517 70	−25	0.063 29
−3	0.476 04	−30	0.038 01
−4	0.437 45	−35	0.022 35
−5	0.401 74	−40	0.012 84
−10	0.259 87		

Substance in solution with high vapor pressure at normal temperatures is often referred to as volatile. Nonvolatile substance in solution has little tendency to escape from the solution. So when we talk about the colligative properties of nonvolatile nonelectrolyte solution, the vapor pressure of solute could be neglected.

1.2.2   Vapor Pressure Lowering-Raoult's Law

Let us first consider a solution with nonvolatile solute dissolved in solvent. An example is the solution of sucrose or cane sugar in water, in which the vapor pressure of sucrose above the solution is almost zero. The vapor pressure of water is not zero and it can be studied as the composition of the solution changes at a given temperature.

We can account for this behavior in terms of the simple model shown in Figure 1-2. The dissolved nonvolatile solute will leads the number of solvent molecules per unit volume to decreases. Thus it lowers the number of solvent molecules at the liquid surface, which proportionately lowers the escaping tendency of the solvent molecules and results in the **vapor pressure lowering** of the solution. Obviously, the more the solute particles are, the greater the lowering in vapor pressure is as shown in Figure 1-3.

> **Question and thinking 1-1**   Why is the vapor pressure of solution lower than that of pure solvent?

In 1887, the French chemist Francois Marie Raoult found that, for nonvolatile non-electrolyte dilute solutions, the relationship between vapor pressure and amount-of-substance fraction of solvent obey the simple equation

$$p = p°x_A \tag{1.7}$$

Where, p is the vapor pressure of solution; $p°$ is the vapor pressure of pure solvent; x_A is amount-of-substance fraction of solvent.

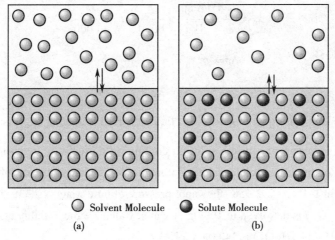

○ Solvent Molecule ● Solute Molecule
(a)   (b)

Figure 1-2   Vapor Pressure of Pure Solvent and Solution
(a) Vapor pressure of pure solvent; (b) Vapor pressure of solution

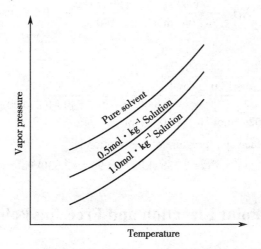

Figure 1-3   Vapor Pressure Curve of Pure Solvent and Solution with Different Concentration

For the dilute solution containing only solute B, $x_A + x_B = 1$. The change of the vapor pressure of the solvent when a nonvolatile solute is added can be expressed as follows

$$\Delta p = p° - p$$
$$= p° - p° x_A = p°(1 - x_A) = p° x_B$$
$$\Delta p = p° x_B \tag{1.8}$$

Which is known as **Raoult's law**, the vapor pressure lowering of a dilute solution that containing nonvolatile, non-electrolyte is directly proportional to the amount-of-substance fraction of the solution and has nothing to do with the nature of solute.

Suppose the mass of solvent is mg, $n_A = \dfrac{m}{M_A}$

For a dilute solution,

$$n_A \gg n_B, \quad \text{so,} \ n_A + n_B \approx n_A$$

$$x_B = \frac{n_B}{n_A + n_B} \approx \frac{n_B}{n_A} = \frac{n_B}{m/M_A} = \frac{n_B}{m} \times M_A$$

$$x_B \approx \frac{n_B}{m} \times \frac{1\,000\text{g}}{1\text{kg}} \times \frac{M_A}{1\,000} = b_B \times \frac{M_A}{1\,000}$$

$$\Delta p = p^\circ x_B = p^\circ \times \frac{M_A}{1\,000} \times b_B$$

$$\Delta p = K b_B \tag{1.9}$$

Where, K is the constant, it is equal to $p^\circ \times \frac{M_A}{1\,000}$. Equation 1.10 is another expression of **Raoult's law**. The vapor pressure lowering of a dilute solution containing nonvolatile, non-electrolyte is directly proportional to the molality of the solute and has nothing to do with the nature of solute.

Sample Problem 1-4 At 293.15K, the vapor pressure of pure water is 2.338 8kPa. Suppose 6.840g of the sucrose ($C_{12}H_{22}O_{11}$) is dissolved in 100.0g of water. Calculate the molality and the vapor pressure of sucrose solution. $MM.(C_{12}H_{22}O_{11})$ is $342.0\text{g} \cdot \text{mol}^{-1}$.

Solution

$$b(C_{12}H_{22}O_{11}) = \frac{6.840\text{g}}{342.0\text{g} \cdot \text{mol}^{-1}} \times \frac{1\,000\text{g} \cdot \text{kg}^{-1}}{100.0\text{g}} = 0.200\,0\text{mol} \cdot \text{kg}^{-1}$$

So, $x(H_2O)$ is

$$x(H_2O) = \frac{\frac{100.0\text{g}}{18.02\text{g} \cdot \text{mol}^{-1}}}{\frac{100.0\text{g}}{18.02\text{g} \cdot \text{mol}^{-1}} + \frac{6.840\text{g}}{342.0\text{g} \cdot \text{mol}^{-1}}} = \frac{5.549\text{mol}}{(5.549 + 0.020\,00)\text{mol}} = 0.996\,4$$

According to Raoult's law, the vapor pressure of the solution is

$$p = p^\circ x_A = 2.338\,8\text{kPa} \times 0.996\,4 = 2.330\text{kPa}$$

1.3 Boiling Point Elevation and Freezing Point Depression

1.3.1 Boiling Point Elevation and Freezing Point Depression of Solution

1. The boiling point and freezing point of pure liquid

A specific definition of **boiling point** is the temperature at which the vapor pressure of a liquid is equal to the external pressure. The **normal boiling point** T_b° is the temperature at which the liquid boils when the external pressure is 1atm. (1atm = 101.3kPa, or standard atmospheric pressure). For example, the normal boiling point of pure water is 373.15K, $T_b^\circ = 100\,°C$.

The boiling point of a liquid is related to the external pressure. The greater the external pressure is, the higher the boiling point is, and vice versa. Therefore, in the actual work, when the hot and unstable substances are extracted or refined, the method of decompression distillation or decompression concentration is often used to reduce the evaporation temperature and prevent the destruction of these substances by high temperature heating. While the heat stable injection and some medical instruments are sterilized, the heat pressure sterilization is often used, which is heated in a closed high pressure sterilizer. The sterilization time is shortened and the sterilization effect is improved by increasing the temperature of water vapor.

Question and thinking 1-2 Why does a small amount of zeolite need to be added to the distillation bottle during evaporation or distillation experiments?

In physics, there is a **superheating** phenomenon in which a liquid is heated to the higher temperature than its boiling point, but without boiling. Superheating is achieved by heating a homogeneous substance in a clean container, free of nucleation sites, while taking care not to disturb the liquid. Superheating can easily cause bumping and danger. Therefore, in laboratory evaporation or distillation, a small amount of zeolite should be added to the distillation bottle.

Freezing point refers to the temperature at which the vapor pressure of a liquid is equal to that of solid under certain external pressure, when solid and liquid phases coexist in equilibrium. The **normal freezing point** T_f° is the temperature at which a substance freezes at 101.3kPa of pressure. For example, the normal freezing point of pure water is 273.15K at 101.3kPa of external pressure; this is also known as the ice point.

The freezing point is related to the external pressure. The freezing point of some liquids increases with the increase of external pressure while some decrease. The values of T_b°, K_b and T_f°, K_f for common pure solvents are listed in Table 1-3.

2. Boiling point elevation and freezing point depression of solution

The normal boiling point of pure liquid or solution is defined as the temperature at which its vapor pressure reaches 101.3kPa. Because the vapor pressure of solution is lower than that of the pure solvent at the same temperatures. As a result, in order for the solution to boil a higher temperature is required than for the pure solvent as shown in Figure 1-4; that is, the boiling point of solution with nonvolatile solute in solvent is higher than that of the pure solvent. This phenomenon is referred to as **boiling point elevation of solution**.

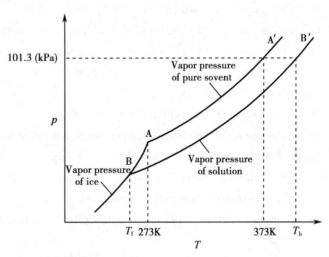

Figure 1-4 Boiling Point Elevation and Freezing Point Depression of Solution

Table 1-3 Boiling-Point Elevation and Freezing-Point Depression Constants of Some Solvents

Solvent	$T_b^\circ/°C$	$K_b/(°C \cdot kg \cdot mol^{-1})$	$T_f^\circ/°C$	$K_f/(°C \cdot kg \cdot mol^{-1})$
Water	100	0.512	0.0	1.86
Acetic acid	118	2.93	17.0	3.90
Benzene	80	2.53	5.5	5.10
Ethanol	78.4	1.22	−117.3	1.99
Carbon tetrachloride	76.7	5.03	−22.9	32.0
Diethyl ether	34.7	2.02	−116.2	1.8
Naphthalene	218	5.80	80.0	6.9

For sufficiently dilute solutions, the elevation of the boiling point ΔT_b is proportional to the concentration of the solution. Because temperature changes are involved (and, as mentioned before, amount-of-substance concentration varies with temperature) it is conventional in discussions of this colligative property to use molality b_B (amount-of-substance of solute per kilogram of solvent) to express the composition of the solution. It is

$$\Delta T_b = K_b b_B \tag{1.10}$$

in Equation 1.10, K_b is the boiling point elevation constant. This equation is quite useful because K_b is a property of the solvent only and does not change from one solute to another if the solution always takes the same solvent. As a result, K_b can be obtained by measuring the boiling point elevation for a known its molality of dilute solutions and tabulated for later use.

Addition of solute causes the boiling point of a solution to increase, but it causes the freezing point to decrease. All we know, water freezes at exactly 0 ℃. At the freezing or melting point, the solid ice and the liquid water have the same vapor pressure. The tendency of the ice to melt is exactly counteracted by the tendency of water to freeze. At the freezing point, ice and water coexist.

What happens when you add a nonvolatile solute to above mixture of ice and water? In many cases, a solute does only dissolve in the liquid solvent and not in the solid solvent. This means that when such a solute is added, the vapor pressure of the solvent in the liquid phase is decreased by dilution, but the vapor pressure of the solvent in the solid phase is not affected. So the equilibrium between the solid and liquid phase is upset. The rate of evaporation for ice is unchanged by the presence of the foreign material, but the rate of evaporation for liquid is changed. So you will find that the ice melting occurs.

To re-establish equilibrium, you must cool the mixture to reach the new balance when the ice and the solution have the same vapor pressure again but below the usual melting point of water as Figure 1-4 shows. For example, the freezing point of a 0.043 mol·kg^{-1} urea solution is roughly -0.079 ℃. It is lower than 0 ℃ obviously. This colligative property is called **freezing point depression of solution**.

The higher the concentration of solute is, the greater the freezing point depresses. For sufficiently small concentrations of solute, the freezing-point depression ΔT_f is related to the molality of solution

$$\Delta T_f = K_f b_B \tag{1.11}$$

Where, K_f is freezing point depression constant that depends only on the properties of the pure solvent as shown in Table 1-3.

Solution always has lower freezing temperature than the pure solvent in Figure 1-4. The phenomenon of freezing-point depression accounts for the fact that seawater, which contains dissolved salts, has a lower freezing point than fresh water. Concentrated salt solutions have still lower freezing points. Spreading salt on an icy road creates a solution with a lower freezing point and causes the ice to melt. Antifreeze added to a car radiator lowers the freezing point of the coolant and keeps the cooling system away from freezing and perhaps cracking the engine block in winter. Figure 1-5 shows the change of temperature in the cooling process of water and solution.

Figure 1-5 (1) shows the ideal cooling curves of water. In the part of the curve where the temperature decreases, the kinetic energy of water also decreases while the potential energy of it keeps the same case as the part a-b in Figure 1-5(1). However, at the phase transition point, where the curve is flat, the temperature keeps constant as the part b-c in Figure 1-5(1). This is because the water has more internal energy than ice. The temperature is just the freezing point of water when water and ice coexist.

However, the experiment can only give the curve (2) instead of curve (1) in Figure 1-5 because of **supercooling** which is the process of lowering the temperature of a liquid below its freezing point without it becoming a solid.

A liquid below its freezing point will crystallize in the presence of a seed crystal or nucleus around which a crystal structure can form creating a solid. However, lacking any such nuclei, the liquid phase can be maintained all the way down to the temperature at which crystal homogeneous nucleation occurs as the part b'-b in Figure 1-5(2).

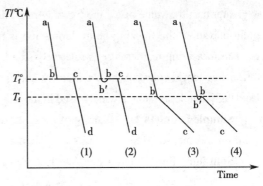

Figure 1-5 The Cooling Curve of Water and Solution
(1) The ideal cooling curve of water; (2) The experimental cooling curve of water; (3) The ideal cooling curve of solution; (4) The experimental cooling curve of solution

The curve (3) in Figure 1-5 is the ideal cooling curve of the solution. Unlike the curve (1), the solution begins to freeze when the temperature drops to T_f, $T_f < T_f^\circ$. With the crystallization of ice, the concentration of solution increases, and the freezing point of the solution decreases. At this point, the freezing point of the solution refers to the temperature (point b, T_f) when the solid solvent begins to crystallize.

Curve (4) is the cooling curve of solution under experimental conditions. It can be seen that proper undercooling will make the observation of the freezing point of the solution relatively easy. The temperature drops to b' below T_f, and then rises to point b until solvent begins to crystallize.

> **Question and thinking 1-3**
> 1. Can you give out the application example of supercooling phenomena on real life or medical research?
> 2. Scientific research results have shown that the anti-drought and anti-cold properties of some plants are related to the laws of vapor pressure lowering and freezing-point depression, how do you explain this phenomenon?

Droplets of supercooled water often exist in stratiform and cumulus clouds, which can be applied for cloud seeding. The most common chemicals used for cloud seeding include silver iodide and dry ice, solid carbon dioxide.

Some plants are able to supercool the fluid in their cells cytosol and vacuole and thereby survive temperatures down to $-40^\circ C$. This is partly achieved through the synthesis of antifreeze proteins that prevent ice nucleation.

A commercial application of supercooling is in refrigeration. For example, there are freezers that cool drinks to a supercooled level so that they form slush when they are opened.

Measurements of the drop in the freezing point, like those of the elevation of the boiling point, can be used to determine molar masses of unknown substances. However, because the K_f value of most solvents is greater than the value of K_b, the freezing-point depression in the same solution is higher than the boiling point elevation, the freezing-point depression is more sensitive than the boiling point elevation and the error in measuring temperature is relatively small. In particular, freezing-point depression is often carried out at low temperatures, which generally does not cause destruction of biological samples. Therefore, freezing-point depression is widely used in medical and biological science experiments.

Sample Problem 1-5 0.638g of urea was dissolved in 250g of water, and the freezing point of the solution was measured to be $-0.079\,℃$. Calculate the relative molar mass of urea.

Solution For solvent water, $K_f = 1.86\,℃ \cdot kg \cdot mol^{-1}$, so

$$\Delta T_f = K_f b_B = K_f \frac{m_B}{m_A M_B}$$

$$M_B = \frac{K_f m_B}{m_A \Delta T_f}$$

Where, m_A and m_B are the mass of solvent and solute, respectively. M_B is the mole mass of the solute ($kg \cdot mol^{-1}$).

$$MM.(CON_2H_4) = \frac{1.86\,℃ \cdot kg \cdot mol^{-1} \times 0.638g}{250g \times 0.079\,℃} = 0.060 kg \cdot mol^{-1} = 60 g \cdot mol^{-1}$$

Table 1-4 The Freezing Point Depression of Some Electrolyte Solution

$b_B / mol \cdot kg^{-1}$	ΔT_f (found) /℃		ΔT_f (calc.) /℃
	NaCl	MgSO$_4$	
0.01	0.036 03	0.030 0	0.018 58
0.05	0.175 8	0.129 4	0.092 90
0.10	0.347 0	0.242 0	0.185 8
0.50	1.692	1.018	0.929 0

1.3.2 Colligative Properties of Electrolyte Dilute Solution

So far we have considered only non-electrolyte solutes, non-dissociating substance. If the solute is electrolyte, what will happen? Are the colligative properties the same as the non-electrolyte? The data on the freezing point depression of NaCl and MgSO$_4$ is shown in Table 1-4.

The experimental values of two solutions are all larger than the calculated values, such as the NaCl solution of $0.10 mol \cdot kg^{-1}$. According to the calculation, it should be $0.185\,8\,℃$, but the test value is $0.347\,0\,℃$. The experimental value is almost 2 times of the calculated value, the theoretical calculation value of the electrolyte solution and the experimental measurement value deviate greatly. Therefore, correction factor must be introduced into the corresponding equations respectively when calculating the colligative properties of dilute electrolyte solutions. Thus, the previous equations should be modified as follows

$$\Delta T_b = i K_b b_B \tag{1.12}$$

$$\Delta T_f = i K_f b_B \tag{1.13}$$

Where, i is called the **Van't Hoff factor** and is defined as follows

$$i = \frac{\text{actual number of particles in solution after dissociation}}{\text{number of formula units initially dissolved in solution}}$$

When the electrolytes solution is high dilutions, i approaches the numbers of ions into which the electrolyte molecule ionizes, such as for NaCl and KNO$_3$, i should be 2; for Na$_2$SO$_4$ and MgCl$_2$, i should be 3. The value of i decreases as the concentration of the electrolyte solute increases. This is mainly because of the attraction between ions which results in departure from the theoretical values for i.

> **Question and thinking 1-4** NaCl and $MgSO_4$ are type AB model of electrolytes. If the correction factor $i=2$, the calculated value is greater than the experimental value, and the greater the solution concentration is, the greater the difference between the calculated value and the experimental value is as shown in Table 1-4. Why?

1.4 Osmotic Pressure of Solution

1.4.1 Osmosis and Osmotic Pressure

When two different kinds of solution with different concentration are mixed in a beaker, a homogeneous solution will be formed finally. This process is called **diffusion**, which is the movement of molecules from a region of higher concentration to one of lower concentration by random molecular motion. In a phase with uniform temperature, diffusion processes tend to lead towards even distributions of molecules.

But if you separate the two solutions with a semipermeable membrane, what will happen?

Semipermeable membrane, also termed as selectively-permeable membrane, partially-permeable membrane or differentially-permeable membrane, is membrane that allows small molecules, such as water molecules, to pass through, but block the passage of large molecules, such as those of proteins or carbohydrates. Semipermeable membranes are fairly common natural and synthetic materials; a sheet of cellophane is an example. There are many kinds of semipermeable membranes, and their permeability is also different.

If dilute solution and pure solvent is separated with a semipermeable membrane as shown in Figure 1-6(a), solvent flows from the left side into the right and raise the height on the right side of the solution as shown in Figure 1-6(b). The driving force of flowing is the tendency of the pure solvent to mix with the solution and dilute it, just as any solutions with different composition tend to mix to a uniform composition when put in contact with each other. Because molecules of solute cannot pass down through the semipermeable membrane, the tendency can be expressed only by solvent molecules passing up. This phenomenon is called **osmosis** which is the process of solvent flow through semipermeable membrane from pure solvent or from dilute solution to more concentrated solution in order to equalize the concentrations of solutes on the two sides of the membrane. There are two conditions for the osmosis: one is the existence of semipermeable membrane, the other is that the number of solvents in the unit volume of the two sides of semipermeable membrane is not equal. On the other hand, there is the concentration difference between two sides of semipermeable membrane.

Along with osmosis and the rising of the height of the solution, the downward pressure increases. Eventually, the height of solution exerts sufficient pressure downward to counteract the flow of solvent molecules across the membrane into the solution. Osmosis then stops like Figure 1-6(c).

When pure solvent and solution are separated by semipermeable membrane, the excess pressure must be exerted on the solution surface to just stop osmosis to happen is called **osmotic pressure**

Π, its unit is Pa or kPa. The osmotic pressure of a solution can be quite large. For a 0.010mol·L^{-1} of non-electrolyte aqueous solution, it has an osmotic pressure of 24.8kPa. Osmotic pressure is another colligative property of solution and plays a vital role in the transport of molecules across cell membranes in living organisms.

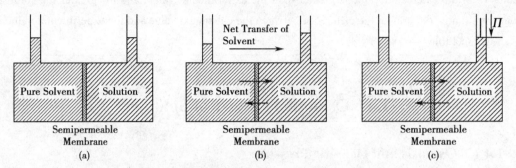

Figure 1-6   Osmosis and Osmotic Pressure
(a) Before osmosis; (b) Osmosis; (c) Osmotic pressure

If the pressure exerted above the solution in Figure 1-6 is greater than the osmotic pressure of the solution, the flowing direction of solvent will change and flow from the solution side to the pure solvent side, this case is called the reverse osmosis. In the same way, the reverse osmosis will also occur when the two solutions with different concentration are separated by a semipermeable membrane, and the external force on the concentrated solution side is greater than the difference in the osmotic pressure between the two sides. According to the reverse osmosis principle, the separation of solute and solvent in solution can be realized by the selection of semipermeable membrane and used widely in the water treatment process.

One of the more interesting uses of osmosis is the desalination of seawater by reverse osmosis. When pure water and seawater are separated by a membrane, the osmosis of water molecules from pure side to seawater side is faster than the reverse one. As osmotic pressure formation, the osmosis rates of water molecules between two sides eventually become equal at an osmotic pressure of about 30atm at 25℃ commonly. If a pressure even greater than 30atm is now applied to the solution side, the reverse osmosis of water molecules becomes favored. As a result, pure water can be obtained from seawater.

1.4.2   The Relationship between Osmotic Pressure of Solution and Concentration, Temperature of Solution

In 1886, Jacobus van't Hoff discovered an important relationship between the osmotic pressure Π, concentration c_B and its absolute temperature T of solution

$$\Pi = c_B RT \tag{1.14}$$

Where, R is the gas constant. Using c_B in mol·L^{-1}, R in kPa·L·mol^{-1}·K^{-1} and T in K in Equation 1.14 gives Π in kPa. For a dilute solution, molarity is approximately equal to the molality, so the Equation 1.14 above can express as follows

$$\Pi \approx b_B RT \tag{1.15}$$

The osmotic pressure of solution is directly proportional to its concentration and it has nothing to do with the nature of solute at the given temperature.

Sample Problem 1-6 2.00g of sucrose ($C_{12}H_{22}O_{11}$) was dissolved in water to 50mL of mixture solution. Calculate the osmotic pressure of the solution at 37.0℃.

Solution The molar mass of sucrose is 342g·mol^{-1}, so

$$c(C_{12}H_{22}O_{11}) = \frac{n}{V} = \frac{2.00g}{342g \cdot mol^{-1} \times 0.050\,0L} = 0.117 mol \cdot L^{-1}$$

According to Equation 1.14 $\Pi = c_B RT$

$$\Pi = 0.117 mol \cdot L^{-1} \times 8.314 kPa \cdot L \cdot K^{-1} \cdot mol^{-1} \times (273.15 + 37.0)\,K = 302 kPa$$

If the solute is electrolyte, those colligative properties just like the non-electrolyte solute mentioned before, the dilute solution must be multiplied a van't Hoff factor i as the follows

$$\Pi = i b_B RT \tag{1.16}$$

Sample Problem 1-7 The normal saline solution used in clinic is 9.0g·L^{-1} of NaCl solution. Calculate the osmotic pressure of it at normal body temperature (37℃).

Solution NaCl dissociated completely in the dilute solution, $i \approx 2$, the molar mass of it is 58.5g·mol^{-1}.

According to $\Pi = i c_B RT$

$$\Pi = \frac{2 \times 9.0g \cdot L^{-1} \times 8.314 kPa \cdot L \cdot K^{-1} \cdot mol^{-1} \times 310.15K}{58.5g \cdot mol^{-1}} = 7.9 \times 10^2 kPa$$

The molar mass of solute can be determined by the colligative properties for the dilute solution. However, the molar mass of larger macromolecule compounds such as protein is often known by determining the osmotic pressure of solution, which is more sensitive than the way to determine the freezing point depression of solution. It can be carried out at normal temperature without affecting the biological samples.

Sample Problem 1-8 1.00g of heme was dissolved in water and prepared 100mL solution. The osmotic pressure of the solution was 0.366kPa at 20℃. Calculate the molar mass of heme.

Solution According to van't Hoff equation

$$\Pi = c_B RT = \frac{m_B}{MM.(B)} RT$$

$$MM.(B) = \frac{m_B RT}{\Pi V}$$

In which, M_B is the molar mass of heme, m_B is the mass of heme, V is the volume of the solution, so

$$MM.(B) = \frac{1.00g \times 8.314 kPa \cdot L \cdot K^{-1} \cdot mol^{-1} \times 293.15K}{0.366 kPa \times 0.100L} = 6.66 \times 10^4 g \cdot mol^{-1}$$

1.4.3 Application of Osmotic Pressure in Medicine

1. Isotonic, hypotonic and hypertonic solution

If two solutions with the same osmotic pressure are separated by a semipermeable membrane, no osmosis will occur. The two solutions are said to be **isotonic** each other. If one of solutions is of lower in the osmotic pressure, it is described as being **hypotonic** with respect to the other solutions. Otherwise it is **hypertonic**. The more concentrated solution is said to be hypertonic with respect to the dilute solution generally.

Osmosis is very important in the regulation of fluid and electrolyte balance in the body. A biological cell consists of cell constituents and cell solution enclosed by cell membrane. When the osmotic pressure of cell solution is equal to that of blood plasma, osmosis does not occur across the cell membrane. The isotonic in medicine indicates that the osmotic pressure of solution is the same as that of blood plasma.

When the concentration of cell solution is different from that of the solution surrounding the cell, osmosis will occur as shown in Figure 1-7.

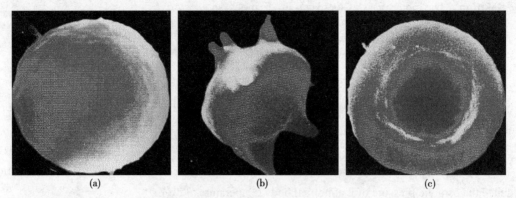

Figure 1-7 The Influence of Osmotic Pressure of a Solution on Red Blood Cells
(a) Cells are placed in 7×10^{-2} mol·L^{-1} NaCl solution; (b) Cells are placed in 2.6mol·L^{-1} NaCl solution; (c) Cells are placed in 0.15mol·L^{-1} NaCl solution

7.0×10^{-2} mol·L^{-1} of NaCl solution are hypotonic comparing with blood plasma. Suppose that a red blood cell is placed in 7.0×10^{-2} mol·L^{-1} of NaCl solution (a hypotonic solution). What will happen? The osmotic pressure in the red blood cell is higher than that of the solution. Therefore, osmosis will take place with the water diffusing into the red blood cell or from dilute to concentrated solution as in Figure 1-7(a). The red blood cell thus enlarges until it bursts. This process of the bursting of a red blood cell because of a hypotonic solution is called **hemolysis.** Thus a hypotonic solution is not usually used for transfusions.

5% of NaCl solution or 10% of glucose solution is an example of a hypertonic solution when compared with blood plasma.

Suppose a red blood cell is placed in hypertonic solution. What will happen? The osmotic pressure in the red blood cell is less than that in the hypertonic solution. Therefore, osmosis will take place with the water diffusing out of the red blood cell as in Figure 1-7(b). The red blood cell thus shrinks. Shrinking of the red blood cell in a hypertonic solution is called **crenation or plasmolysis**.

Usually only isotonic solutions may be safely introduced into the blood stream; solution that is given intravenously to a patient must have an osmotic pressure that is nearly equal to that of blood plasma, to ensure that the blood cells are not adversely affected by osmosis. Hypotonic solutions may cause hemolysis and hypertonic solutions may cause crenation.

Suppose that the red blood cells were placed in 0.15mol·L^{-1} of NaCl solution as in Figure 1-7(c). Will osmosis take place? The answer is no, because there is no difference in osmotic pressure on either side of the semipermeable membrane or they are isotonic. The common name for 0.15mol·L^{-1} of NaCl solution is **physiologic saline solution** which can be given intravenously to a patient without any effect on the red blood cells. 5.0% of glucose solution is also approximately isotonic with body fluids.

In addition to isotonic solution, hypertonic solution is also used in clinic some times. For example, the 2.8mol·L^{-1} of glucose solution, is used to correct hypoglycemia.

2. Osmolarity or osmotic concentration

The osmotic pressure of solution depends upon the number of particles in solution but not their identity. So the osmotic pressure of certain solution is usually expressed in terms of **osmolarity** or **osmotic concentration** c_{os} in medicine. Osmolarity is a measurement of moles of **osmotic active substance** per liter of solution. Its unit is $mol \cdot L^{-1}$. The osmotic active substances refer to the all chemical species that contributes to osmotic pressure of solution. The term comes from the phenomenon of osmosis, and is typically used for osmotic-active solutions.

So, the osmotic pressure of a solution could be expressed also as follows:

$$\Pi = c_{os}RT \tag{1.17}$$

For dilute solutions, osmolarity is expressed in milli-mole per liter in $mmol \cdot L^{-1}$ of solution.

Sample Problem 1-9 Calculate the osmotic concentration of $50g \cdot L^{-1}$ glucose solution and $9.00g \cdot L^{-1}$ NaCl solution (normal saline) in clinic fluids in $mmol \cdot L^{-1}$.

Solution The molar mass of glucose($C_6H_{12}O_6$) is $180g \cdot mol^{-1}$, The osmotic active substance in the solution is glucose molecule.

$$c_{os} = \frac{50.0g \cdot L^{-1}}{180g \cdot mol^{-1}} \times \frac{1\,000mmol}{1mol} = 278mmol \cdot L^{-1}$$

The molar mass of NaCl is $58.5g \cdot mol^{-1}$, The osmotic active substance in the solution of NaCl are Na^+ and Cl^-. The osmotic concentration of $9.00g \cdot L^{-1}$ NaCl solution is

$$c_{os} = \frac{9.00g \cdot L^{-1}}{58.5g \cdot mol^{-1}} \times \frac{1\,000mmol}{1mol} \times 2 = 308mmol \cdot L^{-1}$$

The freezing point of plasma was measured at 0.553℃, and the osmotic concentration of plasma was $297mmol \cdot L^{-1}$. By convention, the solution with the osmotic concentration of $280 \sim 320mmol \cdot L^{-1}$ is regarded as isotonic solution in clinical generally. Table 1-5 shows the concentrations of various substances in normal human plasma, intercellular fluid and intracellular fluid.

Table 1-5 The Osmotic Concentration of Various Substances in Normal Human Plasma, Intercellular Fluid and Intracellular Fluid ($mmol \cdot L^{-1}$)

Osmotic Active Substance	Concentration in Plasma	Concentration in Intercellular Fluid	Concentration in Intracellular Fluid
Na^+	144	137	10
K^+	5	4.7	141
Ca^{2+}	2.5	2.4	
Mg^{2+}	1.5	1.4	31
Cl^-	107	112.7	4
HCO_3^-	27	28.3	10
$HPO_4^{2-}, H_2PO_4^-$	2	2	11
SO_4^{2-}	0.5	0.5	1
Creatine phosphate			45
Carnosine			14
Amino acid	2	2	8
Creatine	0.2	0.2	9
Lactate	1.2	1.2	1.5
Adenosine triphosphate			5

			Continue
Osmotic Active Substance	Concentration in Plasma	Concentration in Intercellular Fluid	Concentration in Intracellular Fluid
Hexose monophosphate			3.7
Glucose	5.6	5.6	
Protein	1.2	0.2	4
Urea	4	4	4
c_{os}	303.7	302.2	302.2

3. Crystalloid osmotic pressure and colloidal osmotic pressure

The total osmotic pressure of plasma is about 773kPa, which are composed of two kinds——crystalloid osmotic pressure and colloidal osmotic pressure.

There are not only many small molecular crystal substances such as NaCl, NaHCO$_3$ and glucose but also large molecular colloidal substances such as protein in plasma. The osmotic pressure formed by small molecular crystal substances is called **crystalloid osmotic pressure** which is about 705.6kPa. The crystalloid osmotic pressure plays important role in regulation balance of fluid and electrolyte on the two sides of cell membrane.

Under normal situations, the concentration of crystal materials at two sides of the cell membrane is equal, and their crystalloid osmotic pressure is equal, the exchange of water keeps dynamic balance, and the cell morphology remains the same basically.

If lack of water in the body, the outside fluids of cells becomes hypertonic, can lead to dehydration and need intake of water. On the other hand, if water intake is too much, the outside fluids of cells become hypotonic, can lead to hemolysis. So, the water containing some of NaCl is better than the pure water to the body with dehydration.

The osmotic pressure formed by large molecular colloidal substances is called **colloid osmotic pressure** or sometimes as the oncotic pressure which is only 2.93~4.00kPa. The colloid osmotic pressure plays important role in regulation balance of fluid and electrolyte at the two sides of blood capillary wall.

In plasma, the colloid osmotic pressure is about 0.5% of the total osmotic pressure. This may be a small percent but because large plasma proteins cannot easily cross through the capillary walls, their effect on the osmotic pressure of the capillary interiors will, to some extent, balance out the tendency for fluid to leak out of the capillaries. In other words, the colloid osmotic pressure tends to pull fluid into the capillaries. In conditions where plasma proteins are reduced, e.g. from being lost in the urine (proteinuria) or from malnutrition, the result of low colloid osmotic pressure can be **edema**, excess fluid buildup in the tissues, which need supply the protein or blood in order to increase the colloid osmotic pressure in plasma.

4. The determination of osmotic pressure of the body fluid

It is more convenient to determine the freezing point of the solution than directly measure the osmotic pressure of the solution. Therefore, the determination of osmotic pressure of many solutions in clinical, such as the pressure of blood, gastric juice, saliva, urine, dialysate, tissue cell culture fluid, is calculated by the freezing point depression of the solution. So the "ice point osmometer" is often used.

The main components of the ice point osmometer include semiconductor refrigerating device, high precision temperature measurement system and supercooled crystal device. The measurement steps of the osmosis pressure are followings, adding the freeze-free liquid in the cold tank and opening the water pipe of the semiconductor refrigerating device; switch on and preheating the machine; correcting the instrument with NaCl standard solution, the osmotic concentration of 300mmol·L^{-1} or 800mmol·L^{-1}; taking 1mL of body fluid sample in the test tube and putting it into the cold tank to determine the osmotic pressure, and the result to be indicated by instrument.

Sample Problem 1-10 The freezing point of human blood was measured to be $-0.56℃$. Calculate the osmotic pressure of it at 37℃.

Solution According to $\Delta T_f = K_f b_B$, so $b_B = \dfrac{\Delta T_f}{K_f}$

$$\Pi = c_B RT \approx b_B RT$$
$$= \dfrac{0.56℃}{1.86℃·kg·mol^{-1}} \times 8.314 kPa·L·K^{-1}·mol^{-1} \times (273.15 + 37)K$$
$$= 7.8 \times 10^2 kPa$$

Key Terms

混合物	mixture
组成标度	composition scale
质量分数	mass fraction
体积分数	volume fraction
质量浓度	mass concentration
摩尔分数	mole fraction
物质的量浓度	amount-of-substance concentration
溶质的质量摩尔浓度	molality of solute
依数性	colligative property
蒸发	evaporation
凝结	condensation
蒸汽压力	vapor pressure
升华	sublimation
蒸气压力下降	vapor pressure lowering
Raoult 定律	Raoult's law
沸点	boiling point
溶液的沸点升高	boiling point elevation of solution
纯液体的冰点下降	freezing point of pure liquid
溶液的冰点下降	freezing point depression of solution
Van't Hoff 系数	Van't Hoff factor
扩散	diffusion
半透膜	semipermeable membrane
渗透	osmosis
渗透压力	osmotic pressure

等渗的	isotonic
低渗的	hypotonic
高渗的	hypertonic
溶血	hemolysis
皱缩	crenation or plasmolysis
渗透浓度	osmolarity or osmotic concentration
晶体渗透压力	crystalloid osmotic pressure
胶体渗透压力	colloidal osmotic pressure
水肿	edema

Summary

The mixture is a system being composed of two or more substances. The composition scale is the relative content of each component in the mixture and can be expressed by mass fraction, volume fraction, mass concentration, mole fraction, amount-of-substance concentration, molarity, etc.

Solutions generally have markedly different properties compared with either the pure solvent or the solute. Some of the properties unique to solutions depend only on the particles number of solute and not their identity. Such properties are called colligative properties. The colligative properties include vapor pressure lowering, boiling point elevation, freezing point depression and osmotic pressure.

If the solute is nonvolatile, the vapor pressure of solution is lower than that of the pure solvent. The value of the vapor pressure lowering is proportional to the amount of solute. The equation that describes that phenomenon is called Raoult's law, $\Delta p = K b_B$.

Boiling point elevation is a colligative property related to vapor pressure lowering. The boiling point is defined as the temperature at which the vapor pressure of liquid equals the atmospheric pressure. Due to vapor pressure lowering, solution will require a higher temperature to reach its boiling point than the pure solvent, $\Delta T_b = K_b b_B$.

Every liquid has a freezing point, the temperature at which a liquid undergoes a phase change from liquid to solid. When solutes are added into solvent, forming solution, the solute molecules disrupt the formation of crystals of the solvent. That disruption in the freezing process results in a depression of the freezing point for the solution relative to the pure solvent, $\Delta T_f = K_f b_B$.

When solution is separated from a volume of pure solvent by semi-permeable membrane that allows only the solvent molecules to pass through, the height of the solution begins to rise. The value of the height difference between the two compartments reflects a property called the osmotic pressure of solution. The amount of osmotic pressure is directly proportional to the concentration of the solute, $\Pi = c_B RT$.

The colligative properties of dilute solutions for the different non-electrolyte substances are same in identical solvent if their molality is same. On the other hand, electrolytes will ionize and each of ions will act as an entity. Therefore, the solution with electrolyte will have the larger changes in colligative properties than the solution with nonelectrolyte if they are equimolal. For electrolyte solutions, the colligative properties should be modified as follows

$$\Delta p = i K b_B;\ \Delta T_b = i K_b b_B;\ \Delta T_f = i K_f b_B;\ \Pi = i b_B RT.$$

where i is the van't Hoff's factor.

Osmotic pressure is important from a biological viewpoint since the physiological membranes such as the red blood cell membranes are semipermeable membranes. In clinic the solutions for transfusions are required to be isotonic. The solution with the osmotic concentration of 280~320mmol·L^{-1} is regarded as isotonic solution in clinical. The hypertonic solution will produce crenation or plasmolysis while hypotonic will result in hemolysis.

The crystalloid osmotic pressure plays important role in regulation balance of fluid and electrolyte on the two sides of cell wall, while the colloid osmotic pressure plays important role in regulation balance of fluid and electrolyte on the two sides of blood capillary wall.

Exercises

1. When 5.50g of biphenyl ($C_{12}H_{10}$), a nonvolatile compound, is dissolved in 100.0g of benzene, the boiling point of the solution exceeds the boiling point of pure benzene by 0.903℃. What is K_b for benzene?

2. When 6.30g of a nonvolatile hydrocarbon of unknown molar mass is dissolved in 150.0g of benzene, the boiling point of the solution is 80.696℃. The boiling point of pure benzene is 80.099℃. What is the molar mass of the hydrocarbon?

3. How many grams of ethylene glycol $C_2H_4(OH)_2$, must be dissolved in 200.0g of water to lower the freezing point to −0.297℃?

4. When 6.00g of a non-ionizable solid is dissolved in 135g of water, the freezing point of this solution is −0.242℃. What is the molecular weight of the solid?

5. The chemical amounts of the major dissolved species in a 1.000L sample of sea water are given below. Estimate the freezing point of the seawater, assuming K_f=1.86℃·kg·mol^{-1} for water. Na$^+$ 0.458mol; Cl$^-$ 0.533mol; Mg^{2+} 0.052mol; SO$_4^{2-}$ 0.028mol; Ca^{2+} 0.010mol; HCO$_3^-$ 0.002mol; K$^+$ 0.010mol; Br$^-$ 0.001mol; Neutral species 0.001mol.

6. Calculate the molality of a solution prepared by dissolving 36.0g of glucose $C_6H_{12}O_6$ in 400.0g of water.

7. Calculate the freezing point of a solution containing 3.60g of glucose $C_6H_{12}O_6$ in 50.0g of water. The molecular mass of glucose is 180.0g·mol^{-1}, the normal freezing point of pure water, T_f°, is 0.00℃, and K_f for water is 1.86℃·kg·mol^{-1}.

8. What is the molecular mass of an organic compound if the solution containing 3.25g of the compound in 125.0g of water has a freezing point of −0.930℃?

9. How many grams of CaCl$_2$ (MM. = 110.9g·mol^{-1}) must be dissolved in 1 000.0g of water to result in 77.4kPa of osmotic pressure at 37℃?

10. 0.243 6g sample of an unknown substance was dissolved in 20.0mL cyclohexane. The density of cyclohexane is 0.779g·mL^{-1}. The freezing point depression ΔT_f was 2.5 degree. Calculate the molar mass of the unknown substance. The freezing point depression constant of cyclohexane is 20.0℃·kg·mol^{-1}.

11. A chemist dissolves 2.00g of a protein in 0.100L of water and finds the osmotic pressure to be 2.127kPa at 25℃. What is the approximate molar mass of the protein?

Supplementary Exercises

1. What are the normal freezing points and boiling points of the following solution? (a) Solution of

21.0g of NaCl in 135mL of water. (b) Solution of 15.4g of urea in 66.7mL of water.

2. If 4.00g of certain non-electrolyte is dissolved in 55.0g of benzene, the resulting solution freezes at 2.36℃. Calculate the molecular weight of the non-electrolyte.

3. A quantity of 7.85 g of a compound having the empirical formula C_5H_4 is dissolved in 301g of benzene. The freezing point of the solution is 1.05℃ below that of pure benzene. What are the molar mass and molecular formula of this compound?

4. Ethylene glycol(EG), $CH_2(OH)CH_2(OH)$, is a common automobile antifreeze. It is cheap, water-soluble and fairly nonvolatile (b.p.197℃). Calculate the freezing point of the solution containing 651g of this substance in 2 505g of water. Would you keep this substance in your car radiator during the summer? The molar mass of ethylene glycol is $62.07g \cdot mol^{-1}$.

5. Solution is prepared by dissolving 35.0g of hemoglobin (Hb) in enough water to make up one liter in volume. If the osmotic pressure of the solution is found to be 10.0mmHg at 25℃, what is the molar mass of hemoglobin.

6. A 0.86 percent by mass solution of NaCl is called "physiological saline" because its osmotic pressure is equal to that of the solution in blood cell. Calculate the osmotic pressure of this solution at normal body temperature (37℃). Note that the density of the saline solution is $1.005g \cdot mL^{-1}$.

Answers to Some Exercises

[Exercises]

1. $2.53℃ \cdot kg \cdot mol^{-1}$
2. $178g \cdot mol^{-1}$
3. 1.98g
4. $342g \cdot mol^{-1}$
5. $-2.04℃$
6. $0.500mol \cdot kg^{-1}$
7. $-0.744℃$
8. $52.0g \cdot mol^{-1}$
9. 1.11g
10. $125g \cdot mol^{-1}$
11. $2.33 \times 10^4 g \cdot mol^{-1}$

[Supplementary Exercises]

1. (a) 102.7℃   (b)101.97℃
2. $118g \cdot mol^{-1}$
3. $127g \cdot mol^{-1}$, $C_{10}H_8$
4. It would be preferable to leave the antifreeze in your car radiator in summer to prevent the solution from boiling.
5. $6.51 \times 10^4 g \cdot mol^{-1}$
6. 763kPa

(王兴坡)

Chapter 2
The Basis of Chemical Thermodynamics

Chemical reactions are always accompanied by the heat absorbed or liberated in reaction. Furthermore, the other forms of energy can be exhibited in photochemical or electrochemical reactions meanwhile. On the other hand, both products and energy used in daily life are all from the chemical reactions mostly in the fields such as medicine, industry and agriculture etc. The aspects referring to the transformation of substance and energy in reactions can be concluded in (1) How much of chemical reaction energy will be transferred or converted? (2) Can a reaction occur spontaneously at a given condition? If it can, what is the reaction degree? The important way to solve these problems is about the thermodynamic principle.

Thermodynamic concerns the macroscopic properties of a system composed by the large number of particles but not focusing on the microscopic structure of substances. The first and the second laws of thermodynamic are the foundation which based on practice and experimental results, not mathematical reasoning. Chemical thermodynamics based on the combination of thermodynamic theory and chemistry was established in which the direction, the limitation and the equilibrium of a chemical reaction are studied generally.

2.1 Thermodynamic System and State Function

2.1.1 System and Its Surroundings

To make meaningful and conventional study in thermodynamics the universe can be thought of two parts, one is the **system** which we are going to observe, measure or have a special interest, the other is defined as **surroundings** including anything else relevant to the system. For example, we observe and investigate 90 ℃ water in a vessel, the system is water and the vessel, but the all parts except this system are regarded as the surroundings.

The concepts of the system and the surroundings in the principle are required for the convenience of thermodynamic study, which can be really existence or not. During the course of a specific study the contents of system and surroundings cannot alter anymore.

According to matter and energy transferred between the system and its surroundings there are three types of system, an **opening system** is that both matter and energy can be transferred between the two parts; the system is **closed system** when matter transfer is not possible; an **isolated system** is neither mechanical nor thermal contact with its surroundings.

> **Question and thinking 2-1**   What system does the following combusting substance belong to?
> (1) Combustion of solid substance in the closed adiabatic oxygen bomb.
> (2) Combustion of gasoline in the cylinder and the pistol is pushed forward and backward.
> (3) Combustion of hydrogen in air.

According to the composition the system can also be a single component system or a multi-component system.

2.1.2 State Functions and Process

Such parameters as coordination in space, speed are needed to study the state of particles in thermodynamics. In the same way, some thermodynamic functions such as temperature T, pressure p, volume V, the amount of substance n, density ρ and viscosity η, etc. are needed to determine the state of a system. These parameters are macroscopic property. When these properties of the system have definite values and do not change with the time, the system is in an equilibrium state. These thermodynamic properties describing the state of a system are also defined as **state functions**. T, p, V, n, and so on are all state functions, and some new state functions will be introduced in this chapter. State functions can be classified as geometric such as volume V, mechanical such as pressure p, thermal such as temperature T and chemical such as quantity n of matter in property. When a certain amount of ideal gas is in a certain equilibrium state, it has the certain values of T, p and V.

Temperature is of particular importance in these thermodynamic state functions. From the macroscopic point of view, temperature indicates the degree of cool or heat degree of a system. It is the indication of the intensity of the random motion of a large number of molecules in the system. If system A and C are at the same temperature, the temperature of system B and C is same, the conclusion that the temperature of the system A and B is the same can be drawn, which is called the zeroth law of thermodynamics. According to this law, system C can be used to measure the standard temperature of material and to determine whether the other substance such as A or B is higher or lower than C in temperature, that is colder or hotter. But the meaning of temperature is not clear yet. The method for giving the specific temperature value is called temperature scale. All of gases, liquids and solids can be used as the materials to measure the standard temperature. The thermodynamic property or function that has a direct and simple relationship with temperature was selected. For example, if the material of measuring standard temperature is gas, volume or pressure usually is adopted; but for the liquid material, volume is adopted usually; and for the metal material, resistance as parameter is usually selected.

Although the relative temperature of objects can be determined with the standard substance, the standard point and the corresponding value need to be specified. This is about the thermometric scale. The Celsius scale is widely used to commemorate by Swedish scientist A. Celsius, where mercury volume is as temperature measurement attribute, the freezing point of water is 0 ℃, the boiling point of water is 100 ℃ under normal pressure. The ideal gas temperature scale is often used in thermodynamics to honor the scientist L. Kelvin, the unit is denoted as K. The relationship between the Celsius scale "℃"

and "K" is

$$T(K) = T(℃) + 273.15$$

There are two types of state functions. The first is intensive property, where the value is proportional to the amount of material in the system. Such properties are additive. For example, the total volume of 50mL of pure water mixed with 50mL of pure water is 100mL. So the volume is an intensive property. The other type of property is the extensive property. For example, the temperature of the water obtained by mixing the pure water of 50℃ with the other part of 50℃ water is still 50℃. So the temperature is an extensive property. The difference of thermodynamic properties is base on whether or not they depend on the amount of substance in the system.

Extensive properties of state function are additive. Their determination requires evaluation of the size of the entire system such as volume and mass. On the other hand, intensive properties are independent of the quantities of particles in system. The examples are temperature, pressure and density etc. The ratio of two state functions possessing extensive property is intensive, for example, molar volume (V/n) and density (m/V).

It should be pointed out that the state functions describing the state of a system are not independent of each other. The relationship of the thermodynamic properties can be described by **state equations**. The task of thermodynamics is to obtain the values of other properties from some variables measured easily. For example, the state equation for ideal gas is $pV = nRT$, R is a constant, the value is $8.314 J·mol^{-1}·K^{-1}$, which indicates that if p, n and V of the perfect gas are known the temperature of the system can be determined from the equation. Likewise if T, n and V are known its pressure value can also be obtained.

Once the state changes of system have taken place, the values of state function may change, the key to understand state functions lies in the fact that the change of state function is completely defined when the initial and the final states are specified independent of the route, the process or the speed of the change. For instance, the temperature change of 50℃ water (initial state) turning to 80℃ (final state) is 30℃, having nothing to do with the detailed process (or path) of the change. When any system changes from original state to returning back again (that is a circular process) the values of alternation of the state functions are always zero. It can also be seen that the change in temperature has nothing to do with the selected temperature scale.

The definition of thermodynamic **process** means that a system goes from state 1 to state 2 (both are equilibrium state). Such as gas compression and expansion, liquid evaporation, chemical reactions and so on are all thermodynamic processes. Generally the processes are classified by **isothermal process**, **isobar process**, **isovolumic process**, **adiabatic process** and **cyclic process**.

Isothermal process indicates that the temperature at final state is the same as that of the initial state. The temperature in the midway may change. The biochemical reactions in human body can be regarded as an isothermal process because usually body temperature is 37℃. The temperature of a system generally equals to that of surroundings in the isothermal process.

At isobar process the pressure at the final state equals to that of at the initial state. For example, the chemical reactions in open beaker or test tube occurring at constant atmospheric pressure can be considered as isobaric processes. In the course of isobar process the pressure of system can be kept constant or varied.

Isovolume process means the volume of system keeps constant, for instance the reactions occurring in the closed and rigid vessel. If there is no exchange of heat between system and its surroundings the course is called adiabatic process. Cyclic process is the basis of thermal machine operation, which refers that the system returns back to its original state through a series of change.

If the above process is slow enough so that each intermediate state of the process is close to the equilibrium state, which is called a quasi-static process that is an ideal but very useful concept in the thermodynamic analysis.

> **Question and thinking 2-2** Are variances of heat and work in the process of friction the state functions?

2.1.3 Heat and Work

1. Heat and work

In thermodynamics there are two ways to exchange the energy between the system and the surroundings. That is heat and work. **Heat** Q is the transferred energy at result of temperature difference and the other form of energy exchange is called **work** W. For example, electrical work and expansion work.

According to the 1970 IUPAC (international union of pure and applied chemistry) heat that is released from the system is given a negative sign ($Q<0$), while heat absorbed by the system is positive ($Q>0$). The work done by the system on the surroundings is negative ($W<0$), if conversely, it is given a positive sign ($W>0$).

It is necessary to emphasize that both heat and work are not state functions. They appear at the process of a state change, or at the boundary between the system and the surroundings, not features of the system itself. For example, the stones on the top of the hill reach its foot along different paths, the amount of work and the heat is different because of different friction. The isothermal expansion of the ideal gas will be used to further illustrate that heat and work are related to the process of the system.

2. Volume work, reversible process and maximum work

As shown in Figure 2-1, an air cylinder filled with a certain amount of ideal gas has a movable excellent thermal conductivity piston. The mass and the friction of the piston are 0. The temperature of gas is T always equaling to that of the surroundings, which means isothermal process. When the ideal gas expands at constant temperature, the piston moves outward a specific distance of l against the outside pressure p_{out}, a certain amount of expansion work is

$$W = -F \times l = -(p_{out} \times A) \times l = -p_{out} \times (A \times l) = -p_{out} \Delta V \tag{2.1}$$

ΔV is the volume change when the gas expands, F is the force done on the cross-sectioned area of the piston with the area A.

In thermodynamics the volume work is the exchanged energy for the variance of the system volume, the other forms of work such as electric or surface work are not involved. The volume work is quite important and useful concept because it is the basis of modern heat engine.

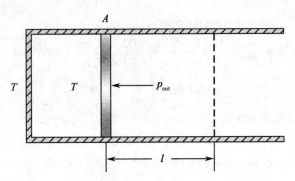

Figure 2-1 Isothermal Expansion of Ideal Gas

Suppose that the initial state of the ideal gas is: $p_{\text{ini.}} = 405.2\text{kPa}$, $V_{\text{ini.}} = 1.00\text{dm}^3$, $T_{\text{ini.}} = 273\text{K}$, the finial state is: $p_{\text{fin.}} = 101.3\text{kPa}$, $V_{\text{fin.}} = 4.00\text{dm}^3$, $T_{\text{fin.}} = 273\text{K}$. The different paths of isothermal expansion from the initial state to the final are illustrated as follows.

Reversible process

(3)

$P_{\text{ini.}}=405.2\text{kPa}$ $P_{\text{out.}}=101.3\text{kPa}$ $P_{\text{fin.}}=101.3\text{kPa}$
$V_{\text{ini}}=1.00\text{dm}^3$ ────────────────▶ $V_{\text{fin.}}= 4.00\text{dm}^3$
$T=273\text{K}$ (1) $T=273\text{K}$

$P_{\text{out.}}=202.6\text{kPa}$ $P_2=202.6\text{kPa}$ $p_{\text{out.}}=101.3\text{kPa}$
──────────────▶ $V_2=2.00\text{dm}^3$ ──────────────▶
(Ⅰ) $T=273\text{K}$ (Ⅱ)

(2)

(1) When the decrease of external pressure is taken in one-step from 405.2kPa to 101.3kPa, the ideal gas in the cylinder will rapidly expand under constant pressure spontaneously, the volume work is

$$W_1 = -p_{\text{out.}} \Delta V = -101.3 \times 10^3 \text{Pa} \times (4-1) \times 10^{-3} \text{m}^3 = -304\text{J}$$

(2) The gas expands to the final state in two steps, the first step of expansion of the gas happens against the external pressure of 202.6kPa, reaching the intermediate equilibrium state $p_2 = 202.6\text{kPa}$, $V_2 = 2.00\text{dm}^3$, $T = 273\text{K}$. The second step of expansion is under the external pressure of 101.3kPa to get the final state. The total expansion work is

$$W_2 = W_\text{I} + W_\text{II}$$
$$= -202.6 \times 10^3 \text{Pa} \times (2-1) \times 10^{-3} \text{m}^3 - 101.3 \times 10^3 \text{Pa} \times (4-2) \times 10^{-3} \text{m}^3 = -405\text{J}$$

(3) If the gas inflation processes infinite times, the pressure of system is only smaller change in dp than external pressure each time, i.e. the quasi static expansion, the volume change is dV, the tiny volume work δW, is

$$\delta W = -p_{\text{out.}} \, \text{d}V = (p_{\text{sys.}} - \text{d}p) \, \text{d}V = -p_{\text{sys.}} \text{d}V$$

The total volume work is W_r.

$$W_3 = W_r = \int_{V_{\text{ini}}}^{V_{\text{fin}}} \delta W = -\int_{V_{\text{ini}}}^{V_{\text{fin}}} p_{\text{sys}} \text{d}V = -\int_{V_{\text{ini}}}^{V_{\text{fin}}} \frac{nRT}{V} \text{d}V = -nRT \ln \frac{V_{\text{fin}}}{V_{\text{ini}}}$$

n, the moles of ideal gas is obtained from the equation.

$$n = \frac{p_1 V_1}{RT} = \frac{405.2 \times 10^3 \text{Pa} \times 1.00 \times 10^{-3} \text{m}^3}{8.314 \text{J} \cdot \text{mol}^{-1} \cdot \text{K}^{-1} \times 273 \text{K}} = 0.178 \text{mol}$$

$$W_r = -0.178 \text{mol} \times 8.314 \text{J} \cdot \text{mol}^{-1} \cdot \text{K}^{-1} \times 273 \text{K} \ln \frac{4.00 \text{dm}^3}{1.00 \text{dm}^3} = -560 \text{J}$$

This example shows that work is not state function. Its value is related to the process that is going on. The third expansion path discussed above is **reversible process** through which the maximum work is done on the surroundings by the ideal gas. The values of the three volume work are figured by shadow area in the Figure 2-2.

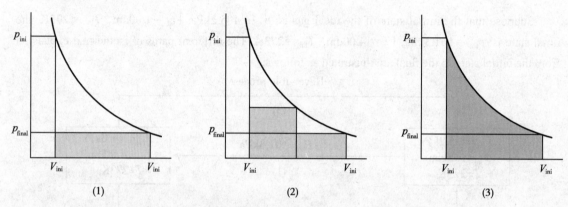

Figure 2-2   Reversible Process and the Maximum Work

The general features of isothermal reversible process are the followings by Figure 2-2.

(1) The volume work done on the surroundings by the system is the maximum.

(2) Through a reversible cycle process both the system and the surroundings return to its original states.

(3) The reversible process is ideal in reality, but such the state changes as the evaporation of liquid at the boiling point or the melting of crystalline at the melting point can be considered reversible approximately.

> **Question and thinking 2-3**
> 1. Discuss the above cases about volume work when the ideal gas in the container was compressed.
> 2. Does thermodynamic "reversible" concept have the same meaning as the "reversible" reaction mentioned in chemistry textbook before?

2.2   Energy Conservation and Heat of Chemical Reaction

The criterion to distinguish reversible or irreversible process is whether both the system and the surroundings can be restored to the original state. The actual process is irreversible. For example, the liquid evaporates at the boiling point, the solid melts at the melting point, energy dissipation, heat conduction, material diffusion, explosion, and life activity, are all irreversible.

2.2.1 Thermodynamic Energy and the First Law of Thermodynamic

1. Thermodynamic energy

Thermodynamic energy or **internal energy** indicates the sum of all forms of energy inside the system including the kinetic energy and potential energy of all molecules, atoms, electrons etc. Due to the complexity of movements of microscopic particles, it is still impossible to determine the absolute value of the thermodynamic energy in a system. But to be sure, a system in a certain state must have a constant value of thermodynamic energy. That is a state function usually denoted by symbol U.

2. The first law of thermodynamics

The first law of thermodynamics is generalized from experience and experiment phenomena of energy transformation in centuries. It can be expressed as "all matter in nature has energy, and energy has various forms, which can be transformed from one form to another and the total value of energy remains constant in the transformation". Energy is neither created nor destroyed, which is called the law of conservation of energy. In any process energy can be transferred by doing work on a system as well as by heating it, so for a closed system heat and work are equivalent way to change thermodynamic energy of a system, the following equation is known as the first law of thermodynamics.

$$\Delta U = Q + W \qquad (2.2)$$

For an open system the variance of thermodynamic energy relates to the amount of the exchanged substance besides heat Q, work W could be volume work, or it could be non-volume work. In the following discussion W will represent only volume work unless otherwise specified. The unit of thermodynamic energy is Joules (J), the first law of thermodynamics relates work to heat, another thermal unit is calories (Cal or Kcal), in food, biological medicine field the unit of Cal or Kcal is stilled used. 1Cal is equal to 4.184J, meaning that work and heat are thermodynamic equivalent in changing the thermodynamic energy of a system.

3. The change in thermodynamic energy and the heat at the constant volume process

The amount of thermodynamic energy in a system cannot be determined but its change can be measured. For a closed system, if a process which keeps constant volume occurs and none of the other types of work exits, according to Equation 2.2,

$$\Delta U = Q_v + W = Q_v + p\Delta V$$

So, $\qquad \Delta U = Q_v \qquad (2.3)$

Equation 2.3 connects the change in thermodynamic energy and the heat exchanged Q_v at the process of constant volume. Many chemical reactions can occur on the condition of constant volume. For example, the bomb calorimeter in Figure 2-3 has been used widely to obtain the change of thermodynamic energy ΔU.

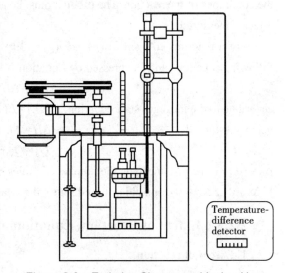

Figure 2-3 Enthalpy Change and Isobar Heat

2.2.2 Enthalpy Change and Isobar Heat

1. Enthalpy of a system

The state of a system changes from state 1 (initial state) to state 2 (final state) through isobar process without any other form of work except volume work, in terms of the first thermodynamic law

$$\Delta U = U_2 - U_1 = Q_p + W$$

namely
$$\Delta U = Q_p - p_{out}\Delta V$$
$$U_2 - U_1 = Q_p - p_{out}(V_2 - V_1)$$

Q_v is the heat exchanged in the process of constant pressure.

if the pressure is constant, $\quad p_1 = p_{out} \quad\quad p_2 = p_{out}$

after substituting, it is $\quad Q_p = (U_2 - U_1) + p_{out}(V_2 - V_1) = (U_2 + p_2V_2) - (U_1 + p_1V_1)$

Term $(U + pV)$ is denoted by symbol H,

$$H \stackrel{\text{def}}{=\!=} U + PV \tag{2.4}$$

$$Q_p = H_2 - H_1$$

Hence, $\quad\quad\quad\quad\quad\quad\quad\quad Q_p = \Delta H \tag{2.5}$

Because both pressure and volume are the state functions the newly introduced thermodynamic function H is also a state function named **enthalpy**.

2. The heat of reaction at constant pressure

The real value of enthalpy of a system cannot be measured, but from Equation 2.4 the change in enthalpy equals to the heat exchange Q_p with the surroundings at constant pressure. The experimental measurement of Q_p is easy and precise with a thermometer. Most of chemical changes and biological reactions take place under normal atmospheric pressure. A great deal of heat information such as dissolving heat, vaporization heat, neutralization heat, reaction heat and so on is based on enthalpy determination. Because most of chemical reactions of interest occur at constant pressure with volume work only, so the heat of chemical reaction is usually expressed by the symbol $\Delta_r H$.

It's important to emphasize that the heat absorbed or released during chemical reactions depends on the conditions of the reaction. The product must be at the same temperature as the reactant when the heat effect is measured.

For a reaction, $aA(g) + bB(g) = dD(g) + eE(g)$, it is assumed that all specials are ideal gas, the following equation can be expressed by Equation 2.4.

$$\Delta H = \Delta U + \Delta(pV)$$

according to the ideal gas state equation, $\Delta(pV) = \Delta(nRT) = \Delta n(RT)$

being substituted $\quad\quad\quad\quad\quad\quad \Delta H = \Delta U + \Delta n(RT)$

or $\quad\quad\quad\quad\quad\quad\quad\quad Q_p = Q_V + \Delta n (RT) \tag{2.6}$

$\Delta n = (d + e) - (a + b)$, meaning the number of moles of gaseous products minus that of gaseous reactants. If $\Delta n = 0$ or $\Delta V \approx 0$ (for solid or liquid state), the relation $Q_p \approx Q_V$ can be considered.

2.2.3 Extent of Reaction, Equation of Thermo-Chemistry and Standard State

1. Extent of reaction

For a reaction equation $\quad\quad\quad\quad aA + bB = dD + eE$

transposition of the above gives $\quad dD + eE - aA - bB = 0$

$$\text{or} \quad \sum_J v_J J = 0 \qquad (2.7)$$

J represents the reactants or the products, v_J is stoichiometric number of substance J. v_J is given negative sign for reactants, but positive sign for products. Extent of the above reaction denoted by ξ can be indicated

$$\xi = \frac{n_A(\xi) - n_A(0)}{-a} = \frac{n_B(\xi) - n_B(0)}{-b} = \frac{n_D(\xi) - n_D(0)}{d} = \frac{n_E(\xi) - n_E(0)}{e}$$

$$\text{or} \quad \xi = \frac{n_J(\xi) - n_J(0)}{v_J} \qquad (2.8)$$

The unit of ξ is mole, the value of ξ is the same both for the reactants and the products.

Sample Problem 2-1 Half of 34g H_2O_2 is catalytic decomposed in 20mins by iodine ion I^-. Two equation forms of this reaction can be expressed as follows.

(1) $H_2O_2 \text{ (aq)} \xrightleftharpoons{I^-} H_2O(l) + 1/2\ O_2(g)$

(2) $2H_2O_2 \text{ (aq)} \xrightleftharpoons{I^-} 2H_2O(l) + O_2(g)$

Calculate the extent of reaction (1) and (2) respectively.

Solution At $t=0$ and $t=20$mins, the moles of reactants and products are listed in the following table.

t / min	$n(H_2O_2)$ /mol	$n(H_2O)$/mol	$n(O_2)$ /mol
0	1.0	0.0	0.0
20	0.50	0.50	0.25

For the reaction (1)

$$\xi = \frac{\Delta n(H_2O_2)}{v(H_2O_2)} = \frac{0.50 - 1.0}{-1} = 0.50 \text{(mol)}$$

Calculation by the product H_2O and O_2 are

$$\xi = \frac{\Delta n(H_2O)}{v(H_2O)} = \frac{0.50 - 0}{1} = 0.50 \text{(mol)}$$

$$\xi = \frac{\Delta n(O_2)}{v(O_2)} = \frac{0.25 - 0}{0.5} = 0.50 \text{(mol)}$$

The values of the extent of reaction are identical for the reactants and products. For chemical reaction (2)

$$\xi = \frac{\Delta n(H_2O_2)}{V(H_2O_2)} = \frac{0.50 - 1.0}{-2} = 0.25 \text{mol}$$

The same results are also obtained according to the products H_2O_2 and O_2. This example demonstrates the equation of chemical reaction should be considered while we calculate the extent of reaction.

2. Thermo-chemical equation and standard state

A thermo-chemical equation can reveal the relationship between the reaction and the heat signed by ΔH exchanged with its surroundings. For example,

(1) $H_2(g) + \dfrac{1}{2} O_2(g) = H_2O(l) \qquad \Delta_r H_{m,298.15}^{\ominus} = -285.8 \text{kJ} \cdot \text{mol}^{-1}$

(2) $2H_2(g) + O_2(g) = 2H_2O(l)$ $\quad \Delta_r H_{m,298.15}^{\ominus} = -571.6 \text{kJ} \cdot \text{mol}^{-1}$

(3) $C(\text{gra.}) + O_2(g) = CO_2(g)$ $\quad \Delta_r H_{m,298.15}^{\ominus} = -393.5 \text{kJ} \cdot \text{mol}^{-1}$

The meaning of sign $\Delta_r H_{m,298.15}^{\ominus}$ is explained as follows:

ΔH: the heat effect of a reaction at constant pressure, negative value means the reaction will release heat to the surroundings and the positive is to absorb heat from environment; here,

r: abbreviation of the word "reaction";

m: meaning $\xi = 1$ mole;

298.15: meaning the temperature of reaction is 298.15K;

$\ominus$: meaning the standard state.

The value of the heat released or absorbed by a reaction is not different when the reaction occurs at different states. To compare the values of reaction heat the standard state must be established. The thermodynamic standard states of solid, liquid, gas and solution are pure solid, pure liquid, perfect gas and ideal solution (concentration is $1\text{mol} \cdot \text{L}^{-1}$ or $1\text{mol} \cdot \text{kg}^{-1}$) respectively at the pressure of 100kPa. For the bio-system, the standard state is that the temperature is at 37°C and the concentration of H^+ is $1 \times 10^{-7} \text{mol} \cdot \text{L}^{-1}$.

Several notes are required to mention: (1) The state of all species involved in the thermo-chemical reaction should be indicated, usually the state of gas, liquid, solid and aqueous solution are denoted by "g" "l" "s" and "aq" respectively. (2) For the solid state different crystalline types, for instance graphite or diamond of carbon element should be illustrated clearly. (3) The temperature and the pressure at which the reaction undergoes should also be denoted in the thermo-equation. (4) ΔH must be given for a reaction followed by the moles of reaction extent in thermo-chemical equation.

2.2.4 Hess's Law and Calculation of Reaction Heat

Hess's law can be stated that the overall heat of a reaction is equal to the sum of heat of reaction for individual step in the reaction measured at the same temperature. Hess's law is particularly useful when measurement of a specific heat of reaction is unfeasible or impossible. It is understandable that the heat of reaction equals to the enthalpy change during the course of a reaction under constant pressure. ΔH is independent of the reaction pathway.

1. Calculation of reaction heat from known thermo-chemical equations

For instance the direct measurement of the heat of the reaction $C(\text{gra.}) + \frac{1}{2} O_2(g) = CO(g)$ is impossible because the formation of some carbon dioxide is inevitable. By Hess's law the heat of the reaction can be calculated from the thermodynamic data of carbon and carbon dioxide.

Sample Problem 2-2 At 298.15K the standard molar enthalpy change of the following reactions has been known:

(1) $\quad\quad\quad\quad C(\text{gra.}) + O_2(g) = CO_2(g) \quad\quad\quad \Delta_r H_{m,1}^{\ominus} = -393.5 \text{kJ} \cdot \text{mol}^{-1}$

(2) $\quad\quad\quad\quad CO(g) + \frac{1}{2} O_2(g) = CO_2(g) \quad\quad\quad \Delta_r H_{m,2}^{\ominus} = -282.99 \text{kJ} \cdot \text{mol}^{-1}$

How much is the $\Delta_r H_{m,3}^{\ominus}$ of the reaction (3) $C(\text{gra.}) + \frac{1}{2} O_2(g) = CO(g)$?

Solution Take the reactants C(gra.) and O_2(g) as the initial state, the final state is the product CO_2(g), the course of the reaction may go through step (1) or step (3), (2)

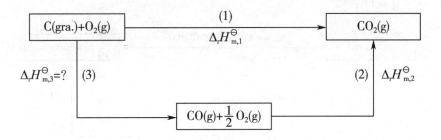

In terms of Hess's law, because

$$\Delta_r H^\ominus_{m,1} = \Delta_r H^\ominus_{m,3} + \Delta_r H^\ominus_{m,2} \quad so, \quad \Delta_r H^\ominus_{m,3} = \Delta_r H^\ominus_{m,1} - \Delta_r H^\ominus_{m,2}$$

$$C(gra.) + O_2(g) = CO_2(g) \qquad \Delta_r H^\ominus_{m,1} = -393.5 kJ \cdot mol^{-1}$$

$$-) \quad CO(g) + \frac{1}{2}O_2(g) = CO_2(g) \qquad \Delta_r H^\ominus_{m,2} = -282.99 kJ \cdot mol^{-1}$$

$$C(gra.) + \frac{1}{2}O_2(g) = CO(g) \qquad \Delta_r H^\ominus_{m,1} = -393.5 kJ \cdot mol^{-1} + 282.99 kJ \cdot mol^{-1}$$

$$= -110.51 kJ \cdot mol^{-1}$$

Sample Problem 2-3 Monoxide carbon is usually used as deoxidizer in smelting metal. For the following reactions the data of $\Delta_r H^\ominus_{m,1}$ and is given as follows.

(1) $Fe_2O_3(s) + 3CO(g) = 2Fe(s) + 3CO_2(g)$ $\Delta_r H^\ominus_{m,1} = -26.7 kJ \cdot mol^{-1}$

(2) $CO(g) + \frac{1}{2}O_2(g) = CO_2(g)$ $\Delta_r H^\ominus_{m,2} = -283.0 kJ \cdot mol^{-1}$

Calculate the value of $\Delta_r H^\ominus_{m,3}$ for the reaction (3) $2Fe(s) + \frac{3}{2}O_2(g) = Fe_2O_3(s)$

Solution From the above equations the equation(3) can be deduced from reaction(1) and reaction(2) with the relation $3 \times (2) - (1) = (3)$, so $\Delta_r H^\ominus_{m,3} = 3 \times \Delta_r H^\ominus_{m,2} - \Delta_r H^\ominus_{m,1}$

$$= -822.3 kJ \cdot mol^{-1}$$

2. Calculation of reaction heat from standard molar enthalpy of formation

At constant pressure for a given reaction $aA + bB = dD + eE$, the heat of reaction is the enthalpy change between products and reactants whose value is independent of the particular way in which the reaction take place, but it is dependant of the temperature, the activity and the states of the reactants and products. For this reason, **the standard molar enthalpy of formation $\Delta_f H^\ominus_m$** of substance B is defined, meaning the formation of 1 mole of substance B from the stable elements at standard state, its unit is $kJ \cdot mol^{-1}$. By the definition the values of $\Delta_f H^\ominus_m$ of the stable elements are set equal to zero such as oxygen (O_2), sulfur (rhombic crystal), carbon (graphite) etc. For instance the $\Delta_f H^\ominus_m$ of H_2O (l, 298.15K) is $-285.8 kJ \cdot mol^{-1}$, which means the standard molar enthalpy change of the reaction $H_2(g, 298.15K, p) + \frac{1}{2}O_2(g, 298.15K, p) = H_2O(l, 298.15K, p)$ is $-285.8 kJ \cdot mol^{-1}$. Note that the stiochiometric number of H_2O is one. The data $\Delta_f H^\ominus_m$ of some compounds at 298.15K are listed in the Appendix 3 of this textbook.

Therefore it is possible for any reaction to construct the two pathways that one is direct path from the elemental compounds of reactants to the products, the other is the procedure of intermediate reactants involved illustrated as follows.

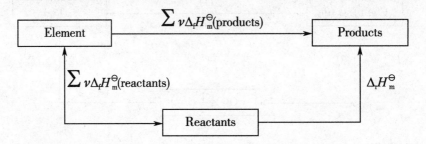

in terms of Hess's law,

$$\sum v\Delta_f H_m^\ominus(\text{products}) = \sum v\Delta_f H_m^\ominus(\text{reactants}) + \Delta_r H_m^\ominus$$

The value of is readily calculated from v

$$\Delta_r H_m^\ominus = \sum v\Delta_f H_m^\ominus(\text{products}) - \sum v\Delta_f H_m^\ominus(\text{reactants}) \qquad (2.9)$$

In Equation 2.9, v is the stoichiometric number of individual substance because enthalpy $\Delta_r H_m^\ominus$ is extensive property.

Sample Problem 2-4 Calculate the standard molar enthalpy of reaction below referring to standard molar enthalpy of formation of the related substance at 298.15K

$$2NH_3(g) + CO_2(g) = CO(NH_2)_2(s) + H_2O(l)$$

Solution The values $\Delta_f H_m^\ominus$ of reactants and products of the reaction are known as follows.

$\Delta_f H_m^\ominus(NH_3,g) = -45.9 kJ \cdot mol^{-1}$; $\Delta_f H_m^\ominus(CO_2,g) = -393.5 kJ \cdot mol^{-1}$; $\Delta_f H_m^\ominus(H_2O,l) = -285.8 kJ \cdot mol^{-1}$; $\Delta_f H_m^\ominus\{CO(NH_2)_2,s\} = -333.1 kJ \cdot mol^{-1}$. According to Equation 2.9,

$$\Delta_r H_m^\ominus = \Delta_f H_m^\ominus\{CO(NH_2)_2,s\} + \Delta_f H_m^\ominus(H_2O,l) - 2 \times \Delta_f H_m^\ominus(NH_3,g) - \Delta_f H_m^\ominus(CO_2,g)$$
$$= -333.1 kJ \cdot mol^{-1} - 285.8 kJ \cdot mol^{-1} + 2 \times 45.9 kJ \cdot mol^{-1} + 393.5 kJ \cdot mol^{-1}$$
$$= -133.6 kJ \cdot mol^{-1}$$

While the temperature of reaction T is close to 298.15K, heat of reaction can be considered as constant, thus

$$\Delta_r H_{m,T}^\ominus \approx \Delta_r H_{m,298.15}^\ominus$$

3. Calculation of heat of reaction from standard molar enthalpy of combustion

It is very difficult for the most of organic compounds to measure the heat of formation from the stable elements, but the heat liberated from the combustion reaction with oxygen have been studied carefully at constant volume by Figure 2-3 and the data of reaction can be obtained.

The term **standard molar enthalpy of combustion** refers to the amount of heat liberated per mole of substance B burned or oxidized completely. The sign is $\Delta_c H_m^\ominus$ and its unit is $kJ \cdot mol^{-1}$. From the definition the conclusion that the values $\Delta_c H_m^\ominus$ of $CO_2(g)$, $H_2O(l)$, $SO_2(g)$, $N_2(g)$ etc. are zero is naturally obtained. Complete combustion or oxidization means such elements as C, H, S, N contained in organic compounds will be transformed to the corresponding substance mentioned above. The data $\Delta_c H_m^\ominus$ of some organic compounds is given in the Appendix 3.

The relation between $\Delta_c H_m^\ominus$ of reactants or products and $\Delta_r H_m^\ominus$ of reaction is illustrated as follows

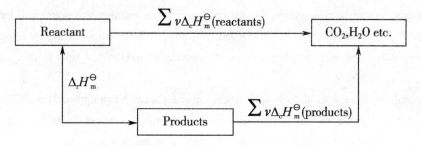

According to Hess's law,
$$\sum v\Delta_c H_m^\ominus (\text{reactants}) = \Delta_r H_m^\ominus + \sum v\Delta_c H_m^\ominus (\text{products})$$
The enthalpy change of a reaction is calculated,
$$\Delta_r H_m^\ominus = \sum v\Delta_c H_m^\ominus (\text{reactants}) - \sum v\Delta_c H_m^\ominus (\text{products}) \tag{2.10}$$

Note the change of the signs of reactants and products compared to Equation 2.9 and the stoichiometric numbers.

Sample Problem 2-5 For the reaction $CH_3CHO(l) + H_2(g) = CH_3CH_2OH$ (l) at 298.15K, it is difficult to determine the value of $\Delta_r H_m^\ominus$. Try to calculate it from the standard molar enthalpy of combustion.

Solution From Appendix 3, $\Delta_c H_m^\ominus(CH_3CHO, l) = -1\,166.9 kJ\cdot mol^{-1}$, $\Delta_c H_m^\ominus(CH_3CH_2OH, l) = -1\,366.8 kJ\cdot mol^{-1}$. According to the definition of standard molar enthalpy of combustion:
$$\Delta_c H_m^\ominus(H_2, g) = \Delta_f H_m^\ominus(H_2O, l) = -285.8 kJ\cdot mol^{-1}$$
$$\Delta_r H_m^\ominus = \Delta_c H_m^\ominus(CH_3CHO, l) + \Delta_c H_m^\ominus(H_2, g) - \Delta_c H_m^\ominus(CH_3CH_2OH, l)$$
$$= -1\,166.9 kJ\cdot mol^{-1} - 285.8 kJ\cdot mol^{-1} + 1\,366.8 kJ\cdot mol^{-1} = -85.9 kJ\cdot mol^{-1}$$

Heat of the reaction can also be calculated by $\Delta_f H_m^\ominus$:
$$\Delta_f H_m^\ominus(CH_3CHO, l) = -192.2 kJ\cdot mol^{-1}; \; \Delta_f H_m^\ominus(H_2, g) = 0 kJ\cdot mol^{-1}; \; \Delta_f H_m^\ominus(CH_3CH_2OH, l) = -277.6 kJ\cdot mol^{-1}. \text{ So,}$$
$$\Delta_r H_m^\ominus = \Delta_f H_m^\ominus(CH_3CHO, l) + \Delta_f H_m^\ominus(H_2, g) - \Delta_f H_m^\ominus(CH_3CH_2OH, l)$$
$$= -277.6 kJ\cdot mol^{-1} - 0 kJ\cdot mol^{-1} + 192.2 kJ\cdot mol^{-1} = -85.4 kJ\cdot mol^{-1}$$

2.3 Entropy and Gibbs's Free Energy

Although the first law of thermodynamics can solve the problems whether the heat of a reaction is absorbed or liberated and the amount of heat in the reaction, for example, heat of 85.9kJ is absorbed from the surroundings as alcohol is decomposed into methanol and hydrogen at 298.15K, $C_2H_5OH(l) = CH_3CHO(l) + H_2(g)$, the adverse reaction $CH_3CHO(l) + H_2(g) = C_2H_5OH(l)$ will release the same amount of heat to the surroundings. There is no answer that which reaction above is spontaneous according to the first law. Problems referred to the direction and the equilibrium of a chemical reaction need to be resolved by the second law of thermodynamics.

2.3.1 Spontaneous Process and Its Characteristics

1. The characteristics of spontaneous process

Spontaneous process of a system occurs naturally without any drive from the surroundings. The spontaneous process can be observed in natural phenomena, for instance, (a) water will move by itself only from the higher level to the lower level; (b) heat always flows from the higher temperature heat reservoir to the lower temperature heat reservoir; (c) electricity tends to flow only from a polar of higher potential to one of the lower, etc.

The general characteristics of these spontaneous processes can be summarized as follows:

(a) Spontaneous process tends to proceed in unidirection, the adverse direction cannot take place without energy supplied by the outside work like the flow of water or heat.

(b) All spontaneous processes can do work on the surroundings, for instance, the flow of water or gas and heat can cause the engine or machine to run, the spontaneous oxidation-reduction reaction can supply electric work and so on.

(c) Any spontaneous process will always lead to an equilibrium state in the end. At the equilibrium state the process cannot continue to take any more and the power to supply work disappears.

2. Impel of a spontaneous reaction

Spontaneity of a reaction is very important for chemical theory and practice. It is waste of time to study the rate and the mechanism of the reaction that cannot occur in thermodynamic principle. In the nineteenth century many experiments failed eventually in the preparation of diamonds from graphite as reactant. Later the thermodynamic theory proved that graphite cannot transform to diamond at the common temperature and pressure, the change must take place at the pressure higher than 1.5×10^9 Pa.

At 70's of the nineteen century French Chemist Berthelot Pem and Denmark chemist Thomson J. meanwhile suggested that only exothermic reactions can proceed spontaneously. It is reasonable because the chemical system would be more stable after giving off some energy to the surroundings. Actually lots of spontaneous reactions are heat-liberating ($\Delta H < 0$).

But there are some endothermic processes that are spontaneous, for example the dissolution of KNO_3 in water, the decomposition of $CaCO_3$, etc.

$$CaCO_3(s) = CaO(s) + CO_2(g)$$

The above reaction obtains heat and takes place spontaneously when the temperature is more than 840℃. The more examples of spontaneous endothermic reaction are

$$CoCl_2 \cdot 6H_2O(s) + 6SOCl_2(l) = CoCl_2(s) + 6SO_2(g) + 12HCl(g)$$

$$N_2O_4(g) = 2NO_2(g)$$

Therefore heat of reaction is not exclusive factor which influences the degree of spontaneity in a chemical process. After studying these examples we found the "**mixture**" or "**disorder**" (more degree of freedom or more numbers of particles) of a system increases through a spontaneous process without exception. Increase of disorder in a system is another drive for spontaneous reaction. Especially, that occurs in the isolated system the degree of disorder is the exclusive factor to determine the spontaneity.

2.3.2 Entropy of A System

1. Entropy and its change

Disorder is a macroscopic state of a system which can be described by the state function "**entropy**" denoted by the letter "S". Entropy is extensive property and the change in entropy of one system depends only on the initial and final state, so we can design a reversible process which initial and final states are the same as the practical irreversible process. The formula of entropy change in isotherm reversible process is deducted

$$\Delta S = \frac{Q_r}{T} \quad \text{or} \quad dS = \frac{\delta Q_r}{T} \tag{2.11}$$

Q_r is the reversible heat effect in the process, it is suggested from Equation 2.11 that the lower temperature of a system is, the less entropy is. When a system liberates heat to the surroundings the entropy in it decreases and vice versa.

The thermodynamic state function entropy relates "the disorder" or "the chaotic" of the system, from the microscopic view German scientist Boltsman revealed that the relation between the microscopic number of possible configurations Ω and the entropy of the system is $S = k\ln\Omega$. This statistical definition allows the entropy to be further understood as a measure of the disorder in a system. When the thermodynamic temperature of a system almost reaches 0K the random motion of molecules almost cease and the entropy equals nearly to the minimum value. **The third law of thermodynamics** indicates that entropy of any perfectly crystalline at the absolute zero of temperature is zero. Based on the third law and Equation 2.11 the entropy has a measurable value for a system, termed as **conventional entropy**.

The standard molar entropy denoted by $S_m^\ominus$ is the conventional entropy of 1mol substance at standard state. The unit of entropy is $J \cdot K^{-1} \cdot mol^{-1}$. Note that the value of standard molar entropy of the stable elementals is not zero, which is different from that of standard molar enthalpy of forming, because generally the stable element compositions is not prefect crystalline and not under the condition of 0K.

The standard molar entropy value of ions in aqueous solution is based on that $S_m^\ominus$ of hydrogen ion is given the value of zero.

Some rules of S referring to the same substance are (1) S (gas) > S (liquid) > S (solid); (2) S (higher temperature) > S (lower temperature); (3) S (lower pressure) > S (higher pressure), pressure can hardly take effect on S of liquid or solid substance; (4) $\Delta S_{mix.} > 0$ is always correct for mixing process of different matter.

The standard molar entropy is different among reactants and products for a reaction. Therefore the entropy change in reaction may be calculated from

$$\Delta_r S_m^\ominus = \sum \nu S_m^\ominus (\text{products}) - \sum \nu S_m^\ominus (\text{reactants}) \tag{2.12}$$

For the extensive property of entropy, the stoichiometric number ν must be multiplied with the corresponding substance. If the temperature variance is limit for a reaction the entropy change at the temperature T may be considered close to that at the temperature of 298.15K.

$$\Delta_r S_{m,T}^\ominus \approx \Delta_r S_{m,298.15}^\ominus \tag{2.13}$$

2. Principle of entropy increase

In terms of Equation 2.11 for adiabatic reversible process, $Q = 0$, $\Delta S = 0$, or the entropy of the system does not change. For the adiabatic irreversible process the entropy of the system increases. Isolated system neither exchanges matter nor energy with its surroundings, therefore the entropy of isolated system will increase in irreversible process or remain constant in the course of reversible process. This is the famous **principle of entropy increase**, one of the expressions for **the second law of thermodynamics**, formulated in mathematic as

$$\Delta S_{iso.} \geq 0 \qquad (2.14)$$

It is important to point out that the principle of entropy increase refers to an isolated system. In fact, the most of systems cannot be regarded as being isolated. In this case the total entropy change (the sum of the entropy change in the system and in the surroundings) is comply with Equation 2.14.

$$\Delta S_{iso.} = \Delta S_{sum.} = \Delta S_{sys.} + \Delta S_{sur.} \qquad (2.15)$$

But, ">" relation applies to irreversible process, "=" relation applies to reversible process or equilibrium state. The entropy of an isolated system never decrease because the value of total entropy is more than zero in the spontaneous process.

Sample Problem 2-6 The volume of 1mol perfect gas increases five times due to isothermal expansion. What is the change of entropy (1) in the reversible isothermal process and (2) the expansion to vacuum, respectively?

Solution (1) In reversible process by Equation 2.11,

$$\Delta S_{sys.} = \frac{Q_r}{T} = -\frac{W_{max.}}{T} = \frac{\int_{V_1}^{V_2} pdV}{T} = \frac{nRT \ln \frac{V_2}{V_1}}{T} = nR \ln \frac{5}{1}$$
$$= 1\text{mol} \times 8.314 \text{J} \cdot \text{K}^{-1} \cdot \text{mol}^{-1} \times 1.609 = 13.4 \text{J} \cdot \text{K}^{-1}$$

(2) Because the expansion to vacuum is spontaneous process the entropy change is identical to that of reversible process for property of state function, $\Delta S_{sys.} = 13.4 \text{J} \cdot \text{K}^{-1}$.

Discussion: In the process of isothermal expansion, the total entropy change can be calculated

$$\Delta S_{sur.} = -\frac{Q_r}{T} = -13.4 \text{J} \cdot \text{K}^{-1}$$

$\Delta S_{sum.} = \Delta S_{sys.} + \Delta S_{sur.} = 13.4 \text{J} \cdot \text{K}^{-1} - 13.4 \text{J} \cdot \text{K}^{-1} = 0 \text{J} \cdot \text{K}^{-1}$, it is reversible process.

There is no exchanged heat between system and surroundings in the process of free expansion, $Q = 0$, hence $\Delta S_{sur.} = 0$

$\Delta S_{sum.} = \Delta S_{sys.} + \Delta S_{sur.} = 13.4 \text{J} \cdot \text{K}^{-1} + 0 \text{J} \cdot \text{K}^{-1} = 13.4 \text{J} \cdot \text{K}^{-1} > 0$, it is spontaneous process.

> **Question and thinking 2-4** Whether the following reactions violate the principle of entropy increase or not?
> 1. Heat is converted to work by coal burning power.
> 2. Water reacts with carbon dioxide to produce glucose in photosynthesis.

2.3.3 Gibbs's Free Energy of System

1. Gibbs's free energy

From the above discussion, we knew that the total entropy change in determining the direction of a spontaneous process is applied for the isolated system. But there are few such of systems that can fulfill the requirement of the isolation of energy and material. It's necessary to find the new state function that will indicate the direction and equilibrium position and to deal with the change in energy and entropy of system only.

In the reversible process of isothermal and isobar, $Q_p = \Delta H$, the entropy change in surroundings,

$$\Delta S_{sur.} = \frac{Q_{r,p\,sur.}}{T} = \frac{-Q_{r,p\,sys.}}{T} = -\frac{\Delta H_{sys.}}{T} \qquad (2.16)$$

Equation 2.15 is substituted to Equation 2.16.

$$\Delta S_{\text{sys.}} + \Delta S_{\text{sur}} = \Delta S_{\text{sys.}} - \frac{\Delta H_{\text{sys.}}}{T} \geqslant 0$$

or $\quad\quad\quad\quad \Delta H - \Delta(TS) \leqslant 0$

If the temperature is constant, $\quad \Delta H - T\Delta S \leqslant 0$

$$\Delta(H - TS) \leqslant 0$$

Define the state function G, $\quad\quad G = H - TS \quad\quad\quad\quad\quad\quad (2.17)$

So, $\quad\quad\quad\quad\quad \Delta G \leqslant 0 \quad\quad\quad\quad\quad\quad (2.18)$

The new state function G is named **Gibbs free energy**. It is applicable to the system that is constrained to be at the constant temperature and constant pressure. G is an extensive variable that relates three functions: entropy S, enthalpy H and temperature T. ΔG is the change in Gibbs free energy. If ΔG is negative in a state change of system, the process will occurs spontaneously. Conversely, ΔG with a positive value is associated with a non-spontaneous process. If ΔG has a value of zero, the system is just at equilibrium state.

Under the conditions of constant pressure and constant temperature by Equation 2.17, we have

$$\Delta G = \Delta H - T\Delta S \quad\quad\quad\quad\quad\quad (2.19)$$

Equation 2.19 is the famous Gibbs equation related to enthalpy, entropy and temperature. The spontaneity of a process is discussed as follows.

(a) $\Delta H < 0$ and $\Delta S > 0$, hence $\Delta G < 0$. Such process occurs spontaneously at any temperature.

(b) $\Delta H > 0$ and $\Delta S < 0$, hence $\Delta G > 0$. Such process will not occur.

(c) $\Delta H < 0$ and $\Delta S < 0$, when the temperature is lower than $\Delta H/\Delta S$, the process occurs spontaneously.

(d) $\Delta H > 0$ and $\Delta S > 0$, when the temperature is higher than $\Delta H/\Delta S$, the process occurs spontaneously.

The transition temperature of some reactions to be spontaneous by changing temperature is defined as

$$T_{\text{trans.}} = \frac{\Delta H}{\Delta S} \quad\quad\quad\quad\quad\quad (2.20)$$

2. Gibbs free energy and non-volume work

ΔG can be the indicator of spontaneity for a process at the constant temperature and pressure, meanwhile there is no other types of work except volume work. For a reversible process at the constant temperature and pressure, it is proved that the maximum non-volume work done by a system to the outside world equals to the value of change in Gibbs free energy.

$$\Delta G = W_{\text{f, max.}} \quad\quad\quad\quad\quad\quad (2.21)$$

So, ΔG can be determined by measuring the maximum non-volume work through the isothermal and isobar reversible process.

3. The direction of a reaction judged by Gibbs free energy

ΔG of a process without non-volume work under the conditions of constant temperature and pressure can be calculated in the following expression

$$\Delta_r G = \sum vG(\text{products}) - \sum vG(\text{reactants})$$

Standard molar Gibbs free energy of formation $\Delta_f G_m^\ominus$ whose unit is $kJ \cdot mol^{-1}$ is defined as the free energy which accompanies the formation of 1 mole compound from its most stable element in the standard state, so it is naturally inferred that $\Delta_f G_m^\ominus$ of the stable element equals to zero. Values of conventional substance may be obtained from the Appendix 3 of this textbook.

4. Calculation of $\Delta_r G_m^\ominus$

At 298.15K, the standard Gibbs free energy change in molar of a reaction $\Delta_r G_m^\ominus$ is

$$\Delta_r G_m^\ominus = \sum \nu \Delta_f G_m^\ominus(\text{products}) - \sum \nu \Delta_f G_m^\ominus(\text{reactants}) \tag{2.22}$$

The function G is extensive. The stoichiometric ν should be concerned. As long as T is not beyond 298.15K too much, the free energy change can be calculated by the standard enthalpy and entropy for a reaction at temperature T.

$$\Delta_r G_{m,T}^\ominus \approx \Delta_r H_{m,298.15}^\ominus - T \Delta_r S_{m,298.15}^\ominus \tag{2.23}$$

Similarly, $\Delta_r G_m^\ominus$ of a reaction can be obtained from $\Delta_r G_m^\ominus$ values of relatived reactions by Hess's law.

Sample Problem 2-7 The oxidation of glucose is the important reaction to obtain the energy for human body. Try to calculate $\Delta_r H_m$, $\Delta_r S_m$ and $\Delta_r G_m$ of reaction, $C_6H_{12}O_6(s) + 6O_2(g) = 6CO_2(g) + 6H_2O(l)$ at 37℃, and to determine whether it is spontaneous.

Solution The body temperature is near to 298.15K, the data of 298.15K can be used.

$$\begin{aligned}
\Delta_r H_m &\approx \sum \nu \Delta_f H_m^\ominus(\text{products}) - \sum \nu \Delta_f H_m^\ominus(\text{reactants}) \\
&= 6\Delta_f H_m^\ominus(CO_2, g) + 6\Delta_f H_m^\ominus(H_2O, l) - \Delta_f H_m^\ominus(C_6H_{12}O_6, s) - 6\Delta_f H_m^\ominus(O_2, g) \\
&= 6 \times (-393.5)\text{kJ} \cdot \text{mol}^{-1} + 6 \times (-285.8)\text{kJ} \cdot \text{mol}^{-1} - (-1\,273.3)\text{kJ} \cdot \text{mol}^{-1} - 6 \times 0 \\
&= -2\,802.5 \text{kJ} \cdot \text{mol}^{-1}
\end{aligned}$$

$$\begin{aligned}
\Delta_r S_m &= \sum \nu S_m^\ominus(\text{products}) - \sum \nu S_m^\ominus(\text{reactants}) \\
&= 6S_m^\ominus(CO_2, g) + 6S_m^\ominus(H_2O, l) - S_m^\ominus(C_6H_{12}O_6, s) - 6S_m^\ominus(O_2, g) \\
&= 6 \times 213.8 \text{J} \cdot \text{K}^{-1} \cdot \text{mol}^{-1} + 6 \times 70.0 \text{J} \cdot \text{K}^{-1} \cdot \text{mol}^{-1} - 212.1 \text{J} \cdot \text{K}^{-1} \cdot \text{mol}^{-1} - 6 \times 205.2 \text{J} \cdot \text{K}^{-1} \cdot \text{mol}^{-1} \\
&= 259.5 \text{J} \cdot \text{K}^{-1} \cdot \text{mol}^{-1}
\end{aligned}$$

$$\begin{aligned}
\Delta_r G_m &= \Delta_r H_m - T\Delta_r S_m = -2\,802.5 \text{kJ} \cdot \text{mol}^{-1} - 310\text{K} \times 259.5 \times 10^{-3} \text{kJ} \cdot \text{K}^{-1} \cdot \text{mol}^{-1} \\
&= -2\,883 \text{kJ} \cdot \text{mol}^{-1}
\end{aligned}$$

Here, $\Delta_r H_m < 0$ and $\Delta_r G_m < 0$, the conclusion can be drawn that the reaction is exothermal and spontaneous. Note the unit of entropy ($\text{J} \cdot \text{K}^{-1} \cdot \text{mol}^{-1}$) is different from that of enthalpy ($\text{kJ} \cdot \text{mol}^{-1}$) generally.

Sample Problem 2-8 A reaction is know, $NH_4Cl(s) = NH_3(g) + HCl(g)$

(1) Calculate the value of $\Delta_r G_m^\ominus$ at 298.15K and judge its spontaneity.

(2) What is the minimum temperature at which the reaction can occur spontaneously at standard state?

Solution (1) By the data in Appendix 3, Equation 2.22 and Equation 2.23,

Way 1: $\quad \Delta_r G_m^\ominus = \Delta_f G_m^\ominus(NH_3, g) + \Delta_f G_m^\ominus(HCl, g) - \Delta_f G_m^\ominus(NH_4Cl, s)$

$\quad\quad\quad\quad = -16.4 \text{kJ} \cdot \text{mol}^{-1} - 95.3 \text{kJ} \cdot \text{mol}^{-1} + 202.9 \text{kJ} \cdot \text{mol}^{-1}$

$\quad\quad\quad\quad = 91.2 \text{kJ} \cdot \text{mol}^{-1}$

Way 2: $\quad \Delta_r H_m^\ominus = \Delta_f H_m^\ominus(NH_3, g) + \Delta_f H_m^\ominus(HCl, g) - \Delta_f H_m^\ominus(NH_4Cl, s)$

$\quad\quad\quad\quad = -45.9 \text{kJ} \cdot \text{mol}^{-1} - 92.3 \text{kJ} \cdot \text{mol}^{-1} + 314.4 \text{kJ} \cdot \text{mol}^{-1} = 176.2 \text{kJ} \cdot \text{mol}^{-1}$

$\quad\quad\quad \Delta S_m^\ominus = S_m^\ominus(NH_3, g) + S_m^\ominus(HCl, g) - S_m^\ominus(NH_4Cl, s)$

$\quad\quad\quad\quad = 192.8 \text{J} \cdot \text{K}^{-1} \cdot \text{mol}^{-1} + 186.9 \text{J} \cdot \text{K}^{-1} \cdot \text{mol}^{-1} - 94.6 \text{J} \cdot \text{K}^{-1} \cdot \text{mol}^{-1} = 285 \text{J} \cdot \text{K}^{-1} \cdot \text{mol}^{-1}$

Hence, $\Delta_r G_m^\ominus = \Delta_r H_m^\ominus - T\Delta_r S_m^\ominus$

$\quad\quad\quad\quad = 176.2 \text{kJ} \cdot \text{mol}^{-1} - 298.15\text{K} \times 285 \times 10^{-3} \text{kJ} \cdot \text{K}^{-1} \cdot \text{mol}^{-1} = 91.2 \text{kJ} \cdot \text{mol}^{-1}$

Here, $\Delta_r G_m^{\ominus} > 0$, the reaction cannot occur spontaneously at 298.15K, therefore $NH_4Cl(s)$ is stable at room temperature.

(2) Using Equation 2.20, $T_{trans} = \dfrac{\Delta H}{\Delta S} = 176\,200 J \cdot mol^{-1} / 285 kJ \cdot K^{-1} \cdot mol^{-1} = 618K$

$NH_4Cl(s)$ will not exist anymore when the temperature is higher than 618K.

5. Calculation of Gibbs free energy at other states

For a reaction, $\qquad aA + bB = dD + eE$

the change in Gibbs free energy at non-standard state is given by

$$\Delta_r G_m = \Delta_r G_m^{\ominus} + RT\ln J \qquad (2.24)$$

Where J is the reaction quotient, R is gas constant and T is the temperature of the reaction. Equation 2.24 is called the isothermal equation of chemical reaction.

The expression of J in solution is

$$J = \dfrac{(c_D/c^{\ominus})^d (c_E/c^{\ominus})^e}{(c_A/c^{\ominus})^a (c_B/c^{\ominus})^b} \qquad (2.25)$$

c is the concentration of products or reactants at a given case in molarity. $c^{\ominus}$ is $1 mol \cdot L^{-1}$. For a gaseous reaction

$$J = \dfrac{(p_D/p^{\ominus})^d (p_E/p^{\ominus})^e}{(p_A/p^{\ominus})^a (p_B/p^{\ominus})^b} \qquad (2.26)$$

p means the partial pressure of gaseous species involved in the reaction, its unit is kPa, $p^{\ominus} = 100 kPa$. Note that the composition for solid or liquid substance is not term in J expression.

Sample Problem 2-9 The decomposition reaction of $CaCO_3(s)$ is

$$CaCO_3(s) = CaO(s) + CO_2(g)$$

If the partial pressure of CO_2 is 0.010kPa, calculate the lowest temperature at which the reaction is spontaneous.

Solution The thermodynamic data are as follows.

	$CaCO_3(s)$	$CaO(s)$	$CO_2(g)$
$\Delta_f G_m^{\ominus}/(kJ \cdot mol^{-1})$	$-1\,206.9$	-634.9	-393.5
$S_m^{\ominus}/(J \cdot K^{-1} \cdot mol^{-1})$	92.9	38.1	213.8

$\Delta_r H_m^{\ominus} = \Delta_f H_m^{\ominus}(CaO,s) + \Delta_f H_m^{\ominus}(CO_2,g) - \Delta_f H_m^{\ominus}(CaCO_3,s)$
$\qquad = -634.9 kJ \cdot mol^{-1} + (-393.5) kJ \cdot mol^{-1} - (-1\,206.9) kJ \cdot mol^{-1} = 178.5 kJ \cdot mol^{-1}$

$\Delta S_m^{\ominus} = S_m^{\ominus}(CaO,s) + S_m^{\ominus}(CO_2,g) - S_m^{\ominus}(CaCO_3,s)$
$\qquad = 38.1 J \cdot K^{-1} \cdot mol^{-1} + 213.8 J \cdot K^{-1} mol^{-1} - 92.9 J \cdot K^{-1} \cdot mol^{-1} = 159 J \cdot K^{-1} \cdot mol^{-1}$

0.010kPa of partial pressure for CO_2 is not at the standard state. Assuming solid $CaCO_3$ starts to decompose at temperature T, from Equation 2.23 and Equation 2.24,

$\Delta_r G_{m,T}^{\ominus} \approx \Delta_r H_{m,298.15}^{\ominus} - T\Delta_r S_{m,298.15}^{\ominus} \approx 178.5 kJ \cdot mol^{-1} - T \times 159 \times 10^{-3} kJ \cdot k^{-1} \cdot mol^{-1}$

$$\Delta_r G_m = \Delta_r G_m^{\ominus} + RT\ln J$$

$\qquad = 178.5 kJ \cdot mol^{-1} - T \times 159 \times 10^{-3} kJ \cdot K^{-1} \cdot mol^{-1} + 8.314 \times 10^{-3} kJ \cdot mol^{-1} \cdot K^{-1} \times T\ln \dfrac{0.010 kPa}{100 kPa}$

$\qquad$ If $\Delta_r G_{m,T} < 0$, $\quad T > \dfrac{178.5 kJ \cdot mol^{-1}}{0.235 kJ \cdot mol^{-1} \cdot K^{-1}} = 759K$

This problem demonstrates that decreasing in the partial pressure of CO_2 is advantageous for the decomposition of solid $CaCO_3$.

Key Terms

系统	system
环境	surroundings
开放系统	opening system
封闭系统	closed system
孤立(隔离)系统	isolated system
状态函数	state functions
广度(容量)性质	extensive properties
强度性质	intensive properties
状态方程	state equations
等温过程	isothermal process
等压过程	isobar process
等容过程	isovolumic process
绝热过程	adiabatic process
循环过程	cyclic process
热	heat
功	work
热力学能	Thermodynamic energy
内能	internal energy
热力学第一定律	The first law of thermodynamics
焓	Enthalpy
反应进度	Extent of reaction
盖斯定律	Hess's law
自发过程	Spontaneous process
熵	Entropy
规定熵	conventional entropy
热力学第二定律	the second law of thermodynamics
熵增原理	principle of entropy increase
吉布斯自由能	Gibbs's free energy
化学反应等温方程式	the isothermal equation of chemical reaction

Summary

Thermodynamics is the study on the energy flow between the system and its surrounding. The first law of thermodynamics states that the change in the internal energy ΔU of a process equals the sum of the heat Q absorbed by the system and the work W done on the system by surrounding, i.e. $\Delta U = Q + W$.

The internal energy U, enthalpy H, entropy S and Gibbs free energy G of a system are all the state functions. But Q and W are not. The values of Q and W depend on how the state change of system takes place.

The heat of reaction at constant volume Q_v is equal to ΔU, whereas the heat of reaction at constant pressure Q_p is equal to ΔH.

The potential energy change is one of factors that influence the spontaneity of a reaction. Exothermic process with the negative values of ΔH tends to proceed spontaneously. The thermodynamic quantity associated with randomness is entropy S. An increase in entropy favors a spontaneous change. Second law of thermodynamics states that the entropy of the universe (system and its surrounding) increases whenever a spontaneous change occurs.

The Gibbs free energy change ΔG allows us to determine the combined effects of temperature, enthalpy and entropy changes on the spontaneity of a chemical or physical change. A change is spontaneous only if the Gibbs free energy of the system decreases ($\Delta G < 0$). When ΔH and ΔS have same algebraic sign, the temperature becomes the critical factor in determining the spontaneity.

Under the standard state ($p^\ominus = 100\text{kPa}$, temperature is usually assigned to be 298.15K), $\Delta_r H_m^\ominus$, $\Delta_r S_m^\ominus$ and $\Delta_r G_m^\ominus$ of a chemical reaction can been calculated on the basis of Hess law

$$\Delta_r H_m^\ominus = \sum \nu \Delta_f H_m^\ominus (\text{products}) - \sum \nu \Delta_f H_m^\ominus (\text{reactants})$$

$$\Delta_r H_m^\ominus = \sum \nu \Delta_c H_m^\ominus (\text{reactants}) - \sum \nu \Delta_c H_m^\ominus (\text{products})$$

$$\Delta_r S_m^\ominus = \sum \nu S_m^\ominus (\text{products}) - \sum \nu S_m^\ominus (\text{reactants})$$

$$\Delta_r G_m^\ominus = \sum \nu \Delta_f G_m^\ominus (\text{products}) - \sum \nu \Delta_f G_m^\ominus (\text{reactants})$$

Where $\Delta_f H_m^\ominus$, $\Delta_c H_m^\ominus$, $\Delta_r G_m^\ominus$ and $S_m^\ominus$ represent the molar enthalpy of formation, molar heat of combustion, molar free energy of formation and molar entropy of substances under the standard state, respectively. At the temperature other than 298.15K, it can be considered that

$$\Delta_r H_{m,T}^\ominus \approx \Delta_r H_{m,298.15}^\ominus, \quad \Delta_r S_{m,T}^\ominus \approx \Delta_r S_{m,298.15}^\ominus \text{ and } \Delta_r G_{m,T}^\ominus \approx \Delta_r H_{m,298.15}^\ominus - T\Delta_r S_{m,298.15}^\ominus$$

When a system reaches equilibrium, $\Delta G = 0$, and no useful work can be obtained from it. At any particular pressure, the equilibrium between two phases for a substance (e.g., liquid ↔ solid, or liquid ↔ vapor) can only occur at a temperature. The entropy change can be computed as $\Delta S = \Delta H / T$. The temperature at which the equilibrium occurs can be calculated from $T = \Delta H / \Delta S$.

Exercises

1. What is a state function and what are the fundamental features of a state function? Which are the state functions among the thermodynamic variables $T, p, V, \Delta U, \Delta H, \Delta G, S, G, Q_p, Q_v, Q, W, W_{f,\max}$? Which are the extensive properties? Which are the intensive properties?

2. Calculate the change of thermodynamic energy in the following system.
 (1) The system releases heat of 2.5kJ, meanwhile it does work of 500J to the surroundings.
 (2) The system release heat of 650J and it is done work of 350J by the surroundings.

3. What is the thermo-chemical equation? Why is it necessary to establish the standard state of a substance in thermodynamics? What are the standard states of substances?

4. The following reaction has been investigated:

$$A + B = C + D \qquad \Delta_r H_{m,1}^\ominus = -40.0 \text{ kJ} \cdot \text{mol}^{-1}$$

$$C + D = E \qquad \Delta_r H_{m,2}^\ominus = 60.0 \text{ kJ} \cdot \text{mol}^{-1}$$

Calculate the molar standard enthalpy $\Delta_r H_m^\ominus$ of following reactions, respectively.

(1) C+D=A+B (2) 2C+2D=2A+2B (3) A+B=E

5. 4.0mol of $H_2(g)$ reacted with 0.2mol of $O_2(g)$ and 0.6mol of $H_2O(g)$ was obtained finally at certain temperature, calculate the reaction extents according to the following two expression

(1) $2H_2(g)+O_2(g)=2H_2O(g)$ (2) $H_2(g)+\frac{1}{2}O_2(g)=H_2O(g)$

6. Illustrate the meaning of these signs, respectively: $\Delta_r H_{m,298.15}^{\ominus}$, $\Delta_r H_m^{\ominus}(H_2O,g)$, $\Delta_c H_m^{\ominus}(H_2,g)$, $S_m^{\ominus}(H_2,g)$, $\Delta_r S_{m,T}^{\ominus}$, $\Delta_r G_{m,T}^{\ominus}$, $\Delta_r G_m^{\ominus}(CO_2,g)$.

7. The standard enthalpy changes of the reactions have been known

(1) $C_6H_6(l)+7\frac{1}{2}O_2(g)=6CO_2(g)+3H_2O(l)$ $\Delta_r H_{m,1}^{\ominus}=-3267.6 kJ \cdot mol^{-1}$

(2) $C(gra.)+O_2(g)=CO_2(g)$ $\Delta_r H_{m,2}^{\ominus}=-393.5 kJ \cdot mol^{-1}$

(3) $H_2(g)+\frac{1}{2}O_2(g)=H_2O(l)$ $\Delta_r H_{m,3}^{\ominus}=-285.8 kJ \cdot mol^{-1}$

How much is the standard heat change $\Delta_r H_m^{\ominus}$ of the reaction $6C(gra.)+3H_2(g)=C_6H_6(l)$?

8. Hydrazine N_2H_4 is a liquid fuel used in driving a rocket where N_2O_4 is oxidant and the products of the combustion are $N_2(g)$ and $H_2O(l)$, $\Delta_r H_m^{\ominus}(N_2H_4,l)=50.63 kJ \cdot mol^{-1}$, $\Delta_r H_m^{\ominus}(N_2O_4,g)=9.16 kJ \cdot mol^{-1}$. Write the reaction equation of combustion and calculate $\Delta_r H_m^{\ominus}$ value of combustion reaction.

9. The thermodynamic data of the following reactions at standard state and 298.15K are

(1) $Fe_2O_3(s)+3CO(g) \rightarrow 2Fe(s)+3CO_2(g)$ $\Delta_r H_{m,1}^{\ominus}=-24.8 kJ \cdot mol^{-1}$, $\Delta_r G_{m,1}^{\ominus}=-29.4 kJ \cdot mol^{-1}$

(2) $3Fe_2O_3(s)+CO(g) \rightarrow 2Fe_3O_4(s)+CO_2(g)$ $\Delta_r H_{m,2}^{\ominus}=-47.2 kJ \cdot mol^{-1}$, $\Delta_r G_{m,2}^{\ominus}=-61.41 kJ \cdot mol^{-1}$

(3) $Fe_3O_4(s)+CO(g) \rightarrow 3FeO(s)+CO_2(g)$ $\Delta_r H_{m,3}^{\ominus}=19.4 kJ \cdot mol^{-1}$, $\Delta_r G_{m,3}^{\ominus}=5.21 kJ \cdot mol^{-1}$

Calculate $\Delta_r H_{m,4}^{\ominus}$, $\Delta_r G_{m,4}^{\ominus}$ and $\Delta_r S_{m,4}^{\ominus}$ of reaction $FeO(s)+CO(g) \rightarrow Fe(s)+CO_2(g)$

10. The decomposition of methanol is $CH_3OH(l) \rightarrow CH_4(g)+\frac{1}{2}O_2(g)$

(1) Whether the reaction can occur spontaneously at the standard state and 298.15K?

(2) What temperature can the reaction be spontaneous be higher than at standard state?

11. Calculate $\Delta_r G_m^{\ominus}$, $\Delta_r H_m^{\ominus}$ and $\Delta_r S_m^{\ominus}$ of reaction $H_2O(g)+CO(g)=H_2(g)+CO_2(g)$ and $S_m^{\ominus}(H_2O,g)$ at standard state and 298.15K.

12. A sick man requires 6300kJ of energy per day averagely. How many liters of $50.0 g \cdot L^{-1}$ glucose solution should be injected into him except having 250g milk and 50g bread per day? The values of combustion of milk, bread and glucose are $3.0 kJ \cdot g^{-1}$, $12 kJ \cdot g^{-1}$ and $15.6 kJ \cdot g^{-1}$, respectively.

13. The overall reaction of sugar metabolism is $C_{12}H_{22}O_{11}(s)+12O_2(g)=12CO_2(g)+11H_2O(l)$

(1) Calculate $\Delta_r G_m^{\ominus}$, $\Delta_r H_m^{\ominus}$ and $\Delta_r S_m^{\ominus}$ at standard state and 298.15K referring to the thermodynamic data from the Appendix.

(2) If 30% of Gibbs free energy is transformed into non-volume work in body, how much of work will be obtained in result of metabolism of 1.00mol of solid sugar at 37℃?

Supplementary Exercises

1. State the first law of thermodynamics in words. What equation defines the change in the internal energy in terms of heat and work? Define the meaning of the symbols, including the significance of their algebraic signs.

2. In what way is Gibbs free energy related to equilibrium?

3. Phosgene, COCl$_2$, was used as a war gas during World War I. It reacts with the moisture in the lungs to produce HCl, which causes the lungs to fill fluid, and CO$_2$. Write an equation of the reaction and compute $\Delta_r G_m^\ominus$. It is known that, $\Delta_f G_m^\ominus(\text{COCl}_2, \text{g}) = -210 \text{kJ} \cdot \text{mol}^{-1}$

4. Given the following reactions,
$$4\text{NO(g)} \rightarrow 2\text{N}_2\text{O(g)} + \text{O}_2\text{(g)} \quad \Delta_r G_{m,1}^\ominus = -139.56 \text{kJ} \cdot \text{mol}^{-1}$$
$$2\text{NO(g)} + \text{O}_2\text{(g)} \rightarrow 2\text{NO}_2\text{(g)} \quad \Delta_r G_{m,2}^\ominus = -69.70 \text{kJ} \cdot \text{mol}^{-1}$$
Calculate $\Delta_r G_{m,3}^\ominus$ for the reaction of $2\text{N}_2\text{O(g)} + 3\text{O}_2\text{(g)} \rightarrow 4\text{NO}_2\text{(g)}$.

5. Chloroform, formerly used as anesthetic and now believed to be a carcinogen (cancer-causing agent), has a heat of vaporization $\Delta_r H_m^\ominus = 31.4 \text{kJ} \cdot \text{mol}^{-1}$. Reaction, CHCl$_3$(l) $\rightarrow$ CHCl$_3$(g), has 94.2J $\cdot$ K^{-1} $\cdot$ mol^{-1} of $\Delta_r S_m^\ominus$. At what temperature do we expect CHCl$_3$ to boil?

6. Triglyceride is one of typical fatty acids, its metabolic reaction is
$$\text{C}_{57}\text{H}_{104}\text{O}_6(\text{s}) + 80\text{O}_2(\text{g}) = 57\text{CO}_2(\text{g}) + 52\text{H}_2\text{O(l)}, \quad \Delta_r H_m^\ominus = -3.35 \times 10^{-4} \text{kJ} \cdot \text{mol}^{-1}$$
Calculate $\Delta_f H_m^\ominus$ of Triglyceride.

7. Predict the sign of entropy change for each of the following reactions:
(a) O$_2$(g)(100kPa, 298K) $\rightarrow$ O$_2$(g) (10kPa, 298K)
(b) NH$_4$Cl(s) $\rightarrow$ NH$_3$(g) + HCl(g)
(c) CO(g) + H$_2$O(g) $\rightarrow$ CO$_2$(g) + H$_2$(g)

Answers to Some Exercises

[Exercises]

2. (1) -3.0kJ (2) -300J

4. 20.0kJ $\cdot$ mol^{-1}

5. 0.30mol; 0.60mol

7. 49.2kJ $\cdot$ mol^{-1}

8. $-1\,254$kJ $\cdot$ mol^{-1}

9. -11.0kJ $\cdot$ mol^{-1}; -6.20kJ $\cdot$ mol^{-1}; -6.20J $\cdot$ K^{-1} $\cdot$ mol^{-1}

10. 164.6kJ $\cdot$ mol^{-1}; 162.1J $\cdot$ K^{-1} $\cdot$ mol^{-1}; 116.3kJ $\cdot$ mol^{-1}; 1 015.42K

11. -41.2kJ $\cdot$ mol^{-1}; -28.6kJ $\cdot$ mol^{-1}; -42.26J $\cdot$ K^{-1} $\cdot$ mol^{-1}; 189.1J $\cdot$ K^{-1} $\cdot$ mol^{-1}

12. 6.31L

13. (1) $-5\,640$kJ $\cdot$ mol^{-1}; 513J $\cdot$ K^{-1} $\cdot$ mol^{-1}; $-5\,796$J $\cdot$ mol^{-1}
    (2) $-5\,799.11$kJ $\cdot$ mol^{-1}; $-1\,740$kJ $\cdot$ mol^{-1}

[Supplementary Exercises]

3. -137.9kJ $\cdot$ mol^{-1}

4. 0.16kJ $\cdot$ mol^{-1}

5. $T = \dfrac{\Delta_r H_m^\ominus}{\Delta_r S_m^\ominus} = \dfrac{31.4 \times 1\,000 \text{J} \cdot \text{mol}^{-1}}{94.2 \text{J} \cdot \text{mol}^{-1} \cdot \text{K}^{-1}} = 333\text{K} \approx 60$

6. $K^\ominus = 7.92 \times 10^8$

7. $\Delta_f H_m^\ominus(\text{CO}_2, \text{g}) = -393.5$kJ $\cdot$ mol^{-1}; $\Delta_f H_m^\ominus(\text{H}_2\text{O}, \text{l}) = -285.8$kJ $\cdot$ mol^{-1}; $\Delta_f H_m^\ominus(\text{O}_2, \text{g}) = 0$

(胡 新)

Chapter 3
Chemical Equilibrium

Chemical thermodynamics can predict the spontaneity of a reaction from the flow of energy. However, we still wonder what is the extent of a given reaction? How much product we can obtain from a particular reaction mixture? How can we adjust the conditions of a reaction to obtain more? Take the industrial synthesis of ammonia from nitrogen and hydrogen as an example, we manager to create conditions to maximize yield of ammonia by controlling reactant concentration, pressure, temperature and so on. All the above issues are related to chemical equilibrium. In this chapter we will study the principles about chemical equilibrium condition in a **reversible reaction** and the factors affecting the equilibrium.

3.1 The Extent of Chemical Reaction and Standard Equilibrium Constant

3.1.1 The Standard Equilibrium Constant

Every chemical reaction consists of both a forward and a reverse reaction, even if one of these occurs only to a very slight extent. For any chemical reaction in a closed system at a particular temperature, with passing time, the amount of reactants decreases and the amount of products increases until a final state is reached in which the amounts of reactants and products no longer change with the time. This apparent cessation of chemical activity suggests the reaction reaches a state of equilibrium because the forward and reverse rates of a reversible reaction have become equal. Hence, the state of equilibrium for a given chemical reaction suggests the extent of the chemical reaction to transform the reactant to the products.

If a reaction is carried out at the constant pressure and temperature only with the volume work, the Gibbs free energy change for the reaction equals to zero at the equilibrium state. Thus, at the equilibrium

$$\Delta_r G_m = \Delta_r G_m^{\ominus} + RT\ln K^{\ominus} = 0$$

Which means that
$$\Delta_r G_m^{\ominus} = -RT\ln K^{\ominus} \tag{3.1}$$

Equation 3.1 is also called **isothermal formula of chemical reaction**, where $K^{\ominus}$ is termed as the **standard equilibrium constant**. From Equation 3.1, we can observe that the value of $K^{\ominus}$ only depends on the temperature and the natural properties, independent of the concentrations or partial pressures of reactants and products in the reaction. As the value of $K^{\ominus}$ is larger, the value of $\Delta_r G_m^{\ominus}$ is more negative, signifying that the forward reaction is more spontaneous.

For the general balanced reaction $a\text{A}(aq) + b\text{B}(g) \rightleftharpoons d\text{D}(aq) + e\text{E}(g)$, where a, b, d, and e are the stoichiometric coefficients for A, B, D and E respectively, the standard equilibrium constant $K^\ominus$ is expressed as

$$K^\ominus = \frac{\{[\text{D}]/c^\ominus\}^d \{p_\text{E}/p^\ominus\}^e}{\{[\text{A}]/c^\ominus\}^a \{p_\text{B}/p^\ominus\}^b} \tag{3.2}$$

Where, [A] and [D] represent the equilibrium concentrations in $\text{mol} \cdot \text{L}^{-1}$, p_B and p_E are the equilibrium partial pressures of the gaseous substance B and product E in kPa. The solution at the standard concentration $c^\ominus$ is $1 \text{mol} \cdot \text{L}^{-1}$ and the gas at standard pressure $p^\ominus$ is 100kPa. In Equation 3.2, we use the ratio of the measured equilibrium quantity of the substance to the thermodynamic standard-state quantity of the substance, allowing units to be canceled, hence $K^\ominus$ is dimensionless.

Equation 3.2 shows that the value of standard equilibrium constant is directly related to the extent of the reaction at a given temperature. When the magnitude of equilibrium constant is larger, the reaction reaches the equilibrium with more products and less reactant, and thus the conversion rate of the reactant is higher.

Some useful rules for writing the expression of standard equilibrium constant are as follows.

(1) Pure solids or liquids do not appear in the equilibrium constant expression because their concentrations are constant. For example, decomposition of $CaCO_3(s)$ upon heating in a closed vessel through

$$CaCO_3(s) \rightleftharpoons CaO(s) + CO_2(g)$$

$$K^\ominus = \frac{p(CO_2)}{p^\ominus}$$

Here, both $CaCO_3$ and CaO is pure solid and hence they are no terms in the expression.

(2) When the reaction occurs in a dilute solution, concentration terms for solvent do not appear in the equilibrium constant expressions even if the solvent takes part in the reaction. Because the amount of solvent is large in the dilute solutions, the change in solvent concentration can be negligible after reaction and thus the solvent concentration can be concerned as a constant. For instance, the dissociation of acetic acid in water happens through

$$HAc(aq) + H_2O(l) \rightleftharpoons H_3O^+(aq) + Ac^-(aq)$$

$$K^\ominus = \frac{\{[H_3O^+]/c^\ominus\}\{[Ac^-]/c^\ominus\}}{\{[HAc]/c^\ominus\}}$$

(3) The standard equilibrium constant expression should match the corresponding balanced chemical equation. For example,

$$N_2(g) + 3H_2(g) \rightleftharpoons 2NH_3(g)$$

$$K_1^\ominus = \frac{\{p(NH_3)/p^\ominus\}^2}{\{p(N_2)/p^\ominus\}\{p(H_2)/p^\ominus\}^3}$$

Suppose that for a certain application, we want an equation based on the synthesis of 1 mole of $NH_3(g)$.

$$\frac{1}{2}N_2(g) + \frac{3}{2}H_2(g) \rightleftharpoons NH_3(g)$$

Here,

$$K_2^\ominus = \frac{p(NH_3)/p^\ominus}{\{p(N_2)/p^\ominus\}^{\frac{1}{2}}\{p(H_2)/p^\ominus\}^{\frac{3}{2}}}$$

Comparing $K_1^\ominus$ and $K_2^\ominus$, we find that their values are different but correlated, that is $K_1^\ominus = (K_2^\ominus)^2$. When the coefficient in a balanced equation is multiplied by a factor n, the equilibrium expression for the new reaction is the original expression raised to the nth power.

(4) When we reverse an equation, we invert the value of $K^\ominus$. Take the oxidation of SO_2 to SO_3 as an example. The balanced reaction is

$$2SO_2(g) + O_2(g) \rightleftharpoons 2SO_3(g)$$

Its standard equilibrium constant is $\quad K^\ominus = \dfrac{\{p(SO_3)/p^\ominus\}^2}{\{p(SO_2)/p^\ominus\}^2\{p(O_2)/p^\ominus\}}$

If we consider the decomposition of SO_3,

$$2SO_3(g) \rightleftharpoons 2SO_2(g) + O_2(g)$$

$$K^{\ominus\prime} = \frac{\{p(SO_2)/p^\ominus\}^2\{p(O_2)/p^\ominus\}}{\{p(SO_3)/p^\ominus\}^2} = \frac{1}{K^\ominus}$$

Thus, the standard equilibrium constant for a reverse reaction $K^{\ominus\prime}$ is the reciprocal of the standard equilibrium constant for the forward reaction $K^\ominus$ expressed as $K^{\ominus\prime} = 1/K^\ominus$.

(5) When two individual equations are added or subtracted, a new chemical reaction called **coupling reaction** is obtained. The equilibrium constants for the elementary steps are multiplied or divided to obtain the equilibrium constant for the overall reaction.

For example, suppose we want to describe the standard equilibrium constant for the formation of CO_2 (g) from C(s) and O_2 (g), we have known $K_1^\ominus$ for the oxidation of C to CO and $K_2^\ominus$ for the oxidation of CO to CO_2. When we add these two equations together, we multiply their standard equilibrium constants to get the new standard equilibrium constant for the overall reaction.

$$C(s) + \frac{1}{2}O_2(g) \rightleftharpoons CO(g) \qquad K_1^\ominus$$

$$CO(g) + \frac{1}{2}O_2(g) \rightleftharpoons CO_2(g) \qquad K_2^\ominus$$

$$\overline{\quad C(s) + O_2(g) \rightleftharpoons CO_2(g) \qquad K_3^\ominus = K_1^\ominus \cdot K_2^\ominus \quad}$$

Sample Problem 3-1 Calculate the value of K_{sp} of AgI at 298.15K according to the data from standard Gibbs free energy of formation listed in Appendix 3.

Solution The dissociation of AgI(s) is based on the following chemical reaction:

$$AgI(s) \rightleftharpoons Ag^+(aq) + I^-(aq)$$

$\Delta_f G_m^\ominus / (kJ \cdot mol^{-1}) \qquad -66.2 \qquad 77.1 \qquad -51.6$

$$\Delta_r G_m^\ominus = \Delta_f G_m^\ominus(Ag^+, aq) + \Delta_f G_m^\ominus(I^-, aq) - \Delta_f G_m^\ominus(AgI, s)$$
$$= 77.1 kJ \cdot mol^{-1} + (-51.6 kJ \cdot mol^{-1}) - (-66.2 kJ \cdot mol^{-1})$$
$$= 91.7 kJ \cdot mol^{-1}$$

At equilibrium, $\Delta_r G_m = 0$. Now, we substitute the value of $\Delta_r G_m^\ominus$ into Equation 3.1 to find $K^\ominus$:

$$\ln K^{\ominus} = -\frac{91.7\times10^{3}\,\text{J}\cdot\text{mol}^{-1}}{8.314\,\text{J}\cdot\text{K}^{-1}\cdot\text{mol}^{-1}\times298.15\,\text{K}} = -36.99$$

$$K^{\ominus} = e^{-36.99} = 8.61\times10^{-17}$$

So
$$K_{sp} = K^{\ominus} = 8.61\times10^{-17}$$

Sample Problem 3-2 The standard equilibrium constant $K_1^{\ominus}$ is 85.5 for the reaction $\text{Hb(aq)} + \text{O}_2(\text{g}) \rightleftharpoons \text{HbO}_2(\text{aq})$ at 292K. At equilibrium, the partial pressure of O_2 in the air is 20kPa and the solubility of O_2 in water is $2.3\times10^{-4}\,\text{mol}\cdot\text{L}^{-1}$ at 292K. Calculate the standard equilibrium constant $K_2^{\ominus}$ and $\Delta_r G_m^{\ominus}$ for the reaction $\text{Hb(aq)} + \text{O}_2(\text{aq}) \rightleftharpoons \text{HbO}_2(\text{aq})$.

Solution At 292K, the standard equilibrium constant $K_1^{\ominus}$ is

$$K_1^{\ominus} = \frac{\{[\text{HbO}_2]/c^{\ominus}\}}{\{[\text{Hb}]/c^{\ominus}\}\{p(\text{O}_2)/p^{\ominus}\}} = \frac{\{[\text{HbO}_2]/c^{\ominus}\}}{\{[\text{Hb}]/c^{\ominus}\}\{20/100\}} = 85.5$$

$$\frac{\{[\text{HbO}_2]/c^{\ominus}\}}{\{[\text{Hb}]/c^{\ominus}\}} = 17.1.$$

With $\dfrac{\{[\text{HbO}_2]/c^{\ominus}\}}{\{[\text{Hb}]/c^{\ominus}\}}$ known, we determine the standard equilibrium constant $K_2^{\ominus}$ at 292K:

$$K_2^{\ominus} = \frac{\{[\text{HbO}_2]/c^{\ominus}\}}{\{[\text{Hb}]/c^{\ominus}\}\{[\text{O}_2]/c^{\ominus}\}} = \frac{17.1}{2.3\times10^{-4}} = 7.435\times10^{4}$$

Now, we substitute the value of $K_2^{\ominus}$ into Equation 3.1 to find $\Delta_r G_m^{\ominus}$.

$$\Delta_r G_m^{\ominus} = -RT\ln K_2^{\ominus} = -8.314\,\text{J}\cdot\text{K}^{-1}\cdot\text{mol}^{-1}\times292\text{K}\times\ln(7.435\times10^{4}) = -27.23\,\text{kJ}\cdot\text{mol}^{-1}$$

3.1.2 Predicting Reaction Direction by Standard Equilibrium Constant $K^{\ominus}$

Obviously, reactions do not usually keep all components in their equilibrium states. We know the relationship between $\Delta_r G_m^{\ominus}$ and $K^{\ominus}$ in Equation 3.1, we obtain $\Delta_r G_m$ of a reaction that applies to any starting concentrations at a given temperature T:

$$\Delta_r G_m = -RT\ln K^{\ominus} + RT\ln J = RT\ln\left(\frac{J}{K^{\ominus}}\right) \tag{3.3}$$

Therefore, as shown in Equation 3.3, the direction of a chemical reaction under non-standard state can be predicted by comparing the value of the reaction quotient J at a particular state with $K^{\ominus}$ as follows:

If $J < K^{\ominus}$, then $\Delta_r G_m < 0$, hence the forward reaction is spontaneous. In other words, the reaction will progress to the right, toward products, until new equilibrium is reached.

If $J > K^{\ominus}$, then $\Delta_r G_m > 0$, hence the reverse reaction proceeds spontaneously. Therefore, the reaction will progress to the left, toward reactants.

In the case $J = K^{\ominus}$, then $\Delta_r G_m = 0$, meaning that the reaction is at equilibrium.

Therefore, the value of $K^{\ominus}$ is an equilibrium indicator to show the system has reached the equilibrium. If $J \neq K^{\ominus}$, this means the system has not reached the equilibrium. Any non-equilibrium mixture of reactants and products moves spontaneously toward the equilibrium mixture. The larger difference between the value of J and $K^{\ominus}$ is, the more strongly the reaction favors products or reactants at equilibrium.

3.2 Experimental Equilibrium Constant

As shown in Equation 3.1, the standard equilibrium constant $K^\ominus$ is obtained from thermodynamic data by the experiments. In other words, $K^\ominus$ gotten from thermodynamics is actually from the result of the experiments indirectly. Nowadays, direct determination of the equilibrium constant can be achieved by modern analytical methods. For example, gas chromatography and mass spectroscopy have been applied to measure the partial pressure or concentration of each gaseous component in the mixture system and then we can obtain the related data by experiment directly and further get the equilibrium constant.

The equilibrium constants directly measured by the experiment are termed as **experimental equilibrium constant** or **empirical equilibrium constant**, which have been widely used in textbooks and scientific literatures. The rules for writing the experimental equilibrium constant expression are the same as those for writing the standard equilibrium constant expression, as discussed before.

For any reversible reaction carried out in the solution

$$a\text{A} + b\text{B} \rightleftharpoons d\text{D} + e\text{E}$$

At equilibrium, the expression of equilibrium constant written in the terms of concentrations is K_c and is expressed as

$$K_c = \frac{[\text{D}]^d [\text{E}]^e}{[\text{A}]^a [\text{B}]^b} \tag{3.4}$$

While for equilibrium involving gases, we can express the equilibrium constant expression based on partial pressures instead of concentrations. For the reaction between gaseous substances

$$a\text{A}(g) + b\text{B}(g) \rightleftharpoons d\text{D}(g) + e\text{E}(g)$$

K_p is expressed as

$$K_p = \frac{p_\text{D}^d p_\text{E}^e}{p_\text{A}^a p_\text{B}^b} \tag{3.5}$$

In many cases, K_p has a value different from K_c. However, since gas molar concentration is the number of moles of gas per volume, K_p and K_c are quantitatively related by the ideal gas law ($PV=nRT$). If we know one, we can calculate the other by the change of in amount of gases $\triangle n_{\text{gas}}$. The symbol $\triangle n_{\text{gas}}$ is the number of moles of gas on the product side minus the number of moles of gas on the reactant side in the balanced reaction. Take the oxidation of SO_2 to SO_3 as an example, the balanced reaction is

$$2SO_2(g) + O_2(g) \rightleftharpoons 2SO_3(g)$$

$$K_p = \frac{p(SO_3)^2}{p(SO_2)^2 p(O_2)}$$

Rearranging the ideal gas reaction to $P = nRT/V = cRT$, we express partial pressures P as cRT and convert them to concentrations

$$K_p = \frac{p(SO_3)^2}{p(SO_2)^2 p(O_2)} = \frac{\{[SO_3]RT\}^2}{\{[SO_2]RT\}^2 \times \{[O_2]RT\}} = \frac{[SO_3]}{[SO_2][O_2]} \times \frac{1}{RT}$$

Thus, $K_p = \dfrac{K_c}{RT}$ or $K_p = K_c(RT)^{-1}$. Notice that the exponent of the RT term equals the change of in amount of gases $\triangle n_{gas}$ in the balanced reaction, -1. Therefore, for the general gaseous reaction, we have

$$K_p = K_c (RT)^{\Delta n_{gas}} \quad (3.6)$$

Different from standard equilibrium constant $K^\ominus$, experimental equilibrium constant K_c or K_p may have units as shown in Equation 3.4 and Equation 3.5. When the stoichiometric coefficients $(a+b)=(d+e)$, K_c or K_p is dimensionless. If $(a+b)\neq(d+e)$, K_c or K_p have the units, however their units are usually omitted for simplify.

In a multiple-phase reaction, both molarities and partial pressures appear in the expression, the experimental equilibrium constant expression can be designated only as K, neither a K_c nor a K_p. For example, oxidation of S^{2-} by O_2 is used in removing sulfides from wastewater. The reaction is $O_2(g) + 2S^{2-}(aq) + 2H_2O(l) \rightleftharpoons 4OH^-(aq) + 2S(s)$. And the experimental equilibrium constant is expressed as $K = \dfrac{[OH^-]^4}{p_{O_2}[S^{2-}]^2}$ since both aqueous S^{2-} and OH^- ion, gaseous O_2 are involved in the reaction.

3.3 The Shift of Chemical Equilibrium

Chemical reaction does not cease at the equilibrium position. When the conditions of reaction change, the original equilibrium position will be disturbed until a new equilibrium position will be re-attained at the new conditions. Shifting an original equilibrium to a new one caused by the change in reaction condition is called **the shift of chemical equilibrium**. Three common disturbances are a change in concentration of a component (that appears in J), a change in pressure (caused by a change in volume), or a change in temperature. We'll discuss each of these changes below.

3.3.1 Effect of A Change in Concentration

The effect of a change in concentration can be found from the Equation 3.3. At equilibrium, the reaction quotient J equals $K^\ominus$. When the system at equilibrium is disturbed by increasing the concentrations of reactants or decreasing the concentrations of the products, J changes to become smaller than $K^\ominus$ at this point thus making the value of $\Delta_r G_m$ negative. The forward reaction becomes spontaneous until a new equilibrium position is reached by consuming some of the added reactant and forming more products with $J=K^\ominus$. In contrast, adding the products or removing the reactants makes J temporarily larger than $K^\ominus$, then $\Delta_r G_m > 0$, and hence the reverse reaction proceeds spontaneously until a new equilibrium will be established with $J=K^\ominus$. Be sure to note for any reaction at a given temperature, change in the concentration of the reaction components only change the equilibrium position of a reaction, the value of $K^\ominus$ at a given temperature does not change with a change in concentration.

Sample Problem 3-3 At 690K, $K^\ominus = 0.10$ for the reaction $CO_2(g) + H_2(g) \rightleftharpoons CO(g) + H_2O(g)$. When the initial concentrations of CO_2 and H_2 are $0.50 \text{mol} \cdot L^{-1}$ and $0.050 \text{mol} \cdot L^{-1}$, (1) what is the conversion rate of CO_2 when the reaction reaches equilibrium? (2) If the initial concentrations of CO_2

and H_2 are both 0.50mol·L^{-1}, what is the conversion rate of CO_2 at this equilibrium?

Solution (1) Suppose both [CO] and [H_2O] are $x$mol·L^{-1} when the equilibrium is reached,

$$CO_2(g) + H_2(g) \rightleftharpoons CO(g) + H_2O(g)$$

initial concns (mol·L^{-1})	0.50	0.050	0	0
changes (mol·L^{-1})	$-x$	$-x$	$+x$	$+x$
equal concns (mol·L^{-1})	$0.50-x$	$0.050-x$	x	x

$$K^\ominus = \frac{\{[CO]/c^\ominus\}\{[H_2O]/c^\ominus\}}{\{[CO_2]/c^\ominus\}\{[H_2]/c^\ominus\}} = \frac{x^2}{(0.50-x)(0.050-x)} = 0.10$$

So $x = 0.030\,4$

The conversion rate of $CO_2 = \dfrac{0.0304}{0.50} \times 100\% = 6.08\%$

(2) For the same initial concentration of CO_2 and H_2 at 0.50mol·L^{-1}, suppose [CO] and [H_2O] are $y$mol·L^{-1} at the equilibrium. The calculation in this case is the same as the calculation in part (1).

$$K^\ominus = \frac{\{[CO]/c^\ominus\}\{[H_2O]/c^\ominus\}}{\{[CO_2]/c^\ominus\}\{[H_2]/c^\ominus\}} = \frac{y^2}{(0.50-y)(0.50-y)} = 0.10$$

So $y = 0.12$ mol·L^{-1}

The conversion rate of $CO_2 = \dfrac{0.12}{0.50} \times 100\% = 24\%$

From this example, we can observe that increasing the initial concentration of H_2 leads to enhance of the conversion rate of CO_2 when fixing the initial concentration of CO_2 unchanged.

> **Question and thinking 3-1** Cl_2 gas can be prepared in the laboratory by the oxidation of concentrated hydrochloride acid by the oxidizing agent MnO_2. A good method to effectively collect the Cl_2 gas is to pass the gases through saturated saline solution instead of pure water. Explain the method for collecting of Cl_2 gas by applying the chemical equilibrium principle?

3.3.2 Effect of A Change in Pressure

Changes in pressure have significant effects only on equilibrium systems with gaseous components. Aside from phase changes, a change in pressure has a negligible effect on liquids and solids because they are nearly incompressible.

Changes in the partial pressure by adding or removing a gaseous reactant or product are similar to the effect of change in concentration on the chemical equilibrium position, as described in 3.3.1. Adding an inert gas to the constant-volume equilibrium mixture makes the partial pressure of reacting component unchanged, and hence no effect on the equilibrium position. However, a change in total pressure by changing the volume causes a large shift in the equilibrium position. Like the change in the concentration, a change in pressure does not alter $K^\ominus$.

For the gaseous reaction in a closed system, take the synthesis of NH_3 as the example,

$$N_2(g) + 3H_2(g) \rightleftharpoons 2NH_3(g)$$

When the equilibrium is reached, $K^\ominus = \dfrac{\{p(NH_3)/p^\ominus\}^2}{\{p(N_2)/p^\ominus\}\{(H_2)/p^\ominus\}^3}$

If the equilibrium mixture has its volume reduces to halve of its original value by increasing the external pressure. The total pressure immediately doubles and the partial pressure for NH_3, H_2 and N_2 doubles correspondingly. The reaction quotient J changes to

$$J = \frac{\{2p(NH_3)/p^\ominus\}^2}{\{2p(N_2)/p^\ominus\}\{2p(H_2/p^\ominus)\}^3} = \frac{1}{4}\frac{\{p(NH_3)/p^\ominus\}^2}{\{p(N_2)/p^\ominus\}\{(H_2)/p^\ominus\}^3} = \frac{1}{4}K^\ominus$$

Thus J becomes less than $K^\ominus$, then $\Delta_r G_m < 0$. As a result, the formation of NH_3 continues spontaneously toward the total number of gas molecules decrease.

Conversely, if the equilibrium mixture has its volume enlarges to twice its original value and the partial pressure of all gaseous species will be halved. The reaction quotient J changes to

$$J = \frac{\{\frac{1}{2}p(NH_3)/p^\ominus\}^2}{\{\frac{1}{2}p(N_2)/p^\ominus\}\{\frac{1}{2}p(H_2/p^\ominus)\}^3} = 4\frac{\{p(NH_3)/p^\ominus\}^2}{\{p(N_2)/p^\ominus\}\{(H_2)/p^\ominus\}^3} = 4K^\ominus$$

Thus $J > K^\ominus$, then $\Delta_r G_m > 0$, the reverse reaction (decomposition of NH_3) continues spontaneously toward the total number of gas molecules increase.

In conclusion, for a reaction that contains gases at equilibrium,

$$aA(g) + bB(g) \rightleftharpoons dD(g) + eE(g)$$

when $(a+b) = (d+e)$, it suggests the number of gas does not change in the reaction, and hence the change in total pressure will not shift the equilibrium position. When $(a+b) \neq (d+e)$, the change in total pressure has a great effect on the equilibrium position. If the total pressure is increased, the reaction will shift to the direction that the total number of gas molecules decreases to counteract the increase of pressure. On the contrary, a decrease of total pressure causes the reaction shifts to the direction that the total number of gas molecules increases.

3.3.3 Effect of A Change in Temperature

Unlike a change in concentration or in pressure, the change in temperature alters the value of standard equilibrium constant $K^\ominus$ of a reaction. According to Equation 3.1 and the Gibbs-Helmholtz Equation, $\Delta_r G_m^\ominus = \Delta_r H_m^\ominus - T\Delta_r S_m^\ominus$, we have

$$RT \ln K^\ominus = -\Delta_r H_m^\ominus + T\Delta_r S_m^\ominus$$

Also
$$\ln K^\ominus = -\frac{\Delta_r H_m^\ominus}{RT} + \frac{\Delta_r S_m^\ominus}{R} \tag{3.7}$$

Suppose a reaction reaches the equilibrium at T_1 with $K_1^\ominus$, and then rise the temperature to T_2 to reach its new equilibrium with $K_2^\ominus$. Based on Equation 3.7, we can obtain

$$(1) \ln K_1^\ominus = -\frac{\Delta_r H_m^\ominus}{RT_1} + \frac{\Delta_r S_m^\ominus}{R}$$

$$(2) \ln K_2^\ominus = -\frac{\Delta_r H_m^\ominus}{RT_2} + \frac{\Delta_r S_m^\ominus}{R}$$

Since change in temperature has a negligible effect on the values of $\Delta_r H_m^\ominus$ and $\Delta_r S_m^\ominus$, they can be regarded as constants during calculation. When equation (2) subtracts equation (1), we get

$$\ln\frac{K_2^\ominus}{K_1^\ominus} = \frac{\Delta_r H_m^\ominus}{R}(\frac{T_2 - T_1}{T_1 T_2}) \tag{3.8}$$

Equation 3.8 shows how the change in temperature impacts on the value of $K^\ominus$. If the forward reaction is endothermic with a positive $\Delta_r H_m^\ominus$ ($\Delta_r H_m^\ominus > 0$), a temperature rise ($T_2 > T_1$) favors the forward direction and increase the value of the equilibrium constant. By contrast if the forward reaction is exothermic with a negative $\Delta_r H_m^\ominus$ ($\Delta_r H_m^\ominus < 0$), increasing temperature favors reverse direction and decreases the value of equilibrium constant. In addition, the absolute value of $\Delta_r H_m^\ominus$ is larger, the impact of the change in temperature on the reaction equilibrium will be more remarkable.

Sample Problem 3-4 The industrial synthesis of ammonia is based on the chemical reaction:
$$N_2(g) + 3H_2(g) \rightleftharpoons 2NH_3(g).$$
Calculate (1) the standard equilibrium constant of the reaction at 298.15K. (2) the standard equilibrium constant at 500℃ and then discuss the effect of temperature on the ammonia synthesis.

Solution (1) From Appendix 3, $\Delta_f G_m^\ominus(NH_3, g) = -16.4 kJ \cdot mol^{-1}$ at 298.15K.

For the above chemical reaction:
$$\Delta_r G_m^\ominus = 2 \times \Delta_f G_m^\ominus(NH_3, g) - \Delta_f G_m^\ominus(N_2, g) - 3 \times \Delta_f G_m^\ominus(H_2, g)$$
$$= 2 \times (-16.4 kJ \cdot mol^{-1})$$
$$= -32.8 kJ \cdot mol^{-1}$$

$$\Delta_r G_m^\ominus = -RT \ln K_1^\ominus$$

$$\ln K_1^\ominus = -\frac{\Delta_r G_m^\ominus}{RT} = \frac{32.8 \times 10^3 J \cdot mol^{-1}}{8.314 J \cdot K^{-1} \cdot mol^{-1} \times 298.15 K} = 13.23$$

$$K_1^\ominus = 5.6 \times 10^5$$

(2) From Appendix 3, $\Delta_f H_m^\ominus(NH_3, g) = -45.9 kJ \cdot mol^{-1}$ at 298.15K. So,
$$\Delta_r H_m^\ominus = 2 \times \Delta_f H_m^\ominus(NH_3, g) - \Delta_f H_m^\ominus(N_2, g) - 3 \times \Delta_f H_m^\ominus(H_2, g)$$
$$= 2 \times (-45.9 kJ \cdot mol^{-1}) = -91.8 kJ \cdot mol^{-1}$$

Since $\Delta_r H_m^\ominus$ is independent of temperature, $\Delta_r H_m^\ominus \approx -91.8 kJ \cdot mol^{-1}$ at 500℃ (773.15K). We determine the standard equilibrium constant $K_2^\ominus$ at 773.15K based on Equation 3.8 and substitute the known values of $\Delta_r H_m^\ominus$, $K_1^\ominus$, T_2 and T_1 into the equation to find $K_2^\ominus$:

$$\ln\frac{K_2^\ominus}{5.6 \times 10^5} = \frac{-91.8 \times 10^3 J \cdot mol^{-1}}{8.314 J \cdot mol^{-1} \cdot K^{-1}}\left(\frac{773.15 K - 298.15 K}{773.15 K \times 298.15 K}\right)$$
$$= -22.75$$

So $$K_2^\ominus = 7.4 \times 10^{-5}$$

Comparing $K_1^\ominus$ at 298.15K and $K_2^\ominus$ at 773.15K, we find that because the formation of NH_3 is exothermal, raising the temperature shifts the equilibrium position to the decomposition of NH_3, thereby decreasing the yield of NH_3. Therefore, the yield of NH_3 favors at low temperature from the view of chemical equilibrium.

As discussed above, when an equilibrium system is subjected to a change in temperature, pressure, or concentration of a reacting species, the reaction responds by attaining a new equilibrium that partially offsets the impact of the change. The most remarkable feature of a reaction at the equilibrium is its ability to return to equilibrium after a change in conditions. Such drive to re-attain equilibrium is concluded by the French chemist Henri Le Châtelier (1884) and the **Le Chatelier's principle** states that

"When a chemical reaction at equilibrium is subjected to a disturbance, the shift of equilibrium adjusts to minimize the effect of the disturbance."

> **Question and thinking 3-2** Hemoglobin carries O_2 from lungs to tissue cells, where O_2 is released. The protein is represented as Hb in its unoxygenated form and as HbO_2 in its oxygenated form. One reason that CO is toxic is that it competes with O_2 in binding to Hb: $HbO_2(aq) + CO \rightleftharpoons HbCO(aq) + O_2(g)$. Applying equilibrium principle to explain why artificial respiration is suggested to treat patients with mild CO poisoning while hyperbaric oxygen therapy is recommended for patients with severe CO poisoning?

Key Terms

可逆反应	reversible reaction
化学反应等温方程式	the isothermal equation of chemical reaction
标准平衡常数	standard equilibrium constant
耦合反应	coupling reaction
实验平衡常数	experimental equilibrium constant
经验平衡常数	empirical equilibrium constant
化学平衡的移动	the shift of chemical equilibrium
勒夏特列规则	Le Chatelier's principle

Summary

1. A reversible reaction will reaches equilibrium when the forward reaction rate is equal to the reverse reaction rate. The magnitude of equilibrium constant is an indication of how far a reaction proceeds toward product at a given temperature.

$K^\ominus$ is the standard equilibrium constant obtained from thermodynamics and K_c (or K_p) are experimental equilibrium constant obtained by experimental measurement. For the general balanced reaction equation

$$aA(aq) + bB(g) \rightleftharpoons dD(aq) + eE(g)$$

The standard equilibrium constant is expressed as

$$K^\ominus = \frac{\{[D]/c^\ominus\}^d \{p_E/p^\ominus\}^e}{\{[A]/c^\ominus\}^a \{p_B/p^\ominus\}^b}$$

If the reaction takes place in the aqueous solution, the equilibrium constant expressions based on equilibrium concentrations is expressed as

$$K_c = \frac{[D]^d[E]^e}{[A]^a[B]^b}$$

If it is the gaseous reaction, the equilibrium constant expressions based on partial pressure is expressed as

$$K_p = \frac{p_D^d p_E^e}{p_A^a p_B^b}$$

2. We compare the values of reaction quotient J and $K^\ominus$ to determine the direction in which a reaction will proceed toward equilibrium by the following equation:

$$\Delta_r G_m = -RT \ln K^\ominus + RT \ln J = RT \ln(\frac{J}{K^\ominus})$$

If $J < K^\ominus$, the reaction will progress to the right and more product forms.

If $J > K^\ominus$, the reaction will progress to the left and more reactant forms.

If $J = K^\ominus$, the reaction is at equilibrium.

3. Le Chatelier's principle states that if an outside influence upsets a chemical equilibrium, the reaction undergoes a change in the direction that counteracts the disturbing influence and, if possible, returns the reaction to equilibrium.

An increase in concentration cause the reaction toward the direction that consume some of it, and a decrease in concentration causes the reaction toward the direction that produce some of it.

For a reaction that involves a change in number of moles of gas, an increase in pressure (decrease in volume) causes the reaction shift toward fewer moles of gas, and a decrease in pressure (increase in volume) causes the opposite change.

Although the equilibrium concentrations of components change as a result of concentration and volume changes, $K^\ominus$ does not change. A temperature change, however, does change $K^\ominus$, higher T increases $K^\ominus$ for an endothermic reaction ($\Delta_r H_m^\ominus > 0$) and decreases $K^\ominus$ for an exothermic reaction ($\Delta_r H_m^\ominus < 0$). The effect of temperature on the chemical equilibrium by the following equation,

$$\ln \frac{K_2^\ominus}{K_1^\ominus} = \frac{\Delta_r H_m^\ominus}{R}(\frac{T_2 - T_1}{T_1 T_2})$$

4. The equilibrium condition plays a role in numerous natural phenomena and affects the methods used to produce many important industrial chemicals. Applying the equilibrium principle, we can maximize the yield of products.

Exercises

1. Write standard equilibrium constant expressions and experimental equilibrium constant expressions for the following reversible reactions.

(1) $H_2(g) + 1/2\ O_2(g) \rightleftharpoons H_2O(g)$

(2) $3C_2H_2(g) \rightleftharpoons C_6H_6(l)$

(3) $C_6H_{12}O_6(s) \rightleftharpoons 2C_2H_5OH(l) + 2CO_2(g)$

(4) $HCN(aq) \rightleftharpoons H^+(aq) + CN^-(aq)$

(5) $MnO_2(s) + 4H^+(aq) + 2Cl^-(aq) \rightleftharpoons Mn^{2+}(aq) + Cl_2(g) + 2H_2O(l)$

2. Given the following information under the standard conditions at 298K

(1) $CO_2(g) + H_2(g) \rightleftharpoons CO(g) + H_2O(g)$ $K_1^\ominus = 0.14$

(2) $CoO(s) + H_2(g) \rightleftharpoons Co(s) + H_2O(g)$ $K_2^\ominus = 67$

(3) $CoO(s) + CO(g) \rightleftharpoons Co(s) + CO_2(g)$

(a) Calculate the value of $K_3^\ominus$ at 298K for the reaction (3).

(b) Calculate $\Delta_r G_{m,823}^\ominus$ for the reaction (2) and (3) and judge which one can reduce CoO(s) more effectively under standard condition, CO(g) or $H_2(g)$?

3. At 700℃, $K_c = 9.0$ for the reaction $SO_2(g) + NO_2(g) \rightleftharpoons SO_3(g) + NO(g)$. If a vessel is filled with these four gaseous substances at the equal initial concentrations of $3.0 \times 10^{-3} mol \cdot L^{-1}$ at 700℃, what is the equilibrium concentration of $SO_3(g)$?

4. CO is toxic is that it binds to the blood protein hemoglobin more strongly than oxygen does. For the reaction $HbO_2(aq) + CO(g) \rightleftharpoons HbCO(aq) + O_2(g)$ at 37℃, $K = 200$. It is found that when the concentration of HbCO reaches 2% of the concentration of HbO_2, it will injure a person's intelligence. After smoking, the concentrations of CO and O_2 in the air inhaled in the lung are measured at $1 \times 10^{-6} mol \cdot L^{-1}$ and $1 \times 10^{-2} mol \cdot L^{-1}$, respectively. Predict whether smoking could injure the person's intelligence based on the calculation.

5. When gaseous phosphorus pentachloride is heated, it decomposes to phosphorus trichloride and chlorine gases: $PCl_5(g) \rightleftharpoons PCl_3(g) + Cl_2(g)$. Mix 2.00mol of PCl_5 with 1.00mol of PCl_3 in a closed container and allows the reaction to proceed at 30℃. It is known that the total pressure is 200kPa and the conversion rate of PCl_5 is 0.91 at equilibrium. Calculate $K^\ominus$.

6. For the reaction $N_2O_4(g) \rightleftharpoons 2NO_2(g)$ at 25℃, $K^\ominus = 0.15$. Suppose the reaction starts with a vessel containing only N_2O_4 sample, and the total pressure inside the vessel is 60kPa after equilibrium has been reached. Calculate decomposition rate of N_2O_4 and the initial pressure of N_2O_4.

7. One of the methods to prepare $Cl_2(g)$ is as follows:
$$MnO_2(s) + 4H^+(aq) + 2Cl^-(aq) \rightleftharpoons Mn^{2+}(aq) + Cl_2(g) + 2H_2O(l)$$
(1) Write the expression of the standard equilibrium constant of this reaction.

(2) Calculate $\Delta_r G^\ominus_{m,298.15}$ according to the thermodynamic data in Appendix 3; Determine $K^\ominus$ under standard conditions at 298.15K and predict whether the forward reaction is spontaneous under this condition.

(3) Suppose the concentration of hydrochloric acid is $12.0 mol \cdot L^{-1}$ and the other substances are at the standard state, can this reaction proceed spontaneously at 298.15K?

8. For the reaction $2SO_2(g) + O_2(g) \rightleftharpoons 2SO_3(g)$, the standard equilibrium constant $K^\ominus$ is 910 at 800K. Assuming the effect of temperature on the $\Delta_r H^\ominus_m$ can be neglected, calculate $K^\ominus$ at the 900K.

9. Based on the reaction $AgCl(s) \rightleftharpoons Ag^+(aq) + Cl^-(aq)$, calculate the solubility product constant K_{sp} of $AgCl(s)$ at 298.15K. The corresponding thermodynamic data can be obtained in Appendix 3.

10. According to the following thermodynamic data, calculate the solubility product constant K_{sp} for Ag_2CO_3 at 298.15K and 373.15K respectively, assuming the values of $\Delta_r H^\ominus_m$ and $\Delta_r S^\ominus_m$ do not change with temperature.

	$Ag_2CO_3(s) \rightleftharpoons$	$2Ag^+(aq) +$	$CO_3^{2-}(aq)$
$\Delta_f H^\ominus_m / kJ \cdot mol^{-1}$	−505.8	105.6	−667.1
$S^\ominus_m / J \cdot K^{-1} \cdot mol^{-1}$	167.4	72.7	−56.9

Supplementary Exercises

1. Given the following reactions,
(1) $4NO(g) \rightleftharpoons 2N_2O(g) + O_2(g)$ $K^\ominus = 3.54 \times 10^{-25}$
(2) $2NO(g) + O_2(g) \rightleftharpoons 2NO_2(g)$ $K^\ominus = 6.26 \times 10^{-13}$
(3) $2N_2O(g) + 3O_2(g) \rightleftharpoons 4NO_2(g)$.

Calculate $K^\ominus$ for the reaction (3) at 298.15K.

2. When the reaction of $C(gra.) + O_2(g) \rightleftharpoons 2CO_2(g)$ reaches equilibrium at 298.15K in a container with the capacity of 1L, its $\Delta_r H_m^\ominus$ is $-393.5 kJ \cdot mol^{-1}$, what's the impact on the equilibrium partial pressure of O_2 under the measures that are as follow: (1) Increase the amount of graphite; (2) Increase the amount of CO_2; (3) Increase the amount of O_2; (4) Reduce the reaction temperature; (5) Adding catalyst.

3. For the reaction of $2NO_2(g) \rightleftharpoons N_2O_4(g)$, its $K^\ominus$ is 7.0×10^{-5} at 298.15K. When the reaction reaches equilibrium, the partial pressure of NO_2 is 20.0kPa, calculate the partial pressure of N_2O_4 and the total pressure of the equilibrium system.

4. A reaction that can convert coal to methane (the chief component of natural gas) is
$$C(gra.) + 2H_2(g) \rightleftharpoons CH_4(g)$$
for which $\Delta_r G_m^\ominus = -50.5 kJ \cdot mol^{-1}$. What is the value of $K^\ominus$ for the reaction at 25℃? Does this value of $K^\ominus$ suggest that studying this reaction as a mean of methane production is worth pursuing?

5. A mixture containing 4.562×10^{-3} mol of $H_2(g)$, 7.384×10^{-4} mol of $I_2(g)$, and 1.355×10^{-2} mol of $HI(g)$ in a 1.0L container at 425.4℃ is at equilibrium. If 1.000×10^{-3} mol of $I_2(g)$ is added, what will be the concentration of $H_2(g)$, $I_2(g)$ and $HI(g)$ after the system has again reached equilibrium?

Answers to Some Exercises
[Exercise]
1.

	standard equilibrium constant expression	experimental equilibrium constant expression
(1)	$K^\ominus = \dfrac{p_{H_2O}/p^\ominus}{(p_{H_2}/p^\ominus)(p_{O_2}/p^\ominus)^{\frac{1}{2}}}$	$K_p = \dfrac{p_{H_2O}}{p_{H_2} p_{O_2}^{1/2}}$
(2)	$K^\ominus = \dfrac{1}{(p_{C_2H_2}/p^\ominus)^3}$	$K_p = \dfrac{1}{p_{C_2H_2}^3}$
(3)	$K^\ominus = (p_{CO_2}/p^\ominus)^2$	$K_p = p_{CO_2}^2$
(4)	$K^\ominus = \dfrac{\{[CN^-]/c^\ominus\}\{[H^+]/c^\ominus\}}{\{[HCN]/c^\ominus\}}$	$K_c = \dfrac{[CN^-][H^+]}{[HCN]}$
(5)	$K^\ominus = \dfrac{\{[Mn^{2+}]/c^\ominus\}(p_{Cl_2}/p^\ominus)}{\{[H^+]/c^\ominus\}^4\{[Cl^-]/c^\ominus\}^2}$	$K = \dfrac{[Mn^{2+}] \times p_{Cl_2}}{[H^+]^4[Cl^-]^2}$

2. (a) $K_3^\ominus = 478.6$

(b) $\Delta_r G_{m,823}^\ominus = -28.77 kJ \cdot mol^{-1}$ For the reaction (2) and $\Delta_r G_{m,823}^\ominus = -42.22 kJ \cdot mol^{-1}$ for the reaction (3), so the reduced ability of CO(g) is greater than that of $H_2(g)$ under standard conditions.

3. $[SO_3] = 0.0045 mol \cdot L^{-1}$

4. The concentration of HbCO is larger than 2.20% of the concentration of HbO_2. Therefore, smoking could injure the person's intelligence.

5. $K^\ominus = 11.83$

6. Decomposition rate of $N_2O_4 = 24\%$, and the initial pressure of N_2O_4 is 48.29kPa.

7. (1) $K^\ominus = \dfrac{\{[Mn^{2+}]/c^\ominus\}(p_{Cl_2}/p^\ominus)}{\{[H^+]/c^\ominus\}^4\{[Cl^-]/c^\ominus\}^2}$

(2) $\Delta_r G^\ominus_{m,298.15} = 25.2 \text{kJ}\cdot\text{mol}^{-1}$ and $K^\ominus = 3.83 \times 10^{-5}$, so the forward reaction cannot proceed spontaneously under standard conditions at 298.15K.

(3) $\Delta_r G^\ominus_{m,298.15} = -11.9 \text{kJ}\cdot\text{mol}^{-1}$, so the forward reaction can proceed spontaneously when the concentration of hydrochloric acid is $12.0 \text{mol}\cdot\text{L}^{-1}$.

8. $K^\ominus_{900} = 33.56$

9. $K_{sp} = 1.74 \times 10^{-10}$

10. $K_{sp} = 1.38 \times 10^{-13}$ (at 298.15K) and $K_{sp} = 7.93 \times 10^{-12}$ (at 373.15K)

[Supplementary Exercises]

1. $K^\ominus = 1.10 \times 10^{-2}$

2.

Impact	the equilibrium partial pressure of O_2
(1) Increase the amount of graphite	unchanged
(2) Increase the amount of CO_2;	increases
(3) Increase the amount of O_2	increases
(4) Reduce the reaction temperature	decreases
(5) Adding catalyst	unchanged

3. $p_{N_2O_4} = 26.68 \text{kPa}$, $p_{total} = 46.68 \text{kPa}$

4. $K^\ominus = 7.02 \times 10^5$ (at 298.15K). This value is quite large, suggesting that studying this reaction as a mean of methane production is worth pursuing.

5. $[H_2] = 3.877 \text{mmol}\cdot\text{L}^{-1}$, $[I_2] = 1.0534 \text{mmol}\cdot\text{L}^{-1}$, $[HI] = 14.92 \text{mmol}\cdot\text{L}^{-1}$

（杨　静）

Chapter 4

Rate of Chemical Reaction

The important problem to a given reaction for us is about the probability whether the reaction will occur or not under the certain conditions. Chemical thermodynamics predicts the spontaneity of a reaction successfully. However, many reactions that should happen by thermodynamics principle only take place too slowly to be useful. For example, the very meaningful reaction between NO and CO, harmful gas reactants from automobile exhaust at normal condition is

$$CO(g) + NO(g) \rightarrow CO_2(g) + \frac{1}{2} N_2(g)$$

Which $\Delta_r G_m^\ominus$ is $-334 \text{kJ} \cdot \text{mol}^{-1}$ at 298K by thermodynamics calculation. It means that this reaction has the obvious spontaneity trend. But there is no actual significance because this reaction rate is very slow. Therefore, the pollution problem from automobile exhaust still has not been solved effectively up till now by this simple reaction. An impossible reaction by thermodynamics principle cannot happen certainly as we known, but the reaction processed spontaneity by thermodynamics principle may be possible to happen. So, it is necessary to consider both the trend and rate for a chemical reaction.

Chemical kinetics is the study on the chemical reaction rate, reaction mechanisms and the factors of affecting the reaction rate.

Chemical reaction rate is affected evidently by many factors such as reactant concentrations, the natural properties of a reaction, temperature, pH of a solution and whether a catalyst is used or not. Chemical thermodynamics focuses on the spontaneity of a reaction. Chemical Kinetics solves the actuality for a reaction. The basic theories about the reaction rate and the effect of main factors on reaction rate are introduced in this chapter.

4.1　Chemical Reaction Rate and Its Expression

4.1.1　The Rate of Chemical Reaction

Reaction rate v expresses how fast the concentration of reactant or product changes with the time. It can be defined as the extent of reaction. That is the extent of reaction changes with time in unit volume

$$v \stackrel{\text{def}}{=\!=} \frac{1}{V} \frac{\mathrm{d}\xi}{\mathrm{d}t} \tag{4.1}$$

Where ξ is the extent of a reaction and the unit is in mole. V is the volume of the reaction system.

For a general reaction

$$aA + bB \rightarrow dD + eE$$
$$d\xi = \nu_B^{-1} \cdot dn_B \tag{4.2}$$

In Equation 4.2, n_B is amount of species in mole for a reaction, ν_B is the stoichiometric coefficient of species in a balanced reaction equation. Substituting them into Equation 4.1, we get

$$v \stackrel{\text{def}}{=\!=} \frac{1}{V} \frac{d\xi}{dt} = \frac{1}{V} \frac{dn_B}{\nu_B dt} = \frac{1}{\nu_B} \frac{dc_B}{dt} \tag{4.3}$$

Here, c_B is concentration in molarity for species B in a reaction. The unit of reaction rate is "$\text{mol} \cdot \text{L}^{-1} \cdot \text{time}^{-1}$". The unit "time" will be taken according to the rate of a reaction in second (s), minute (min), hour (h), day (d), and year (a) etc. Here c_B can be the concentration of reactant or product. Chemical reaction rate is always larger than or equal to zero in numerical value.

For a general chemical reaction, we can take reactant A to express the reaction rate.

$$v_A = -\frac{dc(A)}{dt} \tag{4.4}$$

In Equation 4.4, $\frac{dc(A)}{dt}$ is negative number because c (A) will decrease as the time increases. So, the negative sign "$-$" has to be added. Reaction rate also can be expressed by product D.

$$v_D = \frac{dc(D)}{dt} \tag{4.5}$$

c (D) will increase in a reaction as the time increases.

The reaction rate is a consuming rate for reactants or a producing rate for products in fact. The rate may be very different if expressing it by the different components in a reaction. But the reaction rate is always a certain value by Equation 4.3 even if changing the coefficients of an identical reaction equation. The following equation tells the rate relationships between the different components in a reaction:

$$v = -\frac{1}{a} \cdot \frac{dc(A)}{dt} = -\frac{1}{b} \cdot \frac{dc(B)}{dt} = \frac{1}{d} \cdot \frac{dc(D)}{dt} = \frac{1}{e} \cdot \frac{dc(E)}{dt} \tag{4.6}$$

For example,

$$N_2(g) + 3H_2(g) \rightarrow 2NH_3(g)$$

$$v = \frac{1}{\nu_B} \cdot \frac{dc_B}{dt} = -\frac{1}{1} \frac{dc(N_2)}{dt} = -\frac{1}{3} \frac{dc(H_2)}{dt} = \frac{1}{2} \frac{dc(NH_3)}{dt}$$

or

$$v = v(N_2) = \frac{1}{3} v(H_2) = \frac{1}{2} v(NH_3)$$

Therefore, the reaction rate can be indicated by each of components in a balanced reaction equation.

4.1.2 Average Rate and Instantaneous Rate of A Chemical Reaction

Average rate $\bar{v}$ is expressed in term of the change in the quantity of a reactant or product that takes place in a period of time. For instance, an aqueous solution of H_2O_2 as a common antiseptic will decompose into H_2O and O_2 at room temperature:

$$H_2O_2(aq) \xrightleftharpoons{I^-} H_2O(l) + \frac{1}{2} O_2(g)$$

If we know the concentration of H_2O_2 by measuring the amount of O_2 produced at different time during the reaction to take place, the decomposing rate of H_2O_2 can be known. Table 4-1 lists the data of decomposing rate of H_2O_2 from its initial concentration of $0.80 mol \cdot L^{-1}$.

Table 4-1 Average Rate and Instantaneous Rate of H_2O_2 to Decompose at Room Temperature

t/min	0	20	40	60	80
$c(H_2O_2)$/ $mol \cdot L^{-1}$	0.80	0.40	0.20	0.10	0.050
$\bar{v}$ /$mol \cdot L^{-1} \cdot min^{-1}$		0.020	0.010	0.0050	0.0025
v/ $mol \cdot L^{-1} \cdot min^{-1}$		0.014	0.0075	0.0038	0.0019

The first row in Table 4-1 lists the molarity of H_2O_2 at different time. The second row lists the average decomposing rate of H_2O_2 at a time interval that we have chosen between 0~80min. The average decomposing rate of H_2O_2 is

$$\bar{v} = -\frac{\Delta c(H_2O_2)}{\Delta t}$$

Where $\Delta c(H_2O_2)$ is the difference between final concentration and initial concentration of H_2O_2 over a time interval Δt. As shown in the third row, the reaction rate is not constant, the lower the remaining concentration of H_2O_2 is, the slower the reaction proceeds. You would find from Table 4-1 that the average rate can not reflect the exact rate well at any instant of reaction.

The **instantaneous rate** for a chemical reaction indicates the exact rate at any given instant. It is a limiting value of an average rate as Δt approaches zero. That is

$$v = \lim_{\Delta t \to 0} \bar{v} = \lim_{\Delta t \to 0} \frac{-\Delta c(H_2O_2)}{\Delta t} = -\frac{dc(H_2O_2)}{dt} \qquad (4.7)$$

Generally, instantaneous rate is used to expresses the chemical reaction rate.

The instantaneous rate of H_2O_2 in above example can be obtained by plotting $c(H_2O_2)$ vs. t as shown in Figure 4-1. Where the points of A, B, C and D express the concentrations of H_2O_2 at 20th, 40th, 60th and 80th minute respectively. The instantaneous rate at a specified instant during the reaction is the slope of tangent of curve at a pointed time. You may have found that the instantaneous decomposition rate is not constant at any time in Figure 4-1. The different slops indicate the different rate of H_2O_2 to decompose from 0th minute to 80th minute during the decomposition reaction.

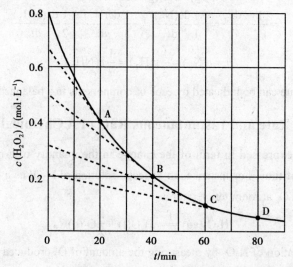

Figure 4-1 Concentration of H_2O_2 vs. Time in the Decomposition Reaction of H_2O_2

4.2 The Effect of Concentration on Reaction Rate

4.2.1 Rate Law of Chemical Reaction

1. Rate law

Many factors will affect the reaction rate. Reactant concentration is one of important factors. A quantitative relationship between reaction rate and reactants concentration is called **rate law** or **rate equation**. For a general reaction,

$$aA + bB \rightarrow Products$$

The rate equation is

$$v = kc^{\alpha}(A)c^{\beta}(B) \tag{4.8}$$

The reaction rate is proportional to the product of reactants concentration raised to the power in rate equation. Where the exponents α and β in Equation 4.8 with respect to $c(A)$ and $c(B)$ must be determined by the experiment only because the coefficients a and b in the balanced reaction may be not equal to α and β, respectively.

Sample Problem 4-1 Reaction rate of NO and H_2 to form N_2 and H_2O at 1 073K is obtained in Table 4-2. Determine the rate law of this reaction from the data.

Table 4-2 Reaction Rate between NO and H_2 (1 073K)

Experiment number	Initial concentration / mol·L^{-1}		Rate of forming N_2 / mol·L^{-1}·min^{-1}
	$c(NO)$	$c(H_2)$	
1	6.0×10^{-3}	1.0×10^{-3}	3.20×10^{-3}
2	6.0×10^{-3}	2.0×10^{-3}	6.38×10^{-3}
3	6.0×10^{-3}	3.0×10^{-3}	9.59×10^{-3}
4	1.0×10^{-3}	6.0×10^{-3}	0.49×10^{-3}
5	2.0×10^{-3}	6.0×10^{-3}	1.98×10^{-3}
6	3.0×10^{-3}	6.0×10^{-3}	4.42×10^{-3}

Solution The reaction is

$$2NO(g) + 2H_2(g) = N_2(g) + 2H_2O(g)$$

Comparing the experiments 1, 2 and 3, the concentrations of NO remain a constant, the reactant rate is proportional to $c(H_2)$. In the same way, comparing experiments 4, 5 and 6, the reactant rate is proportional to $c^2(NO)$ while the concentrations of H_2 is constant. So, the rate law of reaction is

$$v = kc^2(NO)c(H_2)$$

As a reaction goes on, products will be more gradually. Reverse reaction rate becomes quickly more and more. Forward and reverse reaction in a general reaction takes place simultaneously in fact. Thus, we should consider both of forward and reverse reaction rates into account and calculate their difference between them. But it is very complicated to deal with actually. So, we take the initial rate to be the reaction rate as well as to determine the rate law by experiment generally. Reverse reaction rate is neglected so as to make it to be simple greatly when the initial rate is determined at the beginning of a reaction.

Note the following points if we want to get a correct rate law about a chemical reaction:

a) The rate law is gotten only by experiment but not by the balanced reaction equation generally.

b) The concentration of pure solid and pure liquid reactants is no term in rate law. For example,

$$C(s) + O_2(g) \rightarrow CO_2(g)$$

This reaction takes place on the surface of C. The surface area of solid C is a constant due to a certain size. The rate is no relationship to the concentration of solid C here.

c) Concentration of solvent is treated as a constant if the solvent takes part in the reaction and the concentration of other reactants is dilute. For example, Hydrolysis of dilute sucrose solution is

$$C_{12}H_{22}O_{11} + H_2O \rightarrow C_{12}H_{22}O_{11} + C_{12}H_{22}O_{11}$$
$$\text{sucrose} \qquad\qquad \text{glucose} \qquad \text{fructose}$$

The rate equation of this reaction is

$$v = k'c(C_{12}H_{22}O_{11})c(H_2O) = kc(C_{12}H_{22}O_{11})$$

2. Rate constant and reaction order

The proportional constant k in rate law is called **rate constant**. Its numerical value depends on the natural properties of a reaction, temperature and the presence of a catalyst if it is added but not on the reactant concentrations. The significance of rate constant found experimentally is that it equals to the reaction rate in value when the reactant concentrations are $1 \text{mol} \cdot \text{L}^{-1}$. So it is called specific rate too. The larger the magnitude of k is, the faster the reaction undergoes at the given conditions. The unit of k depends on the exponent α and $\beta \cdots$, and the unit of time in rate law.

Chemical reactions are classified by reaction order generally. Here we refer to α as the reaction order with respect to reactant A, and β is the reaction order with respect to reactant B in Equation 4.8. The overall **reaction order** is the sum of $(\alpha + \beta + \cdots)$. If the sun is zero, the reaction is called zero-order reaction. If it is one, first-order reaction; two, second-order reaction; $\cdots$, and so on. The magnitude of reaction order reflects the degree of affecting on rate by reactant concentrations. The larger the reaction order is, the greater the rate is affected in degree by reactant concentrations.

The reaction order must be found by experiment. Its numerical value can be positive integral number, fractions or negative numbers which express that the reactants play a hinder role in the reaction. There is no way to say the exact reaction orders for a reaction with the complicated rate law like the reaction

$$H_2 + Br_2 \rightarrow 2HBr$$

Its rate law is

$$v = kc(H_2) \cdot c(Br_2) / \left[1 + K' \frac{c(HBr)}{c(Br_2)}\right]$$

4.2.2 Characteristics with the Simple Reaction Order

When the overall order is equal to 0, 1 and 2, etc, the reactions is with simple reaction order generally. The main properties of first-, second- and zero-order reaction will be discussed respectively in this section.

1. First-order reaction

The reaction rate is proportional to the reactant concentration by the definition of chemical reaction rate for a **first-order reaction.** Let us take a reaction in which only reactant A decomposes to products

$$aA \rightarrow products$$

For a first-order reaction, the rate law is

$$v = -\frac{dc(A)}{dt} = k\ c(A)$$

By integration operation

$$-\int_{c_0}^{c} \frac{dc(A)}{c(A)} = \int_0^t k\, dt$$

Getting

$$\ln \frac{c_0(A)}{c(A)} = kt \quad \text{or} \quad \lg \frac{c_0(A)}{c(A)} = \frac{kt}{2.303} \tag{4.9}$$

Equation 4.9 is a mathematical relationship between reactant concentration and the time for a first-order reaction. Where $c_0(A)$ is the initial concentration of reactant A. $c(A)$ is the concentration of A at any specified time t during the reaction process. The graph is linear if we plot $\ln c(A)$ against time t. The slope of the line is $-k$. The unit of k in a first-order reaction is $[\text{time}]^{-1}$.

The **half-life** $t_{1/2}$ of a reaction is the time during which the amount of reactant or reactant concentration drops to one-half of its initial magnitude. That is $c(A) = \frac{c_0(A)}{2}$ at this instant. We substitute it into the Equation 4.9 and have the following equation

$$t_{1/2} = \frac{0.693}{k} \tag{4.10}$$

The half-life for a first-order reaction is a constant which is not related to the initial concentration of reactant. The half-life can be used in determining the reaction rate. The larger the half-life is, the slower the reaction undergoes.

The most of decomposition reactions such as decay of radioactive elements, the most of drugs metabolism in the body, rearrangement reaction within the molecule and isomerization reaction, etc. are all the examples of the first-order reaction. Hydrolysis in dilute solution like the hydrolysis of dilute sucrose aqueous solution is second-order reaction originally. But we consider it to be a first-order reaction because the concentration of large amount of water changes so very little during the reaction process as to be regarded as a constant although water is one of reactants. This kind of reactions is called **pseudo-first-order reaction.**

Sample Problem 4-2 It is known that the half-life is equal to 5.26 years for the decay of ^{60}Co. The strength of radioactive substance is expressed by ci [*] (Curie). The γ-ray produced by radioactive ^{60}Co applied in medical linear accelerator often is used in treating cancer. If a hospital bought a machine with 20ci of ^{60}Co, what is the radioactive strength after using it for 10 years?

Solution The decay of radioactive element is a first-order reaction. So,

$$t_{1/2} = \frac{0.693}{k}$$

$$k = \frac{0.693}{t_{1/2}} = \frac{0.693}{5.26a} = 0.132 a^{-1}$$

---

[*] ci is the unit to express the decay times in unit time for the radioactive substance. 1ci is equal to $3.7 \times 10^{10} s^{-1}$.

The initial strength of ^{60}Co is 20ci, $k = 0.132a^{-1}$. Substituting them into Equation 4.9

$$\ln\frac{20ci}{c(Co)ci} = 0.132a^{-1} \times 10a$$

$$c(Co) = 5.3ci$$

Sample Problem 4-3 Metabolism of antibiotic in the human body is a first-order reaction generally. Its concentration in blood at different time is determined after giving an injection of 500mg of antibiotic. The data were shown in the following table. Calculate

(1) The metabolism half-life of this antibiotic.

(2) Ask for the time interval to be injected next time if the lowest effective concentration of antibiotic in blood is $3.7 mg \cdot L^{-1}$?

Time after injection (t/h)	1	3	5	7	9	11	13	15
Concentration in blood ρ / mg $\cdot$ L^{-1}	6.0	5.0	4.2	3.5	2.9	2.5	2.1	1.7
$\ln \rho$	1.79	1.61	1.44	1.25	1.06	0.92	0.74	0.53

Solution (1) We plot lgρ vs. t for this first-order reaction in Figure 4-2. The rate constant k is known from the slope of line. The slope $= -0.087$, $k = 0.087 h^{-1}$, and

$t_{1/2} = 0.693 / k = 0.693 / 0.087 h^{-1} = 8.0 h$

(2) By Figure 4-2, $\ln\rho_0 = 1.87$ at $t = 0$. Substituting ρ_0 and $\rho = 3.7 mg \cdot L^{-1}$ into Equation 4.9. Time interval to take the next time of injection is

$$t = \frac{(\ln \rho_0 - \ln \rho)}{k} = \frac{(1.87 - 1.31)}{0.087 h^{-1}} = 6.4 h$$

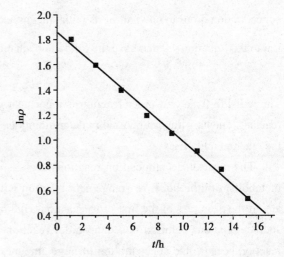

Figure 4-2 Plot of lnρ vs. Time in Blood for the Drug

The result indicates that next injection should carry out after 6.4h from preceding injection on if the drug concentration in blood is wanted to be higher than $3.7 mg \cdot L^{-1}$. Thus, the time interval of injecting this antibiotic in clinic is about 6h generally or four times a day.

2. Second-order reaction

An overall **second-order reaction** has a rate law with the sum of the exponents equal to two. There are two cases for second-order reaction. The first one is

$$(1) \ aA \rightarrow Products$$

The other one is

$$(2) \ aA + bB \rightarrow Products$$

Reaction (2) is the same as the reaction (1) if the initial concentration of reactant A and B is equal and reactant A reacts with B in equal stoichiometric coefficient throughout the reaction. We limit our discussion to a reaction involving only one reactant. That is

$$v = -\frac{dc(A)}{dt} = kc^2(A)$$

Similarly, integrating over time gives the integrated rate law for second-order reaction

$$\frac{1}{c(A)} - \frac{1}{c_0(A)} = kt \tag{4.11}$$

We plot $\frac{1}{c(A)}$ vs. t and get a straight line. The slope of the line is k. The unit of k is concentration$^{-1}\cdot$time^{-1} or $L\cdot mol^{-1}\cdot$time^{-1}. The half-life of second-order reaction is

$$t_{1/2} = \frac{1}{kc_0(A)} \tag{4.12}$$

$t_{1/2}$ of second-order reaction depends on both the rate constant and initial concentration of reactant but not a constant.

Many organic reactions in solution are second-order reaction such as addition reaction, decomposition reaction and substitution reaction.

Sample Problem 4-4 Saponification of ethyl acetate $CH_3COOC_2H_5$ at 298K is a second-order reaction.

$$CH_3COOC_2H_5 + NaOH \rightarrow CH_3COONa + C_2H_5OH$$

If the initial concentration of $CH_3COOC_2H_5$ and NaOH are $0.015\,0 mol\cdot L^{-1}$, Calculate

(1) What is the rate constant and half-life of this reaction if $0.006\,6 mol\cdot L^{-1}$ of NaOH is consumed after 20mins?

(2) What is the instantaneous rate just at 20th min?

Solution (1) Substituting each of data into Equation 4.11 and Equation 4.12

$$k = \frac{1}{t}\left[\frac{1}{c(A)} - \frac{1}{c_0(A)}\right] = \frac{1}{20min} \times \left(\frac{1}{0.015\,0 mol\cdot L^{-1} - 0.006\,6 mol\cdot L^{-1}} - \frac{1}{0.015\,0 mol\cdot L^{-1}}\right)$$
$$= 2.62 L\cdot mol^{-1}\cdot min^{-1}$$

$$t_{1/2} = \frac{1}{kc_0(A)} = \frac{1}{0.015\,0 mol\cdot L^{-1} \times 2.62 L\cdot mol^{-1}\cdot min^{-1}} = 25.5 min$$

(2) $c(A) = 0.015\,0 mol\cdot L^{-1} - 0.006\,6 mol\cdot L^{-1} = 0.008\,4 mol\cdot L^{-1}$ at 20th min. Because of the same initial concentrations for ethyl acetate and NaOH, So

$$v = kc^2(A) = 2.62 L\cdot mol^{-1}\cdot min^{-1} \times (0.008\,4 mol\cdot L^{-1})^2 = 1.85 \times 10^{-4} mol\cdot L^{-1}\cdot min^{-1}$$

3. Zero-order reaction

Reaction rate of **zero-order reaction** have no relationship with the reactants concentration.

$$v = -\frac{dc(A)}{dt} = k\,c^0(A) = k$$

The equation is obtained after integration

$$c_0(A) - c(A) = kt \tag{4.13}$$

We plot c vs. t and a straight line is got. The slope of line is $-k$. The unit of k is $mol\cdot L^{-1}$time^{-1}. The half-life of zero-order reaction is

$$t_{1/2} = \frac{c_0(A)}{2k} \tag{4.14}$$

The common zero-order reaction is that taking place at the active site of solid surface. For example,

NH_3 decomposes on the surface of metal wolfram W acting as a catalyst. NH_3 is absorbed on the surface of W at first, and then it decomposes. The active sites are fixed on the surface of W. Reaction rate is only related to the numbers of active site actually. The increase of concentration of NH_3 does not influence the rate after the active sites are occupied by NH_3 completely. This kind of reaction shows the characteristics of zero-order reaction.

Auxiliary substance developed recently for prolonging and slow-released implantation is added by special meaner to put the drug with short half-life dissolve in the body slowly and regularly. The dissolving rate of drug is relatively constant during a period of considerable length to keep up drug effective in the body for a long time. This situation belongs to zero-order reaction. For example, a hormone medicines sterile tablet made by biological degradation silicone stick in small size containing steroid hormone is embedded under skin. A 4~6 years of validity is maintained with one work. This avoids frequent oral pills and achieves the goal of contraception effective for a long time.

Summarized information about rate law of a reaction with simple order is shown in Table 4-3.

Table 4-3 Characteristic with Simple Reaction Orders

	First-order reaction	Second-order reaction	Zero-order reaction
Rate Law	$\ln\dfrac{c_0(A)}{c(A)} = kt$	$\dfrac{1}{c(A)} - \dfrac{1}{c_0(A)} = kt$	$c_0(A) - c(A) = kt$
Linear Relationship	$\ln c(A) - t$	$1/c(A) - t$	$c(A) - t$
Slope	$-k$	k	$-k$
Half-Life ($t_{1/2}$)	$0.693/k$	$1/[k c_0(A)]$	$c_0(A)/(2k)$
Unit of k	time^{-1}	$\text{conc.}^{-1} \cdot \text{time}^{-1}$	$\text{conc.} \cdot \text{time}^{-1}$

Rate constant for every kind of reaction orders can be obtained from their half-life respectively. Notice that the gotten rate constant by this way is more accurate only for first-order reaction. In addition, the reaction order can be judged by the unit of rate constant.

4.3 Brief Introduction for Rate Theory of Chemical Reaction

The difference in reaction rate for different kinds of reactions is very large. The rate of reactions such as explosion of powders, sensitization of film and the reaction between ions, etc. is too fast to measure by general technique or manners. But the reaction rate in which H_2 reacts with O_2 to H_2O at normal temperature and pressure, the reaction that coal and oil are formed under the depth of stratum, etc. are too slow to be detected. **Chemical kinetics** is the study of how the reaction can takes place and how the reactants convert to the products step by step. The chemical reaction mechanism can be test or explained by collision theory and transition theory that are mature up till now.

4.3.1 Collision Theory and Activation Energy

1. Effective collision and elastic collision

Collision Theory based on the kinetic theory considers that it is necessary for the reactant molecules to have enough energy so as to overcome repelling force from the molecules movement due to the change of valence electron cloud in reactant molecules. And then old chemical bonds between bonded atoms in reactant molecules can be broken by enough energy to form the new bonds in the product molecules that valence electron cloud was be distributed again. Actually, only small fraction of collisions between reactants molecules leads the products to form. The collision is called **elastic collision** in which the collision between reactants molecules does not lead to a reaction and no product as shown in Figure 4-3(a).

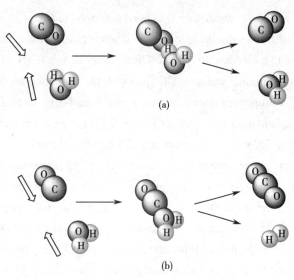

Figure 4-3 Collision between Reactant Molecules
(a) Elastic Collision; (b) Effective Collision

On the other hand, the collision leading a reaction to happen is called **effective collision** as shown in Figure 4-3(b). But how to know which reactants molecules would have the effective collision?

2. Activated molecule and activation energy

Reactant molecules which make effective collision are called **activated molecule**. Only collisions between activated molecules with enough energy to exceed activation energy can lead to the reaction. Collision theory was proposed in 1889 by Swedish chemist Svante Arrhenius. **Activation energy E_a** of a reaction is the minimum energy above the average kinetic energy that molecules must bring their collisions for a reaction to happen. In fact it is an energy threshold that colliding molecules must exceed in order to react. The unit of activation energy is $kJ \cdot mol^{-1}$.

The number of collision between reactants molecules per unit time is called **collision frequency**. Theoretical calculation based on kinetic theory and the facts had proved that the collisions per second between 1 mole of N_2 and 1 mole of O_2 at 101.3kPa and 293.15K is about 10^{27} times. If each time of collisions yields product, it means that the reaction can be over at unimaginable rate, an extreme rate. But in fact, the most of collision does not lead reaction to occur. The gaseous reaction proceeds at a much slower rate perhaps of the order of $10^{-4} mol \cdot L^{-1} \cdot s^{-1}$ generally. However, why every collision between reactants molecules cannot lead to the reaction? The reasonable conclusion is that we should not expect every collision to result in a reaction. Two main factors have to be considered. The first one is that the reactant molecules have to possess the sufficient energy. Another one is that the reactants molecules should take the proper orientation in the space at the moment of their colliding each other so as to form the product molecule. Otherwise, the collision is just elastic collision without products even though the molecules possess enough energy.

Take a gaseous reaction, for instance
$$CO(g) + H_2O(g) \rightarrow H_2(g) + CO_2(g)$$
In which CO reacts with H_2O to CO_2 and H_2, the important change taking place during the effective collision is that the H-O bond in H_2O molecule breaks and O atom from H_2O to be bonded in CO

molecule and another C-O bond is formed to form CO_2 at least. A favorable collision requires the O atom in H_2O molecule to strike the C atom in CO molecule exactly along the direction of C-O bond axis in CO molecule during the collision. May be you find that the proper orientation is the requirement to form the product molecule CO_2 just with the new bond angle. CO_2 and H_2 are formed as a result of collision in proper orientation as shown in Figure 4-3(b). Collision in other orientations but not in an appropriate orientation as shown in Figure 4-3(a) cannot generate the products even if the reactants molecules collide with sufficient energy. Be sure that the effective collision between reactants molecules must need the two requirements, sufficient energy and proper orientation.

Figure 4-4 shows such the distribution for the gaseous molecules in kinetic energy at a given temperature. The kinetic energy E of the gas molecules is laid off along the horizontal axis. The fraction of the total number of molecules having energy within the narrow interval from E to $(E+\Delta E)$ divided by the magnitude of this interval ΔE is laid off along the vertical axis. If the total number of molecules is N and the fraction having energy within the indicated interval is $\Delta N/N$, the quantity laid off along the axis of ordinate will be $\Delta N/(N\Delta E)$. Let us take a column having a width of ΔE. The area of this column is $(\Delta E \cdot \Delta N)/(N\Delta E) = \Delta N/N$. i.e. it equals the fraction of the molecules whose energy is within the interval ΔE. Similarly, the total area under the distribution curve is the sum of the fraction of molecules which is equal to 1. Correspondingly, the ratio of shaded area to the total area under this curve is just equal to the ratio of the sum of activated molecules to the sum of reactant molecules. That is the fraction of activated molecule too, the distribution for reactant molecules in energy tallies with Maxwell-Boltzmann. The fraction of activated molecule f is

$$f = e^{-E_a/RT} \tag{4.15}$$

Where E_a is activation energy of a reaction, R is the universal gas constant and T is the absolute temperature. The magnitude of both E_a and T affects the fraction of effective collision. The less the activation energy is, the more the activated molecule is, the faster the reaction proceeds at a given temperature.

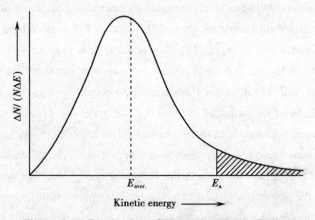

Figure 4-4 Distribution of Molecular Kinetic Energy

The activation energy of various reactions is very different and it is the main factor to affect the reaction rate. For some reactions, it is rather low, maybe less than $40 kJ \cdot mol^{-1}$. It means that a considerable part of collision between the reactant molecules would results in a reaction easily. Such kind of reaction rate is very fast. An example whose activation energy is negligibly small is ionic reaction in solution. For others, it is very high, maybe above 400kJ·mol^{-1}. Such kind of reaction rate is very slow. An example is that of ammonia synthesis from N_2 and H_2. Finally, the reaction will proceed at a moderate rate if the activation energy is in the range of 40~400kJ·mol^{-1}. The activation energy of many chemical reactions is nearly equal to the energy to break the chemical bonds between bonded atoms in reactant molecules generally. We know that the reaction rate will increase as the reactant concentrations become large. Because increasing the reactant concentrations just make the number of activated molecule to be more in fact.

4.3.2 Brief Introduction of Transition State Theory

1. Activated complex

Transition state theory focuses on the procedure of a reaction and it deems that the inner structure of reactant molecule is complicated and multifarious. Their structure will change when reactant molecules move to close proximity to each other but not change suddenly in the moment of collision between reactant molecules. The theory proposed by quantum mechanics and statistical mechanics by Henry Eyring in 1930's of 20th century is based on a hypothetical species believed to exist in a temporary state during reacting. It is an unstable intermediate between the reactants and the products molecules. This temporary state is known as an **activated complex** or **transition state.** The activated complex exists in a very short time because it has a great store of energy. In activated complex, old bonds in reactant molecule have become weaker and new bonds in product have started to form. A transition state can take place in either forward reaction or reverse reaction. The products molecule will form if the bonds between bonded atoms in the part of products molecule continue to be shorten and strengthen. However, the activated complex will decompose back to reactants if the bonds between bonded atoms in the part of reactant molecule become short and strength. So a reaction rate depends on the rate of forming the activated complex generally. For example,

$$N\equiv N-O \; + \; N=O \; \underset{}{\overset{fast}{\rightleftharpoons}} \; N\equiv N\cdots O\cdots N\overset{O}{\diagup} \; \overset{slow}{\longrightarrow} \; N\equiv N \; + \; O-N\overset{O}{\diagup}$$

Reactant   Transition State   Products

(Activated Complex)

O atom in N_2O molecule has been partially removed and partially joined to the NO molecule in transition state. It will decompose and form product molecules N_2 and NO_2 further. You may find that transition state theory give the more reasonable explanation why the different chemical reaction has very different reaction rate.

2. Heat of chemical reaction and activation energy

Figure 4-5 describes both forward and reverse reaction that N_2O and NO react to N_2 and NO_2. This plot shows the variation in the molecule energy during the reaction process. The activated complex is a highly unstable species that possess the maximum amount of

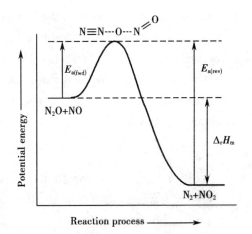

Figure 4-5 A Reaction Process

energy. In the viewpoint of transition theory, The difference between the energy of activated complex and the average energy of reactants is the activation energy for forward reaction $E_{a(fwd.)}$. Similarly $E_{a(rev.)}$ is the activation energy for reverse reaction, the difference between the energy of activated complex and the average energy of products. You should find that E_a is always more than zero by Figure 4-5.

Activation energy difference between $E_{a(fwd.)}$ and $E_{a(rev.)}$ for a reaction is just equal to the reaction heat $\Delta_r H_m$.

$$\Delta_r H_m = E_{a(fwd.)} - E_{a(rev.)} \tag{4.16}$$

The reaction is exothermic if $E_{a(rev.)} > E_{a(fwd.)}$. Conversely, it is endothermic reaction if $E_{a(rev.)} < E_{a(fwd.)}$.

The activation energy just like an **energy barrier** to be overcame for the reactant molecules. If it is smaller, the reaction rate will be fast. In fact, the lifetime of activated complex is too short to be isolated and to determine its structure or geometry unless the advanced laser technique is applied.

> **Question and thinking 4-1** Does the reaction need activation energy to undergo if the average energy of the reactant molecules is more than the average energy of product molecules? Why?

4.4 The Effect of Temperature on Chemical Reaction

4.4.1 The Relationship between Temperature and Rate Constant—Arrhenius Equation

Temperature is the remarkable factor to affect the reaction rate. From the viewpoint of practical experiment, we expect a chemical reaction to go faster at the higher temperature. Svante Arrhenius demonstrated that the rate constant varies with the temperature in 1889. The following formula is called **Arrhenius equation.**

$$k = A e^{-E_a/RT} \tag{4.17}$$

Where, e is the base of natural logarithm, T is absolute temperature in K, R is the universal gas constant $8.314 \text{kJ} \cdot \text{mol}^{-1} \cdot \text{K}^{-1}$ and E_a is the activation energy of a reaction in $\text{kJ} \cdot \text{mol}^{-1}$. A is a constant nearly called **frequency factor** related to the collision frequency and collision orientation of the reactant molecules. It does not vary with the temperature almost. Activation energy E_a only change with the temperature a little generally. Most of case, E_a is considered a constant nearly. So, Both of activation energy and temperature will affect the reaction rate mainly by Equation 4.17.

Arrhenius equation indicates that a rate constant increases as the temperature increases and as the activation energy decreases. The following equation is called Arrhenius equation also after taking natural logarithm of both sides.

$$\ln k = -\frac{E_a}{RT} + \ln A \tag{4.18}$$

We should know the follows further by Arrhenius equation:

(1) The reaction rate will be speed up as the rate constant k increase when the temperature rises because $e^{-E_a/RT}$ becomes large.

(2) For the different reactions at the same temperature, $e^{-E_a/RT}$ and k are small in value if E_a is large.

This indicates that the larger the E_a is, the slower the reaction rate is.

(3) The graph of $\ln k$ vs. $1/T$ is a straight line. The slope is $-E_a/R$ by Equation 4.18. The larger the E_a is, the smaller the slope is, the more the k changes, the larger the reaction rate varies. So the same change in temperature will affect the reaction rate in different degree for the different reactions.

(4) You may find that reaction rate only state the rate of unidirectional reaction, Most of reaction is reversible in fact. Both the rate of forward and reverse reaction will increase in different degree and then the chemical equilibrium will shift when the temperature is changed until it reaches the mew equilibrium position.

> **Question and thinking 4-2** Why a reversible reaction that has been at the equilibrium shift certainly when the temperature is changed? How to understand the shift of a chemical equilibrium from the viewpoint of chemical reaction rate?

k cannot be calculated at a given temperature without A by Arrhenius equation. But it is difficult to know the value of A for a reaction actually. So we derive a useful variation of the equation by writing Equation 4.18 twice at the different temperature T_1 and T_2 from different values of k, that is

$$\ln k_2 = \frac{-E_a}{RT_2} + \ln A$$

$$\ln k_1 = \frac{-E_a}{RT_1} + \ln A$$

$$\ln \frac{k_2}{k_1} = \frac{E_a}{R}(\frac{T_2 - T_1}{T_1 T_2}) \tag{4.19}$$

Equation 4.19 is gotten by taking the second formula at T_2 to minus the first formula at T_1 and then to eliminate the factor A. Here, T_1 and T_2 are two Kelvin temperatures. k_1 and k_2 are the rate constant at corresponding temperatures T_1 and T_2 respectively. Equation 4.19 can be used to calculate activation energy and rate constant of a reaction at a given temperatures.

Sample Problem 4-5 Decomposition of $CO(CH_2COOH)_2$ in aqueous solution has k_{293} of $4.45 \times 10^{-4} s^{-1}$ at 293K and k_{303} of $1.67 \times 10^{-3} s^{-1}$ at 303K. Find E_a of reaction and rate constant k_{313} at 313K.

Solution Substituting $T_1 = 293K$, $k_{293} = 4.45 \times 10^{-4} s^{-1}$, $T_2 = 303K$ and $k_{303} = 1.67 \times 10^{-3} s^{-1}$ into Equation 4.20 to solve for E_a,

$$\ln \frac{1.67 \times 10^{-3} s^{-1}}{4.45 \times 10^{-4} s^{-1}} = \frac{E_a}{8.314 J \cdot mol^{-1} \cdot K^{-1}}(\frac{303K - 293K}{293K \times 303K})$$

$$E_a = 97.6 kJ \cdot mol^{-1}$$

Substituting E_a, k_{303}, T_2 and $T_3 = 313K$ into Equation 4.20, k_{313} at 313K is

$$\ln \frac{k_{313}}{1.67 \times 10^{-3}} = \frac{97.6 kJ \cdot mol^{-1}}{8.314 J \cdot mol^{-1} \cdot K^{-1} \times 10^{-3}}(\frac{313K - 303K}{303K \times 313K})$$

$$k_{313} = 5.76 \times 10^{-3} s^{-1}$$

You can substitute the numbers of E_a, k_{293}, T_1 and T_3 into Equation 4.19 to solve k_{313} also.

4.4.2 The Effect of Temperature on Rate of Chemical Reaction

Energy distribution as shown in Figure 4-6 indicates the effect of increasing the temperature on a chemical reaction. Not only making the average kinetic energy of the molecules in reaction system to increase, but more importantly, the fraction of activated molecules in the system increase with the rise of temperature such as the shade area in Figure 4-6. So the frequency of effective collision is greater at the higher temperature.

Supposing, $E_a = 100 \text{kJ} \cdot \text{mol}^{-1}$ for a reaction, the temperature changes from 298K to 308K. We have

$$\frac{f_{308}}{f_{298}} = \frac{e^{\frac{-E_a}{8.314 \times 308}}}{e^{\frac{-E_a}{8.314 \times 298}}} = 3.7$$

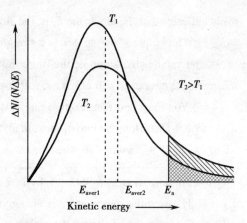

Figure 4-6 Effect of Temperature on the Distribution of Collision Energy

The fraction of activated molecules will increase 3.7 times. And the reaction rate increases 3.7 times also by Equation 4.19. But the average kinetics energy of molecules increases only 3% by the energy calculation. In fact, if the average kinetics energy of molecules in the reaction system increases a little, the fraction of activated molecules would increases more while the temperature rises. That is the reason why the reaction rate will be faster at higher temperature.

> **Question and thinking 4-3** There is a side-reaction in a drug synthesis reaction. It is known that activation energy of side-reaction is less than activation energy of main-reaction, drug synthesis reaction. How to speed up the rate of main-reaction if the frequency factors for both of reactions are equal nearly?

4.5 Brief Introduction of Catalysis

4.5.1 Catalyst and Catalysis

Another effective way to speed up the reaction rate is to use the catalyst besides increasing the reactants concentration or the temperature generally. The substance that is not consumed in a reaction but it affects rate evidently is called **catalyst**. The phenomenon to change the reaction rate under the action of catalyst is known as **catalysis**. For example, H_2 almost does not react with O_2 to H_2O under the normal temperature and pressure, but the reaction will occurs immediately in the presence of a little powder of platinum Pt. In fact, Catalyst makes activated complex in the procedure of reaction appears and its formation needs less energy than that without catalyst. It makes the rate constant larger and rate higher. The reaction rate decreases in the presence of a negative catalyst.

Note that a reaction with catalyst does not yield more product than that without catalyst. Because catalyst would speeds up the forward and reverse reaction to the same extent, it just yields the product more quickly. Catalyst cannot change the direction of a chemical reaction and its equilibrium constant. In other words, it cannot change $\Delta_r G_m$ or $\Delta_r G_m^\ominus$ for a chemical reaction.

Before and after reaction, The form of catalyst maybe changed but the mass and the chemical compositions are not changed because the catalyst participate the reaction and then it will release itself after reaction. So, a little amount of catalyst has the obvious function to the reaction due to regeneration of catalyst time and again in the process of a reaction.

Each catalyst has its own very specific catalysis function. The products may be very different by different catalysts for exactly same reactants. A reaction has its specific selectivity to the catalyst. A simple example is the following reactions:

$$CH_3CH_2OH \xrightarrow[200°C-250°C]{Cu} CH_3CHO + H_2O$$

$$CH_3CH_2OH \xrightarrow[250°C-300°C]{Al_2O_3} CH_3=CH_2 + H_2O$$

Sometimes one of the products can speed up the rate of reaction itself. For example,

$$2KMnO_4 + 3H_2SO_4 + 5H_2C_2O_4 \rightarrow 2MnSO_4 + K_2SO_4 + 8H_2O + 10CO_2$$

The rate is slow at the beginning of the reaction. But the rate will speed up automatically as the reaction proceeds. Product Mn^{2+} here is the catalyst of this reaction. This kind of reaction is called **self catalyzed reaction**.

4.5.2 Brief Introduction of Catalysis Theory

The mechanism of catalytic reaction is very complicated. A basic explanation about it is that a catalyst provides new reaction pathways with lower activation energy and the reaction rate speed up consequently. As shown in Figure 4-7, a catalyst make the one-step mechanism replace two-step mechanism for a reaction. The activation energy E_{a1} and E_{a2} in each step of catalyzed reaction in pathway (2) is much less than the activation energy E_a of an un-catalyzed reaction in pathway (1).

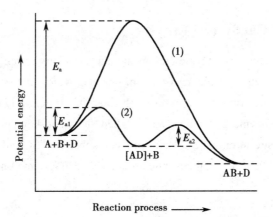

Figure 4-7 Reaction Energy Diagram for a Catalyzed and An Un-catalyzed Reaction

1. Homogeneous catalysis—Middle product theory

Homogeneous catalysis, the catalyst, reactant and product exists together in a same phase. Where D is a catalyst for reaction A + B → AB as shown in Figure 4-7, the middle product is AD in this homogeneous catalytic reaction. A, B, D, AD, and AB are all in the same phase. For example, the overall reaction without catalyst I^- ion is

$$H_2O_2(aq) \rightarrow H_2O(l) + \frac{1}{2}O_2(g) \qquad E_a = 75.3 kJ \cdot mol^{-1}$$

The concrete steps are

(1) $H_2O_2(aq) + I^-(aq) \rightarrow H_2O + IO^-(aq)$ $E_{a1} = 56.6 \text{kJ} \cdot \text{mol}^{-1}$

(2) $IO^-(aq) + H_2O_2(aq) \rightarrow H_2O(l) + O_2(g) + I^-(aq)$ E_{a2}

IO^- is the middle product and I^- ion is the effective catalyst of decomposition reaction of H_2O_2 clearly by comparing their activation energy.

Common catalytic reaction in aqueous solution is catalyzed by acid or base such as the hydrolysis of sucrose and starch. The characteristic of this category of catalytic reaction is that the proton H^+ or OH^- will transfers during the reaction. The new bonds in middle product are easy to be formed and the process only needs less activation energy. That is the reason why the stability of many drugs is related to the acidity of solution for example.

2. Heterogeneous catalysis — Active site theory

The catalysis in which the catalyst is in the different phase from the reaction system is called heterogeneous catalysis. Catalyst is solid in heterogeneous catalysis and the intermediates of reaction are formed on the surface of solid catalyst generally. A key feature of heterogeneous catalysis is that the reactants are in gaseous or aqueous system. Only some atoms on the catalyst surface called active sites attach the reactants. The heterogeneous catalysis reaction is completed in steps, adsorption of reactants, diffusion of reactants along the surface, reaction at the active sites, formation of products adsorbed and detachment of products. For example, Fe is the catalyst in synthetic ammonia reaction. N_2 is adsorbed on the active sites of Fe surface. N_2 is resolved into N atoms due to the bonds in N_2 weakens and breaks at last. H_2 and N atom reacts to NH_3.

$$N_2 + 2Fe \rightarrow 2N \sim Fe$$
$$2N \sim Fe + 3H_2 \rightarrow 2NH_3 + 2Fe$$

4.5.3 Biological Catalyst — Enzyme

The catalytic action known as enzyme-biological catalyst is very specific. The reactant molecules called substrate collide with the specific active site of enzyme, which makes enzyme catalytic reaction occur. Most of enzyme is enormous compared with their substrates. Each biochemical reaction is catalyzed by a particular enzyme generally. This is called "lock-and-key" model. The key "substrate" must fit the lock "active site in enzyme". That is to say, substrate "S" and enzyme "E" reacts to an enzyme-substrate complex "ES", ES will generate product "P" and enzyme further.

$$E + S \rightleftharpoons ES$$
$$ES \rightarrow E + P$$

The mechanism of enzyme catalytic reaction is still to provide a new reaction pathway with low activation energy. Many of enzyme-catalyzed reactions are reversible virtually. The most of them is incredibly efficient.

Catalytic activity of enzyme-catalyzed reaction in human body is about $10^6 \sim 10^{10}$ times as much as uncatalyzed reaction at normal body temperature (37℃). Catalytic activity of enzyme generally depends on the temperature and the pH of the reaction system. The normal body temperature is the most suitable temperature for enzyme-catalyzed reaction. Enzyme is protein with complex construction and large molar mass. The change in pH would vary the charged situation of enzyme molecule with many ionizable groups. Thus, the pH affects the catalytic activity of enzyme. The most suitable pH in human body is neutral nearly.

Question and thinking 4-4 Activation energy of hydrolysis reaction of sucrose catalyzed by H^+ or invertase is $109 kJ \cdot mol^{-1}$ or $48.1 kJ \cdot mol^{-1}$ respectively but $1340 kJ \cdot mol^{-1}$ without catalyst. If we only consider the effects of activation energy and temperature on the reaction rate, how many times of rate is for the reaction catalyzed by H^+ or invertase at 37℃ compared with the identical reaction without catalyst respectively? What temperature does the reaction without catalyst reach if the reaction rate without catalyst is the same as the rate by invertase catalyst at 37℃? What conclusions can you get from this example?

4.6 Brief Introduction of Chemical Reaction Mechanism

4.6.1 Elementary Reaction

The chemical reaction equation indicates the stoichiometric relationship from reactants to products and the other information about the reaction. But we cannot know how the reaction takes place step-by-step by the reaction equation.

Reaction mechanism is a step-by-step description about a chemical reaction in detail. An **elementary reaction** is the reaction consisting of just one step. For example, the reaction mechanism indicates that reaction of CO and H_2O to CO_2 and H_2 is an elementary reaction.

$$CO(g) + H_2O(g) \rightarrow CO_2(g) + H_2(g)$$

Simple reaction consists of an elementary reaction. But this type of reaction is unusual case.

The **molecularity** of an elementary reaction is the sum of the minimum molecule number of reactants needed to collide simultaneously and to produce the reaction. The common elementary reactions include unimolecular, bimolecular and termolecular reaction. For example, the reaction of broking ring of cyclopropane is unimolecular reaction

$$\begin{array}{c} CH_2 \\ / \ \backslash \\ H_2C - CH_2 \end{array} \rightarrow CH_3CH = CH_2$$

The decomposition of $N_2O(g)$ is bimolecular reaction

$$2N_2O(g) \rightarrow 2N_2(g) + O_2(g)$$

And the reaction between $H_2(g)$ and $I(g)$ is termolecular reaction

$$H_2(g) + 2I(g) \rightarrow 2HI(g)$$

Termolecular reaction is extremely rare because of getting out of the situation, the probability of three reactant particles to collide in the proper orientation simultaneously and to react successfully, is very hard. The reaction more than three reactant molecules are not found up till now.

There are many factors to affect the reaction rate as we know. Reactant concentration is one of main factors. Rate law of an elementary reaction can be written directly. Exponent in rate law is just equal to the stoichiometric coefficient in a balanced reaction at the given temperature. Of cause, the reaction order can be known easily. For example, the following is an elementary reaction,

$$NO_2(g) + CO(g) \rightarrow NO(g) + CO_2(g)$$

The rate law of it is

$$v = kc(NO_2)c(CO)$$

This is a second-order reaction.

4.6.2 Complex Reaction

The most of reactions takes place by a series of steps generally. And every step is an elementary reaction in reacting mechanism. They are called **complex reaction**. For example, The overall reaction between $H_2(g)$ and $I_2(g)$ produces $HI(g)$.

$$H_2(g) + I_2(g) \rightarrow 2HI(g)$$

Reaction mechanism known indicates that the reaction occurs in two steps.

(1) $\quad\quad\quad I_2(g) \rightleftharpoons 2I(g) \quad\quad$ (fast)

(2) $\quad\quad H_2(g) + 2I(g) \rightarrow 2HI(g) \quad$ (slow, rate-determining step)

It is important for us to note that the sum of two steps of elementary reactions yields this overall reaction, a complex reaction.

The slowest step in a complex reaction limits the rate of overall reaction process. This step is called **rate-determining step**. The rate of rate-determining step just is the rate of overall reaction commonly.

Keep in mind that the stoichiometric coefficients (a or b) in a balanced reaction are not sure equal to the reaction orders α or β respectively. If we want to get a correct rate law for a complex reaction, we must know the reaction mechanism experimentally and then the rate law of the overall reaction can be obtained by the steps. For example, when $T < 500K$, the above reaction of NO_2 and CO to NO and CO_2 will takes place in two steps but not an elementary reaction as before.

(1) $\quad NO_2(g) + NO_2(g) \rightarrow NO_3(g) + NO(g) \quad\quad$ (slow, rate-determining step)

(2) $\quad NO_3(g) + CO(g) \rightarrow NO_2(g) + CO_2(g) \quad\quad$ (fast)

The reaction mechanism under this condition is a complex reaction. Rate law is

$$v = k\, c^2(NO_2)$$

You would find that the mechanism of a same reaction may be changed at the different conditions by this reaction.

For a common complex reaction

$$aA + bB \rightarrow Products$$

The rate equation of it is

$$v = kc^\alpha(A)c^\beta(B)$$

Where α and β in equation just stand for the exponents determined by experiment.

Notice that we cannot conclude that a reaction must be elementary reaction even if the rate law of this reaction just coincides with the overall reaction equation. Rate law is not ample and necessary evidence to judge whether a reaction is elementary reaction or not. The exact mechanism of a reaction must be determined by the experiment not by its rate law. Rate law only expresses the mathematical relationship quantitatively between rate and concentrations of reactant but not the reaction mechanism.

Key Terms

化学动力学	chemical kinetics
反应速率	reaction rate
平均速率	average rate

瞬时速率	instantaneous rate
速率方程	rate law/rate equation
速率常数	rate constant
反应级数	reaction order
半衰期	half-life
一级反应	first-order reaction
准一级反应	pseudo-first-order reaction
二级反应	second-order reaction
零级反应	zero-order reaction
碰撞理论	collision theory
有效碰撞	effective collision
弹性碰撞	elastic collision
活化能	activation energy
活化分子	activated molecule
过渡态理论	transition state theory
活化络合物	activated complex
能垒	energy barrier
催化剂	catalyst
催化作用	catalysis
均相催化	homogeneous catalysis
多相催化	heterogeneous catalysis
酶	enzyme
底物	substrate
反应机制	reaction mechanism
元反应	elementary reaction
反应分子数	molecularity
复合反应	complex reaction
速率控制步骤	rate-determining step

Summary

1. The reaction rate v is generally defined as the change of reaction extent with the time in unit volume. That is

$$v = \frac{1}{v}\frac{d\xi}{dt}$$

Reaction rate tells us how the concentration of the reactant or product changes with the time. For a normal balanced reaction,

$$aA + bB \rightarrow dD + eE$$

The general reaction rate is

$$v = \frac{1}{v}\frac{d\xi}{dt} = -\frac{1}{a}\frac{dc(A)}{dt} = -\frac{1}{b}\frac{dc(B)}{dt} = \frac{1}{d}\frac{dc(D)}{dt} = \frac{1}{e}\frac{dc(E)}{dt}$$

Reaction rate indicates an instantaneous rate during the reaction process.

2. The rate law $v = kc^{\alpha}(A)c^{\beta}(B)$ indicates the relationship between the reaction rate and the reactant concentrations. The rate constant k in numerical value depends mainly on the nature of the reaction and the temperature. The sum of exponents $(\alpha + \beta + \cdots)$ is referred to the overall reaction order. The unit of k depends on the order of the reaction.

The half-life is the time during which amount of the reactant or its concentration decreases to one-half of its initial one.

3. Two theories about the reaction rate are applied for an elementary reaction. They are collision theory and transition state theory respectively. The fraction of active molecule in the total of reactant molecules determines the reaction rate in collision theory. The magnitude of activation energy is the main factor to affect the rate in transition theory. Activation energy difference between forward and reverse reactions is the enthalpy change or heat of reaction $\Delta_r H_m$ for a reaction.

4. The reaction rate can be increased by raising the temperature according to Arrhenius equation, $k = Ae^{-Ea/RT}$. It indicates that the rate constant increases as the temperature rises and activation energy decreases.

5. Catalysts increase the reaction rate. It provides an alternative reaction pathway with lower the activation energy. A catalyst speeds up the reaction rate for both forward and reverse reaction in the same degree. So it does not change the chemical equilibrium.

Enzymes are proteins with high molar mass that possess efficient activity of catalysis with "lock-and-key" model to substrate. Enzymes are active within relative narrow temperature and pH range.

6. A reaction mechanism is a step-by-step detailed description of a chemical reaction. Reactant is directly converted to products in one step in elementary reaction. An overall reaction in many steps is a complex reaction. An elementary step that occurs much more slowly than all other steps is rate-determining step in a complex reaction.

An elementary reaction may be either unimolecular or bimolecular generally. The termolecular reaction is relatively rare as an elementary reaction.

Reaction order can be obtained from the stoichiometry in a balanced elementary reaction directly. But reaction order has to be determined experimentally for any reaction if the reaction mechanism is not known in advance. We only discuss zero-, first-and second order reaction in this chapter.

Exercises

1. Understand the following concepts.

(1) reaction rate   (2) instantaneous rate   (3) elementary reaction   (4) rate constant
(5) reaction order   (6) half-life   (7) effective collision   (8) activation energy
(9) active complex   (10) rate-determining step   (11) rate law   (12) Arrhenius equation

2. What does the rate constant indicate? What is the unit of rate constant for first-, second-and zero-order reaction respectively if the unit of time is hour and the unit of concentration is $mol \cdot L^{-1}$?

3. What is the relationship between $\Delta_r H_m^{\ominus}$ and activation energy of a reaction?

4. Two reactions will undergo at the same temperature. If $E_{a2} > E_{a1}$, which reaction would be affected by temperature more greatly according to Arrhenius equation?

(1) $A + D \rightarrow E \quad E_{a1}$

(2) $G + J \rightarrow L \quad E_{a2}$

5. If reaction rate of SO_2 is 13.60 mol·L^{-1}·h^{-1} in a moment when SO_2 and O_2 react to SO_3. Calculate the reaction rate of O_2 and SO_3 in this moment respectively.

6. Hydrolysis reaction for the most of agricultural pesticides is first-order reaction. The rate of hydrolysis is important efficiency index for the insecticide. The half-life of deltamethrin at 20℃ is about 23 days. What is the rate constant for deltamethrin to hydrolyze at 20℃?

7. Try to prove that the time needed for a first-order reaction to complete 99.9% is 10 times of the half-life of the reaction nearly.

8. If the reaction rate is 0.014 mol·L^{-1}·s^{-1} when c is 0.50 mol·L^{-1}, Calculate the rate constant for the first-, second-and zero-order reaction respectively?

9. Activation energy of a decomposition reaction is 14.40 kJ·mol^{-1}. It is known that rate constant at 553 K equals $3.5 \times 10^{-2} s^{-1}$, How to adjust the temperature if the decomposition ratio is 90% in 12 minutes?

10. Thermal decomposition of gaseous acetaldehyde is a second-order reaction. It is known that 27.8% of it decomposes after 300 seconds at 500℃ and 36.2% of it decomposes after 300 seconds at 510℃ if the initial concentration of acetaldehyde is 0.005 mol·L^{-1}. Calculate the activation energy of the reaction and the rate constant at 400℃.

11. Somebody bought a drug that period of validity is 2 years kept in refrigerator at 3℃. But he put it for two weeks in a room at 25℃ because of his negligence. It is known that drug will become invalid if 30% of it decomposes, the decomposition ratio of it is not related to the concentration of drug, and the activation energy of it to decompose is 135.0 kJ·mol^{-1}. To tell whether the drug is invalid by calculation?

12. O_2 reacts with hemoglobinin in the body, Hb (hemoglobin) + O_2 → HbO_2 (oxyhemoglobin). It is first-order reaction with respect to Hb and O_2 respectively. The normal concentration should not be less than 8.0×10^{-6} mol·L^{-1} for Hb and 1.6×10^{-6} mol·L^{-1} for O_2 in lungs. $k = 1.98 \times 10^6$ L·mol^{-1}·s^{-1} at 37℃. Calculate

(1) What is the reaction rate of O_2 and HbO_2 in pulmonary blood respectively?

(2) If the reacting rate of HbO_2 reaches 1.3×10^{-4} mol·L^{-1}·s^{-1} for a patient, we want to keep the concentration of HbO_2 in normal case by oxygen therapy. What is the concentration of O_2 in pulmonary blood?

13. Decomposition of penicillin G is a first-order reaction. The following data is known

T/K	310	316	327
k/h^{-1}	2.16×10^{-2}	4.05×10^{-2}	0.119

Find the activation energy and frequency factor A for decomposition of penicillin G.

14. Milk turns sour in about 4 hours at 28℃. But it can be stored about 48 hours before it sours at 5℃ in refrigerator. If souring rate is inversely proportional to the time. Find the activation energy of turning sour for milk.

15. The activation energy of the reaction, $2HI(g) \rightarrow H_2(g) + I_2(g)$, is 184, 105 and 42 kJ·$mol^{-1}$ with catalyst gold Au, catalyst platinum Pt and no catalyst respectively at 25℃. How many times

of decomposition rate compared with the reaction no catalyst is if Au and Pt are used as catalyst respectively?

16. The activation energy of an enzyme-catalyzed reaction is 50.0kJ·mol^{-1}. Estimate the reaction rate for a patient of running a fever (40℃) compared with the normal person (37℃). Neglect the effect of temperature on enzyme activity.

Supplementary Exercises

1. The major reason why the rate of most chemical reaction increase very rapidly as temperature rises is

(a) The fraction of the molecules with kinetic energy greater than the activation energy increases very rapidly as temperature increases.

(b) The average kinetic energy increases as temperature rises.

(c) The activation energy decreases as temperature increases.

(d) The more collisions take place with particles placed so that reaction can occur.

2. At 1 000℃, cyclobutane C_4H_8 decomposes in a first-order reaction, with the very high rate constant of 87s^{-1}, to two molecules of ethylene C_2H_4.

(a) If the initial C_4H_8 concentration is 2.00mol·L^{-1}, what is the concentration after 0.001 0s?

(b) What fraction of C_4H_8 has decomposed in this time?

3. Many gaseous reactions occur in a car engine and exhaust system. One of these is

$$NO_2(g) + CO(g) \rightarrow NO(g) + CO_2(g)$$

Use the following data to determine the individual and overall reaction order:

Expt.	$\dfrac{c_{NO_2}}{mol \cdot L^{-1}}$	$\dfrac{c_{CO}}{mol \cdot L^{-1}}$	$\dfrac{\text{initial rate}}{mol \cdot L^{-1} \cdot s^{-1}}$
1	0.10	0.10	0.005 0
2	0.40	0.10	0.080
3	0.10	0.20	0.005 0

4. Researchers have created artificial red blood cells. These artificial red blood cells are cleared from circulation by a first-order reaction with a half-life of about 6 hours. If it takes 1 hour to get an accident victim, whose red blood cells have been replaced by the artificial red blood cells, to a hospital, what percentage of the artificial red blood cells will be left when the person reaches the hospital?

5. The following data are obtained at a given temperature for the initial rates from a reaction, $A + 2D + E \rightarrow 2M + G$

Expt.	$\dfrac{c_A}{mol \cdot L^{-1}}$	$\dfrac{c_D}{mol \cdot L^{-1}}$	$\dfrac{c_E}{mol \cdot L^{-1}}$	initial rate
1	1.60	1.60	1.00	v_1
2	0.80	1.60	1.00	$v_1/2$
3	0.80	0.80	1.00	$v_1/8$
4	1.60	1.60	0.50	$2v_1$
5	0.80	0.80	0.50	v_5

(a) What is the order of this reaction with respect to A, D and E?

(b) What is the value of v_5 in terms of v_1?

6. If a possible mechanism for the following gaseous reaction is

(1) $2NO \rightleftharpoons N_2O_2$ (fast, K_1)

(2) $N_2O_2 + H_2 \rightarrow H_2O + N_2O$ (slow, k_2)

(3) $N_2O + H_2 \rightarrow N_2 + H_2O$ (fast, k_3)

(a) Write the overall reaction equation.

(b) Write the rate law for the overall reaction.

(c) Indicate the overall reaction order.

Answers to Some Exercises

[Exercises]

2. h^{-1}, $L \cdot mol^{-1} \cdot h^{-1}$, $mol \cdot L^{-1} \cdot h^{-1}$

3. $\Delta_r H_m = E_{a(fwd)} - E_{a(rev)}$

4. (2)

5. $6.8 mol \cdot L^{-1} \cdot h^{-1}$, $13.60 mol \cdot L^{-1} \cdot h^{-1}$

6. $3.01 \times 10^{-2} d^{-1}$

8. (1) $0.014 mol \cdot L^{-1} \cdot s^{-1}$ (2) $0.028 s^{-1}$ (3) $0.056 L \cdot mol^{-1} \cdot s^{-1}$

9. 313K

10. $194.2 kJ \cdot mol^{-1}$, $2.88 \times 10^{-3} L \cdot mol^{-1} \cdot s^{-1}$

11. decomposition ratio is 41.2%, invalid

12. (1) $2.53 \times 10^{-5} mol \cdot L^{-1} \cdot s^{-1}$ (2) $8.2 \times 10^{-6} mol \cdot L^{-1}$

13. $85.5 kJ \cdot mol^{-1}$, 5.47×10^{12}

14. $75.0 kJ \cdot mol^{-1}$

15. 7.0×10^{13} times and 7.8×10^{24} times

16. 1.2 times

[Supplementary Exercises]

1. (a)

2. $1.83 mol \cdot L^{-1}$, 0.085

3. Reaction is zero order in CO; second order in NO_2 and second order overall respectively

4. The artificial red blood cells will be left 89.3%

5. (a) first-order for A. second-order for B and the order of -1 for E

   (b) $v_5 = \dfrac{1}{4} v_1$

6. (a) $2NO + 2H_2 \rightleftharpoons N_2 + 2H_2O$

   (b) $v = k_2 c(N_2O_2) c(H_2) = K_1 k_2 c^2(NO) c(H_2)$

   (c) 3rd-order reaction

(傅 迎)

Chapter 5
Electrolyte Solutions

An electrolyte is the substances that conduct electricity in aqueous solution or in molten state. Most of body fluids, such as body plasma, gastric juice, tear and urine contains many types of electrolyte ions Na^+, K^+, Ca^{2+}, Mg^{2+}, Cl^-, HCO_3^-, CO_3^{2-}, HPO_4^{2-}, $H_2PO_4^-$, SO_4^{2-}, etc. These electrolyte ions are related to the osmotic balance and acidity, as well play an important role in physiological and biochemical function of nervous and muscular tissues. Therefore, it is extremely significance in medical science to master the fundamental theories, characteristics and changing rules of electrolyte solutions.

5.1 Strong Electrolytic Solutions

5.1.1 Electrolytes and Degree of Dissociation

There are two types of electrolytes. **Strong electrolytes** completely ionize or dissociate in the solutions, and they are strong conductors of electricity. Strong acids such as HCl, H_2SO_4, HNO_3, $HClO_4$, and strong bases such as NaOH, KOH, $Ca(OH)_2$ are strong electrolytes. Most of the salts are strong electrolytes like NaCl. **Weak electrolytes** do not dissociate completely into ions in aqueous solution but partially into ions, they are weak conductors of electricity, such as HAc and $NH_3 \cdot H_2O$. Most of case, they exists almost as molecules. However, strong electrolytes are ionic substances such as NaCl, $CuSO_4$ and strong dipole molecule such as HCl. Ions dissociated by weak electrolytes in aqueous solution can attract each other and then partially combine to form weak electrolyte molecules again, so the dissociation reaction of them is reversible reaction. They will reach the dissociation equilibrium.

The strengths of weak electrolyte can be judged by **degree of ionization** or **dissociation** which was defined as a ratio of the number of ionized molecules and the number of molecules dissolved in an aqueous solution when equilibrium is reached. It is represented by α.

$$\alpha = \frac{\text{the number of ionized molecules}}{\text{the total number of molecules}} \times 100\% \tag{5.1}$$

The degree of dissociation may be derived by all experiments that allow determining the number of dissolved particles in a solution, such as colligative properties, including the freezing point T_f, boiling point T_b and osmotic pressure Π of an electrolytic solution.

Sample Problem 5-1 $0.10 \text{mol} \cdot \text{kg}^{-1}$ of an electrolyte substance HA solution has a freezing point of $-0.19\,°C$. Calculate the degree of dissociation of HA.

Solution The dissociation equilibrium of HA in aqueous solution is

$$HA(aq) \rightleftharpoons H^+(aq) + A^-(aq)$$

Initial $b/\text{mol} \cdot \text{kg}^{-1}$ 0.1
At equilibrium $b/\text{mol} \cdot \text{kg}^{-1}$ $(0.1-0.1\alpha)$ 0.1α 0.1α

When the equilibrium is reached, the total concentration of all components including dissociated ions and weak electrolyte molecules is

$$[(0.1-0.1\alpha)+0.1\alpha+0.1\alpha] \text{ mol} \cdot \text{kg}^{-1} = 0.1(1+\alpha) \text{ mol} \cdot \text{kg}^{-1}$$

Therefore, according to $\Delta T_f = K_f b_B$,

$$0.19\text{K} = 1.86\text{K}\,\text{mol}^{-1} \cdot \text{kg} \times 0.1(1+\alpha) \text{ mol} \cdot \text{kg}^{-1}$$

$$\alpha = 2.2\%$$

The degree of dissociation of HA is 2.2%.

> **Question and thinking 5-1** Can you calculate the equilibrium constant K and deduce the relationship of K and α by sample problem 5-1?

The dissociation degree of an electrolyte can be calculated by the observed colligative data and is called as "apparent" degree of dissociation. For $0.10\,\text{mol} \cdot \text{L}^{-1}$ of an electrolytic solution, if the apparent degree of dissociation is greater than 30%, the electrolyte is belong to strong electrolyte. If the degree of dissociation is between 30% and 5%, it is belong to middle strong electrolyte. If the degree of dissociation is less than 5%, it is belong to weak electrolyte.

Theoretically, strong electrolytes are totally ionized into ions in aqueous solution and hence no molecules exist. The degree of dissociation should be 100%. X-ray experiment had proved that most of salts are ionic crystalline in solid state and no "molecules" exist in the crystalline. But some experimental data had shown that the **degree of apparent dissociation** of strong electrolytes is not 100%. Explanation of this paradox is the task of strong electrolyte theory.

5.1.2 Debye-Hückel's Ion Interaction Theory

In 1923, Debye P. and Hückel E. proposed a theory, **ion interaction theory**, which explained the apparent incomplete ionization of strong electrolytes. If an ion moving in an aqueous solution is subjected to the influence of an external field, the surrounding ions will have to move constantly in order to form the **ion atmosphere** as shown in Figure 5-1. During the movement, a force breaking the ion movement will occur, independent of its sign, and obviously this force will increase with increasing concentration.

Debye and Hückel viewed strong electrolytes as existing only in ionic form in aqueous solution, but the ions in solution do not behave independently of one another. There are electrical interactions between ions. They argued that ions possess an ionic atmosphere, which is an average statistical model, in solution, symmetrically distributed about the ion as the center. Statistically a negative ion is more likely to be surrounded by a preponderance of positive ions and a positive ion surrounded by a preponderance of negative ions. That will be more likely the case in more concentrated solutions. The presence of this opposite charged ionic atmosphere about the central ions causes a lowering of the expected inter-ionic forces of attraction and at the same time affects the expected colligative properties by preventing the ions from behaving entirely independently.

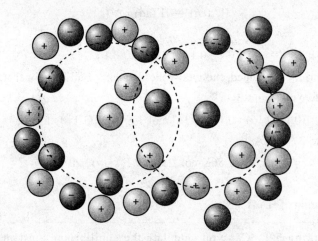

Figure 5-1   Ionic Atmosphere

5.1.3   Activity and Ionic Strength

Because the existence of ionic atmosphere and ion pairs, the effective or "apparent" concentration will be lower than "real" or theoretical concentration. The effective concentration is called **activity**. Activity is expressed as a_B, its unit is 1. For pure liquid or solid substances and solvent in dilute solution such as water, the activity is all equal to 1. We use the term activity in place of concentration, and the two quantities are related by the following equation:

$$a_B = \gamma_B b_B / b^\ominus \qquad (5.2)$$

The activity coefficient γ_B also called **activity factor**, is a factor which varies with the nature of the electrolyte and the concentration of it. $b^\ominus$ is the standard molality (i.e. $1\text{mol}\cdot\text{kg}^{-1}$). The SI units of a_B and γ_B are all 1. Generally, because $a_B < b_B$, thus $\gamma_B < 1$. It is obvious that the more dilute the solution is, the larger the distance between the ions becomes, the weaker the interaction between the ions is, and the chance of appearance of ionic atmosphere and the ionic pair will be less so that the difference between the activity and concentration is less too. So:

a) When the concentration of ions of strong electrolytes is small and the number of ion charges is also small, then activity is close to the concentration, $\gamma \approx 1$.

b) For the neutral molecules in the solution, the difference between the activity and concentration can be omitted and γ can be regarded as 1.

c) For the weak electrolytes, because the concentrations of ions are very low and γ can be regarded as 1. We usually consider the activity coefficients of pure liquid and solid as 1.

Because the positive and negative ions coexist and they can not be separated in the electrolytic solutions, the activity coefficient about the lone ion cannot be determined now. But the experiment can determine of activity coefficients of strong electrolytes. **Mean activity coefficient** is defined as the geometric mean of activity coefficients of positive and negative ions. For the substances like NaCl, KNO_3 electrolyte which general charge of ion is $+1$ and -1, $\gamma_\pm = \sqrt{\gamma_+ \gamma_-}$, where γ_+ is for activity coefficient of the positive ion. γ_- is for activity coefficient of the negative ion. So mean activity is $a_\pm = \sqrt{a_+ \cdot a_-}$.

Activity coefficient reflects the interaction force between ions in the solution. So it is related to the charge and concentration of ions, and then the activity coefficient affects activity. G. N. Lewis introduced a word called **ionic strength** I in 1921. The ionic strength is defined as

$$I \stackrel{\text{def}}{=\!=} \frac{1}{2}\sum b_i Z_i^2 \tag{5.3}$$

Where Z_i is the charge of ion i; b_i is molality of ion i in solution. The unit of ionic strength is $mol \cdot kg^{-1}$. If the solution is dilute, $b_i \approx c_i$.

The ionic strength indicates the strength of interaction between ions. It depends upon the concentrations and charge of ions in solution, but not upon the nature of ions.

According to the Debye-Hückel's theory, the relationship between the activity coefficient and the ionic strength is expressed as follows

$$\lg \gamma_i = -Az_i^2 \sqrt{I}$$

Where, A is the constant. A is $0.509 kg^{1/2} \cdot mol^{-1/2}$ for the aqueous solution at 298K.

The mean activity coefficient $\gamma_\pm$ of an electrolyte, its quantity determined by experiments is equal to the geometric mean of the activity coefficients of the individual ions, that is

$$\lg \gamma_\pm = -A|z_+ \cdot z_-|\sqrt{I}$$

Where, z_+ and z_- is the charge of the positive and negative ions, respectively.

The foregoing equations represent what has been called the Debye-Hückel limiting law; the term "limiting" is used because the derivation is such that the results are applicable only to very dilute solutions approaching the limit of infinite dilution. For NaCl type of electrolytes, results are applicable less than $0.02 mol \cdot L^{-1}$. For higher ionic strength solution, Debye and Hückel's equation should be corrected. In biological systems, the significant effects of ionic strength on the functions of enzyme, hormone and vitamins cannot be neglected.

> **Question and thinking 5-2** Why the correction factor i in calculating the colligative properties of extremely dilute solution of AB type of strong electrolytes such as NaCl and KCl should take 2?

5.2 Acid-Base Theory

Acid and base are the important electrolyte. Scientists studied the properties, composition and structure of acids and bases and proposed many theories. In 1884, S.A. Arrhenius was first to develop a dissociation theory of acid and base. He stated: acids are substances that dissociated in aqueous solution yielding H^+ ion, while bases are substances that yielded OH^- ion. The Arrhenius theory has disadvantages. It was applied only to aqueous solutions because it defined acids and bases in terms of what happens when compounds dissolve in water. Only the compounds that contain the OH^- ion can be classified as Arrhenius bases. This definition is bound to generate many conflicting facts with chemical phenomena. For example, the Arrhenius theory can't explain the fact why NH_4Cl aqueous solution is acidic but NH$_4$Cl itself does not contain H$^+$; as well Na$_2$CO$_3$ or Na$_3$PO$_4$ aqueous solution is basic but Na$_2$CO$_3$ or Na$_3$PO$_4$ does not contain OH$^-$ either. In 1923 J. N. Brønsted and T. M. Lowry independently suggested a different way of describing acid and base, that is the proton theory of acids and bases. Note that Brønsted-Lowry's definitions do not require acids and bases to be only in aqueous solution and it can be applied in non-aqueous solution and even in the gaseous phase. In this section we will discuss the proton theory of acids and bases.

5.2.1 Brønsted-Lowry Theory of Acid and Base

1. The definition of acids and bases

According to the Brønsted-Lowry's theory, **acids** are those donate proton and **bases** are those accept proton. In other words, acids are proton donors and bases are proton acceptors. A great variety of chemical properties and chemical reactions can be correlated by the definitions including the reactions that take place in gaseous phase or in the solvents other than water.

$$\text{Acid} \rightleftharpoons \text{H}^+ + \text{Base}$$
$$\text{HCl} \rightleftharpoons \text{H}^+ + \text{Cl}^-$$
$$\text{HAc} \rightleftharpoons \text{H}^+ + \text{Ac}^-$$
$$\text{H}_2\text{CO}_3 \rightleftharpoons \text{H}^+ + \text{HCO}_3^-$$
$$\text{HCO}_3^- \rightleftharpoons \text{H}^+ + \text{CO}_3^{2-}$$
$$\text{NH}_4^+ \rightleftharpoons \text{H}^+ + \text{NH}_3$$
$$\text{H}_3\text{O}^+ \rightleftharpoons \text{H}^+ + \text{H}_2\text{O}$$
$$\text{H}_2\text{O} \rightleftharpoons \text{H}^+ + \text{OH}^-$$
$$[\text{Al}(\text{H}_2\text{O})_6]^{3+} \rightleftharpoons \text{H}^+ + [\text{Al}(\text{H}_2\text{O})_5\text{OH}]^{2+}$$

In these reactions, the substances on the left side act as acids, those on the right side are bases and protons. That is, acids and bases are not isolated. The remaining part of acid after donating a proton is base which becomes the acid after accepting a proton. Both acid and base can be molecule or ion.

The above equations are called **half reactions of acid-base**. The acid and base at two side in a half reaction are called **conjugate acid-base pairs.** An acid releases a proton and forms its **conjugate base**; a base combines a proton and forms its **conjugate acid**. The two species in the conjugate acid-base pairs differ from each other by a proton. This shows that acid and base are interdependent; they can be transformed into each other.

From the concept of conjugate acid-base pair, we can find that:

(1) Some substances that can either donate or accept a proton are said to be amphoteric, such as H$_2$O and HCO$_3^-$. For example, H$_2$O is not only the conjugate acid of OH$^-$, but also the conjugate base of H$_3$O$^+$.

(2) In the proton theory of acids and bases there is no concept of salt. According to Arrhenius theory, NH$_4$Cl is a salt, it consists of cation acid NH$_4^+$ and anion base Cl$^-$. And for Na$^+$ in Na$_2$CO$_3$ solution, because it can neither donate nor accept proton, it is a non-acid and non-base substance. But CO$_3^{2-}$ is base.

(3) The proton theory of acids and bases broadens the boundary of acids and bases greatly. Acid and base can change into each other and exist interdependently.

2. The essence of acid-base reaction

According to the proton theory, acids and bases exist in pairs or: acid $\rightleftharpoons$ H$^+$ + base. This is a simple expression indicating the conjugate relation between acid and base. Proton cannot exist alone in solutions because of its smallness and large charge density. At the moment of acid's releasing a proton, the proton must combine with a base promptly. So in the actual reactions, the half-reaction that an acid releases a proton must occur at the same moment that a base accepts the proton.

For example, acetate acid HAc dissociates in water. There exist two half-reactions of acid-base.

Acid-base half-reaction 1 $HAc(aq) \rightleftharpoons H^+(aq) + Ac^-(aq)$
$\qquad\qquad\qquad\qquad\qquad$ acid 1 $\qquad\qquad$ base 1

Acid-base half-reaction 2 $H^+(aq) + H_2O(l) \rightleftharpoons H_3O^+(aq)$
$\qquad\qquad\qquad\qquad\qquad\qquad$ base 2 $\qquad$ acid 2

To merge two half-reactions, the overall reaction is

$$HAc(aq) + H_2O(l) \rightleftharpoons Ac^-(aq) + H_3O^+(aq)$$
$\qquad\;\,$ acid 1 $\qquad$ base 2 $\qquad\;\,$ base 1 $\qquad$ acid 2

In above reaction equation, an acid reacts with a base to form a new acid and a new base. Hence, acid 1 and base 1, acid 2 and base 2 are conjugate acid-base pairs respectively. This indicates that the reaction nature of acids and bases in proton theory is a **proton-transfer reaction** process. An acid, acid 1, donates a proton and become its conjugate base, base 1, and the other base, base 2, accepts a proton and becomes its conjugate acid, acid 2. This proton transfer reaction can take place both in aqueous solution and in non-aqueous solution or gaseous phase.

In the proton-transfer reaction, there exists the process of the competition for the proton. The result must be that a strong base captures a proton from a strong acid, and converted to its conjugate acid, a weak acid; and the strong acid donates a proton and converts to its conjugate base, a weak base. The acid-base reaction tends to be direction in which the stronger acid reacts with the stronger base to form the weaker acid and the weaker base. The stronger interaction between acids and bases is, the more completely reaction is. For example,

$$HCl(aq) + NH_3(aq) \rightleftharpoons NH_4^+(aq) + Cl^-(aq)$$

Since HCl is a stronger acid and NH$_3$ is a relatively stronger base, the reaction proceeds from left to right. But for the following reaction

$$Ac^- + H_2O \rightleftharpoons OH^- + HAc$$

HAc is a stronger acid and OH$^-$ is a stronger base, the reaction will proceeds from right to left.

3. Leveling effect and the differentiating effect of solvent

The strength of an acid is related not only to its own characteristics, temperature and concentration, but also closely related to the solvent. For example, HCl, HNO$_3$ and HClO$_4$ are all strong acids while HAc is a weak one in aqueous solution. But HAc is strong acid in liquefied ammonia. The strong acid dissociates completely if their concentrations are not too high, therefore the concentrations of H$_3$O$^+$ are all the same if the concentration of strong acids is same. These acids expressed in HA dissociate in aqueous solution to be expressed as the follows generally.

$$HA(aq) + H_2O(aq) \rightleftharpoons A^-(aq) + H_3O^+(aq)$$

H_2O accepts a proton from HA to form H_3O^+, all of strong acids appear equal or to be leveled to the same strength in aqueous solution. It means that the strongest acid existing in aqueous solution is H_3O^+. This phenomenon leveling all different strength of acids (or bases) to the level of solvated proton such as H_3O^+ is called **leveling effect** of solvent. This due to alkalinity of solvent is enough to accept the all of protons in acid to form hydronium ion H_3O^+. Similarly, OH^- is the strongest base in aqueous solution.

Water is a reasonably good base, moderately protons acceptor. We could choose a different solvent that is poorer base. In this type of solvent system we can differentiate acid strength among the strong acids. This is called the **differentiating effect**. The solvent makes the strong acids appear unequal in acidity. When H_2SO_4, HNO_3 and $HClO_4$ are solved in solvent which is less strongly basic than H_2O, however, they are observed to differ markedly in tendency to give up a proton to the solvent. In pure HAc, a weaker base than H_2O, we do see the difference in acidity, $HClO_4 > H_2SO_4 > HCl > HNO_3$.

It can be concluded that the same substance will appear different acidity or alkalinity in the different solvents. It is related to the ability of solvent to accept the proton. This characteristic of solvent is used widely in non-aqueous acid-base titration.

5.2.2 Autoionization of Water

1. Autoionization and ion-product of water

H_2O is an amphoteric substance that can not only donate proton as an acid, but also accept proton as a base. The **proton self-transfer equilibrium** in H_2O is as follows

$$\text{Half-reaction 1} \quad H_3O^+ \rightleftharpoons H^+ + H_2O$$
$$\text{Half-reaction 2} \quad H_2O \rightleftharpoons H^+ + OH^-$$

overall reaction is $\quad H_2O(l) + H_2O(l) \rightleftharpoons OH^-(aq) + H_3O^+(aq)$

Which H_2O molecule dissociates into H^+ and OH^- ions very slightly in an equilibrium process is known as **autoionization** or **self-ionization of water**. As H_3O^+ ion is a strong acid, OH^- ion is a strong base, so H_2O is both a weak acid and a weak base, Autoionization reaction of H_2O occurs only slightly. The apparent equilibrium constant for the autoionization reaction is

$$K = \frac{[H_3O^+][OH^-]}{[H_2O][H_2O]}$$

Where, $[H_2O]$ can be considered as a constant in dilute solution. So the product of K and $[H_2O]^2$ can be considered as a new constant K_w, the equation of the equilibrium constant can be written as

$$K_w = [H_3O^+][OH^-] \tag{5.4}$$

K_w is called **proton self-transfer constant** or **ion-product constant** for water. Equation 5.4 shows that the product of $[H^+][OH^-]$ is a constant in aqueous solution at a given temperature.

The autoionization reaction of water is an endothermic reaction. So the value of K_w will increases with the temperature rising. For example, K_w is 1.13×10^{-15} at $0°C$, 1.01×10^{-14} at $25°C$, 5.59×10^{-14} at $100°C$. K_w is 1.00×10^{-14} for pure water at 298.15K. Molarity of H^+ and OH^- is all equal to

$1.00 \times 10^{-7} \text{mol} \cdot \text{L}^{-1}$.

$$K_w = [\text{H}_3\text{O}^+][\text{OH}^-] = (1.00 \times 10^{-7})^2 = 1.00 \times 10^{-14}$$

The K_w is applied not only for pure water but also for any dilute aqueous solution. Since the product of $[\text{H}_3\text{O}^+]$ and $[\text{OH}^-]$ is a constant in aqueous solution. If the H_3O^+ ion concentration is known, the concentration of OH^- ion can be calculated based on the Equation 5.6. Hence, the acidity or basicity for an aqueous solution can be expressed by the concentration of H_3O^+ ion.

2. The pH scale

We know that the H_3O^+ and OH^- ion can exist simultaneously in whatever acidic or basic solutions, only the relative content is different. That is: in a neutral solution, $[\text{H}_3\text{O}^+] = [\text{OH}^-] = \sqrt{K_w}$; in an acidic solution, $[\text{H}_3\text{O}^+] > [\text{OH}^-]$; in a basic solution, $[\text{H}_3\text{O}^+] < [\text{OH}^-]$.

In scientific research and industrial production, the H_3O^+ ion concentrations in the solution are often very small, such as $[\text{H}_3\text{O}^+] = 3.9 \times 10^{-8} \text{mol} \cdot \text{L}^{-1}$ in serum. It is not convenient to write the concentration in this way. Hence, the extent of the acidity or basicity of a solution is often expressed by pH. The pH is defined as the negative logarithm of the hydrogen ion activity

$$\text{pH} = -\lg a(\text{H}_3\text{O}^+)$$

In a dilute solution, concentration can be used instead of activity.

$$\text{pH} = -\lg[\text{H}_3\text{O}^+]$$

The basicity of a solution can also be expressed by pOH. pOH is defined as the negative logarithm of the hydroxide ion activity.

$$\text{pOH} = -\lg a(\text{OH}^-) \quad \text{or} \quad \text{pOH} = -\lg[\text{OH}^-]$$

At 25 °C, $[\text{H}_3\text{O}^+][\text{OH}^-] = 1.0 \times 10^{-14}$ in an aqueous solution. Taking the negative logarithm of both sides of the K_w expression gives a very useful relationship among pK_w, pH, and pOH at 25 °C

$$\text{pH} + \text{pOH} = 14$$

The useful range of pH value is usually between 0~14. This is analogous to the H^+ ion concentration between $1 \text{mol} \cdot \text{L}^{-1}$ and $1.0 \times 10^{-14} \text{mol} \cdot \text{L}^{-1}$. If the H^+ ion concentration of a solution is larger than $1 \text{mol} \cdot \text{L}^{-1}$, the H_3O^+ or OH^- ion concentration can be used directly. It is not imperative to use pH. The concept of pH is important in chemistry, medicine and biology. For instance: some biochemical changes in an organism can process normally only in certain range of pH value. The various biological catalysts enzymes are effective only in a given pH value, otherwise its activity will be reduced, even lost. The ranges of pH of various normal body fluids are listed in Table 5-1.

Table 5-1 pH for Several Fluids In Human Body

Body fluids	pH	Body fluids	pH
serum	7.35~7.45	large intestine	8.3~8.4
adult's gastric	0.9~1.5	milk	6.0~6.9
baby's gastric	5.0	tear	~7.4
saliva	6.35~6.85	urine	4.8~7.5
pancreatic	7.5~8.0	cerebrospinal	7.35~7.45
small intestine	~7.6		

5.2.3 The Lewis Acid-Base

Proton theory of acid-base has many advantages, but it has disadvantages, mainly in restricting the acid as the one donating the proton. A more general concept of acids and bases was proposed in 1923 by G.N. Lewis. In Lewis theory of acid-base, also called **electron theory of acid and base**, a Lewis acid is any substance that has a vacant orbital (or the ability to rearrange its bonds to form one) to can accept a pair of nonbonding electrons, a Lewis base is any substance donate pairs of electrons. In other words, a Lewis acid is an electron-pair acceptor, a Lewis base is an electron-pair donor. The product of the reaction of a Lewis acid with a Lewis base is adduct, an acid-base complex

$$A + B \rightarrow A:B$$
$$\text{acid} \quad \text{base} \quad \text{adduct}$$

From the above, we can see that the acid reacts with substance with lone electron pair, so acid is also called electrophilic reagent.

$$F-\underset{\underset{F}{|}}{\overset{\overset{F}{|}}{B}} + [:\ddot{F}:]^- \rightleftharpoons \left[F-\underset{\underset{F}{|}}{\overset{\overset{F}{|}}{B}}\leftarrow F\right]^-$$

$$O\leftarrow \underset{\underset{O}{\downarrow}}{\overset{\overset{O}{\uparrow}}{S}} + [:\ddot{O}:]^{2-} \rightleftharpoons \left[O\leftarrow \underset{\underset{O}{\downarrow}}{\overset{\overset{O}{\uparrow}}{S}}\leftarrow O\right]^{2-}$$

$$Cu^{2+} + 4NH_3 \rightleftharpoons \left[H_3N\rightarrow \underset{\underset{NH_3}{\uparrow}}{\overset{\overset{NH_3}{\downarrow}}{Cu}}\leftarrow NH_3\right]^{2+}$$

From the above, F^- (in KF), O^{2-} (in CaO) and NH_3 are all electron-pair donors so that they are bases, and nucleophilic substances. BF_3, SO_3 and Cu^{2+} (in $CuSO_4$) are all electron-pair acceptors so that they are acids, and also electrophilic substances. The products are called acid-base coordination compounds.

The Lewis theory suggests that acids react with bases to share a pair of electrons, with no change in the oxidation numbers of any atoms. Many chemical reactions can be sorted into one or the other of two classes. Either electrons are transferred from one atom to another, or the atoms come together to share a pair of electrons.

In the Lewis theory, an acid is any ion or molecule that can accept a pair of nonbonding valence electrons. In the preceding section, we concluded that Al^{3+} ions form bonds to six water molecules to give a complex ion.

According to electron theory of acid-base, the acid-base reaction can be divided into four types as following.

Acid-base addition reaction: $Ag^+(aq) + 2NH_3(aq) \rightleftharpoons [Ag(NH_3)_2]^+(aq)$

Alkali substitution reaction: $[Cu(NH_3)_4]^{2+}(aq) + 2OH^-(aq) \rightleftharpoons Cu(OH)_2(s) + 4NH_3(aq)$

Acid substitution reaction: $[Cu(NH_3)_4]^{2+}(aq) + 4H^+(aq) \rightleftharpoons Cu^{2+}(aq) + 4NH_4^+(aq)$

Ambi-substitution reaction: $HCl(aq) + NaOH(aq) \rightleftharpoons NaCl(aq) + H_2O(l)$

One advantage of the Lewis theory is treating an oxidation-reduction reaction as an acid-base reaction. Oxidation-reduction reactions involve a transfer of electrons from one reactant to another. Acids and oxidants are those who can accept electrons in chemical reactions, they are called electrophilic substances. Bases and reductants are electron donor in the reactions, so that they are called nucleophilic substances.

Lewis theory expands the extent of acid and base, and it can be used for the acid-base reactions in organic or non-aqueous solutions. However, the Lewis acid-base is a general concept and up to now no quantitative criterion can be applied for acid-base strength. In 1963, American chemist R.G. Pearson proposed the theory of hard and soft acids and bases (HSAB theory), which can be used to explain the stability of coordination compounds, reaction mechanism and pathways.

5.3 Dissociation Equilibria of Weak Electrolytes

5.3.1 Ionization Constants for Weak Acids and Weak Bases

Weak acids and bases are all weak electrolytes. The proton-transfer processes of weak acids and bases are reversible in aqueous solution. An equilibrium is established when a proton-transfer reaches a certain extent. Consider the proton transfer equilibrium of acetate acid HAc in aqueous solution

$$HAc(aq) + H_2O(l) \rightleftharpoons Ac^-(aq) + H_3O^+(aq)$$

Another example of weak base ammonia NH_3 in aqueous solution is

$$NH_3(aq) + H_2O(l) \rightleftharpoons NH_4^+(aq) + OH^-(aq)$$

If above two reactions reach the equilibrium, this equilibrium is **acid-base dissociation equilibrium**, or simplified as **acid-base equilibrium**. Just the same as all other chemical equilibria, acid-base equilibrium has all the characteristics of chemical equilibria and complies with all common laws of chemical equilibria, including the equilibrium constant.

The equilibrium constant can be defined by thermodynamics, called the standard equilibrium constants, expresses in dimensionless K; it can also be determined directly by experiment, called experimental equilibrium constant, expressed in K. Generally, experimental equilibrium constant is not dimensionless. Considering both the International Chemical Physics Handbook and convergence and habits usage of high school, we use the experimental equilibrium constant K, unless the book is described, and [B] represents the equilibrium concentration of substance B.

For example, the equilibrium constant of HAc can be expressed as

$$K = \frac{[Ac^-][H_3O^+]}{[HAc][H_2O]}$$

For any dilute solutions, the concentration of water, $[H_2O]$, is a constant. So the proceeding equation can be written as follows

$$K_a = \frac{[Ac^-][H_3O^+]}{[HAc]} \tag{5.5}$$

Where, K_a is called the **acid dissociation constant**, or simplified as **acidity constant**.

The value of K_a is a constant at a given temperature. The strength of an acid is related to its ability to donate a proton, the magnitude of which is reflected by the value of its acidity constant, K_a. The greater the value of K_a is, the stronger the acid is, and vice versa. For example: the K_a for HAc, HClO and HCN are 1.75×10^{-5}, 3.9×10^{-8} and 6.2×10^{-10}, respectively, therefore, the strength order of these three acids is: HAc > HClO > HCN. The K_a of some weak acid is very small and pK_a, defined as the negative logarithm of the K_a, is used as a convenience.

Similarly, the base dissociation equilibrium equation of a weak base, NH_3, in aqueous solution is

$$K_b = \frac{[NH_4^+][OH^-]}{[NH_3]} \tag{5.6}$$

K_b is called the **base dissociation constant** or simplified as **basicity constant**. The magnitude of K_b indicates the ability of base to accept the proton. The greater the value of the K_b is, the stronger a base is.

K_a or K_b is a constant related to the temperature, not to the concentrations. The proton-transfer equilibrium constants for several the most common weak acids and bases are listed in Table 5-2. More data will be found in Table 2 of Appendix 3.

Table 5-2 The Proton-Transfer Constants For Weak Acids And Bases (25°C)

Conjugate acid HA	K_a^* (aq)	pK_a (aq)	Conjugate base A^-
H_3O^+			H_2O
HIO_3	1.6×10^{-1}	0.78	IO_3^-
$H_2C_2O_4$	5.6×10^{-2}	1.25	$HC_2O_4^-$
H_2SO_3	1.4×10^{-2}	1.85	HSO_3^-
H_3PO_4	6.9×10^{-3}	2.16	$H_2PO_4^-$
HF	6.3×10^{-4}	3.20	F^-
HCOOH	1.8×10^{-4}	3.75	$HCOO^-$
$HC_2O_4^-$	1.5×10^{-4}	3.81	$C_2O_4^{2-}$
HAc	1.75×10^{-5}	4.756	Ac^-
H_2CO_3	4.5×10^{-7}	6.35	HCO_3^-
H_2S	8.9×10^{-8}	7.05	HS^-
$H_2PO_4^-$	6.1×10^{-8}	7.21	HPO_4^{2-}
HSO_3^-	6×10^{-8}	7.2	SO_3^{2-}
HCN	6.2×10^{-10}	9.21	CN^-
NH_4^+	5.6×10^{-10}	9.25	NH_3
HCO_3^-	4.7×10^{-11}	10.33	CO_3^{2-}
HPO_4^{2-}	4.8×10^{-13}	12.32	PO_4^{3-}
HS^-	1.0×10^{-19}	19.00	S^{2-}
H_2O	1.0×10^{-14}	14.00	OH^-

Acidity increasing ↑ ; Basicity increasing ↓

*: The data of K_a are converted from those of pK_a.

5.3.2 The Relationship between K_a and K_b of A Conjugate Acid-Base Pair

An important relationship exists between the K_a of an acid and K_b of its conjugate base. For example,

$$HA(aq) + H_2O(l) \rightleftharpoons A^-(aq) + H_3O^+(aq)$$

$$K_a = \frac{[H_3O^+][A^-]}{[HA]}$$

And the proton transfer equilibrium for the conjugative base A^- is

$$A^-(aq) + H_2O(l) \rightleftharpoons HA(aq) + OH^-(aq)$$

$$K_b = \frac{[HA][OH^-]}{[A^-]}$$

The sum of the two dissociation reactions is the auto-ionization of water

$$H_2O(l) + H_2O(l) \rightleftharpoons OH^-(aq) + H_3O^+(aq)$$

$$K_w = [H_3O^+][OH^-]$$

$$K_a \cdot K_b = K_w \tag{5.7}$$

So the product of K_a and K_b is the equilibrium constant of water K_w.

Equation 5.7 shows that K_a is inversely proportional to the K_b, demonstrating that the strength of an acid is inversely proportional to that of its conjugate base. In other words, the weaker an acid, the stronger its conjugate base. The weaker a base, the stronger its conjugate acid.

The proton-transfer reactions of a polyprotic acids and polyacid bases proceed in a stepwise style, and are more complicated. With H_2CO_3, its proton-transfer proceeds in two steps. Each step involves the proton-transfer equilibrium

$$H_2CO_3(aq) + H_2O(l) \rightleftharpoons HCO_3^-(aq) + H_3O^+(aq)$$

$$K_{a1} = \frac{[HCO_3^-][H_3O^+]}{[H_2CO_3]} = 4.5 \times 10^{-7}$$

$$HCO_3^-(aq) + H_2O(l) \rightleftharpoons CO_3^{2-}(aq) + H_3O^+(aq)$$

$$K_{a2} = \frac{[CO_3^{2-}][H_3O^+]}{[HCO_3^-]} = 4.7 \times 10^{-11}$$

There exist the proton-transfer equilibria for their conjugate bases, HCO_3^- and CO_3^{2-}, as following

$$HCO_3^-(aq) + H_2O(l) \rightleftharpoons H_2CO_3(aq) + OH^-(aq)$$

$$K_{b2} = \frac{[H_2CO_3][OH^-]}{[HCO_3^-]} = \frac{K_w}{K_{a1}}$$

$$CO_3^{2-}(aq) + H_2O(l) \rightleftharpoons HCO_3^-(aq) + OH^-(aq)$$

$$K_{b1} = \frac{[HCO_3^-][OH^-]}{[CO_3^{2-}]} = \frac{K_w}{K_{a2}}$$

Question and thinking 5-3 H_3PO_4, $H_2PO_4^-$ and HPO_4^{2-} are acids, and their conjugate bases are $H_2PO_4^-$ and HPO_4^{2-}, PO_4^{3-}, respectively. What are the relationships between K_a of each conjugate acid and K_b of its conjugate base?

5.3.3 Shifting An Acid-Base Equilibrium

Acid-Base Equilibrium can be influenced by external factors and may shift; these factors include the concentration, common ion effect and salt effect.

1. The influence of concentration on acid-base equilibrium

The proton transfer equilibrium of weak acid HA in aqueous solution is

$$HA(aq) + H_2O(l) \rightleftharpoons H_3O^+(aq) + A^-(aq)$$

After the equilibrium is reached, if we enlarge the concentration of HA, the equilibrium will be driven to the right, the dissociation of HA, that means the concentration of H_3O^+ and A^- will increase.

Sample Problem 5-2 Calculate the dissociation degree α and the equilibrium concentration $[H_3O^+]$ of $0.100 \text{mol} \cdot L^{-1}$ HAc solution.

Solution: Known: $K_a = 1.75 \times 10^{-5}$ for HAc. Set $[H_3O^+] = x \text{ mol} \cdot L^{-1}$

$$HAc(aq) + H_2O(l) \rightleftharpoons H_3O^+(aq) + Ac^-(aq)$$

	HAc(aq)	H_3O^+(aq)	Ac^-(aq)
Initial / (mol·L^{-1})	c	0	0
At Equilibrium / (mol·L^{-1})	$(c-x)$	x	x

$$K_a = \frac{[H_3O^+][Ac^-]}{[HAc]} = \frac{x^2}{0.100 - x}$$

$$x = [H_3O^+] = 1.32 \times 10^{-3} \text{mol} \cdot L^{-1}$$

So the dissociation degree of $0.100 \text{mol} \cdot L^{-1}$ HAc is

$$\alpha = [H_3O^+]/c(HAc) = 1.32 \times 10^{-3} \text{mol} \cdot L^{-1} / 0.100 \text{mol} \cdot L^{-1} = 1.32 \times 10^{-2} = 1.32\%$$

With the same method as above, the dissociation degree of HAc and $[H_3O^+]$ of several different concentrations can be obtained and are listed in Table 5-3.

Table 5-3 α and $[H_3O^+]$ of Different Concentrations of HAc

c / (mol·L^{-1})	α/ (%)	$[H_3O^+]$ / (mol·L^{-1})
0.020	2.95	5.92×10^{-4}
0.100	1.32	1.32×10^{-3}
0.200	0.935	1.87×10^{-3}

The reason of dilution effect is that the dissociation constant K_a does not change as concentration and the degree of dissociation do. When the solution of a weak acid is diluted, the concentration of HA decreases, the $[H_3O^+]$ also decreases, but the percent dissociation of the weak acid increases, which means the equilibrium move to the direction of dissociation. This is called **law of dilution**.

2. Common ion effect

When a small amount of strong electrolytes containing an ion in common with the equilibrium system, such as solid NaAc, is added into HAc solution, NaAc will be dissociated completely into Na^+ and Ac^- ions. This will make the concentration of Ac^- increase, hence drive the proton-transfer equilibrium of HAc to left direction and decrease the degree of dissociation of acetate acid.

$$HAc(aq) + H_2O(l) \rightleftharpoons H_3O^+(aq) + \boxed{\begin{array}{c} Ac^-(aq) \\ + \\ Ac^-(aq) + Na^+(aq) \leftarrow NaAc(s) \end{array}}$$

$\leftarrow$ The direction of equilibrium moving

When a small amount of strong electrolytes including the same ions, such as solid NH_4Cl or NaOH, is added in $NH_3 \cdot H_2O$ solution, the degree of dissociation of $NH_3 \cdot H_2O$ will decrease.

$$\text{NH}_3(\text{aq}) + \text{H}_2\text{O}(l) \rightleftharpoons \text{OH}^-(\text{aq}) + \boxed{\begin{array}{c}\text{NH}_4^+(\text{aq}) \\ + \\ \text{NH}_4^+(\text{aq}) + \text{Cl}^-(\text{aq}) \leftarrow \text{NH}_4\text{Cl}(s)\end{array}}$$

← The direction of equilibrium moving

In aqueous solution of the weak acid or weak base, when the soluble strong electrolyte is added to the solution and dissociates into the common ion, the degree of dissociation of the weak acid or weak base will be decreased. This phenomenon is known as **common-ion effect**.

Sample Problem 5-3 Adding solid NaAc into $0.10\text{mol}\cdot\text{L}^{-1}$ HAc solution to make its concentration equal to $0.10\text{mol}\cdot\text{L}^{-1}$ (neglect volume change), calculate $[\text{H}_3\text{O}^+]$ and the dissociation degree α of the solution.

Solution Set the dissociated $[\text{H}_3\text{O}^+] = x\text{ mol}\cdot\text{L}^{-1}$:

$$\text{HAc}(\text{aq}) + \text{H}_2\text{O}(l) \rightleftharpoons \text{H}_3\text{O}^+(\text{aq}) + \text{Ac}^-(\text{aq})$$

Initial/(mol·L^{-1})	0.100	0	0.100
At equilibrium/(mol·L^{-1})	$(0.100-x) \approx 0.100$	x	$(0.100+x) \approx 0.100$

Because

$$K_a = \frac{[\text{Ac}^-][\text{H}_3\text{O}^+]}{[\text{HAc}]}$$

$[\text{H}_3\text{O}^+] = K_a \cdot [\text{HAc}]/[\text{Ac}^-] = (1.75 \times 10^{-5} \times 0.100)/0.100\text{mol}\cdot\text{L}^{-1} = 1.75 \times 10^{-5}\text{mol}\cdot\text{L}^{-1}$

$\alpha = [\text{H}_3\text{O}^+]/c(\text{HAc}) = 1.75 \times 10^{-5}\text{mol}\cdot\text{L}^{-1}/0.100\text{mol}\cdot\text{L}^{-1} = 1.75 \times 10^{-4} = 0.0175\%$

For $0.100\text{mol}\cdot\text{L}^{-1}$ HAc, α = 1.32%, $[\text{H}_3\text{O}^+] = 1.32 \times 10^{-3}\text{mol}\cdot\text{L}^{-1}$ (Cf: Sample Problem 5-2). We can see that because of the common ion effect, $[\text{H}_3\text{O}^+]$ and the degree of dissociation of HAc are lowered to the 1/75 of originals. Therefore, the common ion effect can be used to control and adjust the concentration of certain ions and pH of solutions.

> **Question and thinking 5-4** In Sample Problem 5-3, if we add HCl into the solution and make the concentration to $0.100\text{mol}\cdot\text{L}^{-1}$, then will the $[\text{H}_3\text{O}^+]$ and dissociation degree of HAc be the same as adding NaAc? Why?

Using the common ion effect, we can also calculate the pH of a mixed solution of a strong acid and a weak acid or a strong base and a weak base. For example, a mixed solution of HCl and HAc, since the dissociation of HAc is suppressed, $[\text{H}_3\text{O}^+]$ dissociated from HAc can be ignored, and the pH of the solution need only be calculated according to the concentration of strong acid in the mixed solution.

3. Salt effect

If we add strong electrolyte not including ions in common with HAc, such as NaCl, into acetate acid solution, the ionic strength will be increased and the interactions of the ions are increased also therefore the degree of dissociation of HAc will increase moderately. This effect is so called as **salt effect**. For example, when add NaCl into $0.10\text{mol}\cdot\text{L}^{-1}$ HAc solution to make the concentration of NaCl $0.10\text{mol}\cdot\text{L}^{-1}$, the $[\text{H}_3\text{O}^+]$ of the solution will be increased from $1.32 \times 10^{-3}\text{mol}\cdot\text{L}^{-1}$ to $1.82 \times 10^{-3}\text{mol}\cdot\text{L}^{-1}$, and the degree of dissociation of HAc will be increased from 1.32% to 1.82%.

The common ion effect always accompanies along with salt effect, but the salt effect is much smaller than common ion effect so that neglecting the salt effect does not influence the result considerately.

5.4 Calculating pH of Acid-Base Solutions

In order to calculate the pH of a solution, it is necessary to consider the components and their properties from the view of acid-base equilibrium. That means we must make it clear first weather the electrolytes are strong or weak ones, then to analyze which components in the solution are the major ones and which ones are minor enough to be negligible.

It should be kept in mind that there may be some difference between the calculated pH value and the measured one by pH meter because of the influence of ionic strength and activity factor of the solution.

5.4.1 Strong Acid or Strong Base Solutions

Strong acid or strong base are strong electrolytes and completely dissociated in aqueous solutions. For example, the HCl in aqueous solution

$$HCl(aq) + H_2O(l) \rightarrow H_3O^+(aq) + Cl^-(aq)$$

There is equilibrium in water

$$2H_2O(l) \rightleftharpoons H_3O^+(aq) + OH^-(aq)$$

The majority components are H_2O, H_3O^+ and Cl^-. The dissociation of H_2O is very weak and because of intensive inhibition of the common ion effect of HCl, the dissociated H_3O^+ from H_2O can be neglected. So for $0.1 mol \cdot L^{-1}$ HCl solution, the concentration of H_3O^+ is determined mainly by that of HCl. For strong acid HA solutions of moderate concentration, $[H_3O^+] = c(HA)$**; for strong base B, $[OH^-] = c(B)$.

If the concentration of the strong acid or base is very low, however, say, lower than $1.0 \times 10^{-6} mol \cdot L^{-1}$, the dissociated H_3O^+ from H_2O cannot be neglected.

5.4.2 Monoprotic Weak Acid or Monoacid Weak Base Solutions

We can calculate the H_3O^+ and OH^- ion concentrations in a weak acid or base aqueous solution according to the proton-transfer equilibrium constant. There are two kinds of proton-transfer equilibrium in a weak acid, HA, aqueous solution. One is the ionization equilibrium of HA.

$$HA(aq) + H_2O(l) \rightleftharpoons H_3O^+(aq) + A^-(aq)$$

$$K_a = \frac{[H_3O^+][A^-]}{[HA]}$$

Another is the autoionization equilibrium of water

$$H_2O(l) + H_2O(l) \rightleftharpoons H_3O^+(aq) + OH^-(aq)$$

$$K_w = [H_3O^+][OH^-]$$

In a system at equilibrium, the concentrations of H_3O^+, A^-, OH^- and HA are all unknown. Obviously, it is very complex to exactly calculate the H_3O^+ ion concentration. The method of exact calculation is not made here. We may use a reasonable approximate calculation for most cases. The

---

**: H_2SO_4 is considered a strong acid generally, but its $K_{a2} = 1.0 \times 10^{-2}$, so its $[H_3O^+] < 2c(H_2SO_4)$ in fact.

arithmetic process can be simplified, but the result is basically the same as of exact calculation.

(1) The autoionization of water can be neglected when $K_a c_A \geq 20 K_w$. Set initial concentration of HA is c_a, the equilibrium concentration of H_3O^+ is $[H_3O^+]$, the degree of dissociation is α. We will consider only the reaction below,

$$HA(aq) + H_2O(l) \rightleftharpoons H_3O^+(aq) + A^-(aq)$$

Initial concentration	c_a		
At equilibrium	$c_a(1-\alpha)$	αc_a	αc_a

We obtain

$$K_a = \frac{[H_3O^+][A^-]}{[HA]} = \frac{c_a\alpha \cdot c_a\alpha}{c_a(1-\alpha)} = \frac{c_a\alpha^2}{1-\alpha} \quad (5.8)$$

$$K_a = \frac{[H_3O^+]^2}{c_a - [H_3O^+]} \quad (5.9)$$

$$[H_3O^+] = \frac{-K_a + \sqrt{K_a^2 + 4K_a c_a}}{2} \quad (5.10)$$

The above equation is an approximate calculation formula of the H_3O^+ ion concentration in a weak monoprotic acid.

(2) When $\alpha < 5\%$, or $c_a / K_a \geq 500$, the amount of dissociated acid is so much small that we can neglect it, compared with the initial concentration c_a of the acid. That is: $1 - \alpha \approx 1$, $[HA] \approx c_a$,

$$K_a = c_a \alpha^2$$

We obtain

$$\alpha = \sqrt{K_a / c_a} \quad (5.11)$$

$$[H_3O^+] = \sqrt{K_a \cdot c_a} \quad (5.12)$$

Generally speaking, when $K_a \cdot c_a \geq 20K_w$ and $c_a / K_a \geq 500$, calculation error will not be larger than 5%.

The above equation is the simplest formula to calculate the H^+ ion concentration of a weak monoprotic acid. The degree of dissociation is inversely proportional to the square root of concentration by Equation 5.11, which explains **the dilution law** quantitatively.

We give the similar equation for a weak monoacid base solution when $K_b \cdot c_b \geq 20K_w$ and $c_b / K_b \geq 500$ as follows

$$[OH^-] = \sqrt{K_b \cdot c_b} \quad (5.13)$$

Sample Problem 5-4 Calculate the pH of $0.10 \text{mol} \cdot L^{-1}$ HAc solution and the concentrations of HAc, Ac^-, and OH^-, respectively.

Solution For HAc, $K_a = 1.75 \times 10^{-5}$, $c_a = 0.100 \text{mol} \cdot L^{-1}$,

$$K_a \cdot c(HAc) = 1.75 \times 10^{-5} \times 0.100 = 1.75 \times 10^{-6} > 20K_w$$

$c(HAc) / K_a = [0.100 / (1.75 \times 10^{-5})] > 500$, Therefore, we can use the Equation 5.12 to calculate:

$$[H_3O^+] = \sqrt{K_a c_a} = \sqrt{1.75 \times 10^{-5} \times 0.100} \text{mol} \cdot L^{-1} = 1.32 \times 10^{-3} \text{mol} \cdot L^{-1}$$

$$pH = 2.88$$

$[Ac^-] = [H_3O^+] = 1.32 \times 10^{-3} \text{mol} \cdot L^{-1}$, $[HAc] = (0.100 - 1.32 \times 10^{-3}) \text{mol} \cdot L^{-1} \approx 0.100 \text{mol} \cdot L^{-1}$

$$[OH^-] = K_w / [H_3O^+] = 7.58 \times 10^{-12} \text{mol} \cdot L^{-1}$$

It is important to emphasize that only when $K_a c_a$ (or $K_b c_b$) $\geq 20K_w$ or c_a / K_a (or c_b / K_b) > 500, we can use Equation 5.12 or Equation 5.13. Otherwise a larger calculation error will be made, even leading to an absurd conclusion. In very dilute solution $c_a / K_a < 500$, the $[H_3O^+]$ of water dissociation cannot be neglected.

Sample Problem 5-5 Calculate the pH of NaAc solution by adding 4.10g solid NaAc into 500ml water. Known $K_a(HAc) = 1.75 \times 10^{-5}$.

Solution Solid NaAc is completely dissociated when dissolved in water, the pH of the solution is mainly determined by the concentration of Ac^-

$$c(Ac^-) = 4.10g / (82.03g \cdot mol^{-1} \times 0.500L) = 0.100 mol \cdot L^{-1}$$

There exists the following reaction in the solution

$$Ac^-(aq) + H_2O(l) \rightleftharpoons HAc(aq) + OH^-(aq)$$

Known $K_a(HAc) = 1.75 \times 10^{-5}$, $K_b(Ac^-) = K_w / K_a(HAc) = 1.00 \times 10^{-14} / (1.75 \times 10^{-5}) = 5.76 \times 10^{-10}$
As $K_b c_b \geq 20K_w$, $c_b / K_b = [0.100 / (5.75 \times 10^{-10})] > 500$, so

$$[OH^-] = \sqrt{K_b \cdot c} = \sqrt{5.71 \times 10^{-10} \times 0.100} mol \cdot L^{-1} = 7.56 \times 10^{-6} mol \cdot L^{-1}$$

$$pOH = 5.12, \quad pH = 14.00 - 5.12 = 8.88$$

Sample Problem 5-6 The pH value of $0.25 mol \cdot L^{-1}$ HF solution is measured as 1.92, calculate K_a of HF.

Solution pH = 1.92, then $[H_3O^+] = 10^{-1.92} mol \cdot L^{-1} = 0.012 mol \cdot L^{-1}$. There exists the following equilibrium

$$HF(aq) + H_2O \rightleftharpoons F^-(aq) + H_3O^+(aq)$$

Initial (mol·L^{-1})　　　　　　　0.25
At equilibrium (mol·L^{-1})　　(0.25 − 0.012)　　　0.012　　　0.012
Also because of $c_a / K_a = 0.10 / (5.6 \times 10^{-10}) \geq 500$. According to Equation 5.11,

$$K_a = \frac{[H_3O^+][F^-]}{[HF]} = \frac{0.012^2}{0.25 - 0.012} = 6.0 \times 10^{-4}$$

5.4.3 Polyprotic Acid and Polyacid Base Solutions

In a solution of weak polyprotic acid, the dissociation of a polyprotic acid proceeds in a complicated, stepwise fashion. Each dissociation step involves a different dissociation constant expression.

For an example, the two-step dissociation constant of the diprotic acid, H_2S, are $K_{a1} = 8.9 \times 10^{-8}$, $K_{a2} = 1.0 \times 10^{-19}$, respectively. Another example is the three-step dissociation of H_3PO_4 in aqueous solution. At 25℃, $K_{a1} = 6.9 \times 10^{-3}$, $K_{a2} = 6.1 \times 10^{-8}$, $K_{a3} = 4.8 \times 10^{-13}$.

From the data above, we can see: $K_{a1} \gg K_{a2} \gg K_{a3}$, and this is a rule for stepwise dissociation because H_3O^+ from previous step of dissociation will suppress the next step of dissociation by common ion H_3O^+. Therefore, H_3O^+ ions solutions come predominantly from the first step of dissociation for polyprotic acid, H_3O^+ dissociated from the other steps is too minor and can be neglected. Generally, for the polyprotic acid with $(K_{a1} / K_{a2}) > 10^2$ (for most polyprotic acids the ratios are between 10^4 and 10^6), solution may be treated as a monoprotic acid when calculating the $[H_3O^+]$.

Sample problem 5-7 Calculate the concentrations of H^+, HCO_3^-, CO_3^{2-} and OH^-, respectively for the $0.02 mol \cdot L^{-1}$ carbonic acid solution.

Solution When $K_{a1} = 4.5 \times 10^{-7}$, $K_{a2} = 4.7 \times 10^{-11}$, so $(K_{a1}/K_{a2}) > 10^2$, ignore $[H_3O^+]$ from the equilibrium of second step. This means that H_2CO_3 is taken as a monoprotic acid. From the first step of ionization of carbonic acid: $[H_3O^+] = [HCO_3^-]$, $c_a \approx [H_2CO_3]$, let $[H_3O^+]$ from H_2CO_3 to be x, and $[H_3O^+]$ from HCO_3^- be y, there exists the following equilibrium.

$$H_2CO_3(aq) + H_2O(l) \rightleftharpoons HCO_3^-(aq) + H_3O^+(aq)$$

If $[H_2CO_3] = 0.020 mol \cdot L^{-1}$

Initial $(mol \cdot L^{-1})$ 0.020

At equilibrium $(mol \cdot L^{-1})$ $(0.020 - x) \approx 0.020$ $(x - y) \approx x$ $(x + y) \approx x$

As $c_a / K_a \geqslant 500$, by Equation 5.14 can be used

$$K_{a1} = \frac{[HCO_3^-][H_3O^+]}{[H_2CO_3]} = \frac{x^2}{0.020} = 4.5 \times 10^{-7}$$

$$x = 9.5 \times 10^{-5}$$

$$[H_3O^+] = [HCO_3^-] = 9.5 \times 10^{-5} mol \cdot L^{-1}$$

$$HCO_3^-(aq) + H_2O(l) \rightleftharpoons CO_3^{2-}(aq) + H_3O^+(aq)$$

At equilibrium / $mol \cdot L^{-1}$ $(9.5 \times 10^{-5} - y)$ $(9.5 \times 10^{-5} + y)$

$\approx 9.5 \times 10^{-5}$ $\approx 9.5 \times 10^{-5}$

CO_3^{2-} is formed in second step of ionization

$$K_{a2} = \frac{[CO_3^{2-}][H_3O^+]}{[HCO_3^-]} = \frac{(9.5 \times 10^{-5})y}{9.5 \times 10^{-5}} = 4.7 \times 10^{-11}$$

$$y = 4.7 \times 10^{-11}$$

$$[CO_3^{2-}] = 4.7 \times 10^{-11} mol \cdot L^{-1}$$

$$[OH^-] = K_w / [H_3O^+] = 1.0 \times 10^{-14} / 9.5 \times 10^{-5}$$

$$[OH^-] = 1.0 \times 10^{-10} mol \cdot L^{-1}$$

From the above result, it can be seen that $[H_3O^+] = 9.5 \times 10^{-5} mol \cdot L^{-1}$ for the first step of dissociation, and $[H_3O^+] = 4.7 \times 10^{-11} mol \cdot L^{-1}$ for the second step, from H_2O dissociates $[H_3O^+] = 1.0 \times 10^{-10} mol \cdot L^{-1}$. Therefore, the H_3O^+ concentration of the solution is determined by the first step of dissociation of H_2CO_3, ignoring the second step and the dissociation of water is completely reasonable. The pH of the H_2CO_3 solution can be calculated as an approximation of monoprotic weak acid, i.e. when $K_{a1} \cdot c_a \geqslant 20 K_w$ and $c_a / K_{a1} \geqslant 500$

$$[H_3O^+] = \sqrt{K_{a1} \cdot c} = \sqrt{4.5 \times 10^{-7} \times 0.02} mol \cdot L^{-1} = 9.5 \times 10^{-5} mol \cdot L^{-1}, pH = 4.02$$

According to the calculation above, we conclude that

(1) If weak polyprotic acid constants are in the order $K_{a1} \gg K_{a2} > K_{a3}$, the acid is taken as monoprotic acid when calculating the concentration of H^+, and the acidity of it is mainly dependent on K_{a1}.

(2) The concentration of the conjugate base from the second step of ionization of a weak polyprotic acid is nearly equal to the value of K_{a2}, and it is not influenced by the concentration of the weak acid.

(3) The concentrations of conjugate bases from second or third, or forth step of ionizations of weak polyprotic acid are usually quite small.

The ionization of weak polyacid bases is similar to that of weak polyprotic acids. The calculation method is similar to that of weak polyprotic acids.

Sample Problem 5-8 Calculate the pH and the concentration of HCO_3^- and CO_3^{2-} ions of $0.100 \text{mol} \cdot L^{-1}$ Na_2CO_3 solution. Known: $K_{b1} = 2.1 \times 10^{-4}$, $K_{b2} = 2.2 \times 10^{-8}$.

Solution H_2CO_3 is a diprotic acid. CO_3^{2-}-HCO_3^- and HCO_3^--H_2CO_3 are the conjugate acid-base pairs, respectively. There exist the following equilibria

$$CO_3^{2-}(aq) + H_2O(l) \rightleftharpoons HCO_3^-(aq) + OH^-(aq)$$

$$K_{b1} = K_w / K_{a2} = 1.0 \times 10^{-14} / (4.7 \times 10^{-11}) = 2.1 \times 10^{-4}$$

$$HCO_3^-(aq) + H_2O(l) \rightleftharpoons H_2CO_3(aq) + OH^-(aq)$$

$$K_{b2} = K_w / K_{a1} = 1.0 \times 10^{-14} / (4.5 \times 10^{-7}) = 2.2 \times 10^{-8}$$

Due to $K_{b1} / K_{b2} > 10^2$, $c_b / K_{b1} > 500$, and $c_b \cdot K_{b1} > 20 K_w$. So

$$[OH^-] = \sqrt{K_{b1} \cdot c_b} = \sqrt{2.1 \times 10^{-4} \times 0.100} \text{ mol} \cdot L^{-1} = 4.6 \times 10^{-3} \text{ mol} \cdot L^{-1}$$

$$[HCO_3^-] \approx [OH^-] = 4.6 \times 10^{-3} \text{ mol} \cdot L^{-1}$$

$$pOH = 2.34, \quad pH = 14.00 - 2.34 = 11.66$$

$$[CO_3^{2-}] = 0.100 \text{ mol} \cdot L^{-1} - 4.6 \times 10^{-3} \text{ mol} \cdot L^{-1} = 0.095 \text{ mol} \cdot L^{-1}$$

The result shows that predominate species in the solution are still CO_3^{2-} and Na^+ and only about 5% of $0.100 \text{mol} \cdot L^{-1}$ Na_2CO_3 solution dissociate to HCO_3^-.

5.4.4 The pH of Amphoteric Substances Solutions

According to the proton theory of acid and base, the amphoteric substances are the ones which can donate and accept protons. Acid salt, such as HCO_3^-, $H_2PO_4^-$, HPO_4^{2-}, the weak acid weak base salt, such as NH_4Ac, NH_4CN, $(NH_4)_2CO_3$, and amino acid are all amphoteric substances. It is quite complex for amphoteric compounds to be represented at equilibrium as acids and bases. Here we take $NaHCO_3$ as an example. $NaHCO_3$ is dissociated completely in the following way

$$HCO_3^-(aq) + H_2O(l) \rightleftharpoons H_3O^+(aq) + CO_3^{2-}(aq)$$

$$K_a = \frac{[H_3O^+][CO_3^{2-}]}{[HCO_3^-]} = K_{a2}(H_2CO_3) = 4.7 \times 10^{-11} \tag{1}$$

As a base, proton transfer equilibrium of HCO_3^- is

$$HCO_3^-(aq) + H_2O(l) \rightleftharpoons OH^-(aq) + H_2CO_3(aq)$$

$$K_b = \frac{[OH^-][H_2CO_3]}{[HCO_3^-]} = \frac{K_w}{K_{a1}(H_2CO_3)} = 2.2 \times 10^{-8} \tag{2}$$

$$H_2O(l) + H_2O(l) \rightleftharpoons H_3O^+(aq) + OH^-(aq)$$

$$K_w = [H_3O^+][OH^-] \tag{3}$$

It can be seen that the acidity of the amphoteric substances in the aqueous solution all depends on the relative value of the corresponding K_a and K_b, i.e.

When $K_a > K_b$, pH < 7, the solution appears acidic, such as NaH_2PO_4, NH_4F, $HCOONH_4$ and $NH_3^+CH_2COO^-$, etc.

When $K_b > K_a$, pH > 7, the solution appears basic, such as Na_2HPO_4, $NaHCO_3$, $(NH_4)_2CO_3$ and NH_4CN, etc.

When $K_a \approx K_b$, pH $\approx$ 7, the solution appears neutral, such as NH_4Ac, etc.

As can be seen from above equations (1), (2) and (3), since the proton transfer equilibrium is very complicated, in the calculation of [H₃O⁺] in the amphoteric solution, an approximate treatment can be performed according to specific conditions.

When $cK_a \geqslant 20K_w$ and $c \geqslant 20K_a'$ ($K_a' = K_w/K_b$), after derivation and approximate processing, we can get:

$$[H_3O^+] = \sqrt{K_a' K_a} \quad \text{or} \quad pH = \frac{1}{2}(pK_a' + pK_a) \tag{5.14}$$

Equation 5.14 is the equation which can approximately calculate the pH of solutions of amphoteric substances when neglecting the autoionization of water. K_a in the formula is the dissociation constant of the amphoteric substance as the acid, and K_a' is the dissociation constant of the corresponding conjugate acid when the amphipathic substance is the base, and c is the initial concentration of the amphipathic substance.

Sample Problem 5-9 Qualitatively describes the acidity of the Na₂HPO₄ solution.(Known that, for H₃PO₄, $K_{a1} = 6.9 \times 10^{-3}$, $K_{a2} = 6.1 \times 10^{-8}$, $K_{a3} = 4.8 \times 10^{-13}$)

Solution The following equilibrium exists in the Na₂HPO₄ solution:

$$H_2O(l) + HPO_4^{2-}(aq) \rightleftharpoons H_3O^+(aq) + PO_4^{3-}(aq) \quad K_{a3} = 4.8 \times 10^{-13}$$

$$H_2O(l) + HPO_4^{2-}(aq) \rightleftharpoons OH^-(aq) + H_2PO_4^-(aq) \quad K_{b2} = \frac{K_w}{K_{a2}} = \frac{1.0 \times 10^{-14}}{6.1 \times 10^{-8}} = 1.6 \times 10^{-7}$$

HPO_4^{2-} donates a proton in the first dissociation equilibrium and HPO_4^{2-} accepts the proton in the second dissociation equilibrium. K_{a3} and K_{b2} themselves are the acid constants and base constants of the amphoteric substances, because: $K_{b2} > K_{a3}$, the ability of HPO_4^{2-} to accept protons is greater than the ability to give protons, so the solution is basic.

Sample Problem 5-10 At 298.15K, calculate the pH of 0.010mol·L⁻¹ NaHCO₃. Known for H₂CO₃, $K_{a1} = 4.5 \times 10^{-7}$, $K_{a2} = 4.7 \times 10^{-11}$.

Solution Because $K_{a2}c \geqslant 20K_w$, $c \geqslant 20K_a'$, so we can use Equation 5.14 to calculate [H₃O⁺]

$$[H_3O^+] = \sqrt{K_a' \cdot K_a} = \sqrt{K_{a1} \cdot K_{a2}} = \sqrt{4.5 \times 10^{-7} \times 4.7 \times 10^{-11}} \,\text{mol}\cdot\text{L}^{-1} = 4.6 \times 10^{-9}\,\text{mol}\cdot\text{L}^{-1}$$

$$pH = -\lg[H_3O^+] = -\lg(4.6 \times 10^{-9}) = 8.34$$

Sample problem 5-11 Calculate the pH of 0.1mol·L⁻¹ NH₄CN solution. Known for NH₃ K_b is 1.8×10^{-5} and K_a for HCN is 6.2×10^{-10}.

Solution The predominate species in NH₄CN are NH_4^+, CN^- and H₂O, and react as followings

$$NH_4^+(aq) + H_2O(l) \rightleftharpoons NH_3(aq) + H_3O^+(aq) \quad K_a' = K_w/K_b = 5.6 \times 10^{-10}$$

$$CN^-(aq) + H_2O(l) \rightleftharpoons HCN(aq) + OH^-(aq) \quad K_b' = K_w/K_a = 1.6 \times 10^{-5}$$

$$H_2O(l) + H_2O(l) \rightleftharpoons H_3O^+(aq) + OH^-(aq) \quad K_w = 1.0 \times 10^{-14}$$

Because: $K_{a2} c \geqslant 20K_w$, $c \geqslant 20K_a'$, so we can use Equation 5.14 to calculate [H₃O⁺]

$$[H_3O^+] = \sqrt{K_a \cdot K_a'}$$

$$[H_3O^+] = \sqrt{6.2 \times 10^{-10} \times 5.6 \times 10^{-10}}\,\text{mol}\cdot\text{L}^{-1} = 6.0 \times 10^{-10}\,\text{mol}\cdot\text{L}^{-1},\ pH = 9.22$$

Key Terms

强电解质	strong electrolyte
弱电解质	weak electrolyte

解离度	degree of dissociation
表观解离度	apparent degree of dissociation
活度因子	activity factor
离子氛	ion atmosphere
离子强度	ionic strength
质子酸碱理论	Brønsted-Lowry's theory
酸	acid
碱	base
酸碱半反应	half reactions of acid-base
共轭酸碱对	conjugate acid-base pairs
共轭碱	conjugate base
共轭酸	conjugate acid
拉平效应	leveling effect
区分效应	differentiating effect
质子自递平衡	proton self-transfer equilibrium
水的质子自递	autoionization of water
质子自递常数	autoionization constant
离子积	ion-product constant
酸碱电子理论	electron theory of acid and base
酸碱解离平衡	acid-base dissociation equilibrium
酸碱平衡	acid-base equilibrium
酸解离常数	acid dissociation constant
碱解离常数	base dissociation constant
稀释定律	law of dilution
同离子效应	common-ion effect
盐效应	salt effect

Summary

Substances that dissociate or ionize in water to produce cations and anions are electrolytes. The strong electrolytes completely dissociate in aqueous solution, but the apparent dissociation degree is not 100%. The interionic attractions in an aqueous solution prevent the ions from behaving as totally independent particles and ion atmosphere is formed.

An acid is a proton donor, and a base is a proton acceptor in the Brønsted-Lowry theory.

The equilibrium constant for the proton-transfer reaction between two H_2O molecules is called the ion product of water, denoted by K_w, $K_w = [H^+][OH^-]$, which equals 1.0×10^{-14} at 25°C. That between a weak acid and water is called an acidity constant, denoted by K_a. That between a weak base and water is called a basicity constant, denoted by K_b.

A conjugate base is everything that remains of the acid molecule after a proton is lost. A conjugate acid is formed when a proton is transferred to the base. Two substances related in this way, by the expression,

$$\text{Base} + H^+ \rightleftharpoons \text{Acid}$$

They are called a conjugate acid-base pair. For any such pairs, the relation of K_a and K_b is: $K_a \cdot K_b = [H^+][OH^-] = K_w$. Strong acids form weak conjugate bases, and weak acids form strong conjugate bases.

Polyprotic acids dissociate stepwise, with a separate dissociation constant for each step.

The formula of the calculation of $[H^+]$ in a monoprotic weak acid solution is: $[H^+] = \sqrt{K_a \cdot c_a}$ when $K_a \cdot c_a \geq 20K_w$, and $c_a/K_a \geq 500$ or $\alpha < 5\%$. But if not, that is: $K_a = \dfrac{c_a \alpha^2}{1-\alpha}$ and $[H^+] = c_a \alpha$.

Amphoteric substances can act both as an acid and as a base. The $[H^+]$ calculation is given by the common equation of $[H_3O^+] = \sqrt{K_a K_a'}$ or $pH = \dfrac{1}{2} \times (pK_a + pK_a')$.

The extent of the ionization of a weak electrolyte can be reduced by adding a strong electrolyte that provides an ion common to the equilibrium. This phenomenon is called the common-ion effect. The addition of a strong electrolyte not including ions in common slightly increases the degree of ionization of a weak electrolyte. This effect is called the salt effect.

Exercises

1. Indicate the conjugate bases of the following acids: H_2O, H_3O^+, H_2CO_3, HCO_3^-, NH_4^+, $^+NH_3CH_2COO^-$, H_2S and HS^- respectively.

2. Indicate the conjugate acids in the following bases: H_2O, NH_3, HPO_4^{2-}, HAc, NH_2^-, $[Al(H_2O)_5OH]^{2+}$, CO_3^{2-} and $^+NH_3CH_2COO^-$ respectively.

3. Try to explain (1) How many ions are there in the H_3PO_4 solution? List them in their ionic concentration order. Is the concentration of H^+ is three times of PO_4^{3-}? Why? (2) Why the aqueous $NaHCO_3$ solution is more basic than the aqueous NaH_2PO_4 solution?

4. Calculate the proton-transfer equilibrium constant of the following reactions and predict which direction the reactions favor?

    (1) $HNO_2(aq) + CN^-(aq) = HCN(aq) + NO_2^-(aq)$
    (2) $HSO_4^-(aq) + NO_2^-(aq) = HNO_2(aq) + SO_4^{2-}(aq)$
    (3) $NH_4^+(aq) + Ac^-(aq) = NH_3(aq) + HAc(aq)$
    (4) $SO_4^{2-}(aq) + H_2O(l) = HSO_4^-(aq) + OH^-(aq)$

5. pH of gastric juice of normal adult is 1.4, pH of gastric juice of infants is 5.0. How many times is H_3O^+ concentration of gastric juice of adult than that of infant?

6. Calculate the $[H_3O^+]$, $[HS^-]$ and $[S^{2-}]$ in $0.10 \text{mol} \cdot L^{-1}$ H_2S solution. Known: $K_{a1} = 8.9 \times 10^{-8}$, $K_{a2} = 1.0 \times 10^{-19}$.

7. The painkiller morphine ($C_{17}H_{19}NO_3$), mainly extracted from the unripe poppy seed, is a weak base and its $K_b = 7.9 \times 10^{-7}$. Calculate the pH of $0.015 \text{mol} \cdot L^{-1}$ morphine solution.

8. Adding sodium azide (NaN_3) into water can play a bactericidal effect. Calculate the concentrations of each species in $0.010 \text{mol} \cdot L^{-1}$ NaN_3 solution at equilibrium. Known: for Hydrazoic acid HN_3, $K_a = 1.9 \times 10^{-5}$.

9. Salicylic acid (hydroxybenzoic acid, $C_7H_6O_3$) is a diprotic acid, $K_{a1} = 1.06 \times 10^{-3}$, $K_{a2} = 3.6 \times 10^{-14}$.

It is a kind of antiseptic, sometimes used as a painkiller instead of aspirin, but it has a strong acidity which can cause stomach bleeding. Calculate the pH of $0.065 \text{mol} \cdot \text{L}^{-1}$ $C_7H_6O_3$ solution and concentration of each species at equilibrium.

10. Calculate the pH of the following solutions.

(1) the mixture of 100mL $0.10 \text{mol} \cdot \text{L}^{-1}$ H_3PO_4 and 100mL $0.20 \text{mol} \cdot \text{L}^{-1}$ NaOH.

(2) the mixture of 100mL $0.10 \text{mol} \cdot \text{L}^{-1}$ Na_3PO_4 and 100mL $0.20 \text{mol} \cdot \text{L}^{-1}$ HCl.

11. Liquid ammonia, like water, can occur autoionization reaction: $NH_3(l) + NH_3(l) = NH_4^+(aq) + NH_2^-(aq)$. Write the proto-transfer reaction for the acetate acid in liquid ammonia and explain the acidity of acetate acid in liquid ammonia is weaker or stronger than that in water?

12. Lactic acid ($CH_3CHOHCOOH$) is the end product of glycolysis, accumulating in the body can cause body fatigue and lactic acidosis. Known: for lactic acid, $K_a = 1.4 \times 10^{-4}$, pH of a certain sample of yoghurt is 2.45. Calculate the concentration of lactic acid solution.

13. For a $0.2 \text{mol} \cdot \text{L}^{-1}$ HCl solution, (1) To make the pH of the solution equal 4.0, which one, HAc or NaAc, should be added into the solution? (2) If this HCl solution is mixed with equal volumes of $2.0 \text{mol} \cdot \text{L}^{-1}$ of NaAc solution, what is the pH of the mixture? (3) An equal volume of $2.0 \text{mol} \cdot \text{L}^{-1}$ of NaOH solution is added to the solution, what is the pH of the mixture?

14. Quinoline ($C_{20}H_{24}N_2O_2$, $MM. = 324.4 \text{g} \cdot \text{mol}^{-1}$), an important alkaloids, is mainly derived from cinchona bark and is an antimalarial drug. If we know that 1g quinoline can be soluble in 1.90L water, calculate the pH of this saturated solution. Known: $pK_{b1} = 5.1$, $pK_{b2} = 9.7$.

15. Calculate the pH of the following solutions: (1) $0.10 \text{mol} \cdot \text{L}^{-1}$ HCl solution and $0.10 \text{mol} \cdot \text{L}^{-1}$ $NH_3 \cdot H_2O$ solution mixed with equal volume; (2) $0.10 \text{mol} \cdot \text{L}^{-1}$ HAc solution and $0.10 \text{mol} \cdot \text{L}^{-1}$ $NH_3 \cdot H_2O$ solution mixed with equal volume; (3) $0.10 \text{mol} \cdot \text{L}^{-1}$ HCl solution and $0.10 \text{mol} \cdot \text{L}^{-1}$ Na_2CO_3 solution mixed with equal volume.

16. Calculate the pH of the following solutions: (1) $0.20 \text{mol} \cdot \text{L}^{-1}$ H_3PO_4 solution and $0.20 \text{mol} \cdot \text{L}^{-1}$ Na_3PO_4 solution mixed with equal volume; (2) $0.20 \text{mol} \cdot \text{L}^{-1}$ Na_2CO_3 solution and $0.10 \text{mol} \cdot \text{L}^{-1}$ HCl solution mixed with equal volume.

17. Add 6.0g solid NaOH in 1.0L $0.10 \text{mol} \cdot \text{L}^{-1}$ H_3PO_4 solution, after dissolved completely, if the volume of the solution is unchanged, calculate: (1) the pH; (2) osmotic pressure at 37℃; (3) if added 18g glucose into the solution, the osmolarity will be? Is it isotonic solution ($300 \text{mmol} \cdot \text{L}^{-1}$)? Known $MM.(NaOH) = 40.0 \text{g} \cdot \text{mol}^{-1}$; $MM.(C_6H_{12}O_6) = 180.2 \text{g} \cdot \text{mol}^{-1}$.

Supplementary Exercises

1. 125.0mL of $0.40 \text{mol} \cdot \text{L}^{-1}$ propanoic acid, HPr, is diluted to 500.0mL. What will the final pH of the solution be? $K_a = 1.3 \times 10^{-5}$.

2. Ethylamine, $CH_3CH_2NH_2$, has a strong, pungent odor similar to that ammonia. Like ammonia, it is a base. A $0.10 \text{mol} \cdot \text{L}^{-1}$ solution has a pH of 11.86. Calculate the K_b for the ethylamine, and find K_a for its conjugate acid $CH_3CH_2NH_3^+$.

3. Pivalic acid is a monoprotic weak acid. A $0.100 \text{mol} \cdot \text{L}^{-1}$ solution of pivalic acid has a pH = 3.00. What is the pH of $0.100 \text{mol} \cdot \text{L}^{-1}$ sodium pivalate at the same temperature?

4. Answer the following questions.

(1) The weak monoprotic acid HA is 3.2% dissociated in $0.086 \text{mol} \cdot \text{L}^{-1}$ solution. What is the acidity constant, K_a, of HA?

(2) A certain solution of HA has a pH = 2.48. What is the concentration of the solution?

Answers to Some Exercises

[Exercises]

3. (1) The ionic concentration order is: $[H_3O^+] \approx [H_2PO_4^-] > [HPO_4^{2-}] > [OH^-] > [PO_4^{3-}]$, where the concentration of H_3O^+ is not as three times that of PO_4^{3-}.

5. 4 000 times

6. $[HS^-] \approx [H_3O^+] = 9.4 \times 10^{-5} \text{mol} \cdot \text{L}^{-1}$ $[S^{2-}] = 1.2 \times 10^{-13} \text{mol} \cdot \text{L}^{-1}$

7. pH = 10.04

8. $[OH^-] = [HN_3] = 2.3 \times 10^{-6} \text{mol} \cdot \text{L}^{-1}$ $[N_3^-] = [Na^+] = 0.010 \text{mol} \cdot \text{L}^{-1}$
   $[H_3O^+] = 4.3 \times 10^{-9} \text{mol} \cdot \text{L}^{-1}$

9. pH = 2.11 $[C_7H_5O_3^-] \approx [H_3O^+] = 7.8 \times 10^{-3} \text{mol} \cdot \text{L}^{-1}$ $[C_7H_4O_3^{2-}] = 3.6 \times 10^{-14} \text{mol} \cdot \text{L}^{-1}$

10. (1) pH = 9.76   (2) pH = 4.68

12. $0.091 \text{mol} \cdot \text{L}^{-1}$

13. (1) Basic NaAc should be added   (2) pH = 5.70   (3) pH = 13.95

14. pH = 10.1

15. (1) 5.28   (2) 7.01   (3) 8.34

16. (1) 7.21   (2) 10.32

17. (1) 7.21   (2) 644 kPa   (3) $0.35 \text{mol} \cdot \text{L}^{-1}$

[Supplementary Exercises]

1. 2.94

2. 1.9×10^{-11}

3. 9.00

4. (1) 8.8×10^{-5} (2) $0.12 \text{mol} \cdot \text{L}^{-1}$

（马汝海）

Chapter 6
Buffer Solution

The pH is one of the important factors that affect chemical reactions. Many reactions, especially those enzyme-catalyzed reactions in biological body, and body fluids often require a certain pH condition. For example, the normal pH of human blood is in a narrow range between 7.35 and 7.45. The pH value over this range would cause serious illness, acidosis or alkalosis, and even death. How does blood maintain a common pH in contact with countless cellular acid-base reactions? How can a chemist sustain a constant $[H_3O^+]$ in the reaction by reacting or producing H_3O^+ or OH^-? The answers are through the presence of a buffer generally. In this chapter, we will study why buffers are important, how they work, and how to prepare them.

6.1 Buffer Process and Composition of Buffer Solutions

6.1.1 Buffer Process

As we all know, the pure water and some solutions will change the pH when small amount of strong acid or base are added into it, but some solutions that only make the pH varies slightly, such as a solution containing a mixture of acetic acid HAc and sodium acetate NaAc. If small amount of strong acid and strong base are added into this kind of solution even to dilute it, the solution does not change the pH nearly. This solution is called a **buffer solution**.

To understand how a buffer solution works, let's see a solution with the mixture of HAc and NaAc. NaAc dissociate to Na^+ and Ac^- completely in aqueous solution. But weak acid HAc dissociates partly. Because of the common ion effect, $[H_3O^+]$ in solution is lower than that expected for HAc alone in solution. The case is as follows

$$HAc(aq) + H_2O(l) \rightleftharpoons H_3O^+(aq) + Ac^-(aq) \tag{6.1}$$

| Relatively large quantity | Huge excess | Relatively small quantity | Relatively large quantity |

If a small amount of strong acid is added to the mixed solution, the more hydronium ion H_3O^+ reacts with acetate ion Ac^- present.

$$H_3O^+(aq) + Ac^-(aq) \rightarrow HAc(aq) + H_2O(l)$$

The proton-transfer equilibrium by Equation 6.1 will shift and to reach the new equilibrium, thus, solution maintains the pH in stable nearly. Therefore, the conjugate base Ac⁻ is called the anti-acid component.

If a small amount of strong base is added to the mixed solution, the hydroxide ion OH⁻ from the base reacts with H_3O^+ in solution to form H_2O.

$$OH^-(aq) + H_3O^+(aq) \rightarrow 2H_2O(l)$$

The proton-transfer equilibrium shifts to the right and more HAc dissociates to produce some of H_3O^+ by in Equation 6.1. This resists the changes in the pH of the solution significantly. Therefore, the conjugate acid HAc is called the anti-base component.

According to the shift of proton-transfer equilibrium, a solution containing an acid and its conjugate base as the major species in the solution will keep the pH not to variety nearly.

> **Question and thinking 6-1** In HAc aqueous solution, there are HAc molecule and Ac⁻ which come from the dissociation of HAc. Is HAc solution a buffer?

6.1.2 Composition of Buffer Solutions

Usually, a buffer solution is an aqueous solution that contains both members of an acid-base conjugate pair as major species, such as HAc-NaAc solution, NH_4Cl-NH_3 solution and NaH_2PO_4-Na_2HPO_4 solution and so on. An acid-base conjugate pair in buffer solution is also called **buffer system** or **buffer pair**. Table 6-1 shows some common buffer systems.

Table 6-1 Common Buffer Systems

Buffer System	Anti-Base Component	Anti-Acid Component	Proton−Transfer Equilibrium	pK_a(25℃)
HAc-NaAc	HAc	Ac⁻	$HAc(aq) + H_2O(l) \rightleftharpoons Ac^-(aq) + H_3O^+(aq)$	4.756
H_2CO_3-$NaHCO_3$	H_2CO_3	HCO_3^-	$H_2CO_3(aq) + H_2O(l) \rightleftharpoons HCO_3^-(aq) + H_3O^+(aq)$	6.35
H_3PO_4-NaH_2PO_4	H_3PO_4	$H_2PO_4^-$	$H_3PO_4(aq) + H_2O(l) \rightleftharpoons H_2PO_4^-(aq) + H_3O^+(aq)$	2.16
Tris·HCl-Tris	Tris·H⁺	Tris	$Tris·H^+(aq) + H_2O(l) \rightleftharpoons Tris(aq) + H_3O^+(aq)$	8.3(20℃)
$H_2C_8H_4O_4$-$KHC_8H_4O_4$	$H_2C_8H_4O_4$	$HC_8H_4O_4^-$	$H_2C_8H_4O_4(aq) + H_2O(l) \rightleftharpoons HC_8H_4O_4^-(aq) + H_3O^+(aq)$	2.943
NH_4Cl-NH_3	NH_4^+	NH_3	$NH_4^+(aq) + H_2O(l) \rightleftharpoons NH_3(aq) + H_3O^+(aq)$	9.25
$CH_3NH_3^+Cl^-$-CH_3NH_2	$CH_3NH_3^+$	CH_3NH_2	$CH_3NH_3^+(aq) + H_2O(l) \rightleftharpoons CH_3NH_2(aq) + H_3O^+(aq)$	10.66
NaH_2PO_4-Na_2HPO_4	$H_2PO_4^-$	HPO_4^{2-}	$H_2PO_4^-(aq) + H_2O(l) \rightleftharpoons HPO_4^{2-}(aq) + H_3O^+(aq)$	7.21
Na_2HPO_4-Na_3PO_4	HPO_4^{2-}	PO_4^{3-}	$HPO_4^{2-}(aq) + H_2O(l) \rightleftharpoons PO_4^{3-}(aq) + H_3O^+(aq)$	12.32

6.2 Calculating the pH of Buffer Solutions

6.2.1 Henderson-Hasselbalch Equation

In a buffer solution with acid HA and conjugate base A^-, the proton-transfer equilibrium is

$$HA(aq) + H_2O(l) \rightleftharpoons H_3O^+(aq) + A^-(aq)$$

$$K_a = \frac{[H_3O^+][A^-]}{[HA]} \text{ and } [H_3O^+] = K_a \times \frac{[HA]}{[A^-]}$$

Take the negative logarithm of both sides of the equation, make pK_a and pH instead of $-\lg K_a$ and $-\lg[H_3O^+]$, and rearranging the equation to solve for the pH

$$pH = pK_a + \lg\frac{[A^-]}{[HA]} \tag{6.2}$$

In honor of the two scientists who first derived it, this equation is often called the **Henderson-Hasselbalch equation**. $\frac{[A^-]}{[HA]}$ is called **buffer-component ratio**.

In a buffer solution, because of the common ion effect, the equilibrium concentrations of both HA and A^- are virtually the same as their initial concentrations. That is

$$[HA] \approx c(HA) \quad \text{and} \quad [A^-] \approx c(A^-)$$

Using their initial concentrations instead of equilibrium concentrations in Equation 6.2, we rearrange Equation 6.1 here

$$pH = pK_a + \lg\frac{c(A^-)}{c(HA)} \tag{6.3}$$

$c(HA) = n(HA)/V$ and $c(A^-) = n(A^-)/V$

$$pH = pK_a + \lg\frac{n(A^-)/V}{n(HA)/V} = pK_a + \lg\frac{n(A^-)}{n(HA)} \tag{6.4}$$

If the initial concentrations of HA and A^- are identical, we get

$$pH = pK_a + \lg\frac{c(A^-) \times V(A^-)}{c(HA) \times V(HA)} = pK_a + \lg\frac{V(A^-)}{V(HA)} \tag{6.5}$$

Note these important points:

(1) The pH of a buffer depends mainly on the pK_a of HA, as well as on buffer-component ratio. When pK_a is fixed, pH changes with the change of buffer-component ratio. When the ratio equals 1:1, $pH = pK_a$.

(2) The pH of a buffer is related to the temperature of the solution.

(3) Because dilution decreases both the concentrations of HA and A^- in the same proportion, the pH of a buffer remains constant nearly when the solution is diluted.

> **Question and thinking 6-2** If the buffer solution is concentrated solution, is it still a buffer?

Sample Problem 6-1 Prepare for a buffer which results from the mixing of 500mL of $0.200 \text{mol} \cdot L^{-1}$ $NH_3 \cdot H_2O$ with 4.78g of NH_4Cl solid. What is the pH of buffer?

Solution For $NH_3 \cdot H_2O$, $pK_b = 4.75$

for NH_4^+, $pK_a = 14.00 - 4.75 = 9.25$

$$n(NH_4Cl) = \frac{4.78g}{53.5g \cdot mol^{-1}} = 0.0893 mol$$

$$n(NH_3) = 0.200 mol \cdot L^{-1} \times 0.500L = 0.100 mol$$

$$pH = pK_a + \lg \frac{n(NH_3)}{n(NH_4^+)} = 9.25 + \lg \frac{0.100 mol}{0.0893 mol} = 9.30$$

Sample Problem 6-2 Citric acid H_3Cit and its salt is a common buffer system used in cultivating bacteria. Calculate the pH of a solution which results from the mixing of $0.100 mol \cdot L^{-1}$ NaH_2Cit with $0.050 mol \cdot L^{-1}$ NaOH solution by the same volume.

Solution For H_3Cit, $pK_{a1} = 3.13$, $pK_{a2} = 4.76$, $pK_{a3} = 6.40$.

H_2Cit^- reacts with OH^- to produce $HCit^{2-}$. The reaction is as following:

$$H_2Cit^- + OH^- \rightleftharpoons HCit^{2-} + H_2O$$

The residual H_2Cit^- and $HCit^{2-}$ make up a buffer pair.

$$c(H_2Cit^-) = (0.100 - 0.050) mol \cdot L^{-1} / 2 = 0.025 mol \cdot L^{-1}$$

$$c(HCit^{2-}) = 0.050 mol \cdot L^{-1} / 2 = 0.025 mol \cdot L^{-1}$$

$$pH = pK_{a2} + \lg \frac{c(HCit^{2-})}{c(H_2Cit^-)} = 4.76 + \lg \frac{0.025 mol \cdot L^{-1}}{0.025 mol \cdot L^{-1}} = 4.76$$

6.2.2 Correction Formula to Calculate the Accurate pH

Because of ignoring the effect of ionic strength, Henderson-Hasselbalch equation is used to calculate the approximate pH of a buffer actually. Take the activity instead of equilibrium concentration in Equation 6.2. The accurate pH can be got.

$$pH = pK_a + \lg \frac{a(A^-)}{a(HA)} = pK_a + \lg \frac{[A^-] \cdot \gamma(A^-)}{[HA] \cdot \gamma(HA)} = \{pK_a + \lg \frac{\gamma(A^-)}{\gamma(HA)}\} + \lg \frac{[A^-]}{[HA]}$$

$$pH = pK_a' + \lg \frac{[A^-]}{[HA]} \tag{6.6}$$

$\gamma(HA)$ and $\gamma(A^-)$ are the activity coefficients of HA and A^-. $\lg \frac{\gamma(A^-)}{\gamma(HA)}$ is called emendation factor. Both of activity coefficient and emendation factor are related to the charge of weak acid and ionic strength of the solution. pK_a' is the corrected value of pK_a after considering the ionic strength.

6.3 Buffer Capacity and Buffer Range

6.3.1 Buffer Capacity

When small amounts of H_3O^+ or OH^- ion are added into a buffer solution, the pH changes are very small. There is a limit, however, to the amount of protection that a buffer solution can provide. After the buffering agents are consumed, the solution loses its ability to maintain near-constant pH. 1922, Slyke V. derived **buffer capacity**, β, to represent the buffering ability of buffer solution. The buffer capacity is defined as the quantity of strong acid or strong base needed to change the pH of one liter of buffer by 1pH unit.

$$B \stackrel{\text{def}}{=\!=} \frac{\mathrm{d}\,n_{a(b)}}{V\,|\mathrm{dpH}|} \tag{6.7}$$

β is the buffer capacity and has units of $\mathrm{mol\cdot L^{-1}\cdot pH^{-1}}$. $\mathrm{d}n_{a(b)}$ stands for moles of strong acid or strong base which is added to a buffer solution to cause a tiny change in pH. The magnitude of β indicates the relative strength of buffer capacity. The larger the value of β is, the greater the ability of the buffer to resist change in pH is.

What are the factors associated with buffer capacity? Known the buffer capacity and the pH of buffer solution, the variation of the buffer capacity with pH is obtained as shown in Figure 6-1.

Figure 6-1 shows the relation among buffer capacity, concentration, buffer ratio and pH of the solution. From Figure 6-1, buffer capacity depends on the total buffer concentration c_{total} and the buffer-component ratio $[A^-]/[HA]$.

To the same buffer system, if the buffer-component ratio is fixed, the more concentrated the components of a buffer is, the greater the buffer capacity is, and the smaller the pH change is, as shown in Figure 6-1. In other words, you must add more H_3O^+ or OH^- into a high-capacity (concentrated) buffer than to a low-capacity (dilute) buffer to obtain a certain change in the pH.

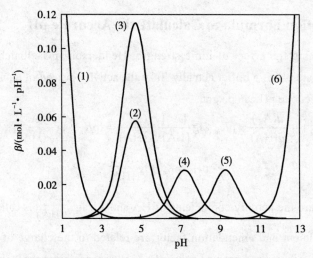

Figure 6-1 The Relation between Buffer Capacity and pH
(1) HCl (2) $0.1\mathrm{mol\cdot L^{-1}}$ HAc+NaOH (3) $0.2\mathrm{mol\cdot L^{-1}}$ HAc+NaOH
(4) $0.05\mathrm{mol\cdot L^{-1}}$ KH_2PO_4+NaOH (5) $0.05\mathrm{mol\cdot L^{-1}}$ H_3BO_3+NaOH (6) NaOH

Buffer capacity is also determined by the buffer-component ratio. Since the buffer ratio also affects the pH, the less the ratio changes is, the less the pH changes is. For a given concentrated buffer solution, the nearer the buffer-component ratio is to 1:1, the greater the capacity is. A buffer has the highest capacity when the buffer ratio is equal to 1 or $[A^-]=[HA]$. That is to say, for a buffer solution with maximum buffer resistance to change in pH and the maximum buffer capacity, the optimum buffer ratio is 1:1.

6.3.2 Buffer Range

The buffer range is the pH range actually in which the buffer is effective. The further the buffer-component ratio is away from 1, the less effective the buffer action is, and the lower the buffer capacity is. In practice, we find that if the $[A^-]/[HA]$ ratio is greater than 10 or less than 0.1, buffering action is poor, and the solution will not be particularly at resisting pH change. Given that $\lg 10 = +1$ and $\lg 0.1 = -1$, we find that buffers have a usable range within ± 1 pH unit of the pK_a of the conjugate acid. That is

$$pH = pK_a + \lg \frac{10}{1} = pK_a + 1 \quad \text{and} \quad pH = pK_a + \lg \frac{1}{10} = pK_a - 1$$

That is, the **buffer range** is the pH range from $pK_a - 1$ to $pK_a + 1$. For example, pK_a for HAc is 4.756, so the buffer range for HAc / Ac$^-$ buffer system is pH = 3.756 to 5.756.

6.4 Preparation of Buffer Solutions

6.4.1 Main Steps to Prepare A Buffer Solution

Buffer solutions have many important applications. In biological systems, buffer action controls the H_3O^+ ion concentration within the necessary limits for the life. For example, human blood has many buffer systems that work together to maintain the pH in a narrow range between 7.35 and 7.45; in chemical analysis, some cations may be separated from each other by selective precipitation under the controlled pH value; many industrial processes such as wine making and paper manufacture are monitored at various stages by a measurement of the pH; a buffer of unusual composition may be required to simulate a cell system or stabilize a fragile biological macromolecule. Even the most sophisticated in many modern medical or biological research applications; automated laboratory frequently relies on personnel versed in chemical techniques with a good knowledge of basic chemistry to prepare a buffer. So it is important to learn how to prepare a buffer commonly.

Several steps are required to prepare a buffer with desired pH:

(1) Choose the optimum buffer system

First, decide the proper chemical composition of the buffer in large extent, and choose the exact conjugate acid-base pairs, the choice are determined by the desired pH. Remember that a buffer is most effective when the buffer component ratio is close to 1. The better way is that the pH of buffer solution is near to pK_a of the conjugate acid. So we should choose an acid whose pK_a is as close as possible to the desired pH, and take its conjugate base together to prepare the buffer. Besides, the chemical behavior of the system should be considered, which must be such that unwanted reactions are minimized.

(2) Determine the buffer concentration

Second, decide how concentrated the buffer should be. As we all know, buffer capacity depends on the concentrations of buffer components. The higher the concentrations of buffer components are, the greater the buffer capacity is. The reasonable total concentration of a buffer is in the range from $0.05 \text{mol} \cdot L^{-1}$ to $0.2 \text{mol} \cdot L^{-1}$ usually. Total concentration is the sum of $[A^-]$ and $[HA]$.

(3) Calculate the buffer components ratio

Next, find the ratio of [A$^-$] / [HA] to give the desired pH by Henderson-Hasselbalch Equation, we can obtain the amount or the volume of buffer components.

(4) Mix the solution and adjust the pH

Due to the effect of ionic strength and other ionic phenomena, the pH of a buffer prepared in this way may vary from the desired pH by as much as several tenths of a pH unit. Therefore, after making up the solution, it is necessary to adjust the pH to the desired value by adding strong acid or strong base while monitoring the solution with pH meter.

Another way to prepare a buffer is to form one of the components during the final mixing step by partial neutralization of the other component. For example, you can prepare a HAc-Ac$^-$ buffer by mixing appropriate amounts of HAc solution and NaOH solution. As the OH$^-$ ions react with the HAc molecules, partial neutralization of the total HAc present produces the Ac$^-$ needed:

$$HAc \text{ (HAc total)} + OH^- \text{ (amt added)} \rightarrow$$
$$HAc \text{ (HAc total - OH}^- \text{ amt added)} + Ac^- \text{ (OH}^- \text{ amt added)} + H_2O$$

Alternatively, you can prepare a HAc-Ac$^-$ buffer by mixing appropriate amounts of NaAc solution and HCl solution.

$$Ac^- \text{ (Ac}^- \text{ total)} + H^+ \text{ (amt added)} \rightarrow Ac^- \text{ (Ac}^- \text{ total - H}^+ \text{ amt added)} + HAc(H^+ \text{ amt added})$$

The following problems illustrate the calculation involving buffer preparation.

Sample Problem 6-3 How to prepare 1 000mL of a buffer solution with pH=4.50?

Solution

(1) Choosing a suitable conjugate acid-base pair: pK_a of HAc is 4.756, which is close to the desired pH, 4.50. Therefore, it is suitable to choose HAc-Ac$^-$ buffer system to prepare this buffer.

(2) Determination of the buffer concentration: The reasonable concentrations of a buffer is in the range from 0.05mol·L^{-1} to 0.2mol·L^{-1}. Therefore, the buffer system which contains 0.10mol·L^{-1} HAc and 0.10mol·L^{-1} NaAc is determined.

(3) Determination of quantities of HAc and NaAc solution: According to Equation 6.5, if $V(Ac^-)$ and $V(HAc)$ represent the volumes of NaAc and HAc respectively,

$$pH = pK_a + \lg \frac{V(Ac^-)}{V(HAc)}$$

$$4.50 = 4.756 + \lg \frac{V(Ac^-)}{1\,000\text{mL} - V(Ac^-)}$$

$$\frac{V(Ac^-)}{1\,000\text{mL} - V(Ac^-)} = 0.55 \quad \text{so, } V(Ac^-) = 355\text{mL}$$

$$V(HAc) = 1\,000\text{mL} - 355\text{mL} = 645\text{mL}$$

That is, preparing 1 000mL of a buffer solution with pH=4.50 by mixing 355mL of 0.10mol·L^{-1} NaAc solution with 645mL of 0.10mol·L^{-1} HAc solution.

Sample Problem 6-4 Phosphate buffer solution with pH closed to 7.0 is commonly used to cultivate enzyme. If we want to prepare a phosphate buffer solution with pH=6.90, what volume of 0.10mol·L^{-1} NaOH solution should be added to 450mL of 0.10mol·L^{-1} H$_3$PO$_4$ solution? Assume no special change in volume after mixing them together. For H$_3$PO$_4$, pK_{a1}=2.16, pK_{a2}=7.21, pK_{a3}=12.32.

Solution According to preparing rule, conjugate acid-base pair $H_2PO_4^- - HPO_4^{2-}$ is the buffer system. The reactions between H_3PO_4 and NaOH after mixing include two steps.

(1) $\qquad H_3PO_4 \text{ (aq)} + NaOH(aq) = NaH_2PO_4(aq) + H_2O(l)$

In this step, H_3PO_4 is neutralized completely to produce NaH_2PO_4, which need 450mL of $0.10 \text{mol} \cdot L^{-1}$ NaOH solution.

$$n(NaH_2PO_4) = 450\text{mL} \times 0.10 \text{mol} \cdot L^{-1} = 45 \text{mmol}$$

(2) $\qquad NaH_2PO_4 \text{ (aq)} + NaOH(aq) = Na_2HPO_4 \text{ (aq)} + H_2O(l)$

In this step, NaH_2PO_4 is neutralized partly to form Na_2HPO_4, which need x mL of $0.10 \text{mol} \cdot L^{-1}$ NaOH solution. So,

$$n(Na_2HPO_4) = 0.10x \text{ mmol} \quad \text{and} \quad n(NaH_2PO_4) = (45 - 0.10x) \text{ mmol}$$

$$pH = pK_{a2} + \lg \frac{n(HPO_4^{2-})}{n(H_2PO_4^-)}$$

$$6.90 = 7.21 + \lg \frac{0.10x \text{ mmol}}{(45 - 0.10x) \text{mmol}}$$

$$\frac{0.10x}{45 - 0.10x} = 0.49 \quad x = 148 \text{mL}$$

$$V(NaOH)_{total} = 450 \text{mL} + 148 \text{mL} = 598 \text{mL}$$

> **Question and thinking 6-3** How many possible buffer systems can be gotten when NaOH solution are added into H_3PO_4 solution. How to get these buffer systems?

In many medical or biological research applications, we usually prepare a buffer solution by the chemical handbook. The compositions of some phosphate buffers and Tris-Tris·HCl buffers widely used in the medical research applications are listed in Table 6-2 and Table 6-3.

Table 6-2 Buffer Solutions Prepared by Mixing $H_2PO_4^-$ with HPO_4^{2-} Solution (25℃)

\multicolumn{5}{c}{50mL $0.1 \text{mol} \cdot L^{-1} KH_2PO_4 + x$ mL $0.1 \text{mol} \cdot L^{-1}$ NaOH dilute to 100mL}
pH	x	β	pH	x	β
5.80	3.6	—	7.00	29.1	0.031
5.90	4.6	0.010	7.10	32.1	0.028
6.00	5.6	0.011	7.20	34.7	0.025
6.10	6.8	0.012	7.30	37.0	0.022
6.20	8.1	0.015	7.40	39.1	0.020
6.30	9.7	0.017	7.50	41.1	0.018
6.40	11.6	0.021	7.60	42.8	0.015
6.50	13.9	0.024	7.70	44.2	0.012
6.60	16.4	0.027	7.80	45.3	0.010
6.70	19.3	0.030	7.90	46.1	0.007
6.80	22.4	0.033	8.00	46.7	—
6.90	25.9	0.033			

Table 6-3　Buffer Solutions Prepared by Mixing Tris with Tris·HCl

Components/mol·kg^{-1}			pH	
Tris	Tris·HCl	NaCl	25℃	37℃
0.02	0.02	0.14	8.220	7.904
0.05	0.05	0.11	8.225	7.908
0.006 667	0.02	0.14	7.745	7.428
0.016 67	0.05	0.11	7.745	7.427
0.05	0.05		8.173	7.851
0.016 67	0.05		7.699	7.382

The formulas of Tris and Tris·HCl are $(HOCH_2)_3CNH_2$ and $(HOCH_2)_3CNH_2·HCl$ respectively. Tris is a weak base with good stability, a high solubility in physiological fluids, which does neither cause precipitations of calcium salts nor effect the activity of enzymes. And its conjugate acid, Tris·HCl has a pK_a (pK_a = 8.3) close to physiological pH. Therefore, Tris-Tris·HCl buffers are widely used in the physiological and biochemical studies. NaCl is added in order to adjust the ionic strength of Tris buffers. If the ionic strength is at 0.16, solution is isotonic clinically.

6.4.2　Standard Buffer Solutions

Standard buffer solutions are used in calibrating the pH meter usually. Table 6-4 lists some data about several standard buffer solutions.

Table 6-4　Standard Buffer Solutions

Solutions	c/mol·L^{-1}	pH(25℃)	Temperature Coefficient/ΔpH·℃$^{-1}$
$KHC_4H_4O_6$	Saturated, 25℃	3.557	−0.001
$KHC_8H_4O_4$	0.05	4.008	+0.001
KH_2PO_4-Na_2HPO_4	0.025, 0.025	6.865	−0.003
KH_2PO_4-Na_2HPO_4	0.008 695, 0.030 43	7.413	−0.003
$Na_2B_4O_7·10H_2O$	0.01	9.180	−0.008

Because pK_a of the acid varies with the temperature, pH of the buffer is also related to the temperature. Temperature coefficient should be considered when the pH meter is used. Temperature coefficient is larger than zero, it means that the pH rises with the increasing of temperature; temperature coefficient is smaller than zero, it means the pH decreases with the increasing of temperature.

Some standard buffer solutions are prepared from a compound, such as potassium hydrogen tartrate $KHC_4H_4O_6$, potassium hydrogen phthalate and borax. How to explain the buffer action of these compounds? On the one hand, these compounds can dissociate in water to large amounts of amphoteric ions. For example, $KHC_4H_4O_6$ dissociate in water completely,

$$KHC_4H_4O_6 \rightarrow K^+ + HC_4H_4O_6^-$$

$HC_4H_4O_6^-$ is an amphoteric ion to form $H_2C_4H_4O_6$ and $C_4H_4O_6^{2-}$. Therefore, it can form two different buffer systems, $H_2C_4H_4O_6$-$HC_4H_4O_6^-$ and $HC_4H_4O_6^-$-$C_4H_4O_6^{2-}$. Because pK_a of $H_2C_4H_4O_6$ is 2.98 and pK_a of $HC_4H_4O_6^-$ is 4.34, their pK_a value is near, and their buffer ranges will be partly overlap, so the buffer ability of it is good. On the other hand, the compound will form a buffer pair after hydrolysis in water. For instance, 1mol of borax $Na_2B_4O_7 \cdot 10H_2O$ in aqueous solution can form 2mol of metaborate acid HBO_2 and 2mol of sodium metaborate $NaBO_2$ which can form a conjugate acid (HBO_2)-base (BO_2^-) pair in $Na_2B_4O_7$ solution. Therefore, $Na_2B_4O_7$ solution has the buffer action and it can be used as a standard buffer solution.

6.5 Main Buffer Systems in Blood

There are various buffer systems in the human blood which can maintain the pH of the blood in the range of 7.35 to 7.45. There are following buffer systems in blood mainly:

In blood plasma, H_2CO_3-HCO_3^-, $H_2PO_4^-$-HPO_4^{2-}, H_nP-$H_{n-1}P^-$ (where H_nP is protein).

In red blood cell, H_2b-Hb^- (where H_2b is hemoglobin), H_2bO_2-HbO_2^- (where H_2bO_2 is oxyhemoglobin), H_2CO_3-HCO_3^-, $H_2PO_4^-$-HPO_4^{2-}.

The primary buffer system in blood is a combination of carbonic acid and its salt, sodium bicarbonate. Carbon dioxide CO_2 is a major acidic product in the metabolism for the body. A person produces averagely 10 to 20mol of CO_2 every day. Blood carries CO_2 from the cells to the lungs, where it is exhaled as waste. There is the equilibrium between dissolved CO_2 and carbonic acid H_2CO_3 in aqueous solution:

$$CO_2 (aq) + H_2O (l) \rightleftharpoons H_2CO_3 (aq)$$

H_2CO_3 and bicarbonate ions HCO_3^- in blood plasma consists of the important buffer system. In the blood of a healthy individual,

$$[HCO_3^-] = 24 \text{mmol} \cdot L^{-1}, [H_2CO_3] = [CO_2(aq)] = 1.2 \text{mmol} \cdot L^{-1}$$

The pH of the blood can be calculated by the Henderson-Hasselbalch equation:

$$pH = pK'_{a1} + \lg \frac{[HCO_3^-]}{[CO_2(aq)]}$$

To consider that the ionic strength of blood is 0.16 commonly and the body temperature is 37°C, pK'_{a1}, corrected value of pK_{a1} of H_2CO_3 is 6.10. So,

$$pH = 6.10 + \lg \frac{0.024 \text{mol} \cdot L^{-1}}{0.0012 \text{mol} \cdot L^{-1}} = 6.10 + \lg \frac{20}{1} = 7.40$$

If the pH of the blood drops below 7.35, a potentially fatal condition is called **acidosis**. If the pH of the blood increases over 7.45, it results in another serious condition, **alkalosis**.

> **Question and thinking 6-4** $[HCO_3^-] / [H_2CO_3]$ ratio in blood plasma is 20:1, it is over an effective buffer ratio. Why it still has the buffer action?

The buffer principle is similar to HAc-NaAc buffer. The equilibrium of $CO_2(aq)$-HCO_3^- buffer in blood is as follows:

$$CO_2(aq) + H_2O(l) \rightleftharpoons H_2CO_3(aq) \rightleftharpoons H^+(aq) + HCO_3^-(aq)$$

When the amount of the acidic substance in blood increases, HCO_3^- ion plays a role of anti-acid component that can react with H_3O^+ and shifts the above equilibrium to the left. The extra CO_2 is released in the lungs and exhaled. HCO_3^- in blood plasma is called "alkali reserve". If extra OH^- is found in the blood system, it reacts with H_2CO_3, which shifts the above equilibrium to the right. The extra HCO_3^- formed is eventually expelled by the kidneys.

Because human body is an open system, despite that the $[HCO_3^-]$ / $[H_2CO_3]$ ratio at pH 7.4 is 20:1, which is over the normal buffer ratio, the buffer action of carbonic acid buffer system is effective in body, CO_2 in the body is transferred to the gas to be removed from the lungs. Thus, our lungs not only get O_2 from the air we inhale, but they also help us to maintain the proper blood pH by expelling waste CO_2 into the air. The complicated set of buffer equilibrium and transport / transfer processes act together to achieve the pH control essential for the proper function of blood chemistry.

Key Terms

缓冲溶液	buffer solution
缓冲作用	buffer action
抗酸成分	anti-acid component
抗碱成分	anti-base component
缓冲系	buffer system
缓冲对	buffer pair
缓冲比	buffer-component ratio
缓冲容量	buffer capacity
缓冲范围	buffer range
酸中毒	acidosis
碱中毒	alkalosis

Summary

A buffer solution is an aqueous solution that contains both components of an acid-base conjugate pair as major species, which scarcely changes its pH even when relatively large quantities of acid or base are added in it. When small amount of H_3O^+ ions are added to the buffer, the conjugate base, anti-acid component can neutralize the strong acid added without change in pH. When small amount of OH^- ions are added to the buffer, the conjugate acid, anti-base component can neutralize the strong base added without changing its pH. The pK_a of the conjugate acid and the buffer-component ratio determine the pH and they are related by the Henderson-Hasselbalch Equation:

$$pH = pK_a + \lg\frac{[A^-]}{[HA]} = pK_a + \lg\frac{c(A^-)}{c(HA)} = pK_a + \lg\frac{n(A^-)}{n(HA)}$$

The buffer capacity, β, can be used to represent the buffering ability of the solution. The larger the value of β is, the greater the ability of the buffer to resist changes in pH is. The buffer capacity depends on both the total concentration of the components and the buffer component-ratio. A buffer has the highest capacity when its components are present in equal concentrations, that is, when $[A^-]$ / $[HA] = 1$. An effective buffer has a pH range from $pK_a - 1$ to $pK_a + 1$.

Buffer solutions are widely used in many fields, especially in biomedical applications. Major steps for the preparation of a desired buffer are: (1) choose the optimum buffer system, (2) determine the buffer concentration, (3) calculate the ratio of buffer components, and (4) adjust the final buffer to the desired pH.

There are various buffer systems in human blood, which can maintain the pH of the blood in the range from 7.35 to 7.45. An important buffer in blood is the carbonic acid with bicarbonate ion system. The pH of blood is 7.40 when the concentration of bicarbonate ion is 20 times that of carbonic acid. Buffers in human body can replenish the components of the buffer solution as they are used up and remove any excess components from the body, which differ from those in the laboratory. One of the functions of the lungs and kidneys is to maintain the $[HCO_3^-] / [H_2CO_3]$ ratio. The complicated set of buffer equilibria in blood and transport/transfer processes act together to achieve the pH control, which is essential for the proper function of blood.

Exercises

1. What is a buffer solution? Illustrate H_2CO_3-HCO_3^- in human blood to explain how a buffer works and how important buffer solutions are in medicine.

2. What is buffer capacity? What are the factors affecting the buffer capacity? Is the buffer capacity of HAc-NaAc buffer system the same as that of H_2CO_3-HCO_3^- buffer system, both at a total concentration of $0.10 mol \cdot L^{-1}$?

3. Which of the following sets of chemicals can be used to prepare buffer solutions?
   (1) $HCl + NH_3 \cdot H_2O$ (2) $HCl + Tris$ (3) $HCl + NaOH$
   (4) $Na_2HPO_4 + Na_3PO_4$ (5) $H_3PO_4 + NaOH$ (6) $NaCl + NaAc$

4. Is it a buffer solution, which is prepared by mixing $0.30 mol \cdot L^{-1}$ pyridine (C_5H_5N, $pK_b = 8.77$) and $0.10 mol \cdot L^{-1}$ hydrochloric acid solution with the same volume? And what is the pH of solution?

5. What is the pH of a solution prepared by dissolving 10.0g of sodium carbonate (Na_2CO_3) and 10.0g of sodium bicarbonate ($NaHCO_3$) in enough water to make 0.250L of solution?

6. Calculate the concentration of methanoic acid (HCOOH) and sodium formate (HCOONa) respectively in a buffer solution at pH 3.90 and a total concentration of $0.400 mol \cdot L^{-1}$. pK_a of HCOOH is 3.75.

7. When 0.20g of solid NaOH is added to 100mL of some buffer solution, the pH of the buffer solution changes to 5.60. If the initial concentration of conjugate acid (HB) of the buffer system, $c(HB)$, is $0.25 mol \cdot L^{-1}$, what is the initial pH of the buffer solution? pK_a of HB is 5.30.

8. Aspirin (acetylsalicylic acid, HAsp, $MM. = 180.2 g \cdot mol^{-1}$) is absorbed from stomach with the form of free acid. A patient takes some antacid to adjust the pH of gastric juice to 2.95, and then takes 0.65g of aspirin. Assume aspirin dissolves immediately, and the pH of gastric juice is invariant. What is the mass of aspirin that the patient can absorb from stomach at once? pK_a of acetylsalicylic acid is 3.48.

9. Calculate the pH of a buffer prepared by mixing $0.10 mol \cdot L^{-1}$ acetic acid (CH_3COOH, HAc) and $0.10 mol \cdot L^{-1}$ NaOH solution with the volume ratio of 3:1.

10. Barbitone($C_8H_{12}N_2O_3$, H_2Bar and $MM. = 184 g \cdot mol^{-1}$) buffer solution is used in some biochemistry experiments. A barbitone buffer solution is prepared by mixing 100mL of barbitone solution which contains 18.4g of barbitone, and 4.17mL of $6.00 mol \cdot L^{-1}$ NaOH solution to make 1 000mL of solution. What is the pH of the solution? pK_{a1} of barbitone is 7.43.

11. Write out the possible buffer systems, the anti-acid components, the anti-base components and

the buffer ranges of the buffer systems when NaOH solution or hydrochloric acid solution are added into the sodium hydrogen citrate (Na_2HCit) solution. If the concentrations of these three solutions are identical, what volume ratio are they mixed with to obtain buffer solutions which have the maximum buffer capacity. For H_3Cit, $pK_{a1}=3.13$, $pK_{a2}=4.76$, $pK_{a3}=6.40$.

12. What volume of $0.10 mol \cdot L^{-1}$ hydrochloric acid solution should be added to 50mL of the following solutions respectively to prepare the solutions at pH=7.00?

(1) $0.10 mol \cdot L^{-1}$ NaOH solution (2) $0.10 mol \cdot L^{-1}$ NH_3 solution (3) $0.10 mol \cdot L^{-1}$ Na_2HPO_4

And judge whether these solutions prepared are buffer solutions, and which one have the best buffer capacity, respectively.

13. What mass of solid NH_4Cl and what volume of $1.00 mol \cdot L^{-1}$ NaOH solution should be used to prepare 1.0L of a buffer solution of pH 9.00? Suppose the overall concentration of the buffer is $0.125 mol \cdot L^{-1}$.

14. Determine the volume of $0.020 mol \cdot L^{-1} H_3PO_4$ and $0.020 mol \cdot L^{-1}$ NaOH solution required to generate 10mL of physiological buffer solution with pH=7.40.

15. If we want to prepare a physiological buffer solution with pH about 7.40 at 20℃, what volume of $0.050 mol \cdot L^{-1}$ hydrochloric acid should be added to 100mL of a buffer solution in which both of the concentration of Tris and Tris·HCl are $0.050 mol \cdot L^{-1}$? And what is the mass of sodium chloride (NaCl) solid should be added into the solution to prepare an isotonic solution compared with blood? pK_a of Tris·HCl is 8.3 at 20℃. Assume the effect of ionic strength is not considered.

16. In the blood plasma of a healthy individual, $[HCO_3^-]=24 mmol \cdot L^{-1}$, $[CO_2(aq)]=1.2 mmol \cdot L^{-1}$. Calculate the pH of blood plasma when $[HCO_3^-]$ of blood reduces to 90% of initial concentration because of diarrhea, and judge whether it results in acidosis. pK'_{a1}, the pK_{a1} of carbonic acid in blood at body temperature (37℃), is 6.10.

Supplementary Exercises

1. How do the anti-acid component and anti-base component of a buffer work? Why are they typically a conjugate acid-base pair?

2. What is the relationship between buffer range and buffer-component ratio?

3. Choose specific acid-base conjugate pairs of suitable for prepare the following buffers (Use Table 6-1 for K_a of acid or K_b of base):

(a) pH ≈ 4.0   (b) pH ≈ 7.0   (c) $[H_3O^+] \approx 1.0 \times 10^{-9} mol \cdot L^{-1}$

4. Choose the factors that determine the capacity of a buffer from among the following and explain your choices.

(a) Conjugate acid-base pair   (b) pH of the buffer   (c) Buffer range
(d) Concentration of buffer components
(e) Buffer-component ratio   (f) pK_a of the acid component

5. What mass of sodium acetate ($NaC_2H_3O_2 \cdot 3H_2O$, $MM. = 136.1 g \cdot mol^{-1}$) and what volume of concentrated acetic acid ($17.45 mol \cdot L^{-1}$) should be used to prepare 500mL of a buffer solution at pH=5.00 that is $0.150 mol \cdot L^{-1}$ overall?

6. Normal arterial blood has an average pH of 7.40. Phosphate ions form one of the key buffering

systems in the blood. Find the buffer-component ratio of a KH_2PO_4 / Na_2HPO_4 solution with this pH. pK_{a2}' of $H_2PO_4^- = 6.80$.

Answers to Some Exercises

[Exercises]

3. (1), (2), (4), (5)

4. pH = 5.53

5. pH = 10.23

6. $c(HCOO^-) = x$ mol·L^{-1} = 0.234 mol·L^{-1}; $c(HCOOH) = 0.166$ mol·L^{-1}

7. pH = 5.45

8. 0.50 g

9. pH = 4.45

10. pH = 6.95

11.

Solution	Buffer system	Anti-acid component.	Anti-base component.	Effective buffer range	Volume ratio at β_{max}
$Na_2HCit + HCl$	$H_2Cit^- - HCit^{2-}$	$HCit^{2-}$	H_2Cit^-	3.76~5.76	2:1
$Na_2HCit + HCl$	$H_3Cit - H_2Cit^-$	H_2Cit^-	H_3Cit	2.13~4.13	2:3
$Na_2HCit + NaOH$	$HCit^{2-} - Cit^{3-}$	Cit^{3-}	$HCit^{2-}$	5.40~7.40	2:1

12. (1) 50 mL; (2) 49.7 mL; (3) 31 mL

13. 6.69 g; 0.045 L (or 45 mL)

14. 38.4 mL; 61.6 mL

15. 77 mL; 1.0 g

16. pH = 7.36

[Supplementary Exercises]

3. (a) HAc and Ac$^-$ (b) $H_2PO_4^-$ and HPO_4^{2-} (c) NH_4^+ and NH_3

4. Choose (d) and (e).

5. 6.53 g; 1.55 mL

6. ≈ 4.00

(章小丽)

Chapter 7
Equilibria of Slightly Soluble Ionic Compounds

Many compounds are quite soluble in water, some compounds dissolve in water to a certain extent, but the compounds are so slightly soluble that they are called "insoluble compounds" sometimes. In a saturated solution of slightly soluble compound, there is equilibrium between the solid compound and free ions from dissolved solid compound. This is heterogeneous equilibrium and it is called dissolution-precipitation equilibrium. Many biological and environmental processes, such as the formation of stalactite and kidney stones, involve the dissolution or precipitation of slightly soluble ionic compound. We will study the aqueous equilibria of slightly soluble ionic compounds in this chapter.

7.1 The Solubility Product Constant and Solubility Product Rule

According to the solubility, electrolytes are generally divided into soluble electrolytes and slightly soluble electrolytes. If the solubility less than $0.1 \text{g} \cdot \text{L}^{-1}$ at 298.15K in water generally, it is referred to slightly soluble electrolytes like AgCl, $CaCO_3$ and PbS and so on. But they dissociate completely once they dissolve in water. They are strong electrolyte but the solubility is small. We only study this kind of compound in this chapter.

7.1.1 Solubility Product Constant

Under the given temperature and pressure conditions, the solvation process of solute in water is dissolving process actually. The process tends toward an equilibrium position spontaneously. In this position, the solution is defined as the **saturation solution**. The concentration of solute is at its maximum value. If we only consider slightly soluble electrolytes, after putting them into water, only a small portion of them will dissociate to free ions but no dissociated molecule is present. The rest part of them in solution exists in solid. This process is called **dissolution**. However, once solute is dissolved, the reverse reaction to form precipitation again will begin to occur. This process is called **precipitation**. Eventually, the saturation solution is formed. The solution reaches the **precipitation-dissolution equilibrium**. The concentration of dissociated solute remains unchanged regardless of the amount of solute added.

A solid of slightly soluble electrolyte exists in equilibrium with its saturated solution. In the case of AgCl, for example, precipitation-dissolution equilibrium is written as follows

$$AgCl(s) \rightleftharpoons Ag^+(aq) + Cl^-(aq)$$

At equilibrium,
$$K = \frac{[Ag^+][Cl^-]}{[AgCl(s)]}$$
$$[Ag^+][Cl^-] = K \cdot [AgCl(s)]$$

As $[AgCl(s)]$ is a constant, $K[AgCl(s)]$ is also a constant, written as K_{sp}

$$K_{sp} = [Ag^+][Cl^-] \tag{7.1}$$

For a saturated solution of slightly soluble ionic electrolyte with the formula A_aB_b, the equilibrium constant for solubility in aqueous solution is called the **solubility product constant**, K_{sp} or simplified as **solubility product**. Like the other equilibrium constants, K_{sp} is temperature dependent. It can be expressed as follows

$$A_aB_b(s) \rightleftharpoons aA^{n+}(aq) + bB^{m-}(aq)$$
$$K_{sp}(A_aB_b) = [A^{n+}]^a \cdot [B^{m-}]^b \tag{7.2}$$

In Equation 7.2, the concentrations of ionic constituents are equilibrium concentrations in molarity. If considering the interaction between ions, the activity should be instead of ionic concentrations. But it will not make difference nearly because the ionic strength is very low. The activity factor is approximate to 1, and the concentration can be used just as activity. Like other equilibrium constant K_{sp} is related to the matter nature and temperature. The solubility product constants of some slightly soluble electrolytes are listed in Table 3 of Appendix 3.

7.1.2 Solubility and Solubility Product Constant

Solubility, S, of a substance is the maximum amount of it to dissolved in a particular solvent at a certain temperature. Both Solubility and solubility product constant can be used to measure the dissolving ability of substance. K_{sp} can be calculated from solubility according to the following equation.

$$A_aB_b(s) \rightleftharpoons aA^{n+}(aq) + bB^{m-}(aq)$$
$$ aS\ mol \cdot L^{-1} \quad bS\ mol \cdot L^{-1}$$
$$K_{sp} = [A^{n+}]^a \cdot [B^{m-}]^b = (aS)^a \cdot (bS)^b = a^a \cdot b^b \cdot S^{(a+b)}$$
$$S = \sqrt[(a+b)]{\frac{K_{sp}}{a^a \cdot b^b}}$$

Sample Problem 7-1 At 298.15K, the solubility of AgCl in an aqueous solution is determined to be $1.91 \times 10^{-3} g \cdot L^{-1}$. What is the value of the K_{sp} for AgCl?

Solution $MM.(AgCl) = 143.4 g \cdot mol^{-1}$. Molarity of its solubility is

$$c = \frac{1.91 \times 10^{-3} g \cdot L^{-1}}{143.4 g \cdot mol^{-1}} = 1.33 \times 10^{-5} mol \cdot L^{-1}$$

The dissociation reaction is
$$AgCl(s) \rightleftharpoons Ag^+(aq) + Cl^-(aq)$$
by
$$K_{sp}(AgCl) = [Ag^+][Cl^-]$$
at 298.15K
$$K_{sp}(AgCl) = (1.33 \times 10^{-5})^2 = 1.77 \times 10^{-10}$$

Sample Problem 7-2 At 298.15K, the molarity of Ag_2CrO_4 in an aqueous solution is given as $6.54 \times 10^{-5} mol \cdot L^{-1}$. What is the value of the K_{sp} for Ag_2CrO_4?

Solution
$$Ag_2CrO_4(s) \rightleftharpoons 2Ag^+(aq) + CrO_4^{2-}(aq)$$
So,
$$[Ag^+] = 2 \times 6.54 \times 10^{-5} mol \cdot L^{-1}; \quad [CrO_4^{2-}] = 6.54 \times 10^{-5} mol \cdot L^{-1}$$
$$K_{sp}(Ag_2CrO_4) = [Ag^+]^2[CrO_4^{2-}] = (2 \times 6.54 \times 10^{-5})^2 \times (6.54 \times 10^{-5}) = 1.12 \times 10^{-12}$$

Sample Problem 7-3 At 298.15K, the value of K_{sp} for $Mg(OH)_2$ in an aqueous solution is 5.61×10^{-12}, What is the solubility for $Mg(OH)_2$?

Solution
$$Mg(OH)_2(s) \rightleftharpoons Mg^{2+}(aq) + 2OH^-(aq)$$

S stands for the solubility of $Mg(OH)_2$ in the saturated solution, that is,

$$[Mg^{2+}] = S, \quad [OH^-] = 2S$$

$$K_{sp}[Mg(OH)_2] = [Mg^{2+}][OH^-]^2 = S(2S)^2 = 4S^3 = 5.61 \times 10^{-12}$$

$$S = \sqrt[3]{\frac{5.61 \times 10^{-12}}{4}} \text{ mol} \cdot L^{-1} = 1.12 \times 10^{-4} \text{ mol} \cdot L^{-1}$$

It is necessary to point out that there are many factors affect solubility of slightly soluble electrolytes, and keep the following points in calculating the K_{sp} or solubility with each other.

(1) Method given above is available to solutions with small ionic strength. In these cases, the activity can be replaced by concentration. For example, solubility and ionic strength of $CaSO_4$ in an aqueous solution are large, so activities of ions cannot be replaced by ions concentrations.

(2) Ions from dissolving precipitations do not hydrolyze or cause other side-reaction. For example, CaS, $CaCO_3$ and $Ba_3(PO_4)_2$, they dissolve in water slightly and form S^{2-}, CO_3^{2-}, PO_4^{3-} anions which will hydrolyze in water further. So the methods to calculate K_{sp} from solubility for them cannot be used.

(3) The methods mentioned above are available for slightly soluble electrolytes in water. They completely ionizes, no soluble molecules exist in solution. If the compounds with strong covalent properties such as Hg_2Cl_2, there is another equilibrium between dissolved molecules $Hg_2Cl_2(aq)$ and aqueous ions $Hg^+(aq)$ and $Cl^-(aq)$ from $Hg_2Cl_2(aq)$ molecules in solution. The precipitation-dissolution equilibrium is not simple. So the methods mentioned above to calculate K_{sp} from solubility cannot be used.

Questions and thinking 7-1 K_{sp} values of PbI_2 and $CaCO_3$ are approximately equal at 25°C. Is the concentration of Pb^{2+} approximately equal to that of Ca^{2+} in saturated solutions? Why?

7.1.3 Solubility Product Rule

At a certain temperature, when A_aB_b reaches the precipitation-dissolution equilibrium, the product of the power of ion concentration in the solution is a constant, K_{sp}. But for any concentration of ion, c, which may differ from the equilibrium concentration, the product is called ion product, I_p.

$$I_p = c^a(A^{n+})c^b(B^{m-})$$

The product of the ion concentrations raised to the power of the stoichiometric coefficients is called **ion product** I_p, the reaction quotient in fact. I_p has the similar form as the K_{sp} expression. The effect of the ion product lies in comparing its value to that attained at the equilibrium, each of precipitation has its own distinct K_{sp} at a given temperature. For a given solution,

(1) If $I_p = K_{sp}$, the solution is saturated, then the solution is at equilibrium.

(2) If $I_p < K_{sp}$, the solution is unsaturated and the solute will continue to dissolve.

(3) If $I_p > K_{sp}$, the solution is supersaturated and precipitation will occur.

These conclusions is called solubility product rule.

7.2 Shifting the Precipitation Equilibrium

7.2.1 Influence Factors of Precipitation Dissolution Equilibrium Shift

1. Common ion effect

The solubility of a solute mainly depends on the temperature of the solution and the nature of solute. Solubility is also affected by the addition of other substances to the solution. The solubility is considerably reduced in a solution if the solution already contains one of its constituent ions as compared to its solubility in a pure solvent. This reduction is called the **common ion effect**. However, the presence of the common ion has no effect on the value of K_{sp}. For example, if a salt such as K_2CrO_4 is dissolved into saturated aqueous solution of Ag_2CrO_4, the solution will dissolve less Ag_2CrO_4 than that in pure water.

Sample Problem 7-4 $K_{sp}(Ag_2CrO_4)$ is 1.12×10^{-12}, what is the solubility of Ag_2CrO_4 in (1) $0.10 \mathrm{mol} \cdot L^{-1}$ of $AgNO_3$ solution, (2) $0.10 \mathrm{mol} \cdot L^{-1}$ of K_2CrO_4 solution.

Solution (1) The solubility of Ag_2CrO_4 is S, the dissociation reaction is

$$Ag_2CrO_4(s) \rightleftharpoons 2Ag^+(aq) + CrO_4^{2-}(aq)$$
$$(2S+0.10) \quad S \; \mathrm{mol} \cdot L^{-1}$$
$$\approx 0.10 \mathrm{mol} \cdot L^{-1}$$
$$K_{sp}(Ag_2CrO_4) = [Ag^+]^2[CrO_4^{2-}]$$

$$S = [CrO_4^{2-}] = \frac{K_{sp}(Ag_2CrO_4)}{[Ag^+]^2} = \frac{1.12 \times 10^{-12}}{0.10^2} \mathrm{mol} \cdot L^{-1} = 1.12 \times 10^{-10} \mathrm{mol} \cdot L^{-1}$$

In pure water, the solubility is $6.54 \times 10^{-5} \mathrm{mol} \cdot L^{-1}$. $AgNO_3$ solution dissolves less Ag_2CrO_4.

(2)
$$Ag_2CrO_4(s) \rightleftharpoons 2Ag^+(aq) + CrO_4^{2-}(aq)$$
$$2S \; \mathrm{mol} \cdot L^{-1} \quad (0.10+S) \approx 0.10 \mathrm{mol} \cdot L^{-1}$$
$$K_{sp}(Ag_2CrO_4) = [Ag^+]^2[CrO_4^{2-}] = (2S)^2 \times 0.10 = 0.40 S^2$$

$$S = \sqrt{\frac{K_{sp}}{0.40}} = \sqrt{\frac{1.12 \times 10^{-12}}{0.40}} \mathrm{mol} \cdot L^{-1} = 1.7 \times 10^{-6} \mathrm{mol} \cdot L^{-1}$$

Similarly, $AgNO_3$ solution dissolves less Ag_2CrO_4 also.

Two results show that in the precipitation equilibrium system of Ag_2CrO_4, more precipitates of Ag_2CrO_4 will form if the reagents containing ion Ag^+ or CrO_4^{2-} are added, and the solubility of Ag_2CrO_4 will decrease. In order to precipitate Ag^+ completely in solution, the common ion effect can be applied by adding a proper excessive precipitant. But sometimes, the more the amount of precipitant is, the higher the solubility is. For example, AgCl can be dissolved by adding excess of HCl due to the coordination reaction.

$$AgCl(s) + Cl^-(aq) \rightarrow [AgCl_2]^-(aq)$$

2. Salt effect

The solubility of precipitation in saturated aqueous solution increased slightly if the soluble salt without the common ions is added, this is called **salt effect**. For example, adding KNO_3 to a saturated aqueous solution of Ag_2CrO_4 promotes the dissolution of solid Ag_2CrO_4. The ionic strength of the solution increases, and activity of ions decrease due to the addition of the soluble salt. The saturation solution becomes unsaturated and the solubility increases.

The common ion effect must be accompanied by the salt effect. But the common ion effect is much more significant than the salt effect. So, the salt effect can be ignored when the two effects coexist.

7.2.2 Formation of Precipitation

1. Formation of precipitation

According to the solubility product rule, if the solution is supersaturated, $I_p > K_{sp}$, precipitation will occur.

Sample Problem 7-5 Determine whether precipitate formed in the following cases:
(1) 10mL 0.020mol·L^{-1} CaCl$_2$ solution mix with 10mL 0.020mol·L^{-1} Na$_2$C$_2$O$_4$ solution;
(2) Lead CO$_2$ gas into 1.0mol·L^{-1} CaCl$_2$ solution until to be saturation.

$$K_{sp}(CaC_2O_4) = 2.32 \times 10^{-9}; \quad K_{sp}(CaCO_3) = 3.36 \times 10^{-9}$$

Solution (1) after mixing, $c(Ca^{2+}) = 0.010 mol·L^{-1}$, $c(C_2O_4^{2-}) = 0.010 mol·L^{-1}$

$$I_P(CaC_2O_4) = c(Ca^{2+}) c(C_2O_4^{2-}) = 0.010 \times 0.010 = 1.0 \times 10^{-4}$$

$I_P(CaC_2O_4) > K_{sp}(CaC_2O_4)$, CaC$_2O_4$ precipitate will occur.

(2) in saturated CO$_2$ aqueous solution, $[CO_3^{2-}] = K_{a2} = 4.7 \times 10^{-11}$

$$I_P(CaCO_3) = c(Ca^{2+})[CO_3^{2-}] = 1.0 \times (4.7 \times 10^{-11}) = 4.7 \times 10^{-11}$$

$I_P(CaCO_3) < K_{sp}(CaCO_3)$, so there will be no precipitate of CaCO$_3$.

2. Fractional precipitation

If two or more ions in solution can react with the same precipitant, the precipitations will occur in orders when this reagent is added into. This phenomenon is called **fractional precipitation**. Fractional precipitation is a technique that separates ions from solution based on their different solubility. If the solubility of precipitation is very different, they can be separated by forming precipitation step by step. For example, I$^-$ and Cl$^-$ ions are in an aqueous solution, add silver nitrate AgNO$_3$ in it slowly. Two precipitates AgCl with K_{sp} of 1.77×10^{-10} and AgI with K_{sp} of 8.52×10^{-17} can form. Because K_{sp} for AgI is smaller than that for AgCl, so AgI will precipitate firstly. If AgNO$_3$ is added to the solution not enough to precipitate Cl$^-$ further, then only AgI precipitation. I$^-$ can be isolated from the original solution by filtering the mixture.

It must be pointed out that, for different types of slightly soluble electrolytes, the precipitating order and the separation results must be determined by calculation according to their solubility and K_{sp} values.

3. Inversion of precipitation

Sometimes, it is necessary to convert one precipitation to another. This process is called **inversion of precipitation**. For example, the main component of boiler fouling is CaSO$_4$ precipitation which has little thermal conductivity. It is difficult to dissolve in acid to remove. However, it can be treated with sufficient Na$_2$CO$_3$ solution. All CaSO$_4$ can be transformed into CaCO$_3$ precipitation which is loose and soluble in acidic solution, and then it can be removed easily. The transformation process is as follows

$$CaSO_4(s) + CO_3^{2-}(aq) \rightleftharpoons CaCO_3(s) + SO_4^{2-}(aq)$$

$$K = \frac{[SO_4^{2-}]}{[CO_3^{2-}]} = \frac{K_{sp}(CaSO_4)}{K_{sp}(CaCO_3)} = \frac{4.93 \times 10^{-5}}{3.36 \times 10^{-9}} = 1.47 \times 10^4$$

The equilibrium constant of the conversion reaction is very large, indicating that the precipitation transformation is quite completely. As the same type of slightly soluble strong electrolyte, it is easier to transform the precipitation from that with larger K_{sp} to that with smaller K_{sp}. However, can this process be reversed? For example, when Ba^{2+} is isolated and identified, it is need to transform $BaSO_4$ into $BaCO_3$. Reverse reaction is as follows

$$BaSO_4(s) + CO_3^{2-}(aq) \rightleftharpoons BaCO_3(s) + SO_4^{2-}(aq)$$

$$K = \frac{[SO_4^{2-}]}{[CO_3^{2-}]} = \frac{K_{sp}(BaSO_4)}{K_{sp}(BaCO_3)} = \frac{1.08 \times 10^{-10}}{2.58 \times 10^{-9}} = \frac{1}{24}$$

According to the equilibrium constant K, $BaSO_4$ can be converted into $BaCO_3$ as long as $c(CO_3^{2-}) > 24\, c(SO_4^{2-})$ in the solution. Adding enough saturated Na_2CO_3 solution into $BaSO_4$, $BaSO_4$ could be transformed into $BaCO_3$ completely.

7.2.3 Dissolving Precipitation

According to the solubility product rule, if $I_p < K_{sp}$, the solute will continue to dissolve. So the precipitation can dissolve by decreasing an ion concentration in a saturated solution of precipitation.

1. Formation of weakly dissociated species.

The weakly dissociated species includes water, weak acid, weak base, complex ion and so on.

(1) The dissolution of metallic hydroxide by acid. OH^- in the hydroxide reacts with H^+ to H_2O, so the metallic hydroxide can be dissolved in acid. For example,

$$Mg(OH)_2(s) \rightleftharpoons Mg^{2+}(aq) + 2OH^-(aq)$$

the direction of equilibrium shift

$+$

$2H^+(aq) + 2Cl^-(aq) \leftarrow 2HCl(aq)$

$\Updownarrow$

$2H_2O(l)$

After adding acid, $c(OH^-)$ becomes much smaller, $I_p < K_{sp}$, causing the precipitation to dissolve. $Mg(OH)_2$ can be also dissolved in NH_4Cl solution because NH_4^+ react with OH^-, this decrease $c(OH^-)$ in the solution, increasing the dissolution of $Mg(OH)_2$.

(2) The dissolution of metallic sulfides by acids. The precipitation ZnS can dissolve in HCl aqueous solution, because S^{2-} can react with H^+ to HS^-, and then HS^- combines with H^+ to form H_2S gas. This reaction makes the equilibrium shift to the right, increasing the dissolution of ZnS.

$$ZnS(s) \rightleftharpoons Zn^{2+}(aq) + S^{2-}(aq)$$

the direction of equilibrium shift

$+$

$2H^+(aq) + 2Cl^-(aq) \leftarrow 2HCl(aq)$

$\Updownarrow$

$H_2S(g)$

(3) The dissolution of $PbSO_4$ type of precipitation. In saturated solutions containing $PbSO_4$

precipitates, adding NH_4Ac to form $Pb(Ac)_2$, a weak electrolyte, it reduces $c(Pb^{2+})$ in the solution and makes the equilibrium shift to the right, increasing the dissolution of $PbSO_4$.

$$PbSO_4(s) \rightleftharpoons SO_4^{2-}(aq) + Pb^{2+}(aq)$$

the direction of equilibrium shift ↓ +

$$2Ac^-(aq) + 2NH_4^+(aq) \leftarrow 2NH_4Ac(aq)$$

$$\updownarrow$$

$$Pb(Ac)_2$$

2. Formation of complex ion.

For example, the AgCl precipitate can be dissolved in ammonia aqueous solution to form complex ion.

$$AgCl(s) \rightleftharpoons Cl^-(aq) + Ag^+(aq)$$

the direction of equilibrium shift ↓ +

$$2NH_3(aq)$$

$$\updownarrow$$

$$[Ag(NH_3)_2]^+(aq)$$

Notice that the formation constant K_f of complex ion is significantly large than K_{sp} of precipitation for the same metallic cation. The complex ion will be described in detail in Chapter 11. As silver ion Ag^+ is formed into complex ion $[Ag(NH_3)_2]^+$, the dissociation reaction of AgCl shifts to the right and form more $[Ag(NH_3)_2]^+$, increasing the dissolution of AgCl.

3. The dissolution of precipitation by redox reaction

Sulfides with large K_{sp} values such as ZnS, PbS and FeS can be dissolved in HCl solution commonly. If sulfides such as HgS or CuS are small in K_{sp} value, they are difficult to be dissolved in HCl solution. If an oxidant is added, these precipitations can be dissolved by the redox reaction. For example, $K_{sp}(CuS)$ is 6.3×10^{-36}, it is soluble in HNO_3 solution. The reaction is as follows:

$$CuS(s) \rightleftharpoons Cu^{2+}(aq) + S^{2-}(aq)$$

$$+ HNO_3(aq)$$

$$\rightarrow S(s) + NO(g) + H_2O$$

The overall reaction is

$$3CuS + 8HNO_3 = 3Cu(NO_3)_2 + 3S\downarrow + 2NO\uparrow + 4H_2O$$

Because S^{2-} ion is oxidized by HNO_3 to form S. The $c(S^{2-})$ is decreased, which result in the dissolution of CuS precipitation.

> **Questions and thinking 7-2** Explain the following phenomena:
> 1. $CaC_2O_4(s)$ dissolve in HCl solution, but not in HAc solution.
> 2. Adding $CaCl_2$ solution to $H_2C_2O_4$ solution, $CaC_2O_4(s)$ is formed. After filtration, ammonia water was added into the filtrate, $CaC_2O_4(s)$ occurs again.

7.3 Applications of Precipitation Dissolution Equilibrium

There are many examples about precipitation-dissolution equilibrium in medicine. The formation of urinary calculus in human body and the formation of bones and tooth decay are all related to precipitation-dissolution equilibrium.

7.3.1 Tooth Decay and Fluoridation

Under the physiological conditions of 37℃ and pH 7.4, mixing Ca^{2+} and PO_4^{3-} together, first, amorphous precipitation calcium phosphate will produce which will convert to octacalcium phosphate afterwards, and finally form the most stable hydroxyapatite (HAP) $Ca_{10}(OH)_2(PO_4)_6(s)$. The formation process of hydroxyapatite consists of a series of reactions and transformation of precipitation. HAP is also called bio-apatite. It is the main component of living bodies including shells, teeth, and fossils of dead bodies. The skeleton of about 55%～75% in the human body belongs to the HAP.

The tooth enamel consists mainly of HAP, which is the hardest substance in the human body. But it is very susceptible to acid. Tooth cavities are caused by acids dissolving the tooth enamel. The acids, which are mostly organic acid formed by the action of specific bacteria on sugars and other carbohydrates present in the plaque adhering to the teeth. The reaction is as follows

$$Ca_{10}(OH)_2(PO_4)_6\,(s) + 8H^+ = 10Ca^{2+} + 6HPO_4^{2-} + 2H_2O$$

The resultant Ca^{2+} and HPO_4^{2-} ions diffuse out of the tooth enamel and are washed away by saliva. As time passes by, tooth cavity forms.

Fluoride ion, presents in drinking water, toothpaste, or other sources, will react with hydroxyapatite to form fluoroapatite (FAP), $Ca_{10}(PO_4)_6F_2$. This mineral, in which F^- has replaced OH^-, is much more resistant to attack by acids because the fluoride ion is a much weaker Brønsted-Lowry base than the hydroxide ion.

$$Ca_{10}(OH)_2(PO_4)_6\,(s) + 2F^- = Ca_{10}F_2(PO_4)_6\,(s) + 2OH^-$$

The above reaction occurs through transforming reaction by precipitation-solubility equilibrium. Because $K_{sp}(HAP) = 6.8 \times 10^{-37}$, $K_{sp}(FAP) = 1.0 \times 10^{-70}$, the equilibrium constant for the transforming reaction is

$$K = \frac{[OH^-]^2}{[F^-]^2} = \frac{K_{sp}(HAP)}{K_{sp}(FAP)} = 6.8 \times 10^{23}$$

The transforming reaction occurs completely because equilibrium constant is very big. In the presence of F^-, a thin layer of FAPs is formed on the enamel surface, preventing acid from dissolving the teeth. Therefore, fluoride can increase enamel caries resistance and promote enamel mineralization. Local high concentration fluorine treatment will produce CaF_2, which has the definite anti-solubility in oral environment, and release F^- at high acidity, which has a long-term anti-caries effect. The higher the pH of plaque is, the more favorable the process of mineralization is. The pH of saliva ranged from 6.35 to 6.85, which increased the pH in plaque and promoted the mineralization. Therefore, in order to prevent tooth decay, instead of brushing your teeth, you should use toothpaste containing fluoride ions.

7.3.2 Urinary Calculus

How does the urinary calculus form? In human bodies, the first step of urine formation is blood entering the kidney, and macromolecules such as proteins and cells are filtered through glomerular filtration. The filtrating fluid is the primitive urine, which enter bladder through a sector of tiny tubing in which contains ions such as Ca^{2+}, Mg^{2+}, NH_4^+, $C_2O_4^{2-}$, PO_4^{3-}, H^+ and OH^-, etc. These ions will react mutually to form precipitates. For example, Ca^{2+} and $C_2O_4^{2-}$ will form calcium oxalate CaC_2O_4, precipitates, which can constitute the urinary calculus. Before the blood pass through glomeruli of kidney, calcium oxalate is supersaturated, $I_p = c(Ca^{2+}) \cdot c(C_2O_4^{2-})$, $I_p > K_{sp}(CaC_2O_4)$, but it is difficult to form precipitation because there exists crystal inhibitors such as protein in blood, calcium oxalate. After the filtration of glomeruli, crystal inhibitors such as protein are filtrated and the filtrated fluids will be inclined to form the precipitation crystalline CaC_2O_4. These phenomena can occur in normal human bodies, but the forming microlith will not clog the tubing and the detention time is very short so that the forming microlith is easy to be washed away by the urine. But the concentration of inhibitor in certain person's urine is too low or renal malfunction, the flow rate is too slow and the detention time will be long, these factors all form urinary calculus. Therefore, in medicine, doctors often use the treatment by accelerating the rate of urination (to lower the detention time), enhancing the amount of the urine (to reduce the concentration of Ca^{2+} and $C_2O_4^{2-}$ to cure and/or prevent the urinary calculus). Usually, drinking sufficient water will help you away from the risk of urinary calculus.

Key Terms

难溶强电解质	insoluble strong electrolyte
同离子效应	common ion effect
盐效应	salt effect
沉淀溶解平衡	precipitation dissolution equilibrium
溶度积常数	solubility product constant
溶解度	solubility
离子积	ion product
饱和溶液	saturation solution
分步沉淀	fractional precipitation
沉淀转化	inversion precipitation

Summary

The solubility product constant K_{sp}, is the equilibrium constant for a slightly soluble ionic compound called precipitation generally in aqueous solution.

$$A_aB_b(s) \rightleftharpoons aA^{n+}(aq) + bB^{m-}(aq)$$
$$K_{sp}(A_aB_b) = [A^{n+}]^a \cdot [B^{m-}]^b$$

The solubility S, of precipitation and its K_{sp} are related by the following equation.

$$S = \sqrt[(a+b)]{\frac{K_{sp}}{a^a \cdot b^b}}$$

I_P is Ionic product, $\quad I_P = c(A^{n+})^a \cdot c(B^{m-})^b$

If 1) $I_P > K_{sp}$, the solute is super-saturated, Precipitation will occur.

2) $I_P < K_{sp}$, the solute is un-saturated. The solute will continue to dissolve.

3) $I_P = K_{sp}$, the solute is saturated. The solution is at equilibrium.

Some metallic cation can be separated by the selective precipitation. This is an important way in qualitative analysis, a procedure for identifying the ions present in a solution.

Common ion effect decreases the solubility of the precipitation. Salt effect slightly increases the solubility of it.

There are some procedures to dissolve a precipitation. The general procedure makes the ions in precipitation-solubility equilibrium to form weak electrolyte, gas, complex ion, another precipitation by adding acid, base, ligand and so on. Sometimes, necessary redox reaction is used to dissolve the precipitation with very small K_{sp}.

Exercises

1. What is the solubility product constant and ionic product of slightly soluble electrolyte? What is the difference and connection between the two terms?

2. Explain the reason why the solubility of $BaSO_4$ in saline is greater than that in pure water, while the solubility of $AgCl$ in saline is lower than in pure water.

3. When added several drops of $2mol \cdot L^{-1}$ aqueous ammonia to 2mL $0.10mol \cdot L^{-1}$ $MgSO_4$ solution, white precipitation $Mg(OH)_2$ was formed, then several drops of $1mol \cdot L^{-1}$ NH_4Cl solution were added to the solution, the precipitation disappeared. Try to explain it.

4. When adding the following substances in a saturated solution of solid AgCl, what effect will it have on the solubility of AgCl?

(1) HCl (2) $AgNO_3$ (3) KNO_3 (4) $NH_3 \cdot H_2O$

5. When the H_2S gas was leaded into the $ZnSO_4$ solution, the precipitation is not completely. However, if NaAc was added to $ZnSO_4$ solution before, and then leading H_2S gas, the precipitation almost completely. Try to explain.

6. Suppose that $Mn(OH)_2$ is dissolved completely in water. It is known, $K_{sp}\{Mn(OH)_2\} = 2.06 \times 10^{-13}$, Try to calculate the values for follow questions:

(1) The solubility of $Mn(OH)_2$ in water.

(2) Solubility of $Mn(OH)_2$ in $0.10mol \cdot L^{-1}$ NaOH solution. Assuming $Mn(OH)_2$ is not the other change in NaOH solution at all.

(3) Solubility of $Mn(OH)_2$ in $0.20mol \cdot L^{-1}$ $MnCl_2$ solution.

7. In the solution of $0.010mol \cdot L^{-1}$ KCl and $0.010mol \cdot L^{-1}$ K_2CrO_4, to add $AgNO_3$ solution dropwise, which one will precipitation at first, AgCl or Ag_2CrO_4? When the second ion begin to precipitation, what is the concentration of the first ion? Neglect the volume change of the solution after adding $AgNO_3$. Known that, $K_{sp}(AgCl) = 1.77 \times 10^{-10}$, $K_{sp}(Ag_2CrO_4) = 1.12 \times 10^{-12}$.

8. About 50% of kidney stones is composed of $Ca_3(PO_4)_2$. Normal daily urine output for a person is 1.4L, in which contains about 0.10g Ca^{2+}. To make $Ca_3(PO_4)_2$ not form precipitation in the urine, which shall the concentration of PO_4^{3-} not be higher than? Doctors always advise patients with kidney stone to

drink more water, can you explain why briefly? $K_{sp}\{Ca_3(PO_4)_2\} = 2.07 \times 10^{-33}$.

9. In a solution containing 0.10mol·L^{-1} Pb^{2+} and 0.10mol·L^{-1} Fe^{2+}, in order to separate Pb^{2+} from Fe^{2+} by forming PbS precipitate, what is the concentration range of S^{2-}? Is it possible to separate Pb^{2+} and Fe^{2+} ions by means of fractional precipitation? Known K_{sp}(PbS) = 8.0×10^{-28}, K_{sp}(FeS) = 6.3×10^{-18}.

10. The concentration of Ca^{2+} in urban water supply is 0.002 0mol·L^{-1}, K_{sp}(CaF$_2$) = 3.45×10^{-11}. Try to calculate:

(1) If NaF is added into this water, what could the highest concentration of F$^-$ be attained before CaF$_2$ precipitate occurs.

(2) If the concentration of F$^-$ in the drinking water is 1mg·L^{-1}, is the concentration of F$^-$ exceeding the standard in the fluorinated water?

11. Mix 500mL of 0.20mol·L^{-1} MgCl$_2$ with 500mL of 0.20mol·L^{-1} NH$_3$·H$_2$O solution. (1) Predict that whether the precipitation forms or not; (2) If there is precipitation, how many grams of NH$_4$Cl are needed to avoid forming of Mg(OH)$_2$ precipitation? Neglect the volume change after adding solid NH$_4$Cl. Known that, $K_{sp}\{Mg(OH)_2\} = 5.61 \times 10^{-12}$, K_b(NH$_3$) = 1.8×10^{-5}, MM.(NH$_4$Cl) = 53.5g·mol^{-1}.

Supplementary Exercises

1. Which of the following compounds are more soluble in acidic solution than in pure water? Write a balanced net ionic equation for each dissolution reaction.

(a) AgBr (b) CaCO$_3$ (c) Ni(OH)$_2$ (d) BaCO$_3$

2. Calculate the solubility of PbCrO$_4$ (K_{sp} = 2.8×10^{-13}) in:

(a) Pure water (b) 1.0×10^{-3}mol·L^{-1} K$_2$CrO$_4$

3. Which has the greater solubility: AgCl with K_{sp} = 1.77×10^{-10} or Ag$_2$CrO$_4$ with K_{sp} = 1.12×10^{-12}? Which has the greater solubility in grams per liter?

4. A saturated solution of an ionic salt MX exhibits an osmotic pressure of 75.4mmHg at 25℃. Assuming that MX is completely dissociated in solution, what is the value of its K_{sp}?

5. Even though Ca(OH)$_2$ is an inexpensive base, its limited solubility restricts its use. What is the pH of a saturated solution of Ca(OH)$_2$? $K_{sp}\{Ca(OH)_2\} = 5.02 \times 10^{-6}$.

6. Ca^{2+}, which causes clotting, is removed from donated blood by precipitation with sodium oxalate Na$_2$C$_2$O$_4$. CaC$_2$O$_4$ is a sparingly soluble salt (K_{sp} = 2.32×10^{-9}). If the desired [Ca^{2+}] is less than 3.00×10^{-8}mol·L^{-1}, what must be the minimum concentration of Na$_2$C$_2$O$_4$ in the blood sample?

7. Prior to having an X-ray exam of the upper gastrointestinal tract, a patient drinks an aqueous suspension of solid BaSO$_4$. (Scattering of X rays by barium greatly enhances the quality of the photograph.) Although Ba^{2+} is toxic, ingestion of BaSO$_4$ is safe because it is quite insoluble. If a saturated solution prepared by dissolving solid BaSO$_4$ in water has [Ba^{2+}] = 1.04×10^{-5}mol·L^{-1}, what is the value of K_{sp} for BaSO$_4$?

8. Clothing washed in water that has a manganese concentration exceeding 0.1mg·L^{-1} (1.8×10^{-6}mol·L^{-1}) may be stained by the manganese. A laundry wishes to add a base to precipitation manganese as the hydroxide, $K_{sp}\{Mn(OH)_2\} = 2.06 \times 10^{-13}$. At what pH is [Mn^{2+}] equal to 1.8×10^{-6}mol·L^{-1}?

Answers to some Exercises

[Exercises]

6. (1) 3.7×10^{-5} mol·L^{-1} (2) $S = 2.1 \times 10^{-11}$ mol·L^{-1} (3) $S = 5.1 \times 10^{-7}$ mol·L^{-1}

7. AgCl precipitates firstly, [Cl$^-$] = 1.6×10^{-5} mol·L^{-1}

8. [PO$_4^{3-}$] ≤ 5.9×10^{-13} mol·L^{-1}

9. $8.0 \times 10^{-27} \sim 6.3 \times 10^{-17}$ mol·L^{-1}.
Pb^{2+} and Fe^{2+} can be completely separated by fractional precipitation.

10. (1) 1.3×10^{-4} mol·L^{-1} (2) concentration of F$^-$ exceeding standard

11. pH 2.82 ~ 9.37

[Supplementary Exercises]

1. (b), (c), and (d) are more soluble in acidic solution.

2. (1) $S = 5.3 \times 10^{-7}$ (2) $S = 2.8 \times 10^{-10}$

3. Ag$_2$CrO$_4$ has both the higher molar and gram solubility.

4. K_{sp}(MX) = 4.00×10^{-6}

5. pH = 12.33

6. 0.077 3 mol·L^{-1}

7. K_{sp}(BaSO$_4$) = 1.08×10^{-10}

8. pH = 10.53

（乔　洁）

Chapter 8
Oxidation-Reduction Reaction and Electrode Potential

Oxidation-reduction reactions or **redox reactions** for short are one of the most important classes of chemical reactions. The energy obtained from redox reactions are involved in a wide variety of important natural processes, including power cells, fuel cells, corrosion of metals, refining of metals, the transmission of nerve impulses in animals and the function of biological membranes,etc.. Redox reaction may be applied to convert chemical energy directly into electrical energy. Conversely, electrical energy may be used to promote chemical reactions. The study of mutual transformation between reaction and electric energy is called **electrochemistry**. Electrochemistry is one of the oldest fields of chemistry, and yet the modern uses of electrochemistry are constantly expanding with the development of many new practical applications.

In this chapter, the principles of redox reaction, the principles of design and construction of cells for the production of electricity, the formations and applications of electrode potentials and the relationship between redox reactions and electrode potentials, as well as some applications of redox reactions in chemical analysis, will be studied.

8.1 Primary Cell and Electrode Potential

8.1.1 Oxidation Number

Oxidation number or **oxidation state** is the apparent charge or charge real of an element in a molecule or componrd ion when all bonding electrons are assumed belong to the element with the lagen electronegativity. Oxidation numbers help us keep track of electrons during redox reaction. A redox reaction can be identified by noting the changes in oxidation numbers.

The rules for assigning oxidation numbers are as follows:

(1) The oxidation number of an atom in its elemental form is always zero, including simple substance (e.g., Mg, Hg, H_2, O_3, P_4).

(2) The oxidation number of a monoatomic ion is the same as the charge on that ion (such as, Na^+ has an oxidation number of $+1$).

(3) The oxidation number of O in most of compounds is -2. The exceptions are peroxides such

as H_2O_2 and Na_2O_2 and superoxides such as KO_2, in which the oxidation numbers are -1 and $-\frac{1}{2}$, respectively.

(4) The oxidation number of H is $+1$ when it bonded to nonmetals atom and -1 when it bonded to active metals (such as, NaH, CaH_2).

(5) The oxidation number of fluorine F in all its compounds is always -1. The oxidation number of the all other halogens X is -1 in all compounds except those combined with O or X having a lower atomic number. For example, the Cl in ClO_2 has an oxidation number of $+4$ and the iodine atom in ICl_3 has an oxidation number of $+3$.

(6) The sum of the oxidation numbers of all elements in a neutral compound is zero. The sum of the oxidation numbers in a polyatomic ion equals the charge of it.

Sample Problem 8-1 What are the oxidation numbers in the following ions? (1) Cr in $Cr_2O_7^{2-}$ (2) Fe in Fe_3O_4

Solution (1) In $Cr_2O_7^{2-}$ ion, the total oxidation numbers must be -2. O at -2, Cr at x,

$$2x + 7 \times (-2) = -2 \qquad x = +6$$

The oxidation number of Cr is $+6$.

(2) In Fe_3O_4, the oxidation numbers must be zero in total. Oxygen at -2, iron at x,

$$3x + 4 \times (-2) = 0 \qquad x = +\frac{8}{3}$$

The oxidation number of Fe is $+\frac{8}{3}$.

So, the oxidation number of elements is very different from their valance.

8.1.2 Redox Half-reaction and Primary Cell

1. Redox reaction

Redox reactions are characterized by the changes in oxidation numbers of some elements in the reactions. Any element which increases in oxidation number is said to be oxidized. Any element which decreases in oxidation number is reduced. In the reaction described as follow

$$Zn(s) + 2H^+(aq) \rightleftharpoons Zn^{2+}(aq) + H_2(g)$$

It is obvious that Zn loses two valence electrons to produce an aqueous Zn^{2+} cation, and the oxidation number of Zn increases from 0 to $+2$. Thus Zn is oxidized in this reaction. The reaction is called **oxidation reaction**. Conversely, two H^+ ions accept the two electrons lost by the Zn to form H_2 molecule, and the oxidation number of H decreases from $+1$ to 0. The H^+ ions are reduced in this reaction. The reaction is called **reduction reaction**. As can be seen from above discussed, the loss of electrons by one species is accompanied by the gain of the electrons by another species. Oxidation and reduction are complementary processes that occur at the same time. In short, the reactions involving exchange of electrons are known as redox reaction also.

In any redox reaction, both oxidation and reduction must occur simultaneously, so the electrons formally have to go from one element to another one. The substance that gains electrons is called either the **oxidizing agent** or **oxidant**. Similarly, a **reducing agent** or **reductant** is a substance that loses electrons, thereby causing another substance to be reduced. Every redox reaction involves both an electron donor as reducing agent and an electron acceptor as oxidizing agent. In the above discussed reaction, $H^+(aq)$ is the oxidizing agent and $Zn(s)$ is the reducing agent.

2. Redox half-reaction and redox electric couple

Although oxidation and reduction always take place simultaneously, it is often convenient to consider them as separate processes. For example, a redox reaction

$$Zn + Cu^{2+} \rightleftharpoons Cu + Zn^{2+}$$

It can be considered to consist of the following, the oxidation reaction that loses electrons is

$$Zn - 2e^- \rightarrow Zn^{2+}$$

And the reduction reaction that gains electrons is

$$Cu^{2+} + 2e^- \rightarrow Cu$$

The above equations that show either oxidation or reduction alone are called **redox half-reaction**. Both half-reactions together are required to represent the whole oxidation-reduction process. In the overall redox reaction, the number of electrons lost in the **oxidation half-reaction** must equal the number of electrons gained in the **reduction half-reaction**.

Half-reactions are useful for showing which reactant loses electrons and which reactant gains them. It can be described as the following general equation.

$$Ox. + ne^- \rightleftharpoons Red. \tag{8.1}$$

Where n is the number of moles of electrons transferred in a half-reaction, "Ox." is **oxidized species** which has relative higher oxidation number of an element, and "Red." is **reduced species** which has relative lower oxidation number of an element. The oxidized and reduced species of the same element in a half-reaction is a **redox electric couple**. In general, the couple can be described as oxidized species / reduced species or, simply, Ox. / Red. such as Cu^{2+} / Cu; Zn^{2+} / Zn.

Every redox half-reaction is a redox electric couple. For example, for the half-reaction,

$$Zn^{2+} + 2e^- \rightarrow Zn$$

The redox electric couple is Zn^{2+} / Zn. Moreover, for some half-reactions which involve H^+ ion and OH^- ion, these ions must also be included in redox electric couple. For example, for the half-reaction

$$MnO_4^- + 8H^+ + 5e^- \rightleftharpoons Mn^{2+} + 4H_2O$$

The oxidized species are MnO_4^- and H^+, and the reduced species is Mn^{2+} and H_2O.

3. Balancing Redox Reaction

A redox reaction must be described by a balanced redox equation. Although some redox reactions can be balanced by inspection, others require systematic balancing procedures. The oxidation number method and the **ion-electron method** or known as the half-reaction method are two common methods to balance the redox equations. Both methods give the same results, but the ion-electron method is more versatile. The following equation will be balanced as an example of the steps of the ion-electron method.

$$KMnO_4 + HCl \rightarrow MnCl_2 + Cl_2 + H_2O$$

(1) Write down the ion equation according to the experimental facts.

$$MnO_4^- + Cl^- \rightarrow Mn^{2+} + Cl_2 + H_2O$$

(2) Divide the total equation into two half-reactions, one for oxidation and the other for reduction.

Reduction half-reaction: $MnO_4^- + H^+ \rightarrow Mn^{2+} + H_2O$

Oxidation half-reaction: $Cl^- \rightarrow Cl_2$

(3) Balance all elements by adjusting stoichiometric coefficients. In acidic solution, add H^+ and H_2O to balance the numbers of H and O atoms. In basic solution, add OH^- and H_2O to balance the numbers of H and O atoms, respectively.

Reduction half-reaction: $MnO_4^- + 8H^+ \rightarrow Mn^{2+} + 4H_2O$

Oxidation half-reaction: $2Cl^- \rightarrow Cl_2$

(4) Balance net charge by adding electrons to the side that is deficient in negative charge.

Reduction half-reaction: $MnO_4^- + 8H^+ + 5e^- \rightleftharpoons Mn^{2+} + 4H_2O$ ①

Oxidation half-reaction: $2Cl^- - 2e^- \rightleftharpoons Cl_2$ ②

(5) Multiply the half-reactions by integers if necessary so that the number of electrons lost in oxidation half-reaction equals the number gained in reduction half-reaction. Then immersing the two half-reactions and simplify by canceling the same species appearing on both sides of the combined equation.

$$① \times 2 \quad 2MnO_4^- + 16H^+ + 10e^- \rightleftharpoons 2Mn^{2+} + 8H_2O$$
$$② \times 5 \quad 10Cl^- - 10e^- \rightleftharpoons 5Cl_2$$

$$2MnO_4^- + 16H^+ + 10Cl^- \rightleftharpoons 2Mn^{2+} + 5Cl_2 + 8H_2O$$

(6) Check and make sure that the number of atoms of each kind and the total charge on the left side is equivalent to the number and the total charge on the right side.

> **Question and thinking 8-1** For the half reaction: $FeS \rightarrow Fe^{3+} + H_2SO_4$, how to complete and balance it?

4. Primary Cell

When a strip of Zn is placed in a solution of $CuSO_4$, the blue color of Cu^{2+}(aq) ions fades and Cu deposits on the surface of Zn strip. At the same time, Zn begins to dissolve. A thermodynamic calculation verifies that the reaction is spontaneous.

$$Zn(s) + CuSO_4(aq) \rightleftharpoons Cu(s) + ZnSO_4(aq) \quad \Delta_r G_m^\ominus = -212.6 kJ \cdot mol^{-1}$$

In the overall reaction, Zn atom transfers two electrons to Cu^{2+} cation directly, so no electrical current generates and no useful work is done in this spontaneous process. The generated energy of this redox reaction converts into heat.

To generate electrical work, the oxidation half-reaction must be separated from the reduction half-reaction so that a flow of electrons can be created through an external circuit. These transformations are illustrated schematically in Figure 8-1. Zn is placed in contact with $ZnSO_4$ solution in one compartment of the cell, and Cu is placed in contact with $CuSO_4$ solution in another compartment. The two metals are connected by conducting wires and a voltmeter. And the solutions in the two beakers are connected by a special device, a **salt bridge**, which is an inverted U tube containing an aqueous gel and a saturated electrolyte solution of KCl or KNO_3, NH_4NO_3 generally. The salt bridge allows a migration

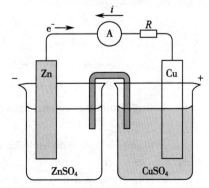

Figure 8-1 Daniel Primary Cell

of ions that maintains the electrical neutrality of the solution*. When the circuit is connected, the meter needle moves. An electrical current flows. Such device which uses a favorable or spontaneous redox reaction to generate electrical energy through an external circuit is called a **primary cell** or **voltaic cell**.

The primary cell discussed above is known as the **Daniel cell**. In a primary cell, each of the two compartments is called a **half cell**. $ZnSO_4$ solution and Zn consists of a half cell and $CuSO_4$ solution and Cu consists of another in the Daniel cell. The two solid metals that are connected by the external circuit are called **electrodes**. The electrodes can be made of materials that participate in the reaction, as in the present example. More typically, the electrodes are made of conducting materials, such as platinum Pt or graphite C, which does not gain or lose mass during the reaction but serves as a surface at which electrons are transferred from it. By definition, the electrode at which oxidation half-reaction occurs is called the **anode**, and the electrode at which the reduction half-reaction occurs is called the **cathode**. In present example, here are the two half cell reactions that occur in cathode and anode.

Anode: $\quad\quad\quad Zn(s) \rightleftharpoons Zn^{2+}(aq) + 2e^-$ $\quad\quad$ Oxidation half-reaction

Cathode: $\quad\quad Cu^{2+}(aq) + 2e^- \rightleftharpoons Cu(s)$ $\quad\quad$ Reduction half-reaction

Electrons from the half-reaction that Zn is oxidized at the anode flow through the external circuit to the cathode, where they are consumed as $Cu^{2+}(aq)$ is reduced. The total reaction that is added together from the two half cell reactions is called **cell reaction**.

Cell reaction: $\quad\quad Zn(s) + Cu^{2+}(aq) \rightleftharpoons Cu(s) + Zn^{2+}(aq)$

In conclusion, any combination of redox electric couples can be used to construct a primary cell and thereby change chemical energy into electrical energy by forcing electrons through an external conductor from anode to cathode. Thus the oxidation half-reaction that loses electrons takes place in the anode, and the reduction half-reaction that gains electrons takes place in the cathode. The cell reaction is exactly the same as the total redox reaction.

A shorthand **cell notation** is often used to describe a primary cell. For example, the notation of the Daniel cell is

$$(-) \; Zn \mid Zn^{2+} (c_1) \parallel Cu^{2+} (c_2) \mid Cu \; (+)$$

The rules for writing the notation of a primary cell are as follows:

(1) By convention, the half cell which is the anode is written on the left, and the anode is indicated by subtraction sign written in parentheses "(−)". The half cell which is the cathode is written on the right, and the cathode is indicated by plus sign written in parentheses "(+)".

(2) The substances involved are denoted by their formulas with their concentrations or pressures written in parentheses following the particular formula.

(3) The salt bridge is indicated by a pair of vertical lines "∥". The electrolyte solution of the half cell must be close to the salt bridge, while the metal electrode of the half cell must be far from it. A single vertical line "∣" indicates contact between two phases like solid and liquid, or liquid and gas, etc. Two or

---

* As zinc (Ⅱ) ions are added to the left beaker because of the loss of electrons by the zinc atoms, negative ions migrate into the beaker from the salt bridge. Meanwhile, positive ions migrate through the salt bridge into the other beaker and neutralize the accumulated negative charge resulting from the depletion of copper (Ⅱ) ions. Thus a complete circuit is achieved, and a continuous flow of electrons through the wires is maintained. Furthermore, the salt bridge can eliminate the liquid junction potential.

more substances in the same phase are separated by commas ",".

(4) If there is not metal as conductor in the half cell, some inert metals, such as graphite and Pt have to be needed. The common **inert electrode** is Pt.

8.1.3 Formation of Electrode Potential

In the Daniel cell, when the circuit is connected, a current occurs. This phenomenon shows that there is a potential difference between the two electrodes. Walther Hermann Nernst, a famous Germany chemist, explained the formation of the electrode potential of metal-metal ion electrode with his **electric double layer theory**.

Consider now what happens when a strip of metal M is dipped into a solution containing the corresponding metal ion M^{n+}. Depending on the nature of M, one of two things may occur. The atoms on M surface may have a tendency to break away from M and go into solution as M^{n+} ions because of the thermal motion of molecules and the action of solvent. When one mole of atom leaves the surface of metal electrode, n moles of electrons are left on M surface. An alternative process is for M^{n+} ion in the solution to tend to withdraw electrons from the electrode, and for the resulting neutral M atoms to stick to its surface. When the two processes are equal in speed, the dynamic equilibrium is set up.

$$M(s) \underset{\text{precipitation}}{\overset{\text{dissolution}}{\rightleftharpoons}} M^{n+}(aq) + ne^- \qquad (8.2)$$

If the tendency of dissolving M dominates in the process, at equilibrium, it will lead to a negative charged electrode. The M^{n+} ions in the solution tend to distribute near M surface because of the electrostatic attraction of the negative charges on M surface. Thus, there is an electric double layer between the excess charges on M surface and M^{n+} ions nearby. This is illustrated in Figure 8-2. Conversely, if the tendency of separating out M dominates in the process, it gives rise to the electrode with positive charges. There is also an electric double layer between the excess M^{n+} ions on M surface and charges nearby. Though the electric double layer is only about 10^{-10}m, it will form a potential difference between the two layers. This potential difference is called **electrode potential**.

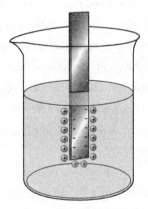

Figure 8-2 Electric Double Layer

The electrode potential can be expressed by the symbol $\varphi_{Ox/Red}$ and its unit is volts.

Obviously, the electrode potential depends on the nature of metal. The more the metal is active, the greater the tendency for the metal dissolving is. And the metal electrode is at a more negative electrode potential. By contrast, the more the metal is inert, the greater the tendency for the metal to deposit. And the metal electrode is at a more positive electrode potential. Furthermore, the electrode potential relates to the temperature and the concentration of the metallic ion in solution. The different electrodes have the different electrode potentials. In the Daniel cell, the potential difference between Zn half cell and Cu half cell causes the electrons to flow if the circuit is completed. In this case, electrons flow from the Zn electrode to Cu electrode. The result is a primary cell in which an electrode potential difference generates a stable current.

There are two half-reactions or two electric couples or two electrodes in a redox reaction. They are equivalent generally.

The potential difference between the two electrodes of a primary cell provides the driving force that pushes electrons through the external circuit. Therefore, this potential difference is called **electromotive force**. The electromotive force of a redox reaction, denoted by the italicized letter E, is also called the cell potential and is measured in volts. For any spontaneous primary cell reaction, the electromotive force will be positive. So in any primary cell, the cathode half-reaction is one with the higher of electrode potential, and the anode half-reaction is lower in electrode potential. The electromotive force of a cell is may be expressed as the difference of two electrode potentials:

$$E = \varphi_{cathode} - \varphi_{anode} \tag{8.3}$$

8.1.4 Standard Electrode Potential

Because the electromotive force of a cell depends on the particular cathode and anode, it would be convenient if we could assign the electrode potential to every electrode. Unfortunately, the electrode potential of a single electrode cannot be measured because electrodes are always paired in a primary cell.

1. Standard hydrogen electrode (SHE)

Although the absolute electrode potential of a single electrode cannot be measured, IUPAC has defined one particular electrode as the reference and defined its electrode potential value as zero volts. Then the electrode potentials of the others can be determined relative to that reference. The reference electrode selected to have zero electrode potential is the **standard hydrogen electrode**, simply, SHE, which is assigned an electrode potential of exactly 0V.

As shown in Figure 8-3, the hydrogen electrode consists of a Pt strip coated with platinum black, the whole bathed in aqueous H^+ solution, and H_2 gas bubbling it continuously. The reaction occurs at the hydrogen electrode

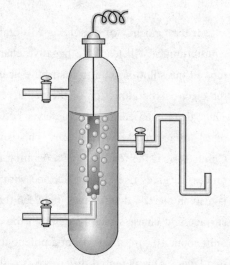

Figure 8-3 Standard Hydrogen Electrode

$$2H^+(aq) + 2e^- \rightleftharpoons H_2(g)$$

when the temperature is 298.15K, the partial pressure of H_2 gas is 100kPa and the concentration of H^+ solution is $1 mol \cdot L^{-1}$ (accurately, the activity is unit), the electrode is the SHE and the value of its electrode potential is defined as 0V.

$$\varphi_{SHE} = 0V$$

2. Measurement of electrode potential

Defined reference value for the SHE makes it possible to determine φ values of other electrodes. The method is to take the SHE and another electrode under test to connect a primary cell. And the measured E values are equal to the φ values of the electrode under test. As an example, Figure 8-4 shows a primary cell taking SHE and Cu^{2+}/Cu electrodes

$(-)\ Pt\ |\ H_2(100kPa)\ |\ H^+(a=1)\ ||\ Cu^{2+}(a)\ |\ Cu(s)\ (+)$

By Equation 8.3, electrode potential of Cu^{2+}/Cu electrode can be known.

$$E = \varphi\ (Cu^{2+}/Cu) - \varphi_{SHE} = \varphi\ (Cu^{2+}/Cu)$$

3. Standard electrode potential

When the electrode operates under standard conditions, the measured electrode potential is called **standard electrode potential** and is denoted $\varphi^{\ominus}_{Ox/Red}$. The unit of $\varphi^{\ominus}_{Ox/Red}$ is volt. The standard electrode potential is the potential developed by a cell in which species in aqueous solution are $1 mol \cdot L^{-1}$ and the partial pressure of gaseous species are 100kPa. Moreover, the temperature is 298.15K most in use.

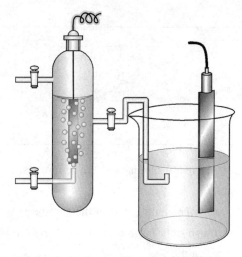

Figure 8-4 Device That Determine the Electrode Potential of Copper Electrode

In the cell to measure the electrode potential of Cu^{2+}/Cu described previously, if electrode is in contact with a $1 mol \cdot L^{-1}$ solution of Cu^{2+},

$(-)\ Pt\ |\ H_2(100kPa)\ |\ H^+(a=1)\ ||\ Cu^{2+}(1mol \cdot L^{-1})\ |\ Cu(s)\ (+)$

The measured value of the electromotive force is the standard electrode potential of Cu^{2+}/Cu electrode.

$$E = \varphi^{\ominus}(Cu^{2+}/Cu) - \varphi_{SHE} = \varphi^{\ominus}(Cu^{2+}/Cu)$$

If measured potential for the cell is 0.341 9V, the standard electrode potential of the Cu^{2+}/Cu electrode is 0.341 9V.

$$\varphi^{\ominus}\ (Cu^{2+}/Cu) = 0.341\ 9V$$

Once the standard electrode potential of an electrode has been measured against the SHE, it may be used in combination with other electrodes in order to determine their standard electrode potentials. Over the years, chemists have carried out many other measurements to determine the standard electrode potentials of electrodes. As a result, the values of standard electrode potentials are listed in reference sources such as the Handbook of Chemistry and Physics. Table 8-1 lists some of the standard electrode potentials.

Table 8-1 Some Representative Standard Electrode Potentials at 298.15K

Half-reaction	$\varphi^{\ominus}/V$
$Na^+ + e^- \rightleftharpoons Na$	-2.71
$Zn^{2+} + 2e^- \rightleftharpoons Zn$	-0.761 8
$Pb^{2+} + 2e^- \rightleftharpoons Pb$	-0.126 2
$2H^+ + 2e^- \rightleftharpoons H_2$	0.000 00
$AgCl + e^- \rightleftharpoons Ag + Cl^-$	0.222 33
$Cu^{2+} + 2e^- \rightleftharpoons Cu$	0.341 9
$I_2 + 2e^- \rightleftharpoons 2I^-$	0.535 5
$O_2 + 2H^+ + 2e^- \rightleftharpoons H_2O_2$	0.695
$Fe^{3+} + e^- \rightleftharpoons Fe^{2+}$	0.771

	Continue
Half-reaction	$\varphi^{\ominus}$ / V
$Ag^+ + e^- \rightleftharpoons Ag$	0.799 6
$Br_2(l) + 2e^- \rightleftharpoons 2Br^-$	1.066
$Cl_2 + 2e^- \rightleftharpoons 2Cl^-$	1.358 27
$Cr_2O_7^{2-} + 14H^+ + 6e^- \rightleftharpoons 2Cr^{3+} + 7H_2O$	1.36
$MnO_4^- + 8H^+ + 5e^- \rightleftharpoons Mn^{2+} + 4H_2O$	1.507

Here is a summary of the definitions and conventions for working with standard electrode potentials table:

(1) The data listed in the standard electrode potentials table are gained from aqueous system, so these data cannot be used for non-aqueous system and solid phase reactions under high temperature.

(2) By convention, the potential associated with each electrode is chosen to be the potential for reduction to occur at that electrode. Thus, standard electrode potentials are tabulated for reduction reactions: $Ox + ne^- \rightleftharpoons Red$, and they are called standard reduction potentials.

(3) The electrode potential is an intensive property and its value is independent of the amount of substance. In other words, changing the stoichiometric coefficient in a half-reaction does not affect the value of the standard electrode potential. For example,

$$Zn^{2+} + 2e^- \rightleftharpoons Zn \qquad \varphi^{\ominus}(Zn^{2+}/Zn) = -0.761\ 8V$$
$$\frac{1}{2}Zn^{2+} + e^- \rightleftharpoons \frac{1}{2}Zn \qquad \varphi^{\ominus}(Zn^{2+}/Zn) = -0.761\ 8V$$

(4) The data listed in the table are gained at 298.15K. There is a little difference for electrode potentials with changes of temperature. But these data can also be used as reference at other temperatures.

4. Standard electrode potential and the relative strengths of oxidizing and reducing agents

The electrode potentials which cause electrons to move in the external circuit between the half cells of a primary cell is a reflection of the relative oxidizing and reducing tendencies of the respective redox couples.

(1) The more positive the standard electrode potential value for an electrode is, the greater the tendency for the oxidized species of the redox couple to be reduced is, so it is the relative strong oxidizing agent. For example, in Table 8-1, MnO_4^- ion is the strongest oxidizing agent and Na^+ ion is the poorest oxidizing agent. But the electrode with the smallest standard electrode potential is the most easily reversed as an oxidation. Thus, Na has a great tendency to transfer electrons to other species. Na is the strongest reducing agent among the substances listed in Table 8-1. In short, the more positive the $\varphi^{\ominus}$ value for an oxidized substance is, the stronger it acts as an oxidizing agent; and the more negative the $\varphi^{\ominus}$ value for a reduced substance is, the stronger it acts as a reducing agent.

(2) In a redox couple, the stronger the oxidized species is, the weaker the corresponding reduced species is. Conversely, the stronger reduced species is, the weaker the corresponding oxidized species is. As an example at the standard conditions, comparing MnO_4^-/Mn^{2+} couple with $Cr_2O_7^{2-}/Cr^{3+}$ couple,

MnO_4^- is stronger than $Cr_2O_7^{2-}$ in their oxidizing ability, while Mn^{2+} is weaker than Cr^{3+} in their reducing ability. The inverse strengths relationship between oxidizing and reducing ability in an electric couple is a conjugate relationship also.

5. Types of electrodes

The electrode is a necessary component of a primary cell. There are four types commonly.

(1) Metal-metal ion electrode

It consists of a metal immersed in its soluble salt solution, such as Zn^{2+}/Zn electrode.

Electrode reaction: $Zn^{2+}(aq) + 2e^- \rightleftharpoons Zn(s)$

Electrode notation: $Zn \mid Zn^{2+}(c)$

(2) Gas electrode

It consists of the gas bubbled over the surface of the inert electrode, while it is bathed in a solution of ions related to the gas. For example, if the gas is chlorine Cl_2, the solution should contain chlorine ions Cl^-.

Electrode notation: $Pt \mid Cl_2(p) \mid Cl^-(c)$

Electrode reaction: $Cl_2(g) + 2e^- \rightleftharpoons 2Cl^-(aq)$

(3) Metal-insoluble salt-anion electrode

It consists of a metal M covered by a layer of insoluble salt MX, the whole immersed in a solution containing X^- ions. A common example is Ag-AgCl(s) electrode, which is silver Ag strip covered by silver chloride AgCl is immersed in Cl^- ions solution.

Electrode notation: $Ag \mid AgCl(s) \mid Cl^-(c)$

Electrode reaction: $AgCl(s) + e^- \rightleftharpoons Ag(s) + Cl^-(aq)$

(4) Redox electrode

It consists of inert electrode immersed in a solution containing a species with two oxidation numbers. For example, Fe^{3+}/Fe^{2+} electrode consists of Pt bathed in the solution of Fe^{2+} ions and Fe^{3+} ions.

Electrode notation: $Pt \mid Fe^{3+}(c_1), Fe^{2+}(c_2)$

Electrode reaction: $Fe^{3+} + e^- \rightleftharpoons Fe^{2+}$

Sample Problem 8-2 The reaction to produce chlorine Cl_2 with potassium permanganate $KMnO_4$ and concentrated hydrochloric acid HCl is as follows

$$2KMnO_4 + 16HCl \rightleftharpoons 2KCl + 2MnCl_2 + 5Cl_2 + 8H_2O$$

According to this reaction, write down the electrode reactions, cell reaction, electrode notation and cell notation, and judge the types of electrodes.

Solution Oxidation half-reaction occurs in the anode, and the reduction half-reaction occurs in the cathode.

Cathode: $MnO_4^- + 8H^+ + 5e^- \rightleftharpoons Mn^{2+} + 4H_2O$

Anode: $2Cl^- - 2e^- \rightleftharpoons Cl_2$

Cell reaction: $2MnO_4^- + 16H^+ + 10Cl^- \rightleftharpoons 2Mn^{2+} + 5Cl_2 + 8H_2O$

Electrode notation of the cathode: $Pt \mid MnO_4^-(c_1), Mn^{2+}(c_2), H^+(c_3)$ — Redox electrode

Electrode notation of the anode: $Pt \mid Cl_2(p) \mid Cl^-(c)$ — Gas electrode

Cell notation: $(-) Pt \mid Cl_2(p) \mid Cl^-(c) \parallel MnO_4^-(c_1), Mn^{2+}(c_2), H^+(c_3) \mid Pt (+)$

> **Question and thinking 8-2** Does a primary cell can be constructed using the same species in both the anode and cathode compartments as long as the concentrations are different? What is the electrode for this cell?

8.2 Electromotive Force and Gibbs Free Energy

A spontaneous electrochemical reaction generates a positive E. At constant temperature and pressure, and non-pV work is involved, the spontaneous direction of a redox reaction is given by Gibbs free energy, a negative $\Delta_r G_m$ is the thermodynamic signpost for a spontaneous reaction. Thus, there is a linkage between electromotive force and Gibbs free energy.

8.2.1 Electric Work and Gibbs Free Energy

The linkage between electromotive force and Gibbs free energy can be made quantitatively. Under the conditions of constant temperature and pressure, the change of Gibbs free energy of a reaction gives the maximum non-volumic work available from that process.

$$\Delta_r G_m = W_{f,\,max.} \tag{8.4}$$

The maximum non-volumic work is the electrical work for a redox reaction generally, so

$$\Delta_r G_m = W_{electrical\ energy,\ max.} \tag{8.5}$$

Moreover, because of

$$W_{electrical\ energy,\ max} = -nFE$$

the relationship between these two important parameters, $\Delta_r G_m$ and E, is

$$\Delta_r G_m = -nFE \tag{8.6a}$$

In Equation 8.6a, the factor n is the number of electrons transferred in the redox process, the constant F is called **Faraday's constant** which is $96\,485 \text{C} \cdot \text{mol}^{-1}$, E is the electromotive force of a cell and $\Delta_r G_m$ is the Gibbs free energy change for the cell reaction. When all substances involved in the reaction are in their standard states, Equation 8.6a can be modified to relate $E^\ominus$ and $\Delta_r G_m^\ominus$.

$$\Delta_r G_m^\ominus = -nFE^\ominus \tag{8.6b}$$

8.2.2 Electromotive Force and the Direction of Spontaneous Redox Reaction

As we all know, a negative $\Delta_r G_m$ is the thermodynamic signpost for a spontaneous reaction. In Equation 8.6, both n and F are positive numbers. Thus, a positive value of E or $E^\ominus$ leads to a negative value of $\Delta_r G_m$ or $\Delta_r G_m^\ominus$. In other word, both a positive value of E and a negative value of $\Delta_r G_m$ indicate a spontaneou reaction. For a common redox reaction, there is a redox reaction needed to decide its spontaneous direction.

$$Ox_1 + Red_2 \rightleftharpoons Red_1 + Ox_2$$

Design a primary cell according to the reaction. Supposing the reactants and products are all in solution, the cell notation is as follows.

$$(-)\ Pt\ |\ Ox_2(aq),\ Red_2(aq)\ ||\ Ox_1(aq),\ Red_1(aq)\ |\ Pt\ (+)$$

Then the electromotive force of the cell is

$$E = \varphi(Ox_1\ /\ Red_1) - \varphi(Ox_2\ /\ Red_2)$$

So, if

$E > 0$, $\Delta_r G_m < 0$, the cell reaction has a tendency to occur forwardly;

$E < 0$, $\Delta_r G_m > 0$, the cell reaction has a tendency to occur reversely;

$E = 0$, $\Delta_r G_m = 0$, the cell reaction is at the equilibrium state.

Under standard conditions, the sign of $E^\ominus$ or $\Delta_r G_m^\ominus$ indicates the direction of spontaneous cell reaction.

Sample Problem 8-3 For a redox reaction, $Cr_2O_7^{2-} + 6Fe^{2+} + 14H^+ \rightleftharpoons 2Cr^{3+} + 6Fe^{3+} + 7H_2O$, To calculate its $\Delta_r G_m^\ominus$ and to judge the spontaneous direction of the reaction under standard conditions by data listed in Table 8-1.

Solution Two half-reactions by redox reaction are as follows

Cathode: $Cr_2O_7^{2-} + 14H^+ + 6e^- \rightleftharpoons 2Cr^{3+} + 7H_2O$ $\varphi^\ominus = 1.36V$

Anode: $Fe^{3+} + e^- \rightleftharpoons Fe^{2+}$ $\varphi^\ominus = 0.771V$

$$E^\ominus = \varphi^\ominus (Cr_2O_7^{2-}/Cr^{3+}) - \varphi^\ominus (Fe^{3+}/Fe^{2+}) = 1.36V - 0.771V = 0.589V$$

For this reaction, $n = 6$, according to Equation 8.6b

$$\Delta_r G_m^\ominus = -nFE^\ominus = -6 \times 96\,485 C \cdot mol^{-1} \times 0.589V = -3.410 \times 10^5 J \cdot mol^{-1} = -341.0 kJ \cdot mol^{-1} < 0$$

So, this redox reaction has a spontaneous tendency forwardly.

Sample Problem 8-4 For a reaction, $Zn(s) + 2H^+(aq) \rightleftharpoons Zn^{2+}(aq) + H_2(g)$, $\Delta_r H_m^\ominus = -153.9 kJ \cdot mol^{-1}$; $\Delta_r S_m^\ominus = -23.0 J \cdot K^{-1} \cdot mol^{-1}$. Calculate the value of standard electrode potential for the electrode reaction, $Zn^{2+}(aq) + 2e^- \rightleftharpoons Zn(s)$ at 298.15K.

Solution $\Delta_r G_m^\ominus = \Delta_r H_m^\ominus - T\Delta_r S_m^\ominus$

$$= -153.9 kJ \cdot mol^{-1} - 298K \times (-23.0/1\,000) kJ \cdot K^{-1} \cdot mol^{-1} = -147.0 kJ \cdot mol^{-1}$$

For the reaction, $n = 2$; by $\Delta_r G_m^\ominus = -nFE^\ominus$

$$E^\ominus = -\Delta_r G_m^\ominus / (nF) = -(-147.0 \times 10^3 J \cdot mol^{-1})/(2 \times 96\,485 C \cdot mol^{-1}) = 0.761\,8V$$

$$E^\ominus = \varphi^\ominus (H^+/H_2) - \varphi^\ominus (Zn^{2+}/Zn) = 0V - \varphi^\ominus (Zn^{2+}/Zn) = 0.761\,8V$$

$$\varphi^\ominus (Zn^{2+}/Zn) = -E^\ominus = -0.761\,8V$$

8.2.3 Standard Electromotive Force and Equilibrium Constant

Equation 8.6b relates $\Delta_r G_m^\ominus$ to $E^\ominus$ for a redox reaction. Moreover, $\Delta_r G_m^\ominus$ is also related to the equilibrium constant $K^\ominus$ by the equation $\Delta_r G_m^\ominus = -RT\ln K^\ominus$. Then we can relate the $E^\ominus$ to the $K^\ominus$ for a redox reaction

$$RT\ln K^\ominus = nFE \quad \text{or} \quad \lg K^\ominus = \frac{nFE^\ominus}{2.303RT} \tag{8.7}$$

Thus, it is a simple to find the equilibrium constant at any temperatures. However, if the temperature is 298.15K, and $R = 8.314 J \cdot K^{-1} \cdot mol^{-1}$, $F = 96\,485 C \cdot mol^{-1}$ to Equation 8.7, we get

$$\lg K^\ominus = \frac{nE^\ominus}{0.059\,16V} \tag{8.8}$$

There are some rules about equilibrium constants of redox reactions.

(1) The equilibrium constant of a redox reaction depends on the nature of oxidizing agent and reducing agent, that is, it depends on the standard electromotive force of a cell. But it is independent of substance concentrations.

(2) $E^\ominus$ is an intensive property. However, multiplying an equation will change the value of n and

hence the value of $\Delta_r G_m^\ominus$. So $\Delta_r G_m^\ominus$ is an extensive property, $K^\ominus$ is also an extensive property.

(3) Commonly, when $n=2$ and $E^\ominus > 0.2V$ or $n=1$ and $E^\ominus > 0.4V$, the equilibrium constant has a very large value ($K^\ominus > 10^6$), which means the reaction proceeds completely.

Sample Problem 8-5 At 298.15K, calculate the equilibrium constant for the following redox reaction, $Zn + Cu^{2+} \rightleftharpoons Cu + Zn^{2+}$.

Solution Design a primary cell according to above reaction

Cathode: $Cu^{2+} + 2e^- \rightleftharpoons Cu$, $\varphi^\ominus(Cu^{2+}/Cu) = 0.341\,9V$

Anode: $Zn \rightleftharpoons Zn^{2+} + 2e^-$, $\varphi^\ominus(Zn^{2+}/Zn) = -0.761\,8V$

$$E^\ominus = \varphi^\ominus(Cu^{2+}/Cu) - \varphi^\ominus(Zn^{2+}/Zn) = 0.341\,9V - (-0.761\,8V) = 1.103\,7V$$

$$\lg K^\ominus = \frac{nE^\ominus}{0.059\,16V} = \frac{2 \times 1.103\,7V}{0.059\,16V} = 37.312\,4$$

$$K^\ominus = 2.053 \times 10^{37}$$

Standard electromotive force can be used to find the equilibrium constant of redox reactions as well as general equilibrium constants, such as dissociation constant of acid or base, K_a or K_b, ion-product constant of water K_w, and solubility-product constant K_{sp} and so on if the general reaction can be designed to be the primary cells.

Sample Problem 8-6 Use the following electrode reactions and standard electrode potentials to determine pK_{sp} of AgCl.

$$Ag^+ + e^- \rightleftharpoons Ag \qquad \varphi^\ominus = 0.799\,6V$$
$$AgCl + e^- \rightleftharpoons Ag + Cl^- \qquad \varphi^\ominus = 0.222\,3V$$

Solution If Ag^+/Ag couple is the cathode and $AgCl(s)/Ag$ couple is the anode. The cell reaction is

$$Ag^+ + Cl^- \rightleftharpoons AgCl(s) \qquad n=1$$

The cell reaction is the reverse process of the precipitation equilibrium of AgCl, so the equilibrium constant of the cell reaction equals the reciprocal of solubility product constant of AgCl, $K^\ominus = \frac{1}{K_{sp}}$. By

$$\lg K^\ominus = \frac{nE^\ominus}{0.059\,16V}$$

$$\lg K^\ominus = \frac{n\{\varphi^\ominus(Ag^+/Ag) - \varphi^\ominus(AgCl/Ag)\}}{0.059\,16V} = \frac{1 \times (0.799\,6V - 0.222\,3V)}{0.059\,16V} = 9.758\,3$$

$$pK_{sp} = -\lg K_{sp} = -\lg(1/K^\ominus) = \lg K^\ominus$$

$$pK_{sp} = 9.757\,8 \qquad K_{sp} = 1.747 \times 10^{-10}$$

Sample Problem 8-7 Use the following electrode reactions and standard electrode potentials to determine K_w.

$$O_2 + 4H^+ + 4e^- \rightleftharpoons 2H_2O \qquad \varphi^\ominus = 1.229V$$
$$O_2 + 2H_2O + 4e^- \rightleftharpoons 4OH^- \qquad \varphi^\ominus = 0.401V$$

Solution According to the electrode potential values, we know that O_2/H_2O couple is the cathode and O_2/OH^- couple is the anode. The cell reaction is

$$H^+ + OH^- \rightleftharpoons H_2O \qquad n=1$$

Obviously, the cell reaction is the reverse process of the dissociation equilibrium of H_2O, so $K^\ominus = \frac{1}{K_w}$.

$$\lg K^\ominus = \frac{nE^\ominus}{0.05916\text{V}} = \frac{n\{\varphi^\ominus(O_2/H_2O)-\varphi^\ominus(O_2/OH^-)\}}{0.05916\text{V}} = \frac{1\times(1.229-0.401)\text{V}}{0.05916\text{V}} = 13.996$$

$$\lg K_w = \lg(1/K^\ominus) = -\lg K^\ominus = -13.996$$

$$K_w = 1.01 \times 10^{-14}$$

8.3 Nernst Equation and Some Factors Affecting the Electrode Potential

The standard electrode potential specifies that all substances must be at standard conditions, that is, concentrations of dissolved solutes must be $1\text{mol}\cdot\text{L}^{-1}$ and gases must be at 100kPa. However, in practice it is unlikely that all of the substances making up an electrode will be at standard conditions, and the electrode potential will not be $\varphi^\ominus$. The electrode potentials under nonstandard conditions can be calculated by Nernst equation derived by Walther Nernst (1864—1941), a German chemist who established many of the theoretical fundations of electrochemistry.

8.3.1 The Nernst Equation of Electrode Potential

The dependence of the electromotive force on concentration can be obtained from the dependence of the Gibbs free energy change on concentration. Recall that the Gibbs free energy change ΔG is related to the standard Gibbs free energy change $\Delta_r G_m^\ominus$.

$$\Delta_r G_m = \Delta_r G_m^\ominus + RT\ln J \tag{8.9}$$

Substituting Equation 8.6a and Equation 8.6b into above equation gives

$$-nFE = -nFE^\ominus + RT\ln J$$

Solving this equation for E gives

$$E = E^\ominus - \frac{RT}{nF}\ln J \tag{8.10a}$$

Here, $E^\ominus$ is the standard electromotive force of a cell, n is the mole number of electrons transferred in the cell reaction and J is the reaction quotient. At $T = 298.15\text{K}$,

$$E = E^\ominus - \frac{0.05916\text{V}}{n}\lg J \tag{8.10b}$$

For a general redox reaction,

$$a\text{Ox}_1 + b\text{Red}_2 \rightleftharpoons d\text{Red}_1 + e\text{Ox}_2$$

The reaction quotient is written as

$$J = \frac{(c_{\text{Red}_1})^d (c_{\text{Ox}_2})^e}{(c_{\text{Ox}_1})^a (c_{\text{Red}_2})^b} \tag{8.11}$$

Then

$$E = E^\ominus - \frac{RT}{nF}\ln\frac{c^d(\text{Red}_1)c^e(\text{Ox}_2)}{c^a(\text{Ox}_1)c^b(\text{Red}_2)} \tag{8.12a}$$

or

$$E = E^\ominus - \frac{0.05916\text{V}}{n}\lg\frac{c^d(\text{Red}_1)c^e(\text{Ox}_2)}{c^a(\text{Ox}_1)c^b(\text{Red}_2)} \tag{8.12b}$$

Equation 8.12 is the **Nernst equation** of electromotive force of the cell.

Equation $E = \varphi(\text{Ox}_1/\text{Red}_1) - \varphi(\text{Ox}_2/\text{Red}_2)$ and equation $E = \varphi_+ - \varphi_-$ is equal for a redox reaction in fact. Similarly, $E^\ominus = \varphi_+^\ominus - \varphi_-^\ominus$.

To use the Nernst equation to half-reactions, we have

$$\varphi_+ - \varphi_- = (\varphi_+^\ominus - \varphi_-^\ominus) - \frac{RT}{nF} \ln \frac{c^d(\text{Red}_1)c^e(\text{Ox}_2)}{c^a(\text{Ox}_1)c^b(\text{Red}_2)}$$

$$= [\varphi_+^\ominus - \frac{RT}{nF} \ln \frac{c^d(\text{Red}_1)}{c^a(\text{Ox}_1)}] - [\varphi_-^\ominus - \frac{RT}{nF} \ln \frac{c^b(\text{Red}_2)}{c^e(\text{Ox}_2)}]$$

Then, obtain

$$\varphi_+ = \varphi_+^\ominus - \frac{RT}{nF} \ln \frac{c^d(\text{Red}_1)}{c^a(\text{Ox}_1)} = \varphi_+^\ominus + \frac{RT}{nF} \ln \frac{c^a(\text{Ox}_1)}{c^d(\text{Red}_1)}$$

$$\varphi_- = \varphi_-^\ominus - \frac{RT}{nF} \ln \frac{c^b(\text{Red}_2)}{c^e(\text{Ox}_2)} = \varphi_-^\ominus + \frac{RT}{nF} \ln \frac{c^e(\text{Ox}_2)}{c^b(\text{Red}_2)}$$

Thus, for a general half-reaction, written as a reduction: $p\text{Ox} + ne^- \rightleftharpoons q\text{Red}$
The Nernst equation is written as

$$\varphi(\text{Ox}/\text{Red}) = \varphi^\ominus(\text{Ox}/\text{Red}) + \frac{RT}{nF} \ln \frac{c^p(\text{Ox})}{c^q(\text{Red})} \tag{8.13a}$$

At 298.15K, substitute corresponding constants to Equation 8.13a, then

$$\varphi(\text{Ox}/\text{Red}) = \varphi^\ominus(\text{Ox}/\text{Red}) + \frac{0.05916}{n} \ln \frac{c^p(\text{Ox})}{c^q(\text{Red})} \tag{8.13b}$$

Equation 8.13 is the Nernst equation of electrode potential, which is one of the most important equations in electrochemistry. In Equation 8.13, n is the mole numbers of electrons transferred in a balanced half-reaction, c_{Ox} and c_{Red} are the concentrations of oxidized species and reduced species of a electric couple respectively; pure liquid and solid substances is no term; gaseous substance is expressed as its partial pressure; the exponents p and q are the coefficients of oxidized species and reduced species in half-reaction respectively; the other relative species in the half-reaction should be considered such as H^+ and OH^- commonly.

From the Equation 8.13a and Equation 8.13b, we can see the following conclusions:

(1)The electrode potential not only depends on the nature of electrode, but also on the temperature and the concentrations or pressures of oxidizing agents, reducing agents and other relative species.

(2) At a certain temperature, if the concentrations of species in half-reaction change, the electrode potential changes consequently.

(3) In general, the electrode potential value depends mainly on the standard electrode potential. And concentrations do not have much effect on it unless the concentration change of species is very large or very small, or the coefficients of species for the half-reaction are very large.

8.3.2 Some Factors Affecting the Electrode Potential

1. Solution acidity

For some half-reactions involving H^+ ion or OH^- ion, the changes of pH of solution will lead to the changes of electrode potential.

Sample Problem 8-8 For a half reaction

$$Cr_2O_7^{2-} + 14H^+ + 6e^- \rightleftharpoons 2Cr^{3+} + 7H_2O \qquad \varphi^{\ominus} = 1.36V$$

Calculate its electrode potential at 298.15K, $c(Cr_2O_7^{2-}) = c(Cr^{3+}) = 1 mol \cdot L^{-1}$, and pH = 6

Solution According to Equation 8.13b

$$\varphi(Cr_2O_7^{2-}/Cr^{3+}) = \varphi^{\ominus}(Cr_2O_7^{2-}/Cr^{3+}) + \frac{0.05916V}{n} lg \frac{c(Cr_2O_7^{2-})c^{14}(H^+)}{c^2(Cr^{3+})}$$

If pH = 6, $c(H^+) = 1 \times 10^{-6} mol \cdot L^{-1}$

$$\varphi(Cr_2O_7^{2-}/Cr^{3+}) = 1.36V + \frac{0.05916V}{6} lg \frac{1 \times (10^{-6})^{14}}{1^2} = 0.532V$$

2. Formation of precipitation

For some half-reactions, if there are some substances that can precipitate oxidized species or reduced species in the system, the formation of precipitation will lead to a marked change in the concentration of oxidized species or reduced species and a consequent change in their electrode potentials.

Sample Problem 8-9 At 298.15K, for a half reaction $Ag^+ + e^- \rightleftharpoons Ag$, $\varphi^{\ominus} = 0.7996V$, calculate its electrode potential when NaCl is added to the electrode solution and the concentration of Cl^- ion is maintained at $1 mol \cdot L^{-1}$. $K_{sp}(AgCl) = 1.77 \times 10^{-10}$.

Solution According to Equation 8.13b,

$$\varphi(Ag^+/Ag) = \varphi^{\ominus}(Ag^+/Ag) + \frac{0.05916V}{n} lg \frac{c(Ag^+)}{1}$$

$$Ag^+ + Cl^- \rightleftharpoons AgCl \qquad [Ag^+][Cl^-] = K_{sp} = 1.77 \times 10^{-10}$$

$$[Ag^+] = K_{sp} / [Cl^-] = 1.77 \times 10^{-10} mol \cdot L^{-1}$$

$$\varphi(Ag^+/Ag) = 0.7996V + 0.05916V \times lg \frac{1.77 \times 10^{-10}}{1} = 0.7996V - 0.577V = 0.223V$$

In fact, when Cl^- ion is added to the electrode solution containing Ag^+ ion, the redox electric couple Ag^+/Ag has been substituted for a new redox electric couple $AgCl/Ag$. The new electrode is a metal-insoluble salt-ion electrode and its electrode reaction is $AgCl + e^- \rightleftharpoons Ag + Cl^-$.

3. Formation of weak acid (base)

Just as formation of precipitation affects electrode potentials, formation of weak acid or weak base will also lead to a marked change in the concentration of oxidized species or reduced species and a consequent change in their electrode potentials.

Sample Problem 8-10 For the cell

$$(-) Sn(s) | Sn^{2+}(1 mol \cdot L^{-1}) \| H^+(1 mol \cdot L^{-1}) | H_2(100kPa) | Pt(s) (+)$$

(a) At the standard conditions, does the reaction $2H^+(aq) + Sn(s) \rightleftharpoons H_2(g) + Sn^{2+}(aq)$ proceed spontaneously?

(b) Judge whether the spontaneous direction will invert when NaAc is added to H^+ ion solution of hydrogen electrode and the concentration of Ac^- ion is maintained at $1 mol \cdot L^{-1}$ and partial pressure of H_2 is 100kPa at equilibrium state.

Solution (a) Assume that the reaction proceed is spontaneously, then

Cathode: $2H^+(aq) + 2e^- \rightleftharpoons H_2(g)$ (reduction) $\varphi^{\ominus}(H^+/H_2) = 0V$
Anode: $Sn^{2+}(aq) + 2e^- \rightleftharpoons Sn(s)$ (oxidization) $\varphi^{\ominus}(Sn^{2+}/Sn) = -0.1375V$
Cell reaction: $2H^+(aq) + Sn(s) \rightleftharpoons H_2(g) + Sn^{2+}(aq)$

$$E^{\ominus} = \varphi_+^{\ominus} - \varphi_-^{\ominus} = \varphi^{\ominus}(H^+/H_2) - \varphi^{\ominus}(Sn^{2+}/Sn) = 0V - (-0.1375V) = 0.1375V > 0$$

The reaction proceed is spontaneously.

(b) There is the following equilibrium in the solution of hydrogen electrode after NaAc is added

$$HAc \rightleftharpoons H^+ + Ac^-$$

$$[H^+] = K_a \frac{[HAc]}{[Ac^-]}$$

$$[HAc] = [Ac^-] = 1 \text{mol} \cdot L^{-1}, \quad K_a = 1.75 \times 10^{-5}$$

$$[H^+] = 1.75 \times 10^{-5} \text{mol} \cdot L^{-1}$$

$$\varphi(H^+/H_2) = \varphi^{\ominus}(H^+/H_2) + \frac{0.05916V}{n} \lg \frac{c^2(H^+)}{p(H_2)/p^{\ominus}}$$

$$= 0V + \frac{0.05916V}{2} \lg \frac{(1.75 \times 10^{-5})^2}{100\text{kPa}/100\text{kPa}} = -0.281V$$

$$E = \varphi_+ - \varphi_- = \varphi(H^+/H_2) - \varphi^{\ominus}(Sn^{2+}/Sn) = -0.281V - (-0.1375V) = -0.1435V < 0$$

The spontaneous direction of the reaction can be inverted because of the decrease of H^+ ion concentration. The cell reaction has a tendency to occur in the opposite direction.

Of course, the electromotive force of the cell can also be calculated according to Equation 8.12b.

$$E = \varphi(H^+/H_2) - \varphi(Sn^{2+}/Sn)$$

$$= [\varphi^{\ominus}(H^+/H_2) - \varphi^{\ominus}(Sn^{2+}/Sn)] - \frac{0.05916V}{n} \lg \frac{c(Sn^{2+})(p(H_2)/p^{\ominus})}{c^2(H^+)}$$

$$= [0V - (-0.1375V)] - \frac{0.05916V}{2} \lg \frac{(100\text{kPa}/100\text{kPa})}{(1.76 \times 10^{-5})^2} = -0.1435V$$

Sometimes, the concentration changes may lead to the inversion of spontaneous direction of a redox reaction. But generally, the electrode potentials depend mainly on the standard electrode potentials but $0.05916 \lg J/n$ does not have much effect on them. Therefore, for a redox reaction under the nonstandard conditions, if $E^{\ominus} > +0.3V$, the cell reaction has a tendency to proceed forwardly; if $E^{\ominus} < -0.3V$, the cell reaction has a tendency to proceed reversely. Under such two conditions, it is difficult to make direction of a reaction inverted by changing concentrations. However, if $-0.3V < E^{\ominus} < +0.3V$, the concentration changes may lead to the inversion of direction of a redox reaction.

Like ΔG, electrode potentials or electromotive force generated by a cell can predict whether a given reaction is feasible and occurs spontaneously or not, but cannot answer the rate or path or mechanism of a redox reaction.

8.4 Measurement of the pH of A Solution by Potentiometry

Potentiometry is an electro-analytical method which is based on the measurement of electromotive force of a cell. Since the electromotive forces of primary cells depend on the species concentrations, measurements of electromotive forces of cells are of considerable importance in analytical chemistry. In many cases, a cell can be devised electromotive force of which depends on concentration of a single ionic species in the solution.

8.4.1 Reference Electrode and Indicator Electrode

Potentiometric measurement system consists of two electrodes. One of the electrodes is known in its electrode potential, and that will remains constant during the measurement. It is called the **reference electrode**. The other electrode must be such that its electrode potential depends on the concentration of the ion being determined. It is called the **indicator electrode**. The electrode combined with a reference electrode and a suitable indicator electrode is called **combination electrode**. For example,

$$(-) M(s) | M^{n+}(c) \| \text{reference electrode} (+)$$

The electromotive force of the cell is given by

$$E = \varphi_{\text{reference}} - \varphi(M^{n+}/M)$$
$$= \varphi_{\text{reference}} - \varphi^{\ominus}(M^{n+}/M) - \frac{RT}{nF}\ln[M^{n+}]$$

Where n, F and R are constants, and $\varphi_{\text{reference}}$ and $\varphi^{\ominus}(M^{n+}/M)$ are also constants at the certain temperature. So we can obtain the concentration of M^{n+} ion when the electromotive force of the cell is measured. This is the principle of potentiometry in the measurement of content of substances.

The standard hydrogen electrode is the most accurate reference electrode because the electrode potentials of most electrodes have themselves been measured against it. But the standard hydrogen electrode is inconvenient for routine, practical measurements in the laboratory. It requires a tank of compressed gas, which is heavy and awkward. Explosive mixtures of H_2 and air may be formed, and catalytic Pt surface is easily poisoned, that is contaminated with absorbed substances that inhibit catalytic activity. Thus, the commonest reference electrodes are the calomel and the silver-silver chloride Ag-AgCl electrodes in the laboratory.

1. The calomel electrode

The **calomel electrode** is a metal-insoluble salt-ion electrode. It consists of two concentric glass tubes, an inner tube and an outer tube. Within the inner tube is a paste-like material known as calomel. Calomel is made by thoroughly mixing metal mercury Hg with mercurous chloride Hg_2Cl_2, a white solid. The outer tube has a porous fibre plug in the tip which acts as the salt bridge to the analyte solution. Potassium chloride KCl solution is in the outer tube. Most commonly, the solution is saturated with KCl crystals. The electrode is then called the **saturated calomel electrode** or SCE, and its single electrode potential is 0.241 2V at 298.15K. A typical SCE available commercially is shown in Figure 8-5.

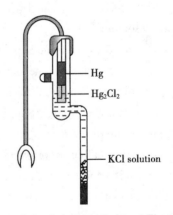

Figure 8-5 Saturated Calomel Electrode

Electrode notation $Pt | Hg_2Cl_2(s) | Hg (l) | Cl^-(c)$
Electrode reaction $Hg_2Cl_2(s) + 2e^- \rightleftharpoons 2Hg(l) + 2Cl^-$

The Nernst equation for this reaction $\varphi = \varphi^{\ominus} - \frac{RT}{2F}\ln c^2(Cl^-)$

At 298.15K, $\varphi = 0.268\,08 - 0.059\,16V \times \lg c(Cl^-)$

The calomel electrode is the most common reference electrode due to its stability, ease of manufacture and convenience of use. It may be fabricated in a variety of sizes and shapes. But it is sensitive to temperature and should not be used above 50 ℃ because Hg_2Cl_2 breaks down, yielding unstable reading.

2. The silver-silver chloride electrode

Ag-AgCl(s) electrode is similar to the SCE in that it is a metal-insoluble salt-ion electrode, is enclosed in glass tube and has a porous fibre tip in contact with the external solution. However, it is different internally. There is only one glass tube and a solution KCl is inside. A Ag wire is coated with a retentive layer of AgCl in this solution from the external lead.

Electrode notation $Ag \mid AgCl(s) \mid Cl^-(c)$

Electrode reaction $AgCl(s) + e^- \rightleftharpoons Ag + Cl^-$

The Nernst equation for this reaction, $\varphi = \varphi^\ominus - \dfrac{RT}{F} \ln c(Cl^-)$

At 298.15K, $\varphi_{AgCl/Ag} = 0.22233\text{V} - 0.05916\text{V} \times \lg c(Cl^-)$

At 298.15K, if the KCl solution are saturated, 1mol·L^{-1} and 0.1mol·L^{-1}, the electrode potentials are 0.1971V, 0.2223V and 0.288V respectively. Ag-AgCl(s) electrode has a superior temperature range, actually usable even above 80 ℃.

The electrode potential of the indicator electrode in the potentiometric measurement is related to the concentration of analyte. For example, pH measurement of solutions is very important in many aspects of chemical analysis. It can be made potentiometrically. Practical pH measurement is glass electrode as its indicator electrode.

A **glass electrode** is shown in Figure 8-6. It consists of a closed-end glass tube that has a very thin glass membrane (about 0.1mm) at the tip. Inside the tube is an internal reference electrode, usually Ag-AgCl(s). It is typically a silver wire coated with silver chloride, the whole dipped into a HCl solution.

When the glass electrode is immersed in the solution whose pH is to be determined, a potential develops due to the fact that the chemical composition inside is different from the chemical composition outside. Specifically, it is the difference in the concentration of the H$^+$ ions on the opposite side of the membrane that causes the potential to develop. The potential is called **membrane potential**. Since the internal H$^+$ concentration is a constant, the membrane potential is directly proportional to the pH of the solution into which it is dipped.

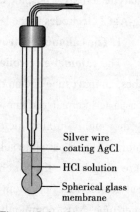

Figure 8-6   Glass Electrode

Glass electrodes are simple to use and maintain. They respond sensitively to H$^+$ ion concentration and provide an accurate measurement of pH values. It is found experimentally that the electrode potential of the glass electrode follows the relation.

$$\varphi_{glass} = K_{glass} + \dfrac{RT}{F} \ln a(H^+) = K_{glass} - \dfrac{2.303RT}{F} pH$$

In the equation, K_{glass} is a constant theoretically, but two different glass electrodes are not likely to give exactly the same K_{glass} because their characteristics change somewhat with time and exposure to solution. So they must be calibrated before each use with standard buffer solutions.

8.4.2 Potentiometric Determination of pH

In pH measurement, a glass electrode connected to a saturated calomel reference electrode are usually employed

$$(-) \text{ glass electrode } | \text{ test solution } | \text{ SCE } (+)$$

The equation relating the electromotive force of this cell to pH is

$$E = \varphi_{SCE} - \varphi_{glass} = \varphi_{SCE} - (K_{glass} - \frac{2.303RT}{F}\text{pH})$$

At the certain temperature, $(\varphi_{SCE} - K_{glass})$ is a constant, K_E, it becomes

$$E = K_E + \frac{2.303RT}{F}\text{pH} \tag{8.14}$$

The Equation 8.14 K_E and pH are unknown hence it cannot be used to evaluate both quantities. It is necessary to assign arbitrarily a pH value to some standard buffer solutions in order to fix a practical scale of pH. The electrodes are dipped into a standard buffer solution of known pH_s, and measure its electromotive force E_s

$$E_s = K_E + \frac{2.303RT}{F}\text{pH}_s \tag{8.15}$$

Combine Equation 8.14 with Equation 8.15, giving

$$\text{pH} = \text{pH}_s + \frac{(E - E_s)F}{2.303RT} \tag{8.16}$$

Where E is the electromotive force of a cell containing the unknown solution and E_s is the electromotive force of a cell containing a standard buffer solution of known or defined pH, that is pH_s. Equation 8.16 is called the **operational definition of pH**.

The pH meters are based on this principle to determine the pH of a solution. However, it is not necessary to calculate the pH of unknown solution according to Equation 8.16 in practice. We just need to dip the combination electrode into a standard buffer solution, await the meter reading electronically adjusted until it reads the correct value, and then dip them into the sample solution and read the pH appearing on a digital display. For precision work, two standard buffer solutions of different pH values are used, and a double calibration is made so that the meter registers both standard values correctly.

8.5 Electrochemistry and Biosensors

A **biosensor** is defined as an analytical device composed of a biological recognition element directly interfaced to a signal transducer which together relates the concentration of an analyte or group of related analytes to a measurable response. It consists of three parts, the **biological recognition element**, the **transducer** or the detector element and the **signal processors**. Typical recognition elements used in biosensors are biological materials (e.g. enzymes, antibodies, nucleic acids, tissue, microorganisms, organelles, cell receptors, natural products etc) or biologically derived materials (e.g. recombinant antibodies, engineered proteins etc) or biomimics (e.g. synthetic catalysts, combinatorial ligands, imprinted polymers etc). The transducer works in a physicochemical way, which may be optical, electrochemical, thermometric, piezoelectric, magnetic or micromechanical, and transforms the signal resulting from the interaction of the analyte with the biological element into another signal that can be more easily measured and quantified. For example, an electrochemical biosensor usually contains three electrodes: a reference electrode, an active electrode and a sink electrode. The target analyte is involved in the reaction that takes place on the active electrode surface, and the ions produced create a potential which is subtracted from that of the reference electrode to give a signal. Then we either measure the current (rate of flow of electrons is now proportional to the analyte concentration) at a fixed potential or the potential can be measured at zero current (this give a logarithmic response). Last part associated with electronics are primarily responsible for the display of the results in a user friendly way.

Generally, biosensors can be mainly classified as two types, **biocatalysis-based biosensors** and **bioaffinity-based biosensors**. In addition, there are other types of classifications such as **microorganism-based biosensors**.

The biocatalysis-based biosensor mainly adopts enzymes as the biological recognition element, catalyzing a signaling biochemical transformation. In the organism, enzymes serve as the extremely specific tools for digesting food substances. This specificity is based on the fact that enzymes can recognize differences of a signal atom in the substance's chemical structure with their exquisite three dimensional structures. The specificity of **enzyme-based biosensor** is based on this specificity of enzymatic action. For example, if an enzyme that is specific for a reaction with glucose is incorporated in biosensors, the sensor has the basis only for glucose and glucose alone. The sensor will not detect a material even if it is related to glucose very closely chemically. The signal should be shown only from the change in the transducer prompted by the interaction of glucose and the glucose-relative enzyme.

The bioaffinity-based biosensor adopts specific binding proteins, nucleic acid, whole cell, cell membrane receptors, antibodies, or antibody-related substances for biomolecular recognition. It can be detect the binding event itself. For example, **immunosensor**, which is one of the bioaffinity-based biosensors, adopts antibodies or antibody-related substances as the biological element. Most antibodies are proteins, which are produced by the immune system of some animals in response to the entry of "foreigner" (the antigen) into the body. Usually, antibodies undergo a physical transformation, in which they tightly bind to the antigen that prompted the response and make it by other elements of the immune system. Moreover, antibodies are very specific which means they recognize and bind to the "foreigner" only. Therefore, antibodies specifically directed against the desired analyte are immobilized on the transducer of the biosensor. Then, the sensor is exposed to the medium of interest (e.g. the blood or other biological fluid). If the specific antigen is present in that medium, it will be bound by the immobilized antibody to form an antigen-antibody complex. This binding event will change some physical properties of the monitored environment at the transducer surface of the sensor. As the result, the change of properties will be signaled by the transducer subsequently.

The first biosensors were reported in the early 1960s. Nowadays, biosensors have been applied to a wide variety of analytical problems in medicine, drug discovery, the environment, food, process industries, security and defence. In all applications, the most important features are specificity, sensitivity and the time required to attain the signal in response to the recognition event. Also, they must be portable, easily to operate and stable. The emerging field of bioelectronics seeks to exploit biology in conjunction with electronics widely and deeply. A key aspect is the interface between biological materials and electronics.

Key Terms

氧化还原反应	oxidation-reduction reaction, redox reaction
电化学	electrochemistry
氧化值，氧化数	oxidation number
氧化反应	oxidation reaction
还原反应	reduction reaction
还原剂	reducing agent, reductant
氧化剂	oxidizing agent, oxidant
氧化还原半反应	redox half-reaction
氧化半反应	oxidation half-reaction
还原半反应	reduction half-reaction
氧化型物质	oxidized species
还原型物质	reduced species
氧化还原电对	redox electric couple
离子-电子法	ion-electron method
盐桥	salt bridge
原电池	primary cell
伏打电池	voltaic cell
丹尼尔电池	Daniel cell
半电池	half-cell
电极	electrode
负极	anode
正极	cathode
电池反应	cell reaction
惰性电极	inert electrode
电池符号	cell notation
双电层理论	electric double layer theory
电极电位	electrode potential
标准氢电极	standard hydrogen electrode, SHE
电动势	electromotive force
标准电极电位	standard electrode potential
法拉第常数	Faraday constant
Nernst 方程式	Nernst equation
电位法	potentiometry
参比电极	reference electrode
指示电极	indicator electrode
甘汞电极	calomel electrode
饱和甘汞电极	saturated calomel electrode, SCE
玻璃电极	glass electrode

膜电位	membrane potential
pH 操作定义	operational definition of pH
生物传感器	biosensor
生物感测元件	biological recognition element
传感器	transducer
信号处理器	signal processor
生物触媒传感器	biocatalysis-based biosensor
生物亲和性传感器	bioaffinity-based biosensor
微生物传感器	microorganism-based biosensor
酶传感器	enzyme-based biosensor
免疫传感器	immunosensor

Summary

Oxidation-reduction reaction or redox reaction is one of the most important chemical reactions, which is characterized by changes in the oxidation numbers of some elements. The species that loses electrons in a redox process is called a reducing agent, and the species that gains electrons is called an oxidizing agent. A redox reaction can be divided into two half-reactions. One of half-reactions describes the oxidation, another describes the reduction. Each half-reaction (Ox. + ne^- ⇌ Red.) involves a redox electric couple, Ox / Red. A redox reaction must be described by a balanced equation by the half-reaction method or ion-electron method.

A spontaneous redox reaction can be designed by a device called a primary cell to generate electrical energy through an external circuit. In a primary cell, there are two electrodes, in which the oxidation half-reaction occurs at the anode or negative electrode and the reduction one at the cathode or positive electrode. The overall redox reaction in a primary cell is called a cell reaction.

The absolute value of electrode potential cannot be measured, but its relative value can be determined by comparing with the standard hydrogen electrode SHE, $\varphi^{\ominus}(H^+ / H_2)$ is 0V by the definition of IUPAC. When an electrode is operated under the standard conditions ($T = 298.15K$, $c = 1 mol \cdot L^{-1}$, $p = 100 kPa$), the measured electrode potential is called the standard electrode potential $\varphi^{\ominus}$. The electrode potential is a reflection of the relative oxidizing and reducing tendencies of the respective redox couple. The larger the standard electrode potential is, the stronger the oxidizing agent is; the lower the standard electrode potential is, the stronger the reducing agent is.

The electrode potential at nonstandard conditions is given by the Nernst equation

$$\varphi(Ox/Red) = \varphi^{\ominus}(Ox/Red) + \frac{RT}{nF} \ln \frac{c^p(Ox)}{c^q(Red)}$$

At 298.15K,
$$\varphi(Ox/Red) = \varphi^{\ominus}(Ox/Red) + \frac{0.05916V}{n} \ln \frac{c^p(Ox)}{c^q(Red)}$$

The potential difference between two electrodes is called electromotive force E of a primary cell.

$$E = \varphi_+ - \varphi_-$$

There is a linkage between electromotive force and Gibbs free energy

$$\Delta_r G_m = -nFE \quad \text{or} \quad \Delta_r G_m^{\ominus} = -nFE^{\ominus}$$

A negative $\Delta_r G_m$ is the thermodynamic signpost for a spontaneous reaction, so, if E or $E^\ominus > 0$, shows a spontaneous process forwardly; if E or $E^\ominus < 0$, shows a spontaneous process reversely. Moreover, the standard electromotive force is related to the equilibrium constant, $\lg K^\ominus = \dfrac{nE^\ominus}{0.059\,16\text{V}}$, many equilibrium constants such as K_a, K_b, K_w, K_{sp} and K_s can be determined if the overall reaction can be designed by two redox half-reactions.

Potentiometry is an electroanalytical method based on Nernst equation and the measurement of electromotive force of a thermodynamic reversible cell. Potentiometric measurement system consists of a reference electrode and an indicator electrode. The saturated calomel electrode (SCE) and AgCl/Ag electrode are used as reference electrodes commonly, and glass electrode is an indicator electrode of H^+ ions concentration in solution. Thus, the pH of a solution can be measured. The operational definition of the pH by IUPAC is given as the following equation

$$\text{pH} = \text{pH}_s + \frac{(E - E_s)F}{2.303RT}$$

Exercises

1. Determine the oxidation numbers of the following underlined elements: $K_2\underline{Cr}O_4$, $Na_2\underline{S}_2O_3$, $Na_2\underline{S}O_3$, $\underline{Cl}O_2$, $\underline{N}_2O_5$, $Na\underline{H}$, $K_2\underline{O}_2$, $K_2\underline{Mn}O_4$.

2. Balance the following redox reactions.
   (1) $MnO_4^-(aq) + H_2O_2(aq) + H^+(aq) \rightarrow Mn^{2+}(aq) + O_2(g) + H_2O(l)$
   (2) $Cr_2O_7^{2-}(aq) + SO_3^{2-}(aq) + H^+(aq) \rightarrow Cr^{3+}(aq) + SO_4^{2-}(aq) + H_2O(l)$
   (3) $As_2S_3(s) + ClO_3^-(aq) + H_2O(l) \rightarrow Cl^-(aq) + H_3AsO_4(sln) + SO_4^{2-}(aq) + H^+(aq)$

3. Explain why chlorine Cl_2 and hydrogen peroxide H_2O_2 can be used as sanitizers? $\varphi^\ominus(Cl_2/Cl^-) = 1.358\text{V}$, $\varphi^\ominus(H_2O_2/H_2O) = 1.776\text{V}$

4. Answer the following questions by their standard electrode potentials.
   (1) Arrange the order of increasing oxidizing ability: $Cr_2O_7^{2-}$, MnO_4^-, MnO_2, Cl_2, Fe^{3+}, Zn^{2+}
   (2) Arrange order of increasing reducing ability: Cr^{3+}, Fe^{2+}, Cl^-, Li, H_2.

5. Write the primary cell notations and judge the spontaneous direction for the following redox reactions under the standard conditions.
   (1) $Zn(s) + Ag^+(aq) \rightleftharpoons Zn^{2+}(aq) + Ag(s)$
   (2) $Cr^{3+}(aq) + Cl_2(g) \rightleftharpoons Cr_2O_7^{2-} + Cl^-(aq)$
   (3) $Fe^{3+}(aq) + I_2(s) \rightleftharpoons IO_3^-(aq) + Fe^{2+}(aq)$

6. Find the proper substances to need the requirement of following conditions by standard electrode potentials.
   (1) Substance which can reduce Co^{2+} to Co, but cannot reduce Zn^{2+} to Zn.
   (2) Substance which can oxidize Br^- to Br_2, but cannot oxidize I^- to I_2.

7. Use the following half-reactions and standard electrode potentials to determine whether or not hydrogen peroxide H_2O_2 will decompose to H_2O and O_2 spontaneously under standard conditions.
   $H_2O_2(aq) + 2H^+(aq) + 2e^- \rightleftharpoons 2H_2O(l)$ $\varphi^\ominus = 1.776\text{V}$
   $O_2(g) + 2H^+(aq) + 2e^- \rightleftharpoons H_2O_2(aq)$ $\varphi^\ominus = 0.695\text{V}$

Chapter 8 Oxidation-Reduction Reaction and Electrode Potential

8. Calculate the electrode potentials for each of the half-reactions by Nernst equation.

(1) $2H^+(0.10 \text{mol} \cdot L^{-1}) + 2e^- \rightleftharpoons H_2(200\text{kPa})$

(2) $Cr_2O_7^{2-}(1.0\text{mol} \cdot L^{-1}) + 14H^+(0.001\ 0\text{mol} \cdot L^{-1}) + 6e^- \rightleftharpoons 2Cr^{3+}(1.0\text{mol} \cdot L^{-1}) + 7H_2O$

(3) $Br_2(l) + 2e^- \rightleftharpoons 2Br^-(0.20\text{mol} \cdot L^{-1})$

9. In a solution, assuming that the concentration of MnO_4^- ion is equal to that of Mn^{2+} ion and other ions are all at the standard conditions, state whether MnO_4^- ion can oxidize I^- ion and Br^- ion under the following conditions: (1) pH = 0.0, (2) pH = 5.5.

10. ClO_2 is a sterilized agent to purify water. Answer the following question.

(1) For the formation reaction of ClO_2: $2NaClO_2(aq) + Cl_2(g) \rightleftharpoons 2ClO_2(g) + 2NaCl(aq)$, calculate the $E^\ominus$, $\Delta_r G_m^\ominus$ and $K^\ominus$.

$$ClO_2(g) + e^- \rightleftharpoons ClO_2^-(aq) \quad \varphi^\ominus = 0.954V$$
$$Cl_2(g) + 2e^- \rightleftharpoons 2Cl^-(aq) \quad \varphi^\ominus = 1.358V$$

(2) Balance the reaction: $ClO_2(g) \rightarrow ClO_3^-(aq) + Cl^-(aq)$

11. Under the standard conditions, when cobalt Co dissolves in $1.0\text{mol} \cdot L^{-1}$ HNO_3 solution, determine whether it will be oxidized to Co^{3+} ion or Co^{2+} ion? If the concentration of HNO_3 in solution is changed, will the above conclusion be changed, too? $\varphi^\ominus (Co^{3+}/Co) = 1.26V$, $\varphi^\ominus (Co^{2+}/Co) = -0.28V$, $\varphi^\ominus (NO_3^-/NO) = 0.96V$.

12. At 298.15K, the measured electromotive force for the following primary cell is 0.420V and the electrode potential of SCE is 0.241 2V. Calculate the pH of gastric juice.

$$(-)\ Pt(s)\ |\ H_2(100\text{KPa})\ |\ \text{gastric juice}\ |\ SCE\ (+)$$

13. For the following oxidization species: Hg_2^{2+}, $Cr_2O_7^{2-}$, MnO_4^-, Cl_2, Cu^{2+}, H_2O_2, in acidic solution, with the increase of pH, which ions will strengthen, weaken and maintain their oxidizing abilities, respectively?

14. Given the concentration cell

$$Cu(s)\ |\ Cu^{2+}(1.0 \times 10^{-4}\text{mol} \cdot L^{-1})\ ||\ Cu^{2+}(1.0 \times 10^{-1}\text{mol} \cdot L^{-1})\ |\ Cu(s)$$

Calculate E and state the cathode and the anode at 298.15K.

15. Given the primary cell

$$(-)\ Zn(s)\ |\ Zn^{2+}(x\ \text{mol} \cdot L^{-1})\ ||\ Cd^{2+}(0.20\text{mol} \cdot L^{-1})\ |\ Cd(s)\ (+)$$

For which $E = 0.388\ 4V$. What is the concentration of Zn^{2+} ion at 298.15K?

16. Use the following half-reactions and standard electrode potentials to determine the solubility product constant K_{sp} for Hg_2SO_4 at 298.15K.

$$Hg_2SO_4(s) + 2e^- \rightleftharpoons 2Hg(l) + SO_4^{2-}(aq) \quad \varphi^\ominus = 0.612\ 5V$$
$$Hg_2^{2+}(aq) + 2e^- \rightleftharpoons 2Hg(l) \quad \varphi^\ominus = 0.797\ 3V$$

17. At 298.15K, for the half-reaction

$$Hg_2Cl_2(s) + 2e^- \rightleftharpoons 2Hg(l) + 2Cl^-(aq) \quad \varphi^\ominus = 0.268V$$

Calculate the concentration of KCl solution when the electrode potential is 0.327V.

18. At 298.15K, the measured E is 0.350V for a primary cell consisting of glass electrode as the anode and SCE as the cathode, and a pH = 6.0 standard buffer solution. If above electrodes and a weak acid HA solution ($c = 0.01\text{mol} \cdot L^{-1}$) consist the primary cell, the measured E is 0.231V. What are the pH and dissociation constant K_a of HA solution?

Supplementary Exercises

1. What is the value of the equilibrium constant at 25℃ for the reaction (refer to the table of standard electrode potentials): $I_2(s) + 2Br^-(aq) \rightleftharpoons 2I^-(aq) + Br_2(l)$?

2. What is $\Delta_r G_m^\ominus$ and $E^\ominus$ at 25℃ of a redox reaction for which $n=1$ and equilibrium constant $K^\ominus = 5 \times 10^3$?

3. Balance the following aqueous skeleton reactions and identify the oxidizing and reducing agents:
   (1) $Fe(OH)_2(s) + MnO_4^-(aq) \rightarrow MnO_2(s) + Fe(OH)_3(s)$ (basic solution)
   (2) $Zn(s) + NO_3^-(aq) \rightarrow Zn^{2+}(aq) + N_2(g)$ (acidic solution)

4. Write the cell notation for the voltaic cells that incorporate each of the following redox reactions:
   (1) $Al(s) + Cr^{3+}(aq) \rightarrow Cr(s) + Al^{3+}(aq)$
   (2) $Cu^{2+}(aq) + SO_2(g) + 2H_2O(l) \rightarrow Cu(s) + SO_4^{2-}(aq) + 4H^+(aq)$

5. A primary cell consists of the SHE as an anode and a Cu^{2+}/Cu electrode. Calculate $[Cu^{2+}]$ when $E_{cell} = 0.25V$.

6. A primary cell consists of Ni^{2+}/Ni and Co^{2+}/Co half cells with the following initial concentrations: $c(Ni^{2+}) = 0.8 mol \cdot L^{-1}$; $c(Co^{2+}) = 0.2 mol \cdot L^{-1}$ (If the volume of solution is the same).
   (1) What is the initial E?
   (2) What is E when $c(Co^{2+})$ reaches $0.4 mol \cdot L^{-1}$?
   (3) What is the equilibrium constant $K^\ominus$?
   (4) What is the value of $[Ni^{2+}]/[Co^{2+}]$ when $E = 0.025V$?

7. A concentration primary cell consists of two hydrogen electrodes. Electrode A has H_2 with 0.9atm of bubbling in $0.1 mol \cdot L^{-1}$ HCl solution. Electrode B has H_2 with 0.5atm if bubbling in $2.0 mol \cdot L^{-1}$ HCl solution. Which electrode is the anode? What is the E? What is the equilibrium constant $K^\ominus$ of this primary cell reaction?

8. In a test of a new reference electrode, a chemist constructs a primary cell consisting of a Zn^{2+}/Zn electrode and a hydrogen electrode under the following conditions: $c(Zn^{2+}) = 0.01 mol \cdot L^{-1}$; $c(H^+) = 2.5 mol \cdot L^{-1}$; $P_{H_2} = 0.3 atm$, Calculate the E at 25℃.

Answers to Some Exercises

[Exercises]

1. $+6, +2, +4, +4, +5, -1, -1, +6$

2. (1) $2MnO_4^-(aq) + 5H_2O_2(aq) + 6H^+(aq) = 2Mn^{2+}(aq) + 5O_2(g) + 8H_2O(l)$
   (2) $Cr_2O_7^{2-}(aq) + 3SO_3^{2-}(aq) + 8H^+(aq) = 2Cr^{3+}(aq) + 3SO_4^{2-}(aq) + 4H_2O(l)$
   (3) $As_2S_3(s) + 5ClO_3^-(aq) + 5H_2O(l) = 5Cl^-(aq) + 2AsO_4^{3-}(aq) + 3SO_4^{2-}(aq) + 10H^+(aq)$

3. According to the standard electrode potentials, $\varphi^\ominus(Cl_2/Cl^-) = 1.358V$, $\varphi^\ominus(H_2O_2/H_2O) = 1.776V$, chlorine and hydrogen peroxide are all strong oxidizing agents, and they can easily oxidize many reducing substances as sanitizers.

4. (1) oxidizing ability: $Zn^{2+} < Fe^{3+} < MnO_2 < Cr_2O_7^{2-} < Cl_2 < MnO_4^-$
   (2) reducing ability: $Cl^- < Cr^{3+} < Fe^{2+} < H_2 < Li$

5. (1) $(-) Zn(s) | Zn^{2+}(aq) \| Ag^+(aq) | Ag(s) (+)$

The cell reaction will proceed spontaneously as written.

(2) $(-)$ Pt(s)$|$Cr$_2$O$_7^{2-}$(aq), Cr^{3+}(aq), H$^+$(aq)$\|$Cl$^-$(aq)$|$Cl$_2$(g)$|$Pt(s) $(+)$

The cell reaction will proceed spontaneously as written.

(3) $(-)$ Pt(s)$|$I$_2$(s)$|$IO$_3^-$(aq), H$^+$(aq)$\|$Fe^{3+}(aq), Fe^{2+}(aq)$|$Pt(s) $(+)$

The cell reaction will proceed spontaneously in the opposite direction.

6. (1) Fe, (2) MnO$_2$

7. H$_2$O$_2$ will spontaneously decompose to H$_2$O and O$_2$ under standard conditions.

8. (1) -0.068V, (2) 0.818V, (3) 1.107V

9. (1) MnO$_4^-$ ion can oxidize I$^-$ ion and Br$^-$ ion; (2) φ(MnO$_4^-$/Mn^{2+})$=0.986$V, MnO$_4^-$ ion can oxidize I$^-$ ion only but can not oxidize Br$^-$ion.

10. (1) $E^{\ominus}=0.404$V, $\Delta_r G_m^{\ominus}=-77\,972J\cdot$mol^{-1}, $K^{\ominus}=4.5\times10^{13}$.

(2) 6ClO$_2$(g)$+$3H$_2$O $\rightleftharpoons$ 5ClO$_3^-$(aq)$+$Cl$^-$(aq)$+$6H$^+$

11. (1) oxidized to Co^{2+} ion (2) the changes of concentration of HNO$_3$ solution can not change above conclusion.

12. pH$=3.02$

13. The oxidizing abilities of Cr$_2$O$_7^{2-}$, MnO$_4^-$ and H$_2$O$_2$ will be strengthened, and the oxidizing abilities of Hg$_2^{2+}$, Cl$_2$ and Cu^{2+} will be maintained.

14. Cu^{2+}(1.0$\times10^{-4}$mol$\cdot$L^{-1})/Cu is the anode, and Cu^{2+}(1.0$\times10^{-1}$mol$\cdot$L^{-1})/Cu is the cathode. $E=0.088\,8$V

15. 0.021mol$\cdot$L^{-1}

16. 5.6×10^{-7}

17. [Cl$^-$]$=0.1$mol$\cdot$L^{-1}

18. pH$=4.0$, $K_a=1.0\times10^{-6}$

[Supplementary Exercises]

1. 1.15×10^{-18}

2. $-21\,100$J$\cdot$mol^{-1}, 0.219V

3. (1) 3Fe(OH)$_2$(s)$+$MnO$_4^-$(aq)$+$2H$_2$O $\rightleftharpoons$ MnO$_2$(s)$+$3Fe(OH)$_3$(s)$+$OH$^-$(aq), MnO$_4^-$(aq) is the oxidizing agent.

(2) 5Zn(s)$+$2NO$_3^-$(aq)$+$12H$^+$ $\rightleftharpoons$ 5Zn^{2+}(aq)$+$N$_2$(g)$+$6H$_2$O, NO$_3^-$(aq) is the oxidizing agent.

4. (1) $(-)$ Al(s)$|$Al^{3+}(c_1)$\|$Cr^{3+}(c_2)$|$Cr(s) $(+)$

(2) $(-)$ Pt(s)$|$SO$_2$(g)$|$SO$_4^{2-}$(c_1), H$^+$(c_2)$\|$Cu^{2+}(c_3)$|$Cu(s) $(+)$

5. 7.8×10^4mol$\cdot$L^{-1}

6. (1) 0.040V (2) 0.028V (3) 6.0 (4) 1.17

7. Electrode A is the anode, 1

8. 0.860V

（喻　芳）

Chapter 9
Atomic Structure and Periodic Law

Today, the research of life science has been developing in microscopic scale, of which the physical quantities are generally quantized. To reveal the laws of motion for microscopic particles has to rely on quantum mechanics. It illuminates that a particle, such as an electron, has wave-particle duality and electron wave is essentially a probability one. The wave function is applied to give mathematical description of electron's wavelike motion in hydrogen atom. Various geometric plots of wave function are used to intuitively present the probability distribution and the magnitude of energy level of electrons. On the basis of explaining the structure of hydrogen atom, that of many-electron atom is clarified. Periodic table reflects the properties of elements varying in a more or less regular fashion. The information about atomic structure is the foundation for understanding the covalent bond, intermolecular forces in Chapter 10 and structures of coordination compounds in Chapter 11. These knowledge are helpful for ulteriorly realizing the biological structures of molecules *in vivo* and the effects of drugs.

9.1 Foundations of Quantum Mechanics and Characteristics of Electronic Motion

The research of atomic structure in quantum mechanics began with hydrogen atom due to only one electron outside the nucleus. Based on the results of studies on the single-electron atom, the structures of complex particles such as many-electron atoms or ions were explored.

9.1.1 Hydrogen Spectrum and Bohr's Model of Hydrogen Atom

In 1909, the British physicist E. Rutherford and his assistant passed a beam of α-particles with high velocity and positive charges through a piece of gold foil with only $10^{-6} \sim 10^{-7}$ m thick. Based on a series of experimental results, he proposed the **nuclear model** in 1911. It stated that the bulk of the atomic mass and the positive charge concentrated into a relatively tiny volume at the center of the atom, known as the "nucleus", and the electrons with an equal amount of negative charge moved around the nucleus. Therefore, the atom was essentially empty. For diverse atoms, the numbers of both positive charge in the nucleus and electrons outside it are different.

Rutherford postulated the velocity of electron moving around the nucleus was high enough to make the centrifugal force generated be equal to the centripetal force of electron due to the attraction of nucleus, to explain why the electrons being not appealed into the nucleus. However, according to Maxwell's classical theory, the electron orbiting a nucleus in high speed will constantly release electromagnetic radiation, causing continuous decrease of moving velocity. As a result, the radius of electrons surrounding the nucleus continually declines, and the electron is finally "sucked" into the nucleus, resulting in "atom collapse". Besides, emitting energy continuously will create a successive atomic spectrum. However, the hydrogen atoms actually produce a series of bright lines in the spectrum. It is obvious that classic physical theories are not suitable for explaining the experimental results of moving particles with high velocity. A new physical theory system has to be developed to accommodate the requirement for the studies in the microscopic fields.

In 1900, to explain the law of radiation (light) from the heated blackbody, the German scientist M. Plank presumed that the blackbody was composed of a series of oscillators in different vibration frequency v. The energy (E) for each oscillator takes a series of specific values which are integer times of that of minimum energy unit ($E = hv$, called energy quantum): $E = nhv$, $n = 1, 2, 3, \cdots$ In the equation, h is **Planck's constant** and equal to 6.626×10^{-34} J·s. According to Plank's theory, the energy of radiation absorbed or emitted by the blackbody may be one of the values of hv, $2hv$, $3hv$, $\cdots$, and nhv. That means the energy is inconsecutive and entitled as quantized. As a result, Plank succeeded in explaining the observed results about the blackbody radiation and opened a new era of dealing with the motion of microscopic particles in quantum theory. His theory also had an impact on the classic physics in the microscopic domain for the first time.

A. Einstein took Planck's quantum theory to produce his solution to the experiment of photoelectric effect successfully and proposed the famous photon theory in 1905. It indicated that light had particle nature in addition to the wave nature. This prompted people for the first time to perceive a certain object by two images: Light is not only a kind of electromagnetic wave characterized by frequency v or wave length λ, but also a beam of particles composed of photons characterized by energy E or momentum p. The wave and particle natures of light are linked quantitatively by h, a very tiny value, that is

$$\lambda = h/p = h/mc$$

These two natures of light are contradictory, but unified. The light shows diverse characteristics under different situation due to its duality. In the process of light being emitted or interacting with entity, the light obviously expresses particle-like nature which was applied to explain the experimental phenomenon about atomic spectroscopy, photoelectric effect and blackbody radiation. On the other hand, when the light is spreading through space, the phenomenon about polarization, interference and diffraction of light should be elucidated from the perspective of wave nature. However, the particle nature of a beam of light as a whole is closely associated with its wave nature.

When white light is scattering, a continuous spectrum in the visible region can be observed. However, a series of bright, but discontinuous spectral lines were emitted when hydrogen atoms were irritated by electric arc, spark or flame. As early as in 1885, J. Balmer discovered the regularity of four visible wavelengths of hydrogen atom as shown in Figure 9-1. Later, other scientists found similar disciplines for a series of ultraviolet and infrared spectral lines respectively. Finally, the wavelength of all these spectral lines could be calculated according to the formula below

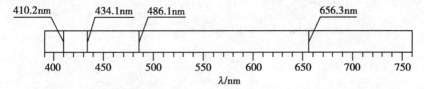

Figure 9-1 Visible Spectral Lines Emitted by Hydrogen Atom

$$\frac{1}{\lambda} = \tilde{R}_H \left(\frac{1}{n_1^2} - \frac{1}{n_2^2} \right) \tag{9.1}$$

Here, λ is the wavelength, $\tilde{R}_H$ is the Rydberg constant, equal to $1.096\,776 \times 10^7 \text{m}^{-1}$ ($1\text{m}^{-1} = 1.986\,48 \times 10^{-25}$ J), both n_1 and n_2 are positive integers (1, 2, 3, ⋯), and n_2 is bigger than n_1.

The Danish scientist N. Bohr proposed three postulations about hydrogen atomic structure in 1913, by synthesizing the results of Plank's quantum theory, Einstein's relativity theory and Rutherford's nuclear model.

(1) The electron moves on a circular orbit around the nucleus and does not emit radiation to avoid "atom collapse". In that case, the electron has certain energy and the atom is always in a state of "energy stability", called **stationary state** for short. The atom in stationary state neither absorbs nor emits radiation. Each stationary state corresponds to a certain **energy level**. The energy level of atom in **ground state** is the lowest. Each of other stationary states is called **excited state**.

(2) When the atom is in a stationary state, the angular momentum L of electron moving around the nucleus is quantized with possible values

$$L = nh / 2\pi \tag{9.2}$$

Where $n = 1, 2, 3, 4, \cdots$.

This is Bohr's quantization rule, and n is called **quantum number**.

(3) Generally, the atom can transit from one discrete stationary state (energy level E_1) to another (energy level E_2). In the process, the electron absorbs or emits radiation, the frequency of which can be calculated according to the formula below

$$h\nu = |E_2 - E_1| \tag{9.3}$$

This is Bohr's frequency rule. The inequation $E_2 > E_1$ indicates atom absorbing light, while $E_2 < E_1$, emitting light.

Bohr deduced the equation of calculating the energy for each stationary state of hydrogen atom according to above first two postulations

$$E = -\frac{R_H}{n^2} \tag{9.4}$$

Where $n = 1, 2, 3, 4, \cdots$.

In Equation 9.4, R_H is a constant and equal to 2.18×10^{-18} J (or 13.6 eV). When $n = 1$, $E = -R_H$, which is the energy of hydrogen atom in the ground state. Corresponding to $n = 2, 3, \cdots$, the energy for each excited state is $-R_H / 4, -R_H / 9, \cdots$, respectively. The energy of hydrogen atom in the stationary state is quantized. Figure 9-2 shows a part of energy levels of hydrogen atom. The frequency of radiation emitted or absorbed when an electron transits from stationary state n_1 to n_2 can be calculated from the following relation which is deduced by combining Equation 9.3 with Equation 9.4

$$h\nu = \left|E_{n_2} - E_{n_1}\right| = R_H \left(\frac{1}{n_1^2} - \frac{1}{n_2^2}\right)$$

or $\quad \dfrac{1}{\lambda} = \dfrac{\nu}{c} = \dfrac{R_H}{hc}\left(\dfrac{1}{n_1^2} - \dfrac{1}{n_2^2}\right)$

Then, the value of $\dfrac{R_H}{hc}$, considered as $\tilde{R}_H$ in Equation 9.1, was calculated to be $1.097\,37 \times 10^7\,\mathrm{m}^{-1}$ and very close to the experimental value.

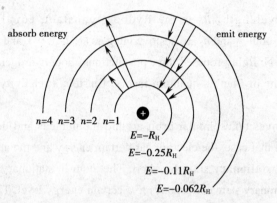

Figure 9-2 Energy Level Diagram of Hydrogen Atom

Bohr had insight into the characteristics of physical quantity quantized in microscopic scales. He applied the idea of "quantization" to explain hydrogen spectrum successfully and constructed the way of studying energy level for atoms with spectral data. But Bohr's theory encountered great difficulties in explaining the fine structure of hydrogen spectrum and the structure of many-electron atoms. Since the concept of planetary-like circle orbit actually belongs to the scope of classic mechanics, Bohr's theory is regarded as old quantum mechanics. Besides, the condition for the quantization of angular momentum in this theory was not properly explained, and the nature of electron was not well understood at that time. In modern quantum mechanics, the method of **describing** the electronic behavior around the nucleus of hydrogen atom was constructed based on the nature of electron and the law of its motion.

9.1.2 Wave-Particle Duality of Electron

According to Einstein's photon theory, light has both wave and particle natures. Light could be described by either wavelength λ, frequency ν as a beam of electromagnetic wave, or energy ($E = h\nu$), momentum ($p = mc$) as particles. According to the equations: $E = mc^2$ and $\nu = c/\lambda$ in photon theory, the formula connecting wave and particle properties of light could be deduced as $\lambda = h/mc$. The limitations of old quantum theory mentioned above also prompt people to further think about the nature of microscopic particles. So, is the nature of microscopic particles such as electrons similar to that of photons?

In 1923, enlightened by dual natures of light, the French physicist L. de Broglie proposed that microparticles, such as electrons, atoms and so on, also had **wave-particle duality**, and concluded **de Broglie relation** which implied the wave nature of microparticle

$$\lambda = \frac{h}{p} = \frac{h}{m\upsilon} \qquad (9.5)$$

In this relation, p is the momentum, m is the mass, υ is the velocity and λ is the wavelength. Wave and particle natures of microparticles were related by Plank's constant h.

In 1927, de Broglie relation was confirmed with the observations of electron beam reflecting on nickel single-crystal in Davisson-Germer experiment of America and electron diffraction in the experiment of the British physicist G. P. Thomson respectively. So, how do we understand the electron wave and reconcile the wave nature of electron with its particle nature?

The electron wave is neither an electromagnetic wave nor a simple harmonic one. M. Born proposed reasonable "statistical interpretation" to reveal the nature of electron wave. For example, in the electron-diffraction experiment, the diffraction images can be obtained by projecting a strong electron beam through crystal onto a photographic film in Figure 9-3(c). If the electron beam is so weak that the electrons were almost being shoot out one by one in Figure 9-3(a) and (b) for enough long time, just as one electron projected for many times, the same diffraction images are acquired. In other words, the position at which the electron appears on the negative after each projection can not be predicted, but the probability of electron appearing at a certain position after many times may be ensured. The images remind us the brighter the spot on the diffraction pattern, the larger the probability of electron showing is, or vice versa. Evidently, the electron wave is a **probability wave**, and the wave intensity of a certain point outside the nucleus is directly proportional to the probability of electron appearing at the point.

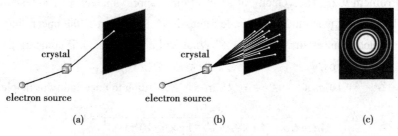

Figure 9-3 Electron-Diffraction Diagram

Sample Problem 9-1 (1) The velocity of electron is $5.9 \times 10^5 \text{m} \cdot \text{s}^{-1}$ under 1V of voltage, electron mass is 9.1×10^{-31}kg, h is 6.626×10^{-34}J·s, please calculate the electron wavelength. (2) The mass of a grain of sand is 1.0×10^{-8}kg and its velocity of motion is $1.0 \times 10^{-2}\text{m} \cdot \text{s}^{-1}$, please calculate the wavelength.

Solution (1) $1\text{J} = 1\text{kg} \cdot \text{m}^2 \cdot \text{s}^{-2}$, $h = 6.626 \times 10^{-34}$J·s. According to de Broglie relation,

$$\lambda = \frac{h}{m\upsilon} = \frac{6.626 \times 10^{-34} \text{kg} \cdot \text{m}^2 \cdot \text{s}^{-1}}{9.1 \times 10^{-31} \text{kg} \times 5.9 \times 10^5 \text{m} \cdot \text{s}^{-1}} = 12 \times 10^{-10} \text{m} = 1\,200\text{pm}$$

(2)
$$\lambda = \frac{6.626 \times 10^{-34} \text{kg} \cdot \text{m}^2 \cdot \text{s}^{-1}}{1.0 \times 10^{-8} \text{kg} \times 1.0 \times 10^{-2} \text{m} \cdot \text{s}^{-1}} = 6.6 \times 10^{-24} \text{m}$$

The above example indicates that the greater the mass of an object, the shorter its wavelength is. Macroscopic object only shows particle properties because its wavelength is too small to be determined. However, the mass of electron is very small, causing its wavelength to be on the same order as atomic size, 1.0×10^{-10}m. Therefore, the de Broglie wavelength of microscopic particles generally can not be neglected. So, is there a trace for the motion of electron outside an atomic nuclear?

9.1.3 Heisenberg's Uncertainty Principle

The macroscopic object, such as no matter a planet or a speck of dust, always has a measurable motion trail. That is, at some point, there are definite both coordinate and velocity or momentum. However, a microscopic particle possesses wave-like motion which is statistic. Therefore, it is expected that it will not have definite position and momentum simultaneously. In 1927, the German physicist W. Heisenberg pointed out that it was impossible to simultaneously acquire accurate both momentum and coordinate of a particle whose wave properties can not be neglected. It was stated that the more precisely the position of some particle was determined, the less precisely its momentum could be known, and vice versa. This is the famous **uncertainty principle**:

$$\Delta x \cdot \Delta p_x \geq \frac{h}{4\pi} \tag{9.6}$$

In Equation 9.6, Δx and Δp_x stand for the uncertainty of coordinate x and that of momentum along x-aixs direction respectively. Because h is Plank's constant, the smaller the Δx is, the bigger the Δp_x is, and vice versa. Uncertainty principle is an inevitable consequence of wave of microparticles. For macroscopic objects, each uncertainty in Equation 9.6 is so small relative to the physical property itself, due to their greater mass and volume, that they can be neglected. Therefore, the macroscopic objects obey the laws of classic mechanics.

Sample Problem 9-2 The velocity of an electron is approximately $6 \times 10^6 \text{m} \cdot \text{s}^{-1}$ when it is near the nucleus. The order of magnitude for atomic radius is 10^{-10}m. What is the uncertainty of coordinate x, Δx, when the velocity uncertainty, Δv is $\pm 1\%$? What will be discovered, if comparing the calculation result with the atomic radius?

Solution $\Delta v = 6 \times 10^6 \text{m} \cdot \text{s}^{-1} \times 0.01 = 6 \times 10^4 \text{m} \cdot \text{s}^{-1}$. According to uncertainty principle,

$$\Delta x \geq \frac{h}{4\pi m \Delta v} = \frac{6.626 \times 10^{-34} \text{kg} \cdot \text{m}^2 \cdot \text{s}^{-1}}{4\pi \times 9.1 \times 10^{-31} \text{kg} \times 6 \times 10^4 \text{m} \cdot \text{s}^{-1}} = 1 \times 10^{-9} \text{m}$$

The uncertainty of coordinate x is at least 10-fold of the order of magnitude for atomic radius. This indicates the accurate position of electron in the atom is unpredictable. Now that it is not possible to determine the velocity and spatial position of an electron simultaneously, how do we study the motion of electron?

9.2 Quantum Mechanical Explanations of Hydrogen Atom

Since the electron wave has the interpretation of probability, the wave behavior of electron should be studied from the viewpoint of statistics. Firstly, it has to be considered how to describe the wave behavior. In classic physics, the functions are usually applied to express the wave behavior. For example, electromagnetic wave can be described by electric field intensity $E(x, y, z, t)$ or magnetic field intensity $H(x, y, z, t)$ which is the function of time t and position (x, y, z). Although the microparticle wave is quite different from the classic one, both have wave properties, such as producing diffraction pattern. So, the wave behavior of microparticle may still be described by a concerned function, called "**wave function**" notated as $\psi(x, y, z)$.

9.2.1 Wave Function and Physical Meanings of Three Quantum Numbers

In 1926, the Austrian physicist E. Schrödinger proposed a partial differential equation, called **Schrödinger's equation**[2*], for the wave behavior of microparticle. The solutions of Schrödinger's equation are a series of reasonable wave functions. Each wave function stands for a certain possible state of electron motion and corresponds to an energy E which is a constant and also acquired from Schrödinger's equation. The energies which are discontinuous or called quantized are in consistent with those calculated by Bohr's model. The state with certain energy is named as stationary state. Among these states, the one with the lowest energy is the ground state. Each of the others is called excited state.

Although wave function is a mathematical expression and seemingly has no clear physical meaning itself, the squared modulus of wave function, $|\psi|^2$, has a definite physical significance. In classic physics, $|E|^2$ or $|H|^2$ are employed to represent the intensity of electromagnetic wave at time t and point (x, y, z). Likewise, $|\psi|^2$ represents the intensity of electron wave, namely **probability density** (the probability of unit volume in maths) of electron appearing at point $p(x, y, z)$ outside the nucleus. For a certain wave function, it has definite mathematical expression, by which along with the value of ψ and that of $|\psi|^2$ at any point $p(x, y, z)$ can be calculated. Then, the ratio of probability for any two points can be obtained. As the probability of electron appearing in the space outside the nucleus is 100%, the probability distribution is acquired. Then, the energy corresponding to the wave function can be calculated. Therefore, the wave function implies the probability distribution of electron in the space outside the nucleus behind its functional expression. Both probability distribution and energy are vital characters of electron motion outside the nucleus.

> **Question and thinking 9-1** If the wave function of H atom is given, the corresponding energy of electron is determined. Why?

Probability distribution of electron outside the nucleus can be demonstrated vividly by the image of **electron cloud**. Figure 9-4 (a) shows the 3D graph of $|\psi|^2$ of hydrogen atom in the ground state, while Figure 9-4 (b), its sectional drawing. The number of small black dot is proportional to the value of $|\psi|^2$ in each location of the drawing. It can be found from Figure 9-4 that the darker the area, the more the number of small black dot is, or the greater the probability density of electron is. It is obvious that electron cloud image reflects the probability distribution of electron, but does not mean that there are many electrons diffusing across the space outside the nucleus.

In quantum mechanics, wave function ψ used to describe the motion of electron outside the nucleus is generally called "**atomic orbital**". It is only a proper and imaginary replacement of the wave function. Theoretically, atomic orbital extends infinitely in space outside the nucleus and the electron cloud disperses across the space. Therefore, the concept of atomic orbital definitely has no meaning of "orbit"

[2*] Schrödinger Equation is a two-step partial differential equation:

$$\frac{\partial^2 \Psi}{\partial x^2} + \frac{\partial^2 \Psi}{\partial y^2} + \frac{\partial^2 \Psi}{\partial z^2} + \frac{8\pi^2 m}{h^2}(E-V)\Psi = 0$$

m is the mass of particle, E is the total energy, V is the potential energy and h is the Plank's constant.

in classic mechanics. Generally, atomic orbital refers to the space in which the probability of electron appearing is about 99% or 95%.

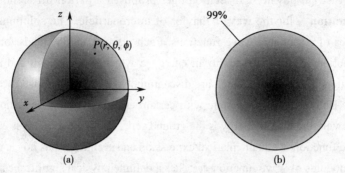

Figure 9-4　Electron Cloud of Hydrogen Atom in the Ground State
(a) Electron cloud; (b) Cross section of electron cloud

Hydrogen atom and hydrogen-like ions, such as He^+, Li^{2+}, have only one electron outside the nucleus. There is no repulsion force between electrons in the single-electron system. The potential energy is just the nucleus-electron attractive energy. As the Schrödinger's equation is relatively simple, accurate solutions of the equation could be acquired. To facilitate the solving of equation, the rectangular coordinate system has to be converted into spherical one. Then, $\psi(x, y, z)$ is transformed into $\psi(r, \theta, \varphi)$ accordingly as shown in Figure 9-5.

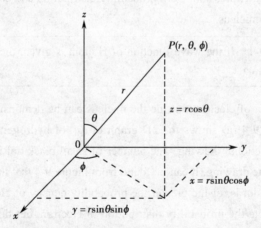

Figure 9-5　Transformation between Rectangular Coordinate System and Spherical Coordinate System

In the process of solving Schrödinger equation, some parameters have to be set and must be integers in order to obtain a series of rational solutions. These parameters are notated as n, l, and m, and also called quantum numbers. The way of acquiring these quantum numbers is more reasonable as compared with the quantum number n in Bohr's theory which was set artificially without any proper explanations in Equation 9.2. This exactly indicates modern quantum theory reveals the quantization characteristics of microparticle more scientifically. Only when the values of three quantum numbers, n, l and m are given, the mathematical form of wave function is determined. So, it is the function $\psi_{n, l, m}(r, \theta, \varphi)$ with three subscripts that really represents a definite state of electron motion.

The rules for taking the values of three quantum numbers and their physical meanings are as follows:

1. Principal quantum number (n)

Principal quantum number is notated as n, n is a positive integer (1, 2, 3, ⋯). It is the dominant factor of orbital's energy. The smaller the n, the lower the electron energy is. When $n=1$, the energy is the lowest. For hydrogen atom or hydrogen-like ion, the energy of orbital only depends on n according to the relation, $E = -\dfrac{Z^2}{n^2} \times 2.18 \times 10^{-18} \text{J}$, in which Z is the atomic nucleus charge. It can be found that the energy is discontinus or quantized due to n being integers.

n also determines the average distance between electron and nucleus or the size of atomic orbital. Therefore, n can be named as **electron shell**. The orbitals with identical n value belong to the same electron shell. $n=1, 2, 3, 4, \cdots$, corresponds to the electron shell of K, L, M, N, ⋯, respectively. The bigger the n value is, the farther the average distance between electron and nucleus is.

2. Orbital angular momentum quantum number (l)

Orbital angular momentum quantum number (or **sublevel**) is notated as l, and determines the shape, angular momentum and magnetic moment of atomic orbital. For each of n, l is an integer from 0 to $(n-1)$, corresponding to each kind of orbital shapes respectively, notated as s, p, d, f, g, ⋯, sequentially in the field of spectroscopy.

In a many-electron atom, the energy of orbital depends on not only n but also l, due to the repulsion force between electrons. Therefore, l is also named as **electron subshell** or **electron sublevel**. When n is given, the energy of orbital in different sublevel rises with the increase of l value. The combination of n and l corresponds to an energy level. For example, the combination of $n=2$ and $l=1$ refers to the 2p subshell or energy level.

3. Magnetic quantum number (m)

Magnetic quantum number m (or **orbital**) determines not only the space orientation of atomic orbital, but also the component of its angular momentum and magnetic moment along the direction of external magnetic field, commonly designated as z direction. For each l value, m is restricted to take $(2l+1)$ values, which are $0, \pm 1, \pm 2, \cdots$, and $\pm l$. It means there are $(2l+1)$ orbitals with different space orientation in each subshell. If $l=1$, m may take the value of $-1, 0, +1$, indicating p-orbitals with three kinds of, or three p-orbitals with diverse, space orientation in the subshell. As the orbital's energy of many-electron atom depends on n and l, and is independent of m, for example, three p-orbitals in a shell have the same energy and are called **equivalent orbitals**.

It can be seen from above discussions, to describe an atomic orbital by $\psi_{n,l,m}(r,\theta,\varphi)$ does not have to rely on complex mathematical expression of wave function, but just a triple of quantum numbers, n, l and m, alternatively. Table 9-1 shows the combination of three quantum numbers n, l, and m. If $n=1$, both l and m should be 0. The combination of three quantum numbers, (1, 0, 0), represents atomic orbital $\psi_{1,0,0}$ or ψ_{1s}, called 1s orbital for short. It indicates only one orbital in K electron shell. If $n=2$, l may take the values of 0 and 1, indicating L shell containing two energy levels. When $l=0$, m only can be 0. The combination (1, 0, 0) stands for $\psi_{2,0,0}$ or ψ_{2s}. When $l=1$, m may be $-1, 0, +1$, indicating that 2p subshell contains three orbitals, $\psi_{2,1,0}$, $\psi_{2,1,1}$ and $\psi_{2,1,-1}$ or ψ_{2p_z}, ψ_{2p_x} and ψ_{2p_y} [3*]. L shell has a total of four

---

[3*] $\psi_{2,1,+1}$ and $\psi_{2,1,-1}$ are complex wave functions. ψ_{2p_x} and ψ_{2p_y} are real wave functions and the results of linear combination of $\psi_{2,1,+1}$ and $\psi_{2,1,-1}$.

atomic orbitals. It is concluded that there are totally n^2 orbitals in each electron shell.

Table 9-1 Combination of n, l, m and the Number of Orbitals, Electrons Full of Each Electron Shell

Principle quantum number n	Orbital angular momentum quantum number l	Magnetic quantum number m	Wave function ψ	Number of orbital in a shell (n^2)	Number of electrons full of a shell ($2n^2$)
1	0	0	ψ_{1s}	1	2
2	0	0	ψ_{2s}	4	8
	1	0	ψ_{2p_z}		
		±1	ψ_{2p_x}, ψ_{2p_y}		
3	0	0	ψ_{3s}	9	18
	1	0	ψ_{3p_z}		
		±1	ψ_{3p_x}, ψ_{3p_y}		
	2	0	$\psi_{3d_{z^2}}$		
		±1	$\psi_{3d_{xz}}, \psi_{3d_{yz}}$		
		±2	$\psi_{3d_{xy}}, \psi_{3d_{x^2-y^2}}$		

9.2.2 Angular Distribution Graphs of Atomic Orbital and Electron Cloud

The probability distribution and energy of electron in a stationary state may be intuitively explained through the diagrams about atomic orbital and electron cloud, which are also very helpful for us to understand the spatial orientation of chemical bonds and the geometric structure of coordination compounds. The $\psi_{n,l,m}(r, \theta, \varphi)$ acquired exactly from Schrödinger equation of hydrogen atom is a function of three independent variables, r, θ and φ. It is very difficult to figure out the graph for the whole $\psi_{n,l,m}(r, \theta, \varphi)$ in 3D coordinate system. Therefore, it is considered to split $\psi_{n,l,m}(r, \theta, \varphi)$ into $R_{n,l}(r)$ and $Y_{l,m}(\theta, \varphi)$, and then map partly the atomic orbital and electron cloud by the two functions respectively.

$$\psi_{n,l,m}(r, \theta, \varphi) = R_{n,l}(r) \cdot Y_{l,m}(\theta, \varphi) \tag{9.7}$$

In Equation 9.7, $R_{n,l}(r)$ is the radial component of wave function and named as radial wave function. It has variable r, the distance between a certain point and a nucleus, and is related with two quantum numbers n and l. $Y_{l,m}(\theta, \varphi)$ is the angular component of wave function, and named as angular wave function. It is a function of angles θ and φ, and related with two quantum numbers l and m. $Y_{l,m}(\theta, \varphi)$ describes the shape and orientation of atomic orbital. The graphs of $R_{n,l}(r)$ and $Y_{l,m}(\theta, \varphi)$ maybe used to observe the probability distribution of electron from the radial and angular aspects respectively. The radial and angular wave functions of hydrogen atomic orbitals in K, L shell and the corresponding energy are listed in Table 9-2.

Firstly, a 3D rectangular coordinate system has to be established to depict the angular distribution graph of atomic orbital and the nucleus is placed in the origin. Draw a straight line from the origin along direction (θ, φ), make the length of the line equal to the value of $|Y|$, link the endpoint of each line to form a contour. Finally, the region of $Y > 0$ in the graph is labeled with sign "+", while that of $Y < 0$, with sign "−". The angular distribution graph reflects the change of $Y_{l,m}(\theta, \varphi)$ with azimuth (θ, φ), not the r. So, it does not mean the distance of electron to the nucleus. It can be found from the subscripts of $Y_{l,m}(\theta, \varphi)$ that the atomic orbitals in different shell should have the same angular distribution graph, only if they have identical l, m value.

Table 9-2   Radial and Angular Wave Functions of Some Hydrogen Atomic Orbitals and Corresponding Energies

Atomic orbital	$R_{n,l}(r)$	$Y_{l,m}(\theta, \varphi)$	Energy/ J
1s	$A_1 e^{-Br}$	$\sqrt{\dfrac{1}{4\pi}}$	-2.18×10^{-18}
2s	$A_2(2-Br)e^{-Br/2}$	$\sqrt{\dfrac{1}{4\pi}}$	$-2.18 \times 10^{-18}/2^2$
$2p_z$		$\sqrt{\dfrac{3}{4\pi}}\cos\theta$	
$2p_x$	$A_3 r e^{-Br/2}$	$\sqrt{\dfrac{3}{4\pi}}\sin\theta\cos\varphi$	$-2.18 \times 10^{-18}/2^2$
$2p_y$		$\sqrt{\dfrac{3}{4\pi}}\sin\theta\sin\varphi$	

*A_1, A_2, A_3 and B are constants.

The sign "+" or "−" in the graph could be regarded as the wave phase corresponding to a fluctuation and should not be interpreted as a sign of charge. In addition, the meaning of sign "+", "−" can be applied to understand the bonding in a molecule. When a covalent bond forms between two atoms, their atomic orbitals overlap. If the lobes provided by the two bonding orbitals have identical sign, their wave functions would have constructive interference to enhance bonding, while, opposite sign, destructive interference to impair bonding.

Since the covalent bond has spatial orientation, the angular distribution of atomic orbital or electron cloud has to be most concerned, in discussing the molecular structure and the change of covalent bond in chemical reaction.

Draw a straight line from the origin along direction (θ, φ) and make the length of the line equal to the value of $|Y|^2$. The contour formed by linking the endpoint of each line is the angular distribution graph of electron cloud. There is no sign "+" or "−" in it because of $|Y|^2 \geq 0$.

1. Angular distribution graphs of s-orbital and s-electron cloud

The angular wave function of s-orbital is a constant which is greater than 0. The length of line from the origin along each direction (θ, φ) is equal because of the same Y value. Linking each end point of the lines makes up a spherical surface. The graph is labeled with sign "+" due to the positive Y value. Figure 9-6(a) shows the sectional graph of s-orbital, while 9-6(b) the 3D one. Figure 9-6(c) shows the angular distribution graph of s-electron cloud which is also a spherical contour without sign "+" or "−".

2. Angular distribution graphs of p-orbitals and p-electron clouds

The angular wave function of p-orbital is related to azimuth θ. For example, the angular wave function of p_z orbital is $Y_{p_z} = \sqrt{3/4\pi}\cos\theta$, and the value of Y_{p_z} changes with azimuth θ as in the following table:

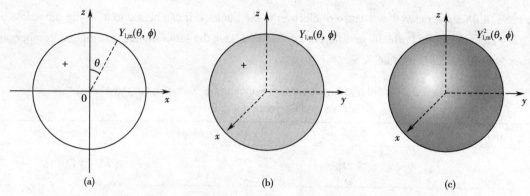

Figure 9-6 Angular Distribution Graphs of s-Orbital and s-Elctron Cloud
(a) Sectional; (b) 3D Graph of s-Orbital; (c) 3D Graph of s-Elctron Cloud

θ	0°	30°	60°	90°	120°	150°	180°
$\cos\theta$	1	0.866	0.5	0	−0.5	−0.866	−1
Y_{p_z}	0.489	0.423	0.244	0	−0.244	−0.423	−0.489

The angular distribution graph of p_z orbital obtained following above drawing procedure consists of two spherical lobes stretching along z-axis. Figure 9-7 shows a sectional drawing of the graph. The **lobe** above x-y plane is labeled with sign "+" because of $Y_{p_z} > 0$, while below the plane, with sign "−" because of $Y_{p_z} < 0$. The two lobes are asymmetric relative to x-y plane. The value of Y function on x-y plane is zero, and the x-y plane is named as **nodal plane**.

As the l value of p-orbital is equal to 1, m may take the values of −1, 0, and +1, indicating p-orbital having three kinds of spatial orientation. The p_z orbital with $m=0$ stretches along z-axis. The p_x with $m=+1$ or −1 and p_y orbitals stretch along x-axis and y-axis respectively, though the shape of the graphs is the same as that of p_z

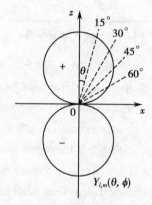

Figure 9-7 Sectional Graph of Angular Distribution of p_z Orbital

orbital. Figure 9-8(a) shows the angular distribution of three p-orbitals, while Figure 9-8(b) that of three p-electron clouds. The lobes in the angular distribution graph of electron cloud have no sign "+" or "−", and the shape is relatively "slimmer" if compared with that of p-orbital.

3. Angular distribution graphs of d-orbitals and d-electron clouds

Figure 9-9 shows the sectional graphs of angular distribution of d-orbitals and their electron clouds. There are four pear-shaped lobes and two nodal planes in the graph of each d-orbital. The d_{xy}, d_{xz} and d_{yz} orbitals have the lobes that are pointing between two of the axes. The planes constituted by two coordinate axes (x-z, y-z and x-y planes) are their nodal planes. $d_{x^2-y^2}$ orbital is stretching along the x-axis and y-axis respectively. Its lobes with sign "+" point to x-axis, while with "−" point to y-axis. Each of the two nodal planes is perpendicular to the x-y plane and stretches along 45° between the two coordinate axes. The graph of d_{z^2} orbital looks very special. It seems to be constituted by two balloons and a tyre which is inlayed between them. Its two nodal planes are located in the direction of θ is 54° 44′,

125° 16' respectively. The angular distribution graph of d-electron cloud is relative "slim" and the lobes have no sign "+" or "−".

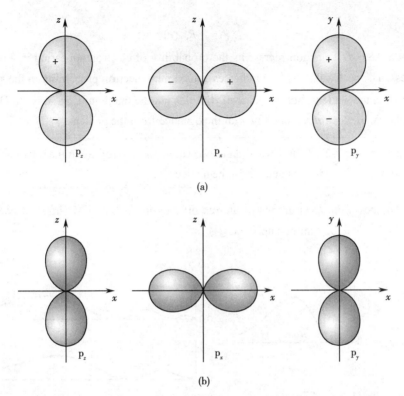

Figure 9-8 Sectional Graphs of Angular Distribution of (a) p-Orbitals (b) p-Electron Clouds

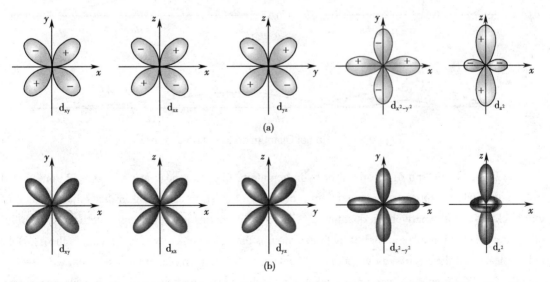

Figure 9-9 Sectional Graphs of Angular Distribution of (a) d-Orbitals and (b) d-Electron Clouds

9.2.3 Radial Distribution Function Graph

The electron probability density along any direction (θ, φ) can be acquired by $Y_{l,m}^2(\theta, \varphi)$. Similarly, the square of radial distribution function, $R_{n,l}^2(r)$, may be related to the probability of electron residing at the distance from the nucleus. The radial distribution function, $D(r)$, is defined as the product of r^2 and $R_{n,l}^2(r)$.

$$D(r) = r^2 R_{n,l}^2(r) \tag{9.8}$$

In Equation 9.8, $D(r)$ function represents the probability of electron appearing in a spherical shell of unit thickness. The peak value of $D(r)$ indicates a maximum electron probability in the spherical shell at the distance from the nucleus, but the value of $R_{n,l}^2(r)$ is not the maximum. So, it is the $D(r)$ that really reflects the variation of electron probability with the distance from the nucleus.

Question and thinking 9-2 Why $R_{n,l}^2(r)$ can not be solely used to represent the relationship between the electron probability and the distance from the nucleus?

Figure 9-10 shows the $D(r)$ graphs of atomic orbitals in K, L and M shells respectively. Some important conclusions can be summed up.

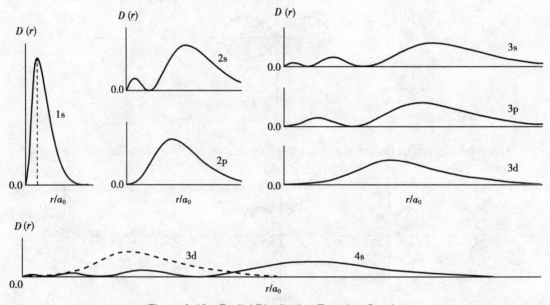

Figure 9-10   Radial Distribution Function Graphs

(1) In the $D(r)$ graph of 1s orbital of hydrogen atom, there is a peak at $r = a_0$ ($a_0 = 52.9$pm, called Bohr radius) which represents the maximum probability of electron appearing in that spherical shell of unit thickness. Or alternatively, in quantum mechanics, Bohr radius just represents where the probability of finding the electron in 1s orbital of hydrogen atom is the maximum. However, in Bohr's theory, Bohr radius means that the electron moves around the nucleus on a circular, planetary-like orbital with $n = 1$.

(2) The $D(r)$ graph has $(n - l)$ peaks when n and l are given simultaneously. Each peak represents a maximum probability of electron appearing at the distance r from the nucleus. If l is given, the greater the n value, the farther the main peak from the nucleus is. This implies that the orbitals actually are distinguished as the "inner" or "outer" one. Besides, the subordinated peak of the curve may be close to the nucleus. This causes the penetration or overlap among diverse orbitals, which is actually the result

of wave behavior of electron. When n is certain, the number of peaks increases upon the decrease of l value. Furthermore, the first peak gets closer to the nucleus, or in other words, penetrates more deeply. This phenomenon is named as orbital penetration. It is more complicated in many-electron atoms when neither n nor l of two orbitals is identical. For example, the first peak of $D(r)$ graph of 4s orbital is even closer to the nucleus than the main peak of 3d orbital.

9.3 Structure of Many-Electron Atom

Since in a many-electron atom, the total energies of electrons include multiple kinetic energies and potential ones which involve in nucleus-electron attractive energies and electron-electron repulsion energies, the whole wave equation of many-electron atoms is very complicated and therefore, should be simplified as that of hydrogen atom, namely single-electron wave equation. Even so, the distance between two electrons is uncertain because of their wave behavior, resulting in the repulsion energy calculated inaccurately. So, it is also difficult to obtain exact solutions from the single-electron wave equation. In that case, it is assumed that electron i moves in an "average" negative field formed by each of other electrons, in addition to the positive field of the nucleus. Now that the repulsion force of electron i acquired from others is an "average" one, it should be independent of the relative position of other electrons. Therefore, the solution of single-electron wave equation only depends on the 3D coordinate of electron i and is named as single-electron wave function ψ_i, which is also defined as the atomic orbital of many-electron atom. The corresponding orbital energy is named as energy E_i. It is obvious that the energy levels of many-electron atom are approximate.

9.3.1 Energy Levels of Many-Electron Atom

According to above assumption, in a many-electron atom, the average repulsion force exerted on the electron i by any other electrons is only related to the position of itself and maybe regarded as an offset of the attraction force between electron i and the nucleus. Or, in other words, the electron i is shielded from the nucleus by each of other electrons. This is called **screening effect**. **Screening constant** σ is generally used to measure the part of the nuclear charge that has been neutralized. So, the amount of positive nucleus charges actually perceived by electron i is called effective nuclear charge Z', and its value is equal to the difference between the positive nuclear charge Z and screening constant.

$$Z' = Z - \sigma$$

Then, the approximate energy of electron i is calculated with Z' instead of Z.

$$E = -\frac{Z'^2}{n^2} \times 2.18 \times 10^{-18} \text{J} \tag{9.9}$$

The energy of electron in a many-electron atom depends on n, Z, and σ. The energy descends with the decrease of n or increase of Z value. Meanwhile, the energy rises with the increase of σ value, causing the screening effect enhanced. The following factors have to be considered to estimate the screening constant σ.

(1) The screening effect of an outer electron to an inner one can be neglected, $\sigma = 0$.

(2) $\sigma = 0.30$ for the screening between two 1s electrons, while $\sigma = 0.35$ for two electrons in the same

shell except $n = 1$.

(3) For an ns or np electron, $\sigma = 0.85$ when screening effect comes from an electron in $(n-1)$ shell, while $\sigma = 1.00$ when that from an electron in more inner shells.

(4) For an nd or nf electron, $\sigma = 1.00$ when screening effect comes from an inner, ns or np electron.

It is concluded that the screening effect mainly comes from inner electrons. When l is fixed, with the increase of n, the electron shell enlarges, which results in the increase of energy level in two aspects, by 1) improving the screening effect and 2) diminishing the attraction of nucleus to the electron.

$$E_{1s} < E_{2s} < E_{3s} < \cdots$$
$$E_{2p} < E_{3p} < E_{4p} < \cdots$$
$$\cdots$$

It was learned that $D(r)$ graph had $(n-l)$ peaks. When n is certain, the smaller the l value is, the more the peak number is (this means the interpenetrating between orbitals) and the closer the first peak gets to the nucleus is (more strongly penetrating of electrons). Then, with the increase of the probability for electron approaching the nucleus, the screening effect from other electrons decreases and the effective nucleus charge improves, finally causing the lower of the energy level. The order of energy levels is as follows.

$$E_{ns} < E_{np} < E_{nd} < E_{nf} < \cdots$$

For two orbitals with different n, l, such as 3d and 4s orbitals, $E_{4s} < E_{3d}$ happens in $_{19}$K and $_{20}$Ca, while $E_{3d} < E_{4s}$ in $_{21}$Sc and subsequent elements of the same period. In other words, the inversion of energy level occurs. The similar phenomenon also appears for 4d and 5s orbitals. Therefore, the energy level or its order in many-electron atom is not fixed. It varies due to the difference in number of nuclear charge or electrons.

The American chemist L. Pauling proposed the order of approximate energy levels of many-electron atom according to the spectroscopic data.

$$E_{1s} < E_{2s} < E_{2p} < E_{3s} < E_{3p} < E_{4s} < E_{3d} < E_{4p} < \cdots$$

In Figure 9-11, the energy levels below are lower, while those above are higher. The oblique arrow threading energy levels indicates the order of approximate energy level by the direction of from bottom to top.

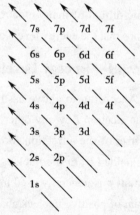

Figure 9-11 Approximate Order of Energy Levels

9.3.2 Electron Spin

In 1925, the Dutch physicists G. Uhlenbeck and S. Goudsmit proposed that electrons had inherent magnetic moment which was independent of its motion in an orbital. The magnetic moment was regarded to form due to the inherent angular momentum of electron, which could be described vividly as "spin" angular momentum. Taking the defining orbital angular momentum and its component in z direction as the reference, the spin angular momentum quantum number, notated as "s" was proposed. s dominates the magnitude of the component of spin angular momentum or magnetic moment along the direction of external magnetic field and can be the value of $+\frac{1}{2}$ or $-\frac{1}{2}$. The spin behavior can also be represented in upward arrow " ↑ " or downward arrow " ↓ ". Two electrons with identical s value are called parallel-spin electrons and their spin behaviors can be notated as " ↑↑ " or " ↓↓ ", meanwhile different s value called antiparallel-spin, notated as " ↑↓ ".

It is worth to be noted that the spin of electron can not be simply regarded as that of a macroscopic object, such as the Earth. The fine-structure of hydrogen atom spectrum can be explained by introducing the spin hypothesis of electron. The electron outside the nucleus has both orbital motion and spin motion, and the state depends on a set of quantum numbers n, l, m, and s. So, does the electron spin affect the structure of many-electron atom?

Sample Problem 9-3 (1) If $n=3$, which values may l and m be? How many orbitals are there in the electron shell? (2) The outermost electron of Na atom is in 3s subshell, please describe its state by quantum numbers n, l, m and s.

Solution (1) When $n=3$, l can be the values of 0, 1 and 2. When $l=0$, m can only be 0. when $l=1$, m can be the values of -1, 0 and $+1$. When $l=2$, m can be the values of -2, -1, 0, $+1$ and $+2$. Thus, there are totally nine atomic orbitals in the electron shell.

(2) The combination of quantum numbers n, l and m for 3s subshell is 3, 0, 0. The motion of electron in 3s subshell may be described as $(3, 0, 0, +\frac{1}{2})$ or $(3, 0, 0, -\frac{1}{2})$.

> **Question and thinking 9-3** Why can two electrons coexist in one orbital only when they are in antiparallel-spin state?

9.3.3 Electron Configuration of Atom

The distribution of electrons of an atom or ion in atomic orbitals is called electron configuration, which is determined in the ground state following three laws.

1. Pauli Exclusion Principle

In 1925, the Austrian physicist, W. Pauli formulated the Pauli Exclusion Principle according to many experimental results, which stated that no two electrons could exist in the same quantum state, identified by four quantum numbers n, l, m, and s. If two electrons of an atom are in the same atomic orbital, that is to say, they must have identical n, l, and m values, their s value should be different due to only two kinds of spin state for electrons. Alternatively, two electrons in one orbital must be in antiparallel-spin state, or an individual orbital has no more than two electrons. For example, the motion of two electrons in 4s orbital of $_{20}$Ca atom can be described by four quantum numbers as $(4, 0, 0, +\frac{1}{2})$ and $(4, 0, 0, -\frac{1}{2})$. As there are n^2 atomic orbitals in any shell, it can accommodate up to $2n^2$ electrons.

2. Aufbau Principle

Aufbau principle, or building-up principle states that in the ground state of an atom or ion, electrons fill atomic orbitals of the lowest available energy levels before occupying higher levels on the premise of Pauli exclusion principle. In this way, the electrons of an atom form the most stable electron configuration possible. If the filling of electrons follows the order of approximate energy level shown in Figure 9-11, the electron configuration of an atom with the lowest energy can be derived. For example, the 1s orbital of $_3$Li atom in the ground state is filled with two electrons before the 2s orbital is occupied by one electron, and its electron configuration is $1s^2 2s^1$. For $_{19}$K atom in the ground state, when K, L and M shells have been filled with 18 electrons, the remaining one electron occupies the 4s orbital, not the 3d one. Its electron configuration is $1s^2 2s^2 2p^6 3s^2 3p^6 4s^1$.

> **Question and thinking 9-4** Must the total energy of atom be the lowest, as long as the electrons are filled in the orbitals of lower available energy levels before occupying higher ones?

3. Hund's Rule

In 1925, the German physicist F. Hund proposed according to many atomic spectrum data that the orbitals of the subshell were each occupied singly with electrons of parallel spin before the occupation of two electrons in one orbital occurred. Hund's rule assumes that the repulsion between the outer electrons is much greater than any other interactions. The electrons in different orbitals are further apart, so that electron-electron repulsion energy may be reduced, and the energy of the atom becomes lower in magnitude. For example, the electron configuration of $_7$N atom in the ground state is $1s^2 2s^2 2p^3$. The states of the outer $2p^3$ electrons are described as follows

$$(2, 1, 0, +\frac{1}{2}), (2, 1, 1, +\frac{1}{2}) \text{ and } (2, 1, -1, +\frac{1}{2})$$

It can also be shown with boxes

$$_7N \quad \boxed{\uparrow\downarrow}_{1s} \quad \boxed{\uparrow\downarrow}_{2s} \quad \boxed{\uparrow|\uparrow|\uparrow}_{2p}$$

The electron configuration of $_6$C atom in the ground state is displayed in following way

$$_6C \quad \boxed{\uparrow\downarrow}_{1s} \quad \boxed{\uparrow\downarrow}_{2s} \quad \boxed{\uparrow|\uparrow|\,}_{2p}$$

For the elements after atomic No. 20, the electron configuration of atoms in the ground state is expressed still in the order of electron shells, though the electrons are filled in the orbitals following the order of approximate energy level. For example, the electron configuration of $_{21}$Sc atom in the ground state is $1s^2 2s^2 2p^6 3s^2 3p^6 3d^1 4s^2$, not the $1s^2 2s^2 2p^6 3s^2 3p^6 4s^2 3d^1$. When $_{21}$Sc atoms are ionized, they lose the electron of 4s orbital first, as it is in the outmost shell. Therefore, the electron configuration of $_{21}$Sc$^+$ ion is $1s^2 2s^2 2p^6 3s^2 3p^6 3d^1 4s^1$.

As the supplement of Hund's rule, when the outer equivalent orbitals of an atom are full (i.e. p^6, d^{10}, f^{14}), or half-full (i.e. p^3, d^5, f^7), or empty (i.e. p^0, d^0, f^0), the atom is in the most stable state, partly due to the most electrons of parallel spin. Therefore, the electron configuration of $_{24}$Cr atom in the ground state is $1s^2 2s^2 2p^6 3s^2 3p^6 3d^5 4s^1$, not the $1s^2 2s^2 2p^6 3s^2 3p^6 3d^4 4s^2$, while that of $_{29}$Cu is $1s^2 2s^2 2p^6 3s^2 3p^6 3d^{10} 4s^1$, not the $1s^2 2s^2 2p^6 3s^2 3p^6 3d^9 4s^2$.

Sample Problem 9-4 Write down the electron configuration of $_{22}$Ti atom in the ground state according to the laws of filling electrons.

Solution According to Pauli Exclusion Principle and Aufbau principle, 1s and 2s orbitals are filled with two electrons respectively. Energy level 2p has three equivalent orbitals ($2p_x$, $2p_y$ and $2p_z$) which are full-filled with six electrons. Then, eight electrons are filled in 3s and 3p orbitals. 4s orbital is preferred by two of the four remaining electrons following the order of approximate energy level. The last two electrons occupy 3d orbital. Hence, the electron configuration of $_{22}$Ti atom in the ground state is $1s^2 2s^2 2p^6 3s^2 3p^6 3d^2 4s^2$.

Atomic core consists of atomic nucleus and inner electron shells possessing the electron configuration of a certain noble gas atom. Generally, atomic core does not change in chemical reactions, and the electron configuration of noble gas atomic structure within the atomic core can be simplified by adding square bracket to the corresponding element symbol. For examples, the electron configuration of $_{20}$Ca atom in the ground state is $1s^2 2s^2 2p^6 3s^2 3p^6 4s^2$, and can be simplified as $[Ar]4s^2$, while that of $_{26}$Fe $1s^2 2s^2 2p^6 3s^2 3p^6 3d^6 4s^2$ as $[Ar]3d^6 4s^2$.

In these examples, the shell structure outside the square bracket is relatively variable, which involves in the change of valence. So, the related electron and shell are named as **valence electron** and **valence shell** respectively. For example, the electron configuration of $_{47}$Ag atom in the ground state is $[Kr]4d^{10}5s^1$. And its valence electron configuration is $4d^{10}5s^1$. Therefore, the structure of atomic valence electron can be seen from the simplified form of electron configuration.

The atom with p-electrons in its outmost shell, in some cases, has $(n-1)d^{10}$ or $(n-2)f^{14}$ structure which is very stable. Because the d or f-electrons in the atom are not valence electrons, they are not included in the valence electron configuration. For example, the electron configuration of $_{32}$Ge atom in the ground state is $[Ar]3d^{10}4s^2 4p^2$, and its valence electron configuration is $4s^2 4p^2$.

The expression of ion electron configuration follows the way of writing that of atom. For example, the electron configurations of Fe^{2+} and Fe^{3+} ion are $[Ar]3d^6$ and $[Ar]3d^5$ respectively.

9.4 Periodic Table and Trends in the Properties of Element

Periodic table presents a lot of information on the elements. Some trends in the properties of elements could be studied by periodic table. The electron configuration of elements deduced by lots of spectroscopic data was applied to explain the periodic trends and reveal the nature of periodic table in microscopic level.

9.4.1 Electron Configuration of Atoms and Periodic Table

1. Energy level groups and periods of elements

It was known initially that the relative magnitude of energy level or the energy difference between energy levels mainly depended on principal quantum number n. However, this viewpoint needs to be reformed after the discovery of energy level inversion. The Chinese chemist Guangxian Xu proposed the concept of energy level group and suggested to calculate the $(n+0.7l)$ value for each energy level. The energy levels with the same integral number of $(n+0.7l)$ value are arranged into one energy level group, and the integer becomes its number. The energy difference between the energy levels in one group is small, while relatively large in the different. Energy level 1s is in energy level group 1. Energy levels

ns, np, along with $(n-1)$d, and $(n-2)$f make up energy level group n. The order of energy-level rising acquired according to $(n+0.7l)$ rule is in consistent with that proposed by L. Pauling, following which the electrons of an atom in the ground state are filled.

Each group of energy level corresponds to one **period** in the periodic table by Table 9-3. Energy level group 1 has only energy level 1s which corresponds to period 1. Thereafter, energy level group n that includes from energy level ns to that of np corresponds to period n. In the period n, the electron configuration of outer shell varies from ns^1 to ns^2np^6. The number of elements is that of electrons full of an energy level group. For example, in period 4, the atomic electron configuration of outer shell starts from $4s^1$ to the end of $4s^24p^6$. However, in addition to 4s and 4p energy levels, this energy-level group contains that of 3d which can accommodate 10 electrons ultimately. Therefore, there are totally 18 elements in period 4. The numbers of elements in periods from one to seven are 2, 8, 8, 18, 18, 32, and 32 respectively. Period 1 is the super short period. Periods 2 and 3 are short periods, and the periods after belong to long period.

Table 9-3   Energy Level Groups and Periods

Energy level	$n+0.7l$	Energy level group	Number of electrons full of an energy level group	Period	Number of elements in a period
1s	1.0	1	2	1	2
2s	2.0	2	8	2	8
2p	2.7				
3s	3.0	3	8	3	8
3p	3.7				
4s	4.0	4	18	4	18
3d	4.4				
4p	4.7				
5s	5.0	5	18	5	18
4d	5.4				
5p	5.7				
6s	6.0	6	32	6	32
4f	6.1				
5d	6.4				
6p	6.7				
7s	7.0	7	32	7	32
5f	7.1				
6d	7.4				
7p	7.7				

Sample Problem 9-5   How many elements are there in Period 7? What is the atomic number of the last element in periodic table?

Solution   According to $(n+0.7l)$ rule, energy level group 7 includes energy level 7s, 5f, 6d and 7p, providing 1, 7, 5, and 3 orbitals respectively. A total of 16 orbitals can accommodate 32 electrons ultimately. So, there are 32 elements in period 7. The atomic number Z of the last element is calculated as follows

$$Z = 2 + 8 + 8 + 18 + 18 + 32 + 32 = 118$$

2. Valence electron configuration and groups

Generally, elements with similar valence electron configuration are arranged in one column, named as a **group**, in periodic table. There are 16 groups in periodic table including 8 representative groups and 8 subgroups. The obvious difference of properties between the elements of representative groups and subgroups can be explained in terms of valence electron configuration.

(1) **Representative groups:** Elements with last electron filled in an s or p-orbital belong to representative groups which consist of Groups ⅠA, ⅡA, ⋯, and ⅧA which is also called Group zero. The inner shells of the representative group elements are full-filled. Their electron configuration of the outmost shell, which is also the valence shell, varies from ns^1 to ns^2np^{1-6}. The number of electrons in the outermost shell is the group number. Hydrogen and helium atoms are special for either of them with only one electron shell, and their electron configurations are $1s^1$ and $1s^2$ respectively. Hydrogen belongs to Group ⅠA. Helium belongs to Group zero and is a noble gas element.

(2) **Subgroups:** Elements with last electron filled in d or f-orbitals belong to subgroups which consist of Groups ⅠB, ⅡB, ⋯, and ⅧB. The character of electron configuration for subgroups elements is that $(n-1)$d or $(n-2)$f orbitals are generally occupied by electrons. $(n-1)$d, $(n-2)$f along with ns belong to valence shell. There are no subgroup elements in Periods 1-3. In Period 4 or 5, there are 10 subgroup elements, considering 3d or 4d valence orbitals. The subgroups after representative groups ⅠA and ⅡA in periodic table are ⅢB ~ ⅦB. Subgroup number equals to the sum of electrons in $(n-1)$d and ns orbitals. Subgroup ⅧB elements occupy three columns in periodic table, and the sum of electrons in $(n-1)$d and ns orbitals varies from 8 to 10. The final two columns belong to Subgroups ⅠB and ⅡB, with 1 and 2 ns electrons, respectively, and either has $(n-1)d^{10}$ valence shell. The number of ns electrons is the subgroup number. In Periods 6 and 7, Subgroup ⅢB elements belong to **lanthanide series** and **actinide series**, either of which has 15 elements. The character of electron configuration is that $(n-2)$ f orbitals are filled with electrons until they are full, and the number of electrons in $(n-1)$d orbitals is 1 or 0 in most cases. The $(n-2)$f orbitals of the elements from subgroups ⅣB to ⅡB are full-filled. In general, their electron configuration of $(n-1)$d and ns orbitals is similar to that of corresponding subgroup elements in Periods 4 and 5.

(3) **Blocks:** Periodic table is divided into five blocks according to the characteristics of valence electron configuration of elements in Figure 9-12.

a. s-block includes representative groups ⅠA and ⅡA. The valence electron configuration of s-block elements is ns^1 or ns^2. Elements in s-block are metals except for hydrogen, and easily lose one or two electrons to become monovalent or divalent ions in chemical reactions. They have no other available

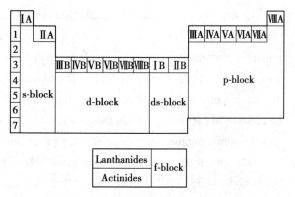

Figure 9-12 Blocks in Periodic Table

oxidation numbers in their compounds. Generally, the oxidation number of hydrogen is +1, but it is -1 in metal hydrides.

b. p-block includes representative groups ⅢA~ⅧA. The valence electron configuration of p-block elements, most of which are nonmetals, is ns^2np^{1-6} except that of helium (i.e. $1s^2$). The elements in group ⅧA consist of noble gas elements. The elements in p-block have variable oxidation numbers and are most active in chemical reactions.

c. d-block includes Subgroups ⅢB~ⅧB. The valence electron configuration of d-block elements, which are all metals, is $(n-1)d^{1-8}ns^2$, $(n-1)d^9ns^1$, or $(n-1)d^{10}ns^0$. They share many physical and chemical properties. Each of d-block elements has multiple oxidation numbers, empty d-orbitals and easily become the central atom of coordination compounds.

d. ds-block includes subgroups ⅠB and ⅡB. The valence electron configuration of ds-block elements, which are all metals, is $(n-1)d^{10}ns^{1-2}$. The $(n-1)d$ orbitals are full-filled. ds-block elements have variable oxidation numbers.

e. f-block includes lanthanide series and actinide series. The valence electron configuration of f-block elements, which are all metals, is $(n-2)f^{0-14}(n-1)d^{0-2}ns^2$. Generally, the number of electrons in n or $(n-1)$ shell of f-block elements is identical, but that in $(n-2)f$ orbitals is different. Elements of lanthanide series and actinide series have similar chemical properties, variable oxidation numbers and share oxidation number +3.

Sample Problem 9-6 Atomic number of a certain element is 25, please write down its electron configuration in the ground state, and indicate which period, group and block the element is in.

Solution There are totally 25 electrons outside the nucleus of the atom. The electron configuration in the ground state is $1s^22s^22p^63s^23p^63d^54s^2$, or $[Ar]3d^54s^2$. The $(n+0.7l)$ values of 3d and 4s energy levels are 4.4 and 4.0 respectively, corresponding to energy level group 4. So, it is in Period 4. The total number of valence electrons is 7, meaning the element is in Subgroup ⅦB, and d-block. It is Mn.

(4) **Transition elements and rare earth elements:** At the beginning, **transition elements** only refer to Subgroup ⅧB elements. Now, the concept of transition elements is broadened to all subgroup elements which belong to d, ds, f-block respectively. Here, lanthanides and actinides are **inner transition elements** because their $(n-2)f$ orbitals are occupied by electrons. Sc and Y are chemically similar to 15 lanthanides, and the 17 elements are known as **rare earth elements**.

All transition elements are metals. Except $_{46}$Pd, they have 1or 2 s-electrons in the outmost shell. Moreover, their $(n-1)d$ orbitals are not full or just exactly full, and their $(n-2)f$ orbitals are not full. So, transition elements have variable oxidation numbers, and their properties are obviously different from those of representative group elements. The electron configuration of element is closely related with its location in periodic table. Generally, it can be figured out from atomic number, and then used to locate the element in periodic table. On the contrary, the location of element in periodic table can be applied to infer its atomic number and electron configuration, and then predict its oxidation number and properties.

The trend of elements in the properties is related with the periodic change of their atomic structures. The periodic trends are demonstrated in effective nuclear charge, atomic radius, ionizing energy, electron affinity energy and electronegativity, and so on.

9.4.2 Periodic Trends in the Properties of Elements

1. Atomic radii

The atomic radius is a measure of the size of its atoms, usually the mean or typical distance from the center of the nucleus to the boundary of the surrounding cloud of electrons. Since the electron cloud distributes across the space outside the nucleus and the boundary is not a well-defined physical entity, an atom does not have a precise radius. For an isolated atom, it is impossible to measure its radius experimentally. The generally observed value of atomic radius is determined by the distance between two adjacent atoms in a crystal, one or two gaseous molecules. Therefore, there are various non-equivalent definitions of atomic radii, such as **covalent radius**, **van der Waals radius**, **metallic radius** and **ionic radius**. When two identical atoms are combined with covalent bond, the covalent radius of the atom is the half distance between the two nuclei. The metallic radius of atom is the half distance between two nuclei of two adjacent atoms in metal crystal. For example, Na_2 molecule is the form of Na element in sodium vapor, and it has a covalent radius. When Na atoms closely stack to each other in metal Na, therefore, it has a metallic radius. If Na^+ and Cl^- ions form NaCl crystal, an ionic radius of Na can be calculated according to the distance between Na^+ and Cl^- ions and the radius of Cl^- ion. When molecules approach to each other by van der Waals forces and achieve a dynamic equilibrium against repulsion force from their electrons, van der Waals radius equals to half distance between nuclei of two adjacent atoms in two neighbor molecules.

Four types of atomic radii of Cl and Na atoms (pm) are as follows.

atom	Covalent radius	Metallic radius	Ionic radius	van der Waals radius
Cl	99		181	198
Na	157	186	99	231

Among the four types of atomic radii, covalent radius and metallic radius are radii of bonding atom, and they are much smaller than van der Waals radius. As covalent bond involves in single, double and triple bonds, the covalent radius is related to the type of covalent bond. For example, there are three covalent radii for C atom: r(single bond) = 77pm, r(double bond) = 67pm, r(triple bond) = 60pm.

The periodic trend of atomic radius is mostly related to the effective nuclear charge and the shell numbers of an atom.

For two adjacent representative elements in a given period, the right atom has one more both nucleus charge and electron that occupies s or p-orbital of outmost shell and increases the screening constant of valance electron 0.30 or 0.35. The effective nucleus charge increases 0.65 at least. Thus, the attraction between nucleus and valance electron improves rapidly and the atomic radius decrease obviously.

For two adjacent transition elements in a given period, the right atom has one more electron that occupies $(n-1)$d or $(n-2)$f orbital and increases the screening constant of electron in outmost shell 0.85 or 1.00. The effective nucleus charge increases at most 0.15. It means the attraction between nucleus and electron in outmost shell increases rarely, and as a result, the atomic radius decreases slowly. The effective nucleus charge of inner transition elements does not increase obviously and the atomic radius is almost unchanged with the increase of atomic number.

The approximate difference between atomic radii of two adjacent elements in a given period is as follows.

Non-transition element (~10pm) > transition element (~5pm) > inner transition element (~1pm).

With the increase of shell number of representative elements down a group, electrons in outermost shell are more far away from nucleus. But, effective nucleus charge increases slowly due to the screening effect from electrons in inner shells. Thus, atomic radius increases obviously down the group in periodic table.

The atomic radius of subgroup elements increases with the addition of shell number down the group. But, the atomic radii of subgroup elements within a group in Periods 5 and 6 are actually close. That is why these elements coexist in mine and are difficult to be isolated.

2. Ionization energy, electron affinity, and electronegativity

Ionization energy is used to valuate the tendency of atom or ion losing their electrons. First ionization energy I_1 is qualitatively defined as the amount of energy required to remove a valence electron of an isolated gaseous atom in the ground state to form a cation, and has the trend in periodic table. Generally, I_1 increases with the decrease of the atomic radius or increase of the effective nuclear charge along each period of the table, from the alkali metals to the noble gases. However, there are several exceptions. For example, three 2p orbitals of N atom are just half-filled with one electron in each, which bring extra stability to N atom according to the supplement of Hund's rule. As a result, I_1 of N atom is higher than that of O atom, and so is for Be and B atoms. With the increase of shell number down the representative group, the outermost shell electron is getting further away from the nucleus, meanwhile the effective nucleus charge increases rarely. This decreases the electron-nucleus attraction and makes the electron in outermost shell ionized more easily, and then I_1 decreases gradually.

Electron affinity of an atom is defined as the amount of energy released when an electron is added to a neutral atom in the gaseous state to form a negative ion. It is used to measure the ability of atom combining with electrons. Generally, the halogen atom releases more energy when combined with electrons, if compared with any other atom across a period. Therefore, halogen atoms are relatively easier to derive electrons. On the contrary, the metal atom releases less energy or even absorbs energy when combined with electrons, so, it is hard to get an electron to form a negative ion.

Either ionization energy or electron affinity just reflects the tendency of atom to lose or get electrons unilaterally. Actually, some atoms neither lose nor get electrons easily, such as C and H atoms. Therefore, it is a limitation to solely use ionization energy or electron affinity to describe the metallicity or non-metallicity of elements. Considering ionization energy and electron affinity comprehensively, in 1932, L. Pauling firstly proposed the definition of **electronegativity**, notated as symbol χ. The electronegativity of F atom is the maximum and the value was nearly to be 4.0. Based on this, χ values of other elements were figured out. Electronegativity can describe the tendency of an atom to attract a shared pair of covalent electrons towards itself. The higher the associated electronegativity value, the more an element attracts electrons towards it.

Table 9-4 shows the value of element electronegativity which have been revised according to thermochemical data and varies periodically with atomic number. It can be seen that the electronegativity of representative group elements gradually increases on passing from left to right along a period and decreases down a group. For subgroup elements, there is no obvious trend of electronegativity. Elements with strong electronegativity are concentrated at the upper-right corner of periodic table, such as F, O, Cl, N, Br, S, C and some other nonmetal elements. Elements with weak electronegativity are located at the lower-left corner of periodic table, such as Cs, Rb, Ba and some other alkaline metal, alkali earth metal elements. The electronegativity value of 2 is generally used to classify an element as a metal or nonmetal approximately.

Table 9-4 Electronegativity of Elements

H 2.18																	He
Li 0.98	Be 1.57											B 2.04	C 2.55	N 3.04	O 3.44	F 3.98	Ne
Na 0.93	Mg 1.31											Al 1.61	Si 1.90	P 2.19	S 2.58	Cl 3.16	Ar
K 0.82	Ca 1.00	Sc 1.36	Ti 1.54	V 1.63	Cr 1.66	Mn 1.55	Fe 1.80	Co 1.88	Ni 1.91	Cu 1.90	Zn 1.65	Ga 1.81	Ge 2.01	As 2.18	Se 2.55	Br 2.96	Kr
Rb 0.82	Sr 0.95	Y 1.22	Zr 1.33	Nb 1.60	Mo 2.16	Tc 1.90	Ru 2.28	Ru 2.20	Pd 2.20	Ag 1.93	Cd 1.69	In 1.73	Sn 1.96	Sb 2.05	Te 2.10	I 2.66	Xe
Cs 0.79	Ba 0.89	La 1.10	Hf 1.30	Ta 1.50	W 2.36	Re 1.90	Os 2.20	Ir 2.20	Pt 2.28	Au 2.54	Hg 2.00	Tl 2.04	Pb 2.33	Bi 2.02	Po 2.00	At 2.20	

The concept of electronegativity has been widely used, for example, to understand polarity of chemical bond, explain some physical or chemical properties of substance, and predict the trend of metallic or non-metallic nature of elements.

9.5 Elements and Health of Human Body

9.5.1 Essential Elements of Human Body and Brief Introduction on Their Biofunctions

Up to now, 118 elements have been registered and named. Among them, the first 92 elements exist naturally. Elements of atomic numbers 93~118 have only been synthesized in laboratories or nuclear reactors. 81 elements have been detected in life entities, and they are biological elements. 11 elements are macroelements because the mass fraction of each of them in human body is greater than 0.05%. The mass fraction of microelement or trace element in human body is less than 0.05%. The amount of an element in human body has a certain relationship with its content in circumstance. In early 1970s, the English geochemist, E.T. Hamilton discovered the similarity in component and content distribution of the elements between human blood and sea water. The elements are divided into essential elements and non-essential elements according to their biofunctions in human body. Essential elements collect 11 macroelements in Table 9-5 and 18 trace elements in Table 9-6. Macroelements include four metal elements, Na, K, Ca and Mg, and each atomic number is less than 21 in periodic table. Most of the essential microelements locate in Period 4.

Table 9-5 Macroelements in human body

Element	Content*/ g	Mass fraction in body/ %	Distribution in human body
O	45 000	64.30	water, component of organic compounds
C	12 600	18.00	component of organic compounds
H	7 000	10.00	water, component of organic compounds
N	2 100	3.00	component of organic compounds
Ca	420	2.00	same as above, bone, teeth, muscle, body fluid
P	700	1.00	same as above, skeleton, teeth, phospholipids, phosphoprotein
S	175	0.25	Sulfur amino acids, hair, nail, skin
K	245	0.35	intracellular fluid
Na	105	0.15	extracellular fluid, bone
Cl	105	0.15	Cerebrospinal fluid, gastrointestinal tract, extracelluar fluid, bone
Mg	35	0.05	Bone, teeth, intracellular fluid, soft tissues

*The content is the mass of element in 70kg human body.

Table 9-6 Essential trace elements in human body

Element	Content*/ mg	Plasma concentration/ μmol·L^{-1}	Distribution in the human tissues	Proof history
Fe	2 800~3 500	10.75~30.45	Red blood cell, liver, bone marrow	17th century
F	3 000	0.63~0.79	bone, teeth	in 1971
Zn	2 700	12.24~21.42	Muscle, bone, skin	in 1934
Cu	90	11.02~23.6	Muscle, connective tissue	in 1928
V	25	0.20	Fat	in 1971
Sn	20	0.28	Fat, skin	in 1970
Se	15	1.39~1.9	Muscle (cardiac muscle)	in 1957
Mn	12~20	0.15~0.55	Bone, muscle	in 1931
I	12~24	0.32~0.63	Thyroid	in 1850
Ni	6~10	0.07	Kidney, skin	in 1974
Mo	11	0.04~0.31	liver	in 1953
Cr	2~7	0.17~1.06	Lung, kidney, pancreas	in 1959
Co	1.3~1.8	0.003	Bone marrow	in 1935
Br	<12			
As	<117		Hair, skin	in 1975
Si	18 000	15.31	Nodus lymphaticus, nail	in 1972
B	<12	3.60~33.76	Brain, liver, kidney	in 1982
Sr	320	0.44	Bone, teeth	

*The content is the mass of element in 70kg human body.

The essential elements involve in various aspects of composition and activity of life.

(1) They are principal components of human tissues. For example, elements H, O, C, N, S and P are major constituents of proteins, nucleic acids, saccharides and fats which are the foundations of lives. Elements Ca, P, and Mg are the important constituents of bone and teeth.

(2) They are constituents of the substances with some special functions. For example, ferrum is a constituent of hemoglobin, Iodine is an essential constituent of thyroid hormone, chrome exists in glucose tolerance factor (GTF), and cobalt is the central atom of vitamin B_{12}. The trace elements, such as Zn, Mo, Mn, and Cu, are the active centres of enzymes, and some of them are activators or inhibitors of enzymes.

(3) They participate to maintain the osmotic pressure of body fluid.

(4) They participate to maintain the acid-base and electrolyte equilibria in body.

(5) They participate to keep the excitability of nerves and muscles.

It should be noted that the definition of "essential elements" or "non-essential elements" is relative.

A non-essential element may turn into an "essential element" in future with the improvement of test facilities and diagnostic methods. For example, arsenicum was considered as a harmful element and until its essentiality for human body was discovered in 1975. Now, many researches show that rubidium is related to life course, but further research is needed to recognize its "essentiality".

Generally, essential trace elements with different valence numbers show different biofunctions on animate objects. For example, Cr^{3+} has strong ability to coordinate with weak organic ligand or inorganic ligand to be functional in vivo. But, Cr^{6+}, mainly as the form of chromate, is poisonous and can permeate into red blood cell. In the process of Cr^{6+} being reduced to Cr^{3+}, the activity of glutathione reductase is inhibited. This prompts the transformation of hemoglobin into methemoglobin and causes the disorder of red blood cell carrying oxygen. Then, the decrease of oxygen in blood leads to various toxic symptoms.

The concentrations of essential elements have proper scale, *in vivo*. It is bad for health if the amount of a certain essential element is either excessive or insufficient relative to the scale.

The biological effects of elements are closely related to their locations in periodic table. Generally, for elements in s and p-blocks, the nutrient effects decrease and toxicities increase on passing from left to right along a period or down a group.

Sample Problem 9-7 Iodine is an essential constituent of thyroxine. Iodine deficiency can cause goiter, cretinism and mental retardation. The simplest and most effective method to prevent iodine deficiency is to use salt with iodine. Please locate iodine in periodic table and find out which hydride is the strongest reductant in that group.

Solution Iodine is in Period 5, Group ⅦA and p-block. The non-metal elements in main Group ⅦA are F, Cl, Br, I, and their hydrides are HF, HCl, HBr, and HI, respectively. Electronegativity of iodine within the elements is the weakest, and I^- is the ion to lost electron most easily. So, HI is the strongest reductant.

9.5.2 Harmful Elements to Human in Environment

Among biohazardous elements, Pb, Ge, Hg, Tl have attracted more and more attention on polluting environment. The mechanisms of toxicity caused by these biohazardous elements involve in inhibiting the functions of essential active groups of biomacromolecules, substituting essential metal ions in biomacromolecules, or transforming their active conformations, and finally destroy the immune system of human body, and produce neurotoxicity or carcinogenesis.

1. Lead(Pb)

In 1994, the first Childhood Lead Poisoning Prevention Meeting warned: "85% of children in industrial district have ingested overdose lead". In China, the concerned Administrations also published that 51.6% of children in cities were lead-poisoned. Either lead or lead compound causes giant toxicity. They mainly destroy the hemopoietic system, nervous system, kidney, and cause irreversible damage to children's intelligence. Among the environmental pollutants, the primary damage to children's health is caused by lead pollutants that mainly come from the exhaust of automobiles of using gasoline with tetra-ethyl lead as an antiknock compound. At present, the leaded gasoline has been forbidden in many Chinese cities. Eating Fe-rich foods may have certain effect in lead-discharge by increasing the iron content in blood, which enhances the iron-hemoglobin affinity and promote the lead-hemoglobin dissociation.

2. Mercury(Hg)

In the Ming Dynasty, Yingxing Song recorded the method of preventing mercury poisoning. Mercury and most of mercury compounds are poisonous. The toxicity of organomercury is larger than that of inorganic mercury. The Minamata disease which originated from methylmercury intoxication burst out in 1956, in Minamata City, Kumamoto Prefecture, Japan and shocked the whole world. The pollution origin is the organomercury in sewage of a local nitrogenous fertilizer factory. It poisoned the fishes in the river and then people through food chain. In 1971, methyl mercury caused 6 530 people poisoned and 459 of them dead in Iraq.

3. Cadmium(Cd)

After World War II, the cadmium poisoning resulted in the prevalence of Itai-Itai disease around Jinzu River and its tributaries, Toyama Prefecture, Japan. 258 people suffered from Itai-Itai disease and 128 of them were dead. The sufferers were from 30 to 70 years old and almost females, in which most were just before or after the menopause and between the ages of 47 and 54. The victims could hardly suffer from the pain of arthrosis and bones of the whole body. The content of cadmium in the bones of the dead was 159 times higher than that of healthy adults. The pollution source is the sewage with cadmium released into river by a mining company, which polluted the rice fields, and resulted in chronic poisoning of local populace due to the "cadmium rice". Vitamin D and calcium agents can be used to treat the disease in addition to some special methods for cadmium excretion.

4. Thallium(Tl)

Thallium which is in Period 6, Group ⅢA and all thallium compounds are poisonous. The reasons of thallium poisoning may involve in iatrogenic, food and environment pollutions, concerned occupation. The typical appearances of thallium poisoning are the shedding of hairs and crinis pubis. Effective expellent for thallium poisoning is Prussian blue.

Key Terms

有核模型	nuclear model
普朗克常数	Planck constant
线状光谱	line spectrum
定态	stationary state
能级	energy level
基态	ground state
激发态	excited state
跃迁	transition

量子数	quantum number
波粒二象性	wave-particle duality
de Broglie 关系式	de Broglie relation
概率波	probability wave
不确定原理	uncertainty principle
波函数	wave function
Schrodinger 方程	Schrodinger's equation
概率密度	probability density
电子云	electron cloud
原子轨道	atomic orbital
主量子数	principal quantum number
电子层	electron shell
轨道角动量量子数	orbital angular momentum quantum number
电子亚层	electron subshell, electron sublevel
磁量子数	magnetic quantum number
简并轨道	equivalent orbital
径向波函数	radial wave function
角度波函数	angular wave function
节面	nodal plane
径向分布函数	radial distribution function
屏蔽作用	screening effect
屏蔽常数	screening constant
有效核电荷数	effective nuclear charge
自旋角动量量子数	spin angular momentum quantum number
电子组态	electronic configuration
Pauli 不相容原理	Pauli exclusion principle
构造原理	building-up principle or Aufbau principle
洪特规则	Hund's rule
原子芯	atomic core
价电子	valence electron
价电子层	valence shell
周期	period
族	group
过渡元素	transition element
内过渡元素	inner transition element
稀土元素	rare earth element
原子半径	atomic radius
共价半径	covalent radius
van der Waals 半径	van der Waals radius
金属半径	metallic radius

离子半径	ionic radius
电负性	electronegativity
生命元素	biological element
常量元素	macroelement
微量元素	microelement
痕量元素	trace element
必需元素	essential element
非必需元素	non-essential element
有毒或有害元素	poisonous element or harmful element

Summary

Rutherford's model gave us the qualitative picture of an essentially empty atom with a heavy positive nucleus in the center and light negative electrons moving around it, but encountered great difficulties with the prediction of continuous emission spectra for atoms. In 1913, N. Bohr proposed a theory of atomic structure by synthesizing the results of Plank's quantum theory and Einstein's relativity theory, providing a quantitative prediction of line spectra. In his theory, an atom can exist for a long time without radiating in certain stationary state with discrete energy which is quantized. The frequency of radiation emitted or absorbed by an atom as a result of a transition between two energy levels is determined by $h\nu = |E_2 - E_1|$. A number of conclusions drawn from his assumptions agreed quantitatively with the experimental observations on hydrogen atom. However, Bohr's theory was a failure for many-electron atoms. The quantum mechanics that gives correct description of electronic behavior in atoms has its roots in a hypothesis put forward by L. de Broglie in 1923. A particle, such as an electron, with momentum p should have associated with it a wave whose wavelength is given by $\lambda = h/p$. This is a consequence of the uncertainty principle of Heisenberg, which really means that we are limited in our ability to know simultaneously where the electron is and where it is going. It leads us to refer to where the electron is likely to be found, not the trace of its motion.

The wave function (ψ) obtained by E. Schrödinger through solving a wave equation is a mathematical description of the electron's wavelike motion in an atom. Each of these different possible waves is called an atomic orbital that has a characteristic energy. The square of the wave function (ψ^2) represents the probability of finding the electron in some small element of volume at various places around the nucleus. The wave functions that describe the orbitals are characterized by the values of principal quantum number (n), orbital angular momentum quantum number (l) and magnetic quantum number (m). n and l determine the size and shape of an orbital respectively, while m determines its orientation in space relative to the other orbitals. n stands for main level and the energy of an orbital only depends on the n value in the case of hydrogen atom. For any given n, sublevel l may take the values of 0, 1, 2, $\cdots$, $(n-1)$ and affects the energy of an orbital to a certain degree in many-electron atoms. The values of n and l determine the energy level. A sublevel with $l=0$ has a spherical (s) orbital, $l=1$, three two-lobed (p) orbitals, and $l=2$, five four-lobed (d) ones.

The distribution of charge in an orbital could be indicated in different ways. The angular distribution of electron can be showed by polar plots of the square of angular wave function $Y_{l,m}(\theta, \varphi)$. A radial distribution plot can show us how the electron occupies the space near the nucleus for a particular energy level. The order of energy levels, $E_{1s} < E_{2s} < E_{2p} < E_{3s} < E_{3p} < E_{4s} < E_{3d} < E_{4p} < \cdots$, in many-electron atoms was deduced by L. Pauling according to spectral data.

In addition to n, l, m, the spin angular momentum quantum number (s) has to be applied to specify electrons. s may take the value of 1/2 or $-$1/2 and arises because the electron behaves as if it is spinning. Pauli Exclusion Principle limits the number of electrons in any given orbital to two with the spins being opposite. The electron configuration of an atom in the ground state is the way the electrons are distributed among the orbitals following Pauling's energy level order and Hund's rule which requires electrons occupying different equivalent orbitals in a spin-parallel manner. For representative group elements, the valence electrons in the outermost shell are responsible for chemical changes. For transition elements, inner d electrons are also involved in chemical reactions. The elements of a group in periodic table have similar chemical behavior because of their similar valence shell.

Many of the properties, such as atomic size and electronegativity, of the elements vary in a more or less regular fashion. For example, atomic size which depends on both outer shell and effective nuclear charge obviously increases down a representative group, while gradually decreases on passing from left to right along a period. Across a transition series, size remains relatively constant. Electronegativity decreases down a representative group and increases on passing from left to right along a period.

Exercises

1. What are the wave properties of electron? What are the differences between electron wave and electromagnetic wave?

2. "The 1s electron moves on a spherical orbit." Is it true or false? Why?

3. What is the de Broglie wavelength of electron with velocity of $7 \times 10^5 \mathrm{m \cdot s^{-1}}$?

4. The mass of a bullet and its velocity are 10g and $1\,000 \mathrm{m \cdot s^{-1}}$, respectively. Explain its particle properties, and why its movement obeys the rules of classical mechanics according to de Broglie relation and uncertainty principle. (The velocity uncertainty of a bullet, Δv_x, equals to $10^{-3} \mathrm{m \cdot s^{-1}}$)

5. Why can one atomic orbital only be occupied by up to 2 electrons?

6. Write down the notation of energy levels or orbitals for
   (1) $n=2$, $l=1$ (2) $n=3$, $l=2$ (3) $n=5$, $l=3$
   (4) $n=2$, $l=1$, $m=-1$ (5) $n=4$, $l=0$, $m=0$

7. The electron configuration of nitrogen atom is $2s^2 2p^3$. Describe the motion of its valence electrons by the combination of four quantum numbers.

8. Which is reasonable among the subshells listed below? How many orbitals does a reasonable subshell contain?
   (1) 2s   (2) 3f   (3) 4p   (4) 5d

9. Fill the correct answers in the blanks below. Each atom involved is in the ground state.

Atomic number	Electron configuration	Valence electron configuration	Period	Group
49				
	$1s^2 2s^2 2p^6$			
		$3d^5 4s^1$		
			6	IIB

10. Write down the electron configuration and number of unpaired electrons for the atoms or ion listed below without the help of periodic table.

(1) The seventh element in Period 4

(2) The noble gas element in Period 4

(3) The most stable ion of element No. 38

(4) The representative group element whose 4p orbitals are half-filled

11. Write down the electron configuration of Ag^+, Zn^{2+}, Fe^{3+} and Cu^+ ions.

12. The atom of an element has 24 electrons. Which period, group, block is the element in? What is the difference of atomic radius between the element and either of its two neighbors in periodic table?

13. The first ionization energy I_1 of representative group elements increases on passing from left to right along a period. However, why the I_1 of $_{15}$P is higher than that of $_{16}$S in Period 3?

14. Rearrange the following atoms in the order of decrease of electronegativity and explain the reasons.

$$As, F, S, Ca, Zn$$

15. The characteristic of valence electron configuration of each atom in the ground state is listed below. Which group the element is in? or which element it is?

(1) It has two electrons in p-orbitals.

(2) It has two electrons with quantum numbers $n=4$ and $l=0$, and six electrons with quantum numbers $n=3$ and $l=2$.

(3) Its 3d orbitals are half-filled, and 4s orbital has one electron.

16. Copper is needed to participate in the process of metabolism and transportation of ferrum in human body. Ceruloplasmin can catalyze the oxidation of Fe^{2+} into Fe^{3+} ion, and transport ferrum to bone marrow. Try to explain why the oxidation easily happens according to the theory of atomic structure.

17. Selenium is closely related to health. It exists in the form of selenium-enzyme or selenium-protein in vivo. The deficiency of selenium may cause Keshan disease, Kashin-Beck disease and Cataract, etc. Predict the oxide of selenium with highest oxidation state according to its location in periodic table.

Supplementary Exercises

1. An electron in hydrogen atom undergoes a transition from level $n=5$ to $n=3$. What is the frequency of the emitted radiation?

2. What is the wavelength of a neutron obtained from a nuclear pile, traveling at a speed of $3.90 \times 10^3 \text{m} \cdot \text{s}^{-1}$?

3. How many subshells are there in M shell? How many orbitals are there in f subshell?

4. Among the following electron configurations, which is reasonable according to the theories about atomic structure? Explain why the others are not expressed correctly.

(1) $1s^2 2s^1 2p^6$ (2) $1s^2 2s^2 2p^6 3s^1 3d^6$ (3) $1s^2 2s^2 2p^8$ (4) $1s^2 2s^1 2p^6 3s^2 3d^{10}$

5. The electron configuration of thallium atom in the ground state is $[Xe]4f^{14}5d^{10}6s^2 6p^1$. Give the group and period of this element. Classify it as a representative group, a d-transition, or an f-transition element.

6. With the aid of periodic table, please rearrange the order of the following elements in increasing electronegativity.

(1) Sr, Cs, Ba   (2) Ca, Ge, Ga   (3) P, As, S

Answers to Some Exercises

[Exercises]

1. Since an electron with high velocity does not have definite position and momentum simultaneously, it is impossible to depict its motion trail like that of a macroscopic object. The wave-like motion of electron is statistic and can be expressed through probability. In modern quantum mechanics, the squared modulus of the wave function, $|\psi|^2$ that is probability density, is used to represent the intensity of electron wave. Therefore, the electron wave is a probability wave. The electromagnetic waves are synchronized oscillations of electric and magnetic fields that propagate at the speed of light through a vacuum. They are not probability waves, but energy waves.

2. It is false. Since there is not classical, planetary-like circle orbital in an atom, the motion of an electron can not be described through a trace, but by wave function. The "1s" here means the wave function whose shape is spherical. The 1s electron can be found everywhere in the space outside the nucleus. But the probability of appearing in a spherical shell at different distance from the nucleus is diverse.

3. The de Broglie wavelength of electron with velocity of $7 \times 10^5 \text{m} \cdot \text{s}^{-1}$ is 1 040pm.

4. According to de Broglie relation,

$$\lambda = \frac{h}{m\upsilon} = \frac{6.626 \times 10^{-34} \text{ kg} \cdot \text{m}^2 \cdot \text{s}^{-1}}{10 \times 10^{-3} \text{kg} \times 1\,000\text{m} \cdot \text{s}^{-1}} = 6.626 \times 10^{-35} \text{m}$$

The de Broglie wavelength of the bullet is so small that its wave properties can be neglected.

According to uncertainty principle,

$$\Delta x \geqslant \frac{h}{4\pi m \Delta \upsilon_x} = \frac{6.626 \times 10^{-34} \text{kg} \cdot \text{m}^2 \cdot \text{s}^{-1}}{4\pi \times 10 \times 10^{-3} \text{kg} \times 10^{-3} \text{m} \cdot \text{s}^{-1}} = 5.3 \times 10^{-30} \text{m}$$

The position uncertainty of the bullet is so small that it can be neglected. Therefore, the bullet can fly precisely along the trajectory.

5. According to Pauli Exclusion Principle, no two electrons could exist in the same quantum state, identified by four quantum numbers n, l, m, and s. If more than two electrons are in one orbital, at least two electrons have the same quantum state, meaning the break of Pauli Exclusion Principle. Therefore, one atomic orbital only can be occupied by up to 2 electrons in antiparallel-spin state.

6. (1) 2p energy level, (2) 3d energy level, (3) 5f energy level, (4) $2p_x$ or $2p_y$ orbital, (5) 4s orbital.

7. $(2, 0, 0, +1/2)$, $(2, 0, 0, -1/2)$, $(2, 1, -1, +1/2)$, $(2, 1, 0, +1/2)$, $(2, 1, 1, +1/2)$

8. (1) reasonable, and it only has one orbital. (2) unreasonable, since $n = 3$, l must be smaller than 3. (3) reasonable, and it has three orbitals. (4) reasonable, and it has five orbitals.

9.

Atomic number	Electron configuration	Valence electronic configuration	Period	Group
49	$[Kr]4d^{10}5s^25p^1$	$5s^25p^1$	5	III A
10	$1s^22s^22p^6$	$2s^22p^6$	2	0
24	$[Ar]3d^54s^1$	$3d^54s^1$	4	VI B
80	$[Xe]4f^{14}5d^{10}6s^2$	$5d^{10}6s^2$	6	II B

10. (1) $[Ar]3d^54s^2$ and five unpaired electrons, (2) $[Ar]3d^{10}4s^24p^6$ and no unpaired electrons, (3) $[Kr]5s^2$ for the atom, $[Kr]5s^0$ for its most stable ion and no unpaired electron, (4) $[Ar]3d^{10}4s^24p^3$ and

three unpaired electrons.

11. Ag^+: $[Kr]4d^{10}$, Zn^{2+}: $[Ar]3d^{10}$, Fe^{3+}: $[Ar]3d^5$, Cu^+: $[Ar]3d^{10}$

12. The element is in Period 4, Group ⅥB, d-block and belongs to the transition elements. The difference of atomic radius between the element and either of its two neighbors in periodic table is approximately 5pm.

13. For the elements from left to right along a period, the atomic radius decreases, and the effective nuclear charge increases, therefore, to ionize the outermost electrons needs more energy. According to the supplement of Hund's rule, the valence electrons in 3p orbitals of $_{15}P$ are more stable due to half-filled, if compared to those of $_{16}S$. As a result, the I_1 of $_{15}P$ is higher than that of $_{16}S$.

14. The positions of the five elements in periodic table are as follows. Since the electronegativity of elements increases on passing from left to right along a period and decreases down a group, the arrangement of elements in the order of electronegativity decreasing is F, S, As, Zn, Ca.

Group Period	ⅡA	ⅡB	ⅤA	ⅥA	ⅦA
2					F
3				S	
4	Ca	Zn	As		

15. (1) The valence electron configuration of the element is ns^2np^2. The element is in Group ⅣA.
(2) The valence electron configuration is $3d^64s^2$. The element is Fe which is in Period 4, Group ⅧB.
(3) The valence electron configuration is $3d^54s^1$. It is element Cr which is in Period 4, Group ⅥB.

16. The electron configuration of Fe^{2+} is $[Ar]3d^6$. After losing one electron, Fe^{2+} is changed into Fe^{3+} whose 3d orbitals are half-filled. According to the supplement of Hund's rule, Fe^{3+} is more stable. This means Fe^{2+} is easily to be oxidized into Fe^{3+}.

17. Element Se is in Period 4, Group ⅥA, p-block of periodic table. Its valence electron configuration is $4s^24p^4$. It can lose up to 6 electrons. Since the oxidation number of oxygen is generally -2, the oxide of Se with the highest oxidation state is SeO_3.

[Supplementary Exercises]

1. The frequency of radiation emitted is $2.34 \times 10^{14} s^{-1}$.

2. The wavelength of a neutron traveling at a speed of $3.90 \times 10^3 m \cdot s^{-1}$ is 0.102nm.

3. There are 3 subshells in M shell. There are 7 orbitals in f subshell.

4. All the electron configurations are possible except the one in answer (3).

5. Thallium is a representative group element in Period 6, Group ⅢA of periodic table.

6. (1) Cs, Ba, Sr   (2) Ca, Ga, Ge   (3) As, P, S

（钮因尧）

Chapter 10
Covalent Bond and Intermolecular Forces

Atoms can interact with one another in several ways to form aggregates. Simple elements and compounds exist in the form of molecule or crystal. We define a bond as a force that holds a group of two or more atoms together and makes them function as a unit. There are various types of weak forces between molecules known as intermolecular forces. Their strength is weaker than that of chemical bonds. The chemical structures and intermolecular forces play a very important role in determining the nature of substances. The types of chemical bonds are ionic bond, covalent bond (including coordination bond) and metallic bond. Among them, the covalent bond exists in more than 90% of the known compounds. On the other hand, the stability or strength of a chemical bond is mainly related to the bonding energy. The bonding energy is between tens and hundreds of kilojoules per mole generally. To understand the behavior of substances, we must understand the nature of chemical bonding and the factors determining molecular structure.

Molecular bonding and structure play the central role in determining the course of chemical reactions, many of which are vital to our survival. For example, most reactions in biological systems are very sensitive to the structures of the participating molecules. Molecules that act as drugs must have exactly the right structure to perform their functions correctly. In this chapter, we will introduce the covalent bonding theory and molecular geometry as well as intermolecular forces.

10.1 Valence Bond Theory

The classical covalent bond theory was proposed as early as in 1916 by American Chemist G. N. Lewis, who suggested that covalent bond is formed by both bonding atoms to provide the outer layer lone electron to form shared electron pair. This shared electron pair would comprise the single bond holding the atoms together. In this arrangement, each bonded atom generally reaches the stable electron configuration of a noble gas. Covalent bond theory of Lewis revealed initially the distinction between covalent bond and ionic bond. However, it regarded electrons as stationary negative charge, and thus could not explain why two negative charged electrons were mutually matching instead of exclusive. Nor could it explain why covalent bond was in orientation, and why some covalent molecules with less than 8 outermost electrons of the central atoms (such as BF_3) or more than 8 (such as PCl_5) were still fairly stable. The Lewis theory has problems with odd-electron species such as NO and O_2, and situations in which it is not possible to represent a molecule through a single structure, resonance.

For a more complete explanation of chemical bond formation, we look to quantum mechanics. In fact, the quantum mechanical study of chemical bonding also provides a means for understanding molecular geometry. In 1927, German Chemist W. Heitler and F. London applied quantum mechanics to deal with molecular structure of H₂ and revealed the nature of covalent bond. L. Pauling and J.C. Slater etc. established the modern **valence bond (VB) theory** which was developed and based on the research by W. Heitler and F. London. In 1932, American Chemist R. S. Muiliken and Germany Chemist F. Hund also put forward **molecular orbital (MO) theory**. Both of them are modern theories of covalent bond. Neither theory perfectly explains all aspects of bonding, but each has contributed something to our understanding of many observed molecular properties. Let us start our discussion of valence bond theory by considering the formation of a H₂ molecule from two H atoms.

10.1.1 Formation of Hydrogen Molecule

The Lewis theory describes the H-H bond in terms of pairing of the two electrons on the H atoms. In the framework of valence bond theory, the covalent H-H bond is formed by the overlap of the two 1s orbitals in the H atoms. By overlap, we mean that the two orbitals share a common region in space.

What happens to two H atoms as they move toward each other and form a bond? Figure 10-1 shows the energy of interaction of two H atoms. It is assumed that the energy of free H atoms is zero for brief discussion. Initially, when two H atoms are far apart, the interaction between them can be neglected. As the atoms approach each other, each electron is attracted by the nucleus of the other atom. At the same time, the electrons repel each other, as do the nuclei. While the atoms are still separated, attraction is stronger than repulsion, so that the potential energy of the system decreases even to be negative as the atoms approach to each other. This trend continues until at a particular internuclear distance, 74pm. The potential energy reaches its lowest value, $-436 kJ \cdot mol^{-1}$. This is the specific state in which the two H atoms substantial overlap the 1s orbitals and form a stable H₂ molecule. This stable state is a ground

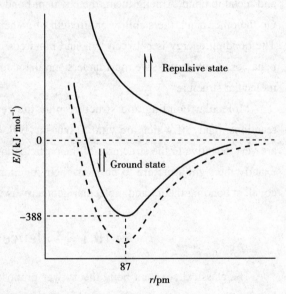

Figure 10-1   A Plot of Energy versus Internuclear Distance for Two H Atoms

state of H₂ molecule. If the distance between nuclei decreases further, the potential energy of the system will rise steeply and finally becomes positive as a result of enhanced electron-electron and nucleus-nucleus repulsion.

If two H atoms have electrons with the same spin, as the distance between two H atoms decreases, the electron probability density in the internuclear region is nearly zero and repulsion between electrons is greater than the attraction of the electrons for the nucleus of the other H atom. So the electrons cannot pair to form a bond. This unstable state is repulsive state as shown in Figure 10-2.

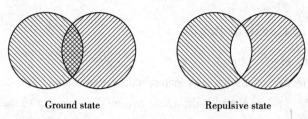

Ground state   Repulsive state

Figure 10-2 The Electron Probability Density in the Internuclear Region under the Ground State and Repulsive State of Hydrogen Molecule

10.1.2 Main Points of Valence Bond Theory

The concept of overlapping atomic orbitals is applies equally well to diatomic and polyatomic molecules other than H_2. The main points of valence bond theory can be summarized as follows:

(1) Only the unpaired electrons with opposite spin can be paired (overlap effectively), it enhanced electron probability density between the two nuclei, reduces system energy thus forming the stable covalent bond.

(2) After the two unpaired electrons with opposite spin are paired, they cannot pair with other unpaired electrons further. Therefore, the number of covalent bond formed in an atom is confined to its number of unpaired electron, which is called the saturation feature of covalent bond.

(3) The more the atomic orbitals of two electrons overlap is, the higher the electron density between the two nuclei is, and the more stable the covalent bond is, which is called maximum overlap principle of atomic orbital. The extent of overlap depends on shapes and directions of the orbitals. An s orbital is spherical, the electron density in each direction is the same, but for p or d electrons, the electron density is the greatest along the stretching direction of orbital. Thus, whenever possible, a bond involving p or d orbitals will be oriented in the direction that maximizes overlap. This is the direction feature of covalent bond. For example, in the formation of HCl molecule, the 1s orbital of H atom overlaps with the half-filled $3p_x$ orbital of Cl along x axis of that orbital in Figure 10-3 (a). Any other direction would result in less overlap, thus, it will not form covalent bond in Figure 10-3 (b) and 10-3 (c).

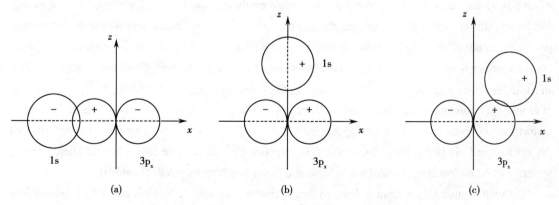

Figure 10-3 Bonding in HCl Represent by Atomic Orbital Overlap

10.1.3 Types of Covalent Bonds

1. Sigma (σ) bond and pi (π) bond

In this section, we focus on the mode by which orbitals can overlap by "end-to-end' or "side-by-side" to see the detailed makeup of covalent bonds. These two modes give rise to the two types of covalent bonds which are sigma (σ) bonds or pi (π) bonds respectively as shown in Figure 10-4.

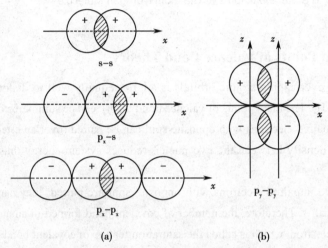

Figure 10-4 σ Bonds and π Bonds
(a) σ bonds; (b) π bond

Since the atomic orbitals are different in shape, they may mutually overlap in different ways. Bonds formed by end-to-end overlapping the atomic orbitals between two bonded atoms are called σ bonds. σ bonds are symmetrical with respect to rotation about the bond axis. By this definition, common forms of σ bonds are s-s, s-p_x and p_x-p_x, where x is defined as the axis of the bond. Single bonds are always σ bonds.

Bonds formed in side-by-side overlapping fashion between p or d orbitals are called π bonds. A π bond is not symmetrical about the bond axis. In a π bond, the electron density is in two separate regions (lobes), one above and the other below the plane of the molecule. Both parts make one π bond. Because the common place of side-by-side overlap fashion is much less than that of end-to-end overlap fashion, π bond is weaker than σ bond, making it much more easily broken. From the perspective of quantum mechanics, this is explained by significantly less overlap between the component *p*-orbitals due to their parallel orientation. Although π bond is weaker than sigma bond, π bond is a component of multiple bonds, together with σ bond. π bond makes the second bond in double bond, and the second and third in triple bond. For example, in N_2, the electron configuration of N atom is $1s^2 2s^2 2p_x^1 2p_y^1 2p_z^1$. When two N atoms were combined with each other to form a N_2 molecule, each of N atoms provides a $2p_x$ orbital to form a σ bond by end-to-end overlapping of the two $2p_x$ orbitals. At the same time, the two parallel $2p_z$ orbitals and the two parallel $2p_y$ orbitals overlap side-by-side to form two π bonds. Therefore, the structure of N_2 molecule consists of one σ bond and two π bonds as shown in Figure 10-5.

Rotation around single bond is free, while rotation around double bond is not free. If one doubly bonded carbon were to be rotated relative to the other, the p orbitals could no longer overlap side-by-side. Free rotation about single bonds and lack of free rotation around double bonds are important factors in the properties of many molecules, including physiologically important molecules like proteins.

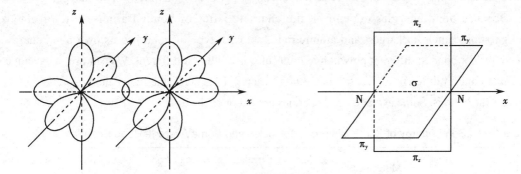

Figure 10-5 Schematic Diagram of N_2 Molecule Formation

2. Normal covalent bond and coordinate covalent bond

According to the origin of a shared electron pair between two bonded atom, covalent bonds can be classified into **normal covalent bond** and **coordinate covalent bond**. In a normal covalent bond, each of bonded atoms contributes one electron to produce a shared pair, such as covalent bonds in H_2, O_2 and HCl. A shared electron pair originally from a bonded atom only is called a coordinate covalent bond. The properties of a coordinate covalent bond do not differ from those of a normal covalent bond because both electrons are alike whatever their origin. It is just named differently to show that the electron pair comes from one atom. To be distinguished from normal covalent bond, coordination bond is represented by a sign of " $\rightarrow$ ", the arrow is pointing toward the atom which accepts the electron pair. For example, carbon monoxide CO contains one coordinate bond and two normal covalent bonds between C and O atom. The two 2p orbitals in O atom were combined with the two 2p orbitals in C atom to form one σ bond and one π bond, respectively. And the third bond is a coordinate bond where the O atom donates the lone-pair of electrons to one unoccupied 2p orbital of C atom. So in CO molecule, C atom is an electron acceptor and O atom is an electron donor. The formation of coordination bond can be expressed as

$$:\!\ddot{C}\!\cdot + \cdot\ddot{O}\!: \longrightarrow :\!C\!\equiv\!O:$$

Coordination bonding theory will be further studied in Chapter 11.

10.1.4 Covalent Bond Parameters

The nature of chemical bond is characterized as **bond parameters**. They include bond energy, bond length and bond angle.

1. Bond energy

The **bond energy** (*BE*) is used to measure the strength of a chemical bond. Bond energy is commonly given in the unit of $kJ \cdot mol^{-1}$, equivalent to molecular **dissociation energy** (*DE*) in diatomic molecule. At 100kPa and 298.15K, the energy needed to dissociate 1mol ideal gaseous molecule AB into gas-phase A atom and B atom is called dissociation energy of AB molecule.

For example,　　　$H_2(g) \rightarrow 2H(g)$　　　$BE(\text{H-H}) = DE(\text{H-H}) = 436 kJ \cdot mol^{-1}$

Bond energy and dissociation energy in polyatomic molecule is different. For example, the first dissociation energy of O-H bond of H_2O molecule is $502 kJ \cdot mol^{-1}$, and $423.7 kJ \cdot mol^{-1}$ is needed to cleave the remaining O-H bond. The average bond energy of O-H bond in H_2O is $463 kJ \cdot mol^{-1}$.

Because bond energies depend on the characteristics of bonded atoms — their electron configurations, nuclear charges, and atomic radii, so, each type of bond has its own bond energy. A same covalent bond in different polyatomic molecules is a little bit different. We can use the average of the same bond in different molecules as the bond energy. In general, the larger the bond energy is, the stronger the bond is. Some average bond energies are list in Table 10-1.

Table 10-1   Bond Energy of Some Diatomic Molecules and Some Average Bond Energies BE / $kJ \cdot mol^{-1}$

Molecules	Bond energy	Molecules	Bond energy	Covalent bond	Average bond energy	Covalent bond	Average bond energy
H_2	436	HF	565	C—H	413	N—H	391
F_2	165	HCl	431	C—F	460	N—N	159
Cl_2	247	HBr	366	C—Cl	335	N=N	418
Br_2	193	HI	299	C—Br	289	N≡N	946
I_2	151	NO	286	C—I	230	O—O	143
N_2	946	CO	1 071	C—C	346	O=O	495
O_2	493			C=C	610	O—H	463
				C≡C	835		

2. Bond length

In molecular geometry, **bond length** or bond distance is the average distance between the nuclei of two bonded atoms. Bond lengths are equal to the sums of the radii of the atoms joined by the bond. Bond length can be measured by means of X-ray diffraction, or spectroscopic technique. The experiment results show that the bond length of an identical bond in different molecules is almost the same. And therefore, an average bond length can be used. For example, the length of C-C single bond in diamond is 154.2pm. It is 153.3pm in ethane, 154pm in propane, and 153pm in cyclohexane. Therefore, the length of C-C single bond is set to be 154pm.

Bond length is inversely related to bond strength and the bond dissociation energy, as a stronger bond is also a shorter bond. When the two same atoms were combined with each other to form bonds, the bond length of single bond is longer than that of double bond, which is longer than that of triple bond. For example, the bond lengths of C=C and C≡C are 134pm, and 120pm respectively.

3. Bond angle

Bond angle is formed between two adjacent bonds on an atom. It is an important parameter which reflects molecule geometry. For example, the bond angle of H_2O is $104°45'$, which shows that the molecular geometry of H_2O is bent; but the bond angle of CO_2 is $180°$, which shows that the molecular geometry of CO_2 is linear. In general, bond length and bond angle can determine molecule geometry.

10.2 The Polarity of Bond

Bond polarity depends on the electronegativity difference (ΔEN) between the bonded atoms. When the two bonded atoms have the same values of the electronegativity, the electrons are shared equally and the centers of positive charge and negative charge in a molecule are coincided exactly. Such a covalent bond is known as the **non-polar covalent bond**. The simplest example of it is the bond in diatomic molecules with identical atoms, such as H_2 and O_2, etc. However, in the covalently bonded HCl molecule, the H and Cl atoms do not share the bonding electrons equally because H and Cl are different in electronegativity. The bond in HCl is called a **polar covalent bond**. This unequal distribution of electron density gives the bond partially negative and positive poles. Such a polar covalent bond is depicted by a polar arrow (⊢⟶) pointing toward the negative pole or by δ^+ and δ^- symbols, where the lowercase Greek letter delta (δ) represents a partial charge.

$$\overset{\delta^+ \quad \delta^-}{H-Cl} \qquad \overset{\longrightarrow}{H-Cl}$$

The greater the difference in electronegativity between bonded atoms is, the more polar the bond is. If the difference in electronegativity is large enough, electrons transferred from the less electronegativity to the more electronegativity atom, and the bond become ionic bond. Bonds would vary from nonpolar to slightly polar to very polar even to mainly ionic depending on the difference in electronegativity between the bonded atoms. Some examples are list in Table 10-2.

Table 10-2 Relation between the Electronegativity Difference of Bonded Atoms and Bonding Types

Substances	NaCl	HF	HCl	HBr	HI	Cl_2
ΔEN	2.23	1.80	0.98	0.78	0.48	0
Bonding type	Ionic bond		Polar covalent bond			Non-polar covalent bond

10.3 Valence Shell Electron Pair Repulsion Model

Although valence bond theory gives a clear picture of chemical bond formation and successfully explains the bond nature, the orientation and the saturation of covalent bonds, it cannot be used to predict the molecular geometry of a molecule—that is, the three-dimensional arrangement of the atoms. In 1940 N.V. Sidgwick and others proposed the **valence-shell electron-pair repulsion model (VSEPR)**. This model is useful for predicting the molecular geometries of AX_m or ions formed between main group elements. Because the lone pair will affect the shape of molecule, so, when we apply this model to predict the shape of molecule, we often write AX_m into AX_mE_n further for convenient. Where m and n are integers, A is the central atom with the lower group number and the lower electronegativity, X is a surrounding atom or group of atoms bonded to the central atom A, and E is a lone pair of electrons. In VSEPR model, we focus on the all of electrons pairs in valence shell for the central atom A in a structure. The main idea of this model is that the structure around a bonded atom is determined by minimizing repulsion between electron pairs. It means that the bonding pairs and lone pairs around the central atom are positioned as far apart as possible to keep the minimal repulsive force each other. The existence of lone pairs will increases the repulsive force, The repulsion force order between two electron pairs is:

Lone-pair *vs* lone-pair > lone-pair *vs* bonding-pair > bonding-pair *vs* bonding-pair

Lone electron pairs occupy the positions around the central atom but no contribution to the shape of molecule. It affects molecular bond angles that change the basic type of molecular shape.

The steps of predicting molecular geometry by applying VSEPR are as follows:

1. Determine the valence electron-pairs number (VPN) around the central atom A, which is calculated by the following equations.

$VPN = \frac{1}{2}$ [the number of valence shell electrons of A + all of valence electrons number of X ±ionic charge (negative charge +; positive charge −)]

(1) The number of valence shell electrons of A equals to the group number of it. A halogen atom provides seven electrons. An element as central atom in group oxygen provides 6 electrons.

(2) Usually H, O, S and halogen are the surrounding atoms. As the surrounding atoms, an H or halogen atom provides one electron, but O and S atom are regarded not to provide the electron.

(3) The number of negative charges should be added for anions while that of positive charges should be subtracted for cations in calculating the valence shell electrons.

(4) When calculating the number of electron pairs, if there is one electron remaining, it is taken as one pair.

(5) Multiple bonds, such as double bond and triple bond, are treated as a single bonding group.

2. Predict molecular geometry According to the number of valence shell electron pair around A, find the corresponding valence shell electron pair geometry from Table 10-3. It should be noted that the valence shell electron pair geometry of A refers to the spatial arrangement of valence shell electron pair around A, but molecular geometry refers to spatial arrangement of X in molecule, not including lone electron pair. If there is no lone electron pair, the molecular geometry is just the same as the valence shell electron pair geometry. If there is lone electron pair, the molecular geometry will be "distorted".

Table 10-3 Ideal Geometries of Molecule and Valence Shell Electron Pairs for Some Simple Molecules and Ions

VPN	Valence electron pair geometry	Class of molecule	Bonding pairs	Lone pairs	Molecular geometry	Examples
2	Linear	AX_2	2	0	Linear	$HgCl_2$, CO_2
3	Trigonal planar	AX_3	3	0	Trigonal planar	BF_3, NO_3^-
		AX_2E	2	1	Bent	$PbCl_2$, SO_2
4	Tetrahedral	AX_4	4	0	Tetrahedral	SiF_4, SO_4^{2-}
		AX_3E	3	1	Trigonal pyramidal	NH_3, H_3O^+
		AX_2E_2	2	2	Bent	H_2O, H_2S

VPN	Valence electron pair geometry	Class of molecule	Bonding pairs	Lone pairs	Molecular geometry	Examples
5	Trigonal bipyramidal	AX_5	5	0	Trigonal bipyramidal	PCl_5, PF_5
		AX_4E	4	1	Distortion tetrahedral	$SF_4, TeCl_4$
		AX_3E_2	3	2	T-shaped	ClF_3
		AX_2E_3	2	3	Linear	I_3^-, XeF_2
6	Octahedral	AX_6	6	0	Octahedral	SF_6, AlF_6^{3-}
		AX_5E	5	1	Square pyramidal	BrF_5, SbF_5^{2-}
		AX_4E_2	4	2	Square planar	ICl_4^-, XeF_4

Sample Problem 10-1 How to predict the molecular geometry of SO_4^{2-} ion?

Solution The charge number of SO_4^{2-} is -2, there are 6 valence electrons in central atom S, and the surrounding O atom does not provide electrons, therefore the number of valence shell electron pair of S atom is $(6+2)/2=4$, its valence shell electron pair geometry is regular tetrahedron. As a result, the number of the surrounding atoms is also 4, this shows that there is no lone electron pair in valence shell electron pair, therefore the molecular geometry of SO_4^{2-} is regular tetrahedron.

Sample Problem 10-2 How to predict the molecular geometry of H_2S molecule?

Solution S is the central atom of H_2S molecule. There are 6 valence electrons in S atom, bonded with two H atoms which provides one electron respectively, therefore the number of valence shell electron pair of S atom is $(6+2)/2=4$, its valence shell electron pair geometry is tetrahedron. Because there are 2 surrounding atoms, so this indicates that there are two lone electron pairs in the valence shell electron pair, the molecular geometry of H_2S molecule is bent.

Sample Problem 10-3 How to predict the molecular geometry of HCHO molecule?

Solution C is the central atom of HCHO molecule. There are 4 valence electrons in C atom, bonded with two H atoms which provides one electron respectively, and the number of valence shell electron pair of C atom is $(4+2)/2=3$. Since there are no lone pairs of electrons in C atom, the molecular geometry of HCHO molecule is trigonal planar. But we all know that there is one $C=O$ double bond in HCHO. So VSEPR theory can be applied to the simple molecules with multiple bonds.

The effect of a double bond on shapes is similar to the effect of a single bond. As far as the shape predicted by VSEPR is concerned, multiple bonds are no different compared with single bonds. The strong repulsion of multiple bonds can affect on the bond angle of molecule and change the molecular geometry. In HCHO molecule, $C=O$ is a double bond, therefore, $\angle HCH < \angle HCO$ and the molecular geometry is not plane equilateral triangle.

> **Question and thinking 10-1** The constituent atoms of BF_3 and ClF_3 are $1:3$. Whether the molecular geometry of them is identical or not? Why?

10.4 Hybridization Theory of Atomic Orbitals

The VSEPR model provides a relatively simple and straightforward method for predicting the geometry of molecules. But it does not explain how the shapes come from the interactions of atomic orbitals. Valance bond theory successfully describes the formation of covalent bond and explains the direction and saturation properties of it. Based on valance bond theory, L. Pauling proposed hybridization theory in 1931 to explain the experimental fact that the molecular geometry of CH_4 is tetrahedral. Hybridization theory complements and develops modern valence bond theory in explanation of molecular geometry and atomic bonding properties.

10.4.1 Main Points of Hybridization Theory

The main points of Hybridization theory are as follows:

(1) In the bonding process, quantum-mechanical calculations show that the different types of atomic orbitals called wave-function which have similar energy in the same atom can combine linearly, reassign energy and determine spatial orientations to form new atomic orbitals. The spatial orientations of these new orbitals lead to more stable bonds and are consistent with observed molecular shapes. The process of orbital mixing is called **hybridization**, and the new atomic orbitals are called **hybrid orbitals**. The number of hybrid orbitals obtained is equal to the number of atomic orbitals that participate in the hybridization process.

(2) The value of angular wave function of hybrid orbital in a certain direction is enlarged comparing with that of atomic orbital that will be more beneficial to have a maximum overlap between orbitals, so the bonding ability of hybrid orbital is stronger than that of atomic orbital.

(3) The bond angle between hybrid orbitals is trying to get the largest distribution in the space, so that mutual repulsive energy between the hybrid orbitals can be minimized, in this way the bond is getting stable. The bond angles between different types of hybrid orbitals are different, so that molecules have different spatial configurations after forming bond.

It is important to note that the concept of hybridization is not applied to isolated atoms. It is a theoretical model used only to explain covalent bonding. You can imagine hybridization as a process in which atomic orbitals mix, hybrid orbitals form, and electrons enter them with parallel spins by Hund's rule to create the stable bonds. In truth, though, hybridization is a mathematically derived result from quantum mechanics that accounts for the molecular shapes we observe.

10.4.2 Types of Hybrid Orbitals

1. s-p hybridization and s-p-d hybridization

According to the type of atomic orbitals participating in hybridization, there are two common types of hybridization, s-p hybridization and s-p-d hybridization.

(1) s-p hybridization

If hybridization takes place between ns orbital and np orbitals which are in the similar energy level, it is termed as s-p hybridization. As the numbers of p orbitals participating in hybridization may be different, s-p type of hybridization can be divided into three fashions: sp, sp^2 and sp^3 hybridization.

a. sp hybridization

One ns orbital is combined with one np orbital to form two sp hybrid orbitals. This kind of hybridization is called sp hybridization. Each sp hybrid orbital contains 1/2 s orbital component and 1/2 p orbital component. The two sp orbitals lie on a straight line, so the angle between the two hybrid orbitals is 180°. After two sp hybrid orbitals overlap with other atomic orbitals from other bonded atoms, it can form linear geometry molecule. Process of sp hybridization and shape of sp hybrid orbital are shown in Figure 10-6.

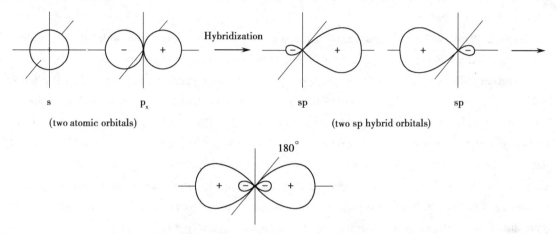

Figure 10-6　Schematic Diagram of sp Hybrid Orbitals Formation

Sample Problem 10-4　How to explain the molecular geometry of $BeCl_2$?

Solution　The central atom in $BeCl_2$ is Be atom. The valence electron configuration of Be atom is $2s^2$. In the course of forming $BeCl_2$ molecule, one electron in Be atom's 2s orbital should be excited to the empty 2p orbital, its valence electron configuration is $2s^1 2p^1$. 2s orbital and 2p orbital with one-electron respectively are mixed to form two sp hybrid orbitals. The two hybrid orbitals will be arranged as far apart as possible from each other with the bond angle 180°. The two unhybridized 2p orbitals in Be atom lie perpendicular to each other and to the bond axes. Two equivalent σ_{sp-p} bonds are formed by overlap of the two sp hybrids with the 3p orbitals of two Cl atoms. Therefore, the spatial geometry of $BeCl_2$ molecule is linear as shown in Figure 10-7.

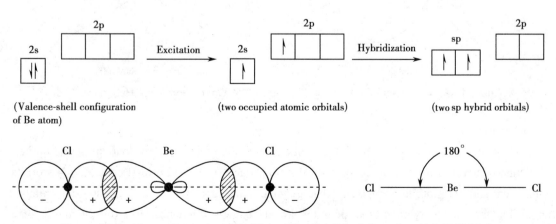

Figure 10-7　Linear Geometry of $BeCl_2$ and Spatial Orientation of sp Hybrid Orbitals

b. sp² hybridization

One ns orbital is combined with two np orbitals to form three sp² hybrid orbitals. This kind of hybridization is called sp² hybridization. Each sp² hybrid orbital contains 1/3 s orbital component and 2/3 p orbital component. In order to diminish the repulsive energy among the bonds, the three sp² orbitals lie in a plane, and the angle between any two of them is 120°. Figure 10-8(a) shows it. After the three sp² hybrid orbitals overlap with the other three atomic orbital from bonded atoms to form three bonds, it can form flat triangular molecule.

Sample Problem 10-5 Experimental evidence has shown that boron trifluoride BF_3 molecule is planar with all the F-B-F angles equal to 120°. How to explain the geometry of BF_3 molecule?

Solution The central atom in BF_3 is B atom. The valence electron configuration of B atom is $2s^2 2p^1$. In the course of forming BF_3 molecule, one electron in B atom's 2s orbital should be excited to the empty 2p orbital, its valence electron configuration is $2s^1 2p^2$. The 2s orbital combines two 2p orbitals with single-electron respectively to form three sp² hybrid orbitals. The three sp² hybrid orbitals lie in a plane and the angle between any two of them is 120°. The third 2p orbital in B atom remains empty and unhybridized. When they respectively overlap with 2p orbital of F atom, three equivalent σ_{sp^2-p} bonds are formed. The unhybridized 2p orbital is perpendicular to the trigonal bonding plane. Therefore, the geometry of BF_3 molecule is equilateral triangle as shown in Figure 10-8 (b).

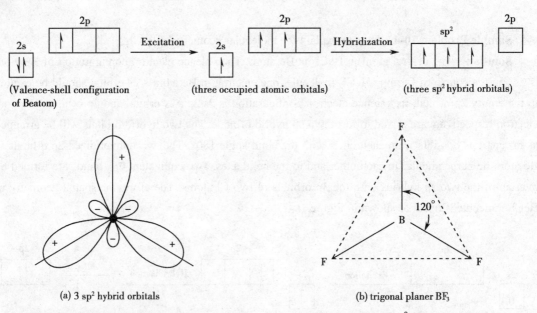

(a) 3 sp² hybrid orbitals (b) trigonal planer BF_3

Figure 10-8　Molecular Geometry of BF_3 and Spatial Orientation of sp² Hybrid Orbitals

c. sp³ hybridization

One ns orbital and all three np orbitals of the central atom mix and form four sp³ hybrid orbitals. This kind of hybridization is called sp³ hybridization. Each sp³ hybrid orbital contains 1/4 s orbital component and 3/4 p orbital component. The geometric arrangement of those four hybrid orbitals is tetrahedral, the four sp³ hybrid orbitals point towards the corners of tetrahedron at 109°28′ to each other as shown in Figure 10-9(a).

Sample Problem 10-6 Experimental evidence has shown that the bond angles in methane are 109° 28′ giving the overall shape of tetrahedron. How to explain the geometry of CH₄ molecule?

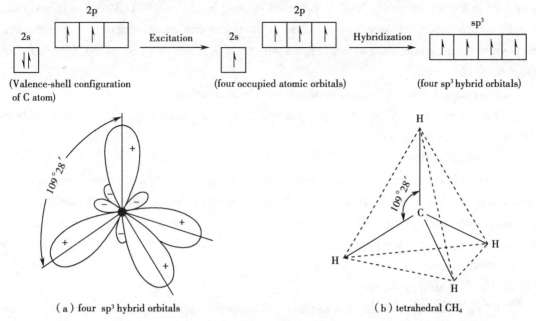

(a) four sp³ hybrid orbitals (b) tetrahedral CH₄

Figure 10-9 Molecular Geometry of CH₄ and Spatial Orientation of sp³ Hybrid Orbitals

Solution The central atom in methane is C atom. The valence electron configuration of C atom is $2s^22p^2$. In the course of forming methane molecule, one electron in C atom's 2s orbital should be excited to empty 2p orbital, its valence electron configuration is $2s^12p^3$. Mixing the 2s orbital with all three 2p orbitals generate four sp³ hybrid orbitals, the four sp³ hybrid orbitals lie in tetrahedral. Four sp³ hybrid orbitals of C atom overlap with 1s orbitals of four H atoms respectively to form four equivalent σ_{sp^3-s} C-H bonds. All the HCH angles are 109° 28′. CH₄ molecule is the most frequently cited molecule to have tetrahedral shape as shown in Figure 10-9(b).

Three types of s-p hybridization mentioned above are summarized in Table 10-4.

Table 10-4 Three Types of s-p Hybridization

The type of hybridization	sp	sp²	sp³
Hybridized atomic orbitals	1 ns + 1 np	1 ns + 2 np	1 ns + 3 np
Number of hybrid orbitals	2 × sp hybrid orbitals	3 × sp² hybrid orbitals	4 × sp³ hybrid orbitals
Angle between hybrid orbitals	180°	120°	109° 28′
Molecular Geometry	Linear	Trigonal planar	Tetrahedral
Examples	BeCl₂, C₂H₂	BF₃, BCl₃	CH₄, CCl₄

Question and thinking 10-2 The structural and geometry of organic molecules, such as C₂H₆, C₂H₄ and C₂H₂, can also be explained by hybridization. Whether the hybridization type of C in the three molecular is the same or not? Whether the π bond is formed between two C atom?

(2) s-p-d hybridization

If hybridization takes place among $(n-1)$ d orbital, ns orbital and np orbital or ns, np, nd orbitals which are in the similar energy level, it is termed as s-p-d hybridization. This type of hybridization is quite complicated; they usually exist in the compound which contains transition element. How the atomic orbitals mix to form hybrid orbitals will be discussed in more detail in Chapter 11 about coordination compound.

2. Equivalent hybridization and nonequivalent hybridization

According to the energy levels of hybrid orbitals, hybridization may be classified into equivalent and nonequivalent hybridization.

(1) Equivalent hybridization

The hybridization, which generates hybrid orbitals with equal original orbital proportions and equal energy level, is called **equivalent hybridization**. Usually, if the number of valence electrons of the central atom equals to the number of its hybrid orbitals, its hybridization is equivalent hybridization. For instance, hybridization of central atoms of $BeCl_2$, BF_3 and CH_4 are sp, sp^2 and sp^3 equivalent hybridization, respectively.

(2) Nonequivalent hybridization

The hybridization, which generates hybrid orbitals with unequal original orbital proportions and unequal energy levels, is called **nonequivalent hybridization**. Usually, in nonequivalent hybridization, some hybrid orbitals are occupied by lone pair electrons.

Now we take the formation of NH_3 molecule and H_2O molecule as examples see Figure 10-10.

Sample Problem 10-7 By experimental measurement, the bond angles of three N-H bonds in NH_3 molecule are $107°$, and the molecular geometry of NH_3 is triangular pyramid. How to explain the geometry of NH_3?

Solution Valence shell electron configuration of the central atom N is $2s^2 2p^3$. Mixing the 2s orbital with all three 2p orbitals generate four sp^3 nonequivalent hybrid orbitals, the four sp^3 hybrid orbitals lie in a tetrahedron. Because one of sp^3 hybrid orbitals is occupied by a lone pair of electrons, only three half-filled sp^3 hybrid orbitals are involved in bond formation. The three hybrid orbitals form $\sigma_{sp^3 - s}$ N-H bonds. Repulsion among the lone-pair electrons and other three bonding pairs makes the H-N-H bond angles from $109°28'$ to $107°$, so the geometry of NH_3 is triangular pyramid, just as VSEPR theory predicts.

Sample Problem 10-8 Obtained from experiment, there are two O-H bonds in H_2O molecule, and their bond angle is $104°45'$ (smaller than the bond angle of CH_4, equivalent sp^3 hybridization), the molecular geometry is bent. How to explain the spatial geometry of H_2O?

Solution Valence shell electron configuration of the central atom O is $2s^2 2p^4$. In the process of forming H_2O molecule, one 2s orbital is combined with three 2p orbitals to form four sp^3 nonequivalent hybrid orbitals. Two half-filled sp^3 hybrid orbitals are involved in bonding to form two $\sigma_{sp^3 - s}$ bonds, and the other two sp^3 hybrid orbitals accommodates the lone pairs on O respectively. Repulsion between the lone-pair electrons and the bonding pairs makes the H-O-H bond angles from $109°28'$ to $104°45'$ (smaller than the bond angle of NH_3). The geometry of H_2O molecule is bent.

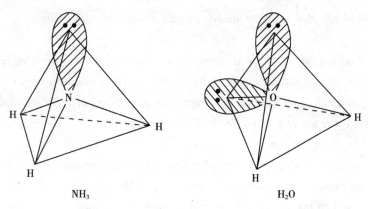

Figure 10-10 Structural Diagram of NH$_3$ Molecule and H$_2$O Molecule

10.5 Molecular Orbital Theory

The modern valence bond theory, which is based on the orbital overlap between bonded atoms, is straightforward and easy to understand. And it illustrated the nature of covalent bonds. The hybridization theory successfully explains the geometries of covalent molecules and has been widely applied. However, the theory suggests that the bonding electron pairs in a molecule only move in a narrow region between the two bonded atoms. The molecule is not regarded as a whole. And none of the above theories adequately explains magnetic and spectral properties as well the other properties. So there is still some limitation for the theories to observe the properties of molecules. Consider O$_2$ molecule, the electron configuration of O atom is $1s^2 2s^2 2p_x^2 2p_y^1 2p_z^1$, according to valence bond theory, two O atoms should be bonded with one σ bond and one π bond, and O$_2$ should be diamagnetic as all electrons are paired. But experiments have shown that O$_2$ has two unpaired electrons and it is paramagnetic. Furthermore, valence bond theory cannot resolve some paradoxes, such as why a single-electron bond exists in H$_2^+$ or why a three-electron bond exists in O$_2$ molecule. To address these questions, American Chemist R.S. Milliken and German Chemist F. Hund proposed another quantum mechanical approach called **molecular orbital (MO) theory** in 1932. This theory is based on the molecular integrity and can explain the structure of polyatomic molecules well, so it is an important part of modern covalent bonding theory. It starts with a simple picture of molecules, but it quickly becomes complex in details. We can provide only an overview here.

10.5.1 The Central Themes of Molecular Orbital Theory

1. The central themes of MO

(1) The theory assigns the electrons in a molecule to series of orbitals that belong to the molecule as a whole. These orbitals are called molecular orbitals (MOs), which result from allowed interaction between the atomic orbitals. An atomic orbital is associated with only one atom, while a molecular orbital with the entire molecule. A molecular orbital is also a mathematical function which describes a region of space in a molecule where the electrons are most likely to be found. And a molecular orbital can accommodate just two electrons, and the electrons must have opposing spins. Furthermore, atomic

orbitals are labeled by s, p, d⋯, while molecular orbitals are labeled correspondingly by σ, π, δ, ⋯., and so on.

(2) Molecular orbital can be obtained through linear combination of atomic orbitals (LCAO). The number of molecular orbitals formed equals to the number of those involved atomic orbitals combined linearly. In the H_2 molecule, for example, two singly occupied 1s atomic orbitals combine to form two molecular orbitals. There are two ways for the orbital combination to occur—an additive way and a subtractive way. The energy of the additive combination is lower than that of 1s atomic orbitals of H atom. The molecular orbital with energy lower than the energy of the atomic orbitals that were combined is called a **bonding molecular orbital.** A bonding molecular orbital not only has lower energy but also greater stability than the atomic orbitals from which it was formed. The energy of subtractive combination is higher than that of 1s atomic orbital of H atom. The molecular orbital with energy higher than the energy of the atomic orbitals that were combined is called an **antibonding molecular orbital**. An antibonding molecular orbital has higher energy and lower stability than the atomic orbitals from which it was formed. Since not any electrons in molecule can occupy the central region between the nuclei, where there is a node, region of zero electron density, and can't contribute to bonding. The two nuclei therefore repel each other. The molecular orbital whose energy does not obviously change if compared with that of atomic orbital combined is called **nonbonding orbital**.

(3) A molecular orbital has a specific size, shape and energy. Both the bonding and antibonding molecular orbitals of H_2 are sigma (σ) molecular orbitals because they are cylindrically symmetrical around the bond axis. The bonding molecular orbital is denoted by σ_{1s}, that is, a σ molecular orbital derived from 1s atomic orbitals. Antibonding orbitals are denoted with a superscript star "*", the antibonding molecular orbital derived from 1s atomic orbital is σ_{1s}^*. That is, σ_{1s} has lower energy and greater stability than the 1s orbitals on H atom from which it was formed and σ_{1s}^* has higher energy and lower stability than the 1s orbitals from which it was formed. As the names "bonding" and "antibonding" suggest, placing electrons in a bonding molecular orbital yields a stable covalent bond, whereas placing electrons in an antibonding molecular orbital results in an unstable state. In the H_2 molecule, we consider only the interaction between 1s orbitals; with more complex molecules, we need to consider additional atomic orbitals as well. Nevertheless, for all of s orbitals, the process is the same as for 1s orbitals. For p orbitals, the process is more complex because they can interact with each other in two different ways. For example, two 2p orbitals can approach each other and combine to produce a sigma bonding(σ) and a sigma antibonding(σ^*) molecular orbital, as shown in Figure 10-12. Alternatively, the two p orbitals can combine to generate a bonding pi molecular orbital (π) and an antibonding pi molecular orbital(π^*). In a pi molecular orbital (bonding or antibonding), the electron density is concentrated above and below a line joining the two nuclei of the bonding atoms. The type of interaction between atomic orbitals can be further categorized by the molecular-orbital symmetry labels σ and π. When electrons are filled in σ molecular orbitals, they are called σ-electrons; when electrons are filled in π molecular orbitals, they are called π-electrons.

(4) Just as we can write the electron configuration for an atom, we can write one for a molecule. The symbol of each occupied molecular orbital is written in parentheses, with the number of electrons in it as a superscript outside, for example, the electron configuration of H_2 is $(\sigma_{1s})^2$. The electron

configuration in molecular orbitals also obeys Pauli's exclusion principle, energy minimum principle and Hund's rule. Before filling molecular orbitals with electrons, the molecular orbital's energy level order should be known. At present it is confirmed mainly by molecular spectrum experiments.

(5) In molecular orbital theory, **bond order** representing the intensity of bonds

$$\text{Bond order} = \frac{\text{the number of bonding electrons} - \text{the number of antibonding electrons}}{2}$$

In general, the higher the bond order, the stronger the bond is. If bond order is greater than zero, the molecule is more stable than the separate atoms, so it will form. For H_2, the bond order is $\frac{1}{2} \times (2-0) = 1$. If bond order is zero, the atoms cannot combine to form a molecule.

2. Rules for making molecular orbitals by the combination of atomic orbitals

In order to construct the molecular orbital effectively, every combining atomic orbital is required to obey to the following three principles:

(1) Symmetry-matching principle

Only symmetry-matching atomic orbitals can make up molecular orbital. What kind of two atomic orbitals is symmetry matching? We can carry on two symmetry operations on their angle distribution, namely rotation and reflection. Rotation means to rotate around the bonded axis where x is defined as the axis of the bond; reflection means to reflect on a plane with the bonded axis, xy plane or xz plane, like look into a mirror. If rotating or reflecting, the math symbol of atomic orbital lobe does not change, it is symmetric; while if the math symbol of atomic orbital lobe changes, it is antisymmetric. When two atomic orbitals are symmetric or antisymmetric on rotation or reflection, they symmetrically match and can combine and form the molecular orbital effectively. Otherwise if one is symmetric and the other is antisymmetric, they do not match symmetrically and cannot be combined to form molecular orbital. For example, as shown in Figure 10-11, the atomic orbitals in (a) and (b) are combined by additive way to form the σ bonding molecular orbital. Atomic orbitals in (c) are combined by subtractive way to form the σ antibonding molecular orbital. The two p_y orbitals can combine to generate a bonding pi molecular orbital in (d) or an antibonding pi molecular orbital in (e).

While (f) and (g) show that along the regarding xz plane one atomic orbital is symmetric but the other is antisymmetric, therefor they do not symmetrically match, molecular orbital cannot be obtained.

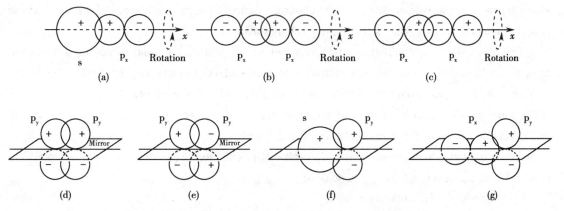

Figure 10-11 Drawing of Atomic Orbital Symmetry-Matching

Several simple linear combinations of atomic orbitals are rotational symmetric (along x axis) s-s, s-p_x, and p_x-p_x combinations forming σ molecular orbitals, and plane reflectional antisymmetric (on xz plane or xy plane) p_y-p_y or p_z-p_z combinations forming the π molecular orbitals. Figure 10-12 is drawing of two symmetry-matching atomic orbitals combining to form the molecular orbital.

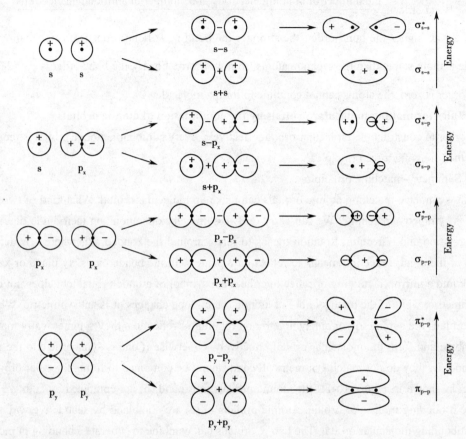

Figure 10-12 Drawing of Two Symmetry-Matching Atomic Orbital Combining to Form Molecular Orbital

(2) Principle of approximate energy

Only the atomic orbitals with approximate energy can effectively make up molecular orbital among symmetry-matching atomic orbitals. And the closer their energy, the more efficiency of the combination between atomic orbitals is. That is called energy approximate principle.

This principle is very important in deciding whether two different kinds of atomic orbitals can compose molecular orbital or not. For example, the energy of 1s orbital of H atom is $-1\,312\text{kJ}\cdot\text{mol}^{-1}$, the energy of 1s, 2s and 2p orbitals of F atom is $-67\,181, -3\,870.8$ and $-1\,797.4\text{kJ}\cdot\text{mol}^{-1}$ respectively. When forming HF molecule, in view of symmetry matching, 1s orbital of H may combine with any of 1s, 2s or 2p orbital of F to form molecule orbital, but according to energy approximate principle, 1s orbital of H only effectively combine with 2p orbital of F. Therefore, H atom and F atom are combined to form HF molecule through a σ_{s-p_x} single bond.

(3) Principle of orbital maximum overlap

When two symmetry-matching atomic orbitals carry on the linear combination, the more they overlap, and the lower the combined molecular orbital energy is, the stronger the chemical bond is formed. That is called the principle of orbital maximum overlap.

Among three principles above, symmetry-matching principle is the most important, which decides whether atomic orbital can be combined into molecular orbital or not. On the premise of conforming to symmetry-matching principle, principles of energy approximate and orbital maximum overlap decide the combinational efficiency of atomic orbitals.

10.5.2 Application of Molecular Orbital Theory

Each molecular orbital has its corresponding energy level, if they are arranged in order, a molecular orbital energy-level diagram would be obtained.

1. Molecular orbital diagram of homonuclear diatomic molecules

We are now ready to study the ground-state electron configuration of molecules containing second-period elements. We will consider only the simplest case, that of homonuclear diatomic molecules, or diatomic molecules containing the same atoms. Because the energy of 2s orbital is significantly different from 2p orbital, there are two models of energy level orders. If the orbital energy difference between 2s and 2p is too large ($>1\,500\,\text{kJ}\cdot\text{mol}^{-1}$), orbital interaction would be weak, so orbital linear combination occurs in s-s or p-p, the energy order of homonuclear diatomic molecular orbitals is

$$\sigma_{1s} < \sigma_{1s}^* < \sigma_{2s} < \sigma_{2s}^* < \sigma_{2p_x} < \pi_{2p_y} = \pi_{2p_z} < \pi_{2p_y}^* = \pi_{2p_z}^* < \sigma_{2p_x}^*$$

The molecular orbital energy-level diagram of the O_2 and F_2 molecules belongs to this model as shown in Figure 10-13(a).

If energy difference between 2s and 2p is relatively smaller ($<1\,500\,\text{kJ}\cdot\text{mol}^{-1}$), 2s and 2p orbital interactions affect the way in which atomic orbitals combine, which makes σ_{2p_x} molecular orbital energy surpass that of π_{2p_y} and π_{2p_z} molecular orbitals. The energy order for homonuclear diatomic molecular orbitals is

$$\sigma_{1s} < \sigma_{1s}^* < \sigma_{2s} < \sigma_{2s}^* < \pi_{2p_y} = \pi_{2p_z} < \sigma_{2p_x} < \pi_{2p_y}^* = \pi_{2p_z}^* < \sigma_{2p_x}^*$$

The molecular orbital energy-level diagrams of Li_2, B_2, C_2 and N_2 belong to this model as shown in Figure 10-13(b).

Sample Problem 10-9 Do H_2^+ and He_2 molecules exist?

Solution H_2^+ is formed by an H atom and an H nucleus with only one 1s electron, its ground-state electron configuration in molecular orbitals is $(\sigma_{1s})^1$, which means H_2^+ ion is bonded with one single-electron bond, and its bond order is 1/2. H_2^+ has a relatively weak bond, but it should exist.

The ground-state electron configuration of helium He atom is $1s^2$. Two He atoms have four electrons. In He_2, with two electrons in σ_{1s} MO and two electrons in σ_{1s}^* MO, both the bonding and antibonding orbitals are filled. Stabilization from the electron pair in the bonding MO is canceled by destabilization from the electron pair in the antibonding MO. Because the bond order of He_2 molecule is zero, $(2-2)/2$, we predict that a covalent He_2 molecule does not exist.

Sample Problem 10-10 Use the MO theory to write the electron configuration of N_2 and find its bond order.

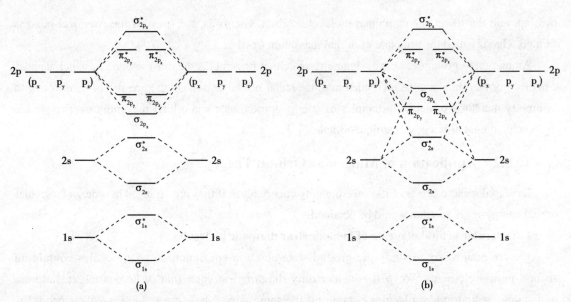

Figure 10-13 The Two Possible Molecular Orbital Energy-Level Diagrams for Period 2 Homonuclear Diatomic Molecules.

(a) MO energy-levels for O_2 and F_2; (b) MO energy-levels for Li_2, Be_2, B_2, C_2 and N_2

Solution The electron configuration of N atom is $1s^2 2s^2 2p^3$. There are fourteen electrons of N_2 filling in corresponding molecular orbital according to Figure 10-13 (b), therefore the formula of N_2 orbital is

$$N_2[(\sigma_{1s})^2(\sigma_{1s}^*)^2(\sigma_{2s})^2(\sigma_{2s}^*)^2(\pi_{2p_y})^2 = (\pi_{2p_z})^2(\sigma_{2p_x})^2]$$

The electrons on inner atomic orbitals are noble and not to take part in bonding when the molecule is forming. The ground-state electron configuration of N_2 in molecular orbital can be written as follows:

$$N_2[KK(\sigma_{2s})^2(\sigma_{2s}^*)^2(\pi_{2p_y})^2 = (\pi_{2p_z})^2(\sigma_{2p_x})^2]$$

Here each K represents two electrons on the atomic orbital of K electron shell.

In this case, the force of $(\sigma_{2s})^2$ bonding counterbalances that of $(\sigma_{2s}^*)^2$ antibonding, which does not work for bonding. $(\sigma_{2p_x})^2$ produce one σ bond and two π bonds are formed by $(\pi_{2p_y})^2$ and $(\pi_{2p_z})^2$ respectively. Therefore there are one σ bond and two π bonds in N_2. The electrons fill in bonding orbitals which decrease molecular energy, therefore N_2 is very stable. Its bond order is $(10-4)/2=3$.

Sample Problem 10-11 Why is O_2 paramagnetic? Determine the bond order and the chemical activity of O_2.

Solution The electron configuration of O atom is $1s^2 2s^2 2p^4$. Thus, there are 16 electrons in O_2. Using the order of increasing energies of the molecular orbitals discussed above, the ground-state electron configuration of O_2 is

$$O_2[KK(\sigma_{2s})^2(\sigma_{2s}^*)^2(\sigma_{2p_x})^2(\pi_{2p_y})^2 = (\pi_{2p_z})^2(\pi_{2p_y}^*)^1 = (\pi_{2p_z}^*)^1]$$

According to Hund's rule, the last two electrons enter $\pi_{2p_y}^*$ and $\pi_{2p_z}^*$ equivalent orbitals with parallel spins. Ignoring the $(\sigma_{2s})^2$ and $(\sigma_{2s}^*)^2$ because their net effects on bonding are zero, $(\sigma_{2p_x})^2$ form one σ bond; the bonding force of $(\pi_{2p_y})^2$ cannot completely counter-balance with the antibonding force of $(\pi_{2p_y}^*)^1$, and they form a three-electron π bond due to the same spatial location; $(\pi_{2p_z})^2$ and $(\pi_{2p_z}^*)^1$ form another one. So, O_2 molecule has one σ bond and two three-electron π bonds. There are two single electrons in two three-electron π bonds respectively. The bond order of O_2 is 2, $(10-6)/2$. And it is paramagnetic because of single electrons in the molecule.

In each three-electron π bond of O_2, two electrons are in the bonding orbital, and one electron locates in antibonding orbital. The decreased energy of three-electron π bond is only a half of that of single bond, so it is greatly weaker than the normal π bond. In fact, the molecular bond energy of O_2 is only 495kJ·mol^{-1}, which is lower than that of a general double bond. Because O_2 includes three-electron π bond with weak binding force, its chemical nature is quite active and it may release electron to turn into molecular ion O_2^+.

2. Molecular orbital diagram of heteronuclear diatomic molecules

Heteronuclear diatomic molecules, those composed of two different atoms, have asymmetric MO diagrams because the atomic orbitals of two bonded atoms have unequal energies. Atoms with greater effective nuclear charge pull their electrons closer to the nucleus and thus have more stable atomic orbitals and higher electronegativity values.

For the heteronuclear diatomic molecules or ions of Period 2 elements, atomic nuclear charge is the primary factor impacting molecular orbital energy level, if total nuclear number of heteronuclear diatomic molecule is less than or equal to two times of N atomic number (i.e. fourteen), the diagram of its orbital energy-level is coincident with the energy-level order shown in Figure 10-13(b); if more than twice of 14, then the energy-level diagram is coincident with Figure 10-13(a). Let's apply MO theory to the bonding in NO and HF.

Sample Problem 10-12 Please compare the stability of nitrogen monoxide NO molecule and nitrogen monoxide ion NO$^+$ by MO theory.

Solution Because the total electrons number of N and O is fifteen, the electron configuration of NO is

$$NO[(\sigma_{1s})^2(\sigma_{1s}^*)^2(\sigma_{2s})^2(\sigma_{2s}^*)^2(\sigma_{2p_x})^2(\pi_{2p_y})^2=(\pi_{2p_z})^2(\pi_{2p_y}^*)^1]$$

Here $(\sigma_{2p_x})^2$ forms one σ bond; $(\pi_{2p_z})^2$ forms one π bond; $(\pi_{2p_y})^2$ and $(\pi_{2p_y}^*)^1$ forms into one three-electron π-bond. The bond order is $(10-5)/2=2.5$. After NO loses one electron and becomes NO$^+$ ion, the $\pi_{2p_y}^*$ orbital is empty. Then there is one σ bond and two π bond in NO$^+$ and the bond order is $(10-4)/2=3$. Therefore, NO$^+$ ion is more stable than NO molecule.

Sample Problem 10-13 Using MO theory to analyze the bonding in HF molecule.

Solution To form the MOs in HF, we should know which atomic orbitals will combine. The high effective nuclear charge of F means that its electrons are held so tightly that its 1s, 2s, and 2p orbitals have lower energy than the 1s of H. According to three principles of atomic orbital linear combination in MO theory, the comprehensive analysis can make sure that 1s atomic orbital of H interacts with 2p$_x$ atomic orbital of F through end to end overlap to form one bonding molecular orbital 3σ and one antibonding molecular orbital 4σ (heteronuclear diatomic molecular orbitals in different periods are usually represented by 1σ, 2σ, 3σ, ⋯ 1π, 2π, ⋯). The two 2p orbitals of F are called nonbonding molecular orbitals. Because they are not involved in bonding, they have the same energy as that of the isolated atomic orbitals. The molecular orbital energy-level diagram for HF is shown in Figure 10-14. Here 1σ, 2σ and two 1π are nonbonding orbitals. The bond order of HF is one, and there is only one σ bond in the molecule. The electron configuration of HF molecule is $1\sigma^2 2\sigma^2 3\sigma^2 1\pi^4$.

216　　Chapter 10　Covalent Bond and Intermolecular Forces

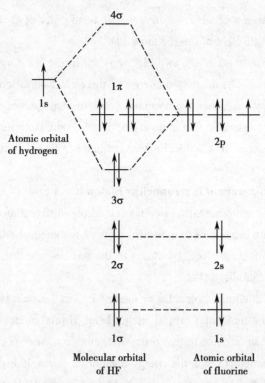

Figure 10-14　The Molecular Orbital Energy-Level Diagram of HF

10.6　Delocalized π Bond and Free Radical

10.6.1　Delocalized π Bond

π bonds are divided into localized π bonds and delocalized π bonds. A **localized bond** is a double-center bond, while a **delocalized bond** is a multi-center bond. Many compounds have delocalized π bonds (Π). A delocalized π bond with m electrons and n center atoms can be denoted as Π_n^m.

A delocalized π bond can be formed if two following conditions are satisfied:

1. The bonded atoms containing delocalized π bond form σ bond by hybrid orbital, and make up basic molecular structure. They are on identical plane. Each of them provides one p orbital which is vertical to the plane and the orbitals are mutually parallel, which guarantee the p orbitals overlap side-by-side in maximum.

2. The electrons number in delocalized π bond is less than doubled number of bonding p orbitals. According to orbital theory, the number of atomic orbitals combined can produce equivalent number of molecular orbitals. If the number of p electrons is equal to double of molecular orbitals, both bonding and antibonding orbitals are fully occupied, and the reduced energy balances the energy rising, so the bond cannot be formed effectively.

Sample Problem 10-14　How to analyze the π bond in ozone O_3 molecule?

Solution　The geometry of O_3 molecule is bent, as is shown in Figure 10-15. Here the two sp^2 hybrid orbitals of central O atom were combined with two $2p_x$ orbitals of the other two O atoms respectively to form two $\sigma_{sp^2\text{-}p}$ bonds, and the third sp^2 hybrid orbital is occupied by lone-pair electrons. Additionally, there is a pair of electrons in unhybridized p orbital of central O atom and each of the terminal O provide one half-filled $2p_z$ orbitals. The three unhybridized 2p orbitals were combined to forms a delocalized π bond with three O atoms center and four electrons are vertical to molecular plane. It is usually indicated by Π_3^4. O_3 molecule is diamagnetic, which means there is no single electron.

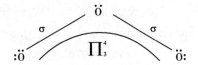

Figure 10-15 Delocalized π Bonds in Ozone Molecule

Question and thinking 10-3 Whether the delocalized π bond is formed in benzene C_6H_6 molecule? Compared with the classical structural formula of benzene, what are different properties if the delocalized π bond is formed?

10.6.2 A Brief Introduction of Free Radical

In Chemistry, the molecule like NO, atom like H, ion like O_2^- and atomic group like ·OH which contain the single-electron are called **free radical**. Single electron tends to pair with other electron, it is easy to lose or obtain the electrons, displaying rather active chemical characteristics. Organism can produce free radicals and remove free radicals. Excessive free radicals produced in organism can cause diseases, but free radicals can also cure diseases under some certain conditions.

Active oxygen free radicals transformed from O_2 are active participants in the process of physiology, pathology and aging. Two electrons of the highest energies in ground state of O_2 are filled up in two degenerate anti-bonding orbitals respectively and spin parallelly. The algebraic sum of their spin angular momentum quantum numbers (total spin angular momentum) S is $1/2 + 1/2 = 1$, and the spin multiplicity is equal to $2 \times S + 1 = 3$. Therefore, O_2 in ground state is called triplet oxygen, usually indicated by $^3O_2$. When $^3O_2$ is activated, electrons in two π^* orbital will spin conversely, $S = 1/2 + (-1/2) = 0$ and spin multiplicity $= 2 \times S + 1 = 1$. Thus singlet oxygen is formed, indicated by $^1O_2$. Electrons of $^1O_2$ on π^* orbital is arranged as follows.

$$^1O_2: \quad \underset{\pi^*_{2p_y}}{\uparrow\downarrow} \quad \underset{\pi^*_{2p_z}}{\quad} \quad \text{or} \quad \underset{\pi^*_{2p_y}}{\quad} \quad \underset{\pi^*_{2p_z}}{\uparrow\downarrow}$$

The energy of singlet oxygen is higher than that of triplet oxygen and it lacks of electrons in outer orbital. Besides, it is strongly oxidative, which can have reaction with all kinds of biological systems, such as viruses, cells. Therefore, singlet oxygen is active oxygen. Organisms have a series of biocatalysts (enzymes), so it is easy to form singlet oxygen in organism than in pure chemical system. For example, in leucocytes, $^3O_2$ changes into $^1O_2$ through several procedures. During this period, the movement of electrons in two π^* orbital of $^3O_2$ can be simplified as

$$\underset{^3O_2}{\uparrow\ \uparrow} \xrightarrow{+e^-} \underset{\cdot O_2^-}{\uparrow\downarrow\ \uparrow} \xrightarrow{+e^-} \underset{O_2^{2-}}{\uparrow\downarrow\ \uparrow\downarrow} \xrightarrow{-2e^-} \underset{^1O_2}{\uparrow\downarrow\ \_\_}$$

Super oxygen molecule $\cdot O_2^-$ is produced in the process. Since there is one single electron in its π^* orbital, it is a super oxygen anion free radical. Active oxygen, $^1O_2$ and its precursor O_2^- produced in leucocytes are both bacteria killers, which is favorable to diseases cure.

The existing free radical can promote favorable reactions to organism on one hand, and also can induce unfavorable reactions. Under certain circumstances, $^3O_2$ can react with biomolecules in human organs and form active oxygen like $\cdot O_2^-$, H_2O_2 and $\cdot OH$ (hydroxyl free radical). Here $\cdot O_2^-$ and $\cdot OH$ are active oxygen free radicals. When they react with molecule without single electron, they will change the molecule to free radical no matter whether they obtain or lose an electron. For example,

$$H_2O_2 + \cdot O_2^- \rightarrow O_2 + \cdot OH + OH^-$$

The produced free radical $\cdot OH$ will react with biomolecule in organ

$$\cdot OH + RH \rightarrow H_2O + R\cdot$$

Here the produced organic free radical $R\cdot$ can continue to react with other substances and produce new free radicals. Therefore, cell damages brought by $\cdot OH$ are rather serious. Free radicals can be passed on through this way. If there are excessive active oxygen free radicals in cells, they will damage cells and cause all kinds of diseases.

The only way to terminate free radicals is disproportion, namely the same two free radicals pass on electrons reciprocally. One radical can be reduced by obtaining an electron, and the other one can be oxidized by losing an electron. Excessive free radicals in organ can be removed by superoxide dismutase SOD, catalase CAT and glutathione peroxidase GSH-PX, etc. For example, the reaction of SOD removing $\cdot O_2^-$ is

$$\cdot O_2^- + \cdot O_2^- + 2H \xrightarrow{SOD} O_2 + H_2O_2$$

While catalase can remove H_2O_2

$$2H_2O_2 \xrightarrow{CAT} O_2 + 2H_2O$$

Excessive free radicals in organ can be removed not only by self-protection of some enzyme but also by taking some natural or synthetic antioxidant (also called a free radical inhibitor) which can eliminate the free radicals and protect the organism.

> **Question and thinking 10-4** Recent studies suggest that free radicals may be involved in cancer and even aging. What is free radical? What is the mechanism of free radical damaging the body? What measures can be taken to reduce the damage to the body?

10.7 Intermolecular Forces

In the above sections, we have discussed the covalent bond which is a strong force between the bonded atoms inside the molecule. Under certain conditions, substances usually exist in three different states, which are gas, liquid and solid, due to the different distances between the molecules. And the distance between molecules depends on the intermolecular forces. Generally intermolecular forces are relatively weaker and they are categorized into the two types, van der Waals forces and hydrogen bond. Intermolecular forces are closely related to the polarity and polarizability of molecules.

10.7.1 Polarity and Molecular Polarizability

1. Molecular Polarity

Knowing the molecular shape is key to understanding the physical and chemical behavior of a substance. One of the most far-reaching effects of molecular shape is about molecular polarity, which can influence melting point, boiling point, solubility, reactivity and even biological function. Molecule can be divided into two categories, polar molecule and non-polar molecule.

Let's take the simple molecule as examples to know molecular polarity. In diatomic molecules, all homonuclear diatomic molecules such as H_2, O_2 and F_2, the distribution of electron charge density between the two identical bonded atoms is even and the centers of positive charge and negative charge coincide, these kind of molecules are **non-polar molecule**. And the only bond in a homonuclear diatomic molecule is non-polar because two bonded atoms are no difference in electronegativity. While for heteronuclear diatomic molecules such as HF and HCl, due to the two bonded atoms are different in electronegativity, the only bond is polar. The distribution of electron charge density isn't even and the centers of positive charge and negative charge don't coincide. These kind of molecules are **polar molecule**.

In larger molecules, both shape and bond polarity determines the molecular polarity, an uneven distribution of charge over the whole molecule or large portion of it, even if polar bonds are present, the molecule will not necessarily be a polar one. For examples, the C-O bond in CO_2 molecule and C-H bond in CH_4 molecule are polar bonds. However, since the geometries of CO_2 and CH_4 are linear and tetrahedral respectively, both of them are non-polar molecules because of the coincidence of the center of positive charge with that of the negative charge. The molecule geometry of H_2O is bent and that of NH_3 is triangular pyramid. They are polar molecules because the centers of positive charge and negative charge in the both of molecules don't coincide.

A quantitative measure of the molecular polarity is its **electric dipole moment** ($\vec{\mu}$) given in the unit called a debye (D), which is derived from SI units of charge (coulomb, C) and length (m): $1D = 3.34 \times 10^{-30} C \cdot m$. Electric dipole moment $\vec{\mu}$ is product of charge q and distance d between positive and negative charge centers.

$$\vec{\mu} = q \cdot d$$

The electric dipole moment is a vector quantity, which means that it has both magnitude and direction. Measured values of electric dipole moment of some molecules are shown in Table 10-5. Non-polar molecule has a net dipole moment of zero, the larger the electric dipole moment is, the stronger the polarity of the molecule is.

Table 10-5 Some Dipole Moment $\vec{\mu} / 10^{-30} C \cdot m$

Molecule	$\vec{\mu}$	Molecule	$\vec{\mu}$	Molecule	$\vec{\mu}$
H_2	0	BF_3	0	CO	0.40
Cl_2	0	SO_2	5.33	HCl	3.43
CO_2	0	H_2O	6.16	HBr	2.63
CH_4	0	HCN	6.99	HI	1.27

2. Molecular Polarizability

No matter whether molecules are polar or non-polar, when they are induced by the external electric field, the center of positive and negative charge in molecules will always be changed. As shown in Figure 10-16, the centers of positive and negative charges of non-polar molecule are at the same position ($\vec{\mu} = 0$), but relative displacement occurs when it is induced by external electric field, such deformation can create molecular dipole; on the other hand, the centers of positive and negative charges of polar molecule are at different positions, so polar molecule has **permanent dipole**, which is increased or decreased, according to the direction of electric field. The phenomenon that the molecule deformation by an external electric field may cause dipole or the dipole moment of molecule increase is known as **polarizability**. This polarizability resulting dipole in molecule is said to be an **induced dipole**, its $\Delta\vec{\mu}$ value is in Figure 10-16.

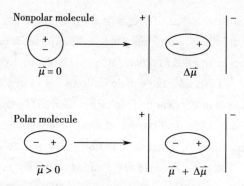

Figure 10-16 Schematic Diagram of Molecular Polarization Affected by External Electric Field

The polarizability of molecule not only occurs under external electric field but also occurs when the molecular interact, which is one important reason why interaction forces exist between molecules.

10.7.2 Van der Waals' Forces

There are weak forces between molecules whose strength is only 1/10 to 1/100 of chemical bond. They are named van der Waals forces because they were first proposed by the Dutch physicist van der Waals. All these forces are attractive in nature and play important roles in determining physical properties of substances, such as melting point, boiling point and solubility. Such forces can be divided into orientation, induction and dispersion forces in terms of their causes and characteristics.

1. Orientation force

The molecules with permanent dipoles attract each other electrostatically; the positive end of molecule attracts the negative end of another molecule and so on, leading to an alignment of molecules. Permanent dipole orientation arising from inter-molecular attractions is known as **orientation force** (dipole-dipole force). Orientation force occurs between polar molecules which possess dipole moment. Their origin is electrostatic, and they can be understood in terms of Coulomb's law. The larger the dipole moment is, the greater the force is. Figure 10-17 shows how two polar molecules interact.

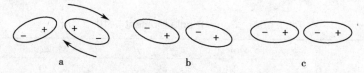

Figure 10-17 Schematic Diagram of Interactions between Two Polar Molecules

2. Induction force

Induction force occurs between the polar molecule and non-polar molecule or the polar molecule. When the polar molecule and non-polar molecule approaches, the permanent dipole of polarity molecule is like an external electric field, which makes the non-polar molecule polarize, then come into being induced dipole. The induced dipole attracts permanent dipole, as shown in Figure 10-18. Induction force results from the interaction of the dipole moment of a polar molecule and the induced

Figure 10-18 Schematic Diagram of Interaction between Polar Molecule and Nonpolar Molecule

dipole moment of a non-polar molecule. The greater the dipole moment and molecule polarizability is, the stronger the induction interaction is. When the two polar molecules approach to each other, each molecule will have induced dipole due to mutual polarization by their permanent dipoles. Induction force is a kind of additional orientation force.

3. Dispersion force

Dispersion force is a weak intermolecular force arising as a result of instantaneous dipoles in atoms or molecules. Dispersion forces are exhibited by non-polar molecules because of the correlated movements of the electrons in interacting molecules. Because the electrons in adjacent molecules "flee" as they repel each other, electron density in a molecule becomes redistributed in proximity to another molecule. This is frequently described as the formation of instantaneous dipoles that attract each other (Figure 10-19). Dispersion forces are present between all chemical groups, and usually represent the main part of the van der Waals forces.

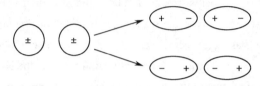

Figure 10-19 Schematic Diagram of Dispersion Force

Dispersion force usually increases with molar mass for the following reason. Molecule with the bigger molar mass tends to have more electrons, and its dispersion force increases in strength with the number of electrons. Furthermore, the bigger molar mass often means the bigger atoms whose electron cloud distribution is more easily to be disturbed because its outer electrons are less tightly held by nucleus.

To sum up, dispersion force exists between non-polar molecules. Both induction force and dispersion force exist between non-polar molecules and polar molecules. Orientation force, induction force and dispersion force exist between polar molecules. Table 10-6 lists three forces distributed in some molecules.

Table 10-6 Distribution of Van der Waals forces/kJ·mol^{-1}

Molecule	Orientation force	Induction force	Dispersion force	Total energy
Ar	0.000	0.000	8.49	8.49
CO	0.003	0.008	8.74	8.75
HI	0.025	0.113	25.86	26.00
HBr	0.686	0.502	21.92	23.11
HCl	3.305	1.004	16.82	21.13
NH$_3$	13.31	1.548	14.94	29.80
H$_2$O	36.38	1.929	8.996	47.31

Van der Waals forces do not belong to the scope of chemical bonds. They have the following characteristics: they are electrostatic attractions, in most covalent bonds, they are quiet weak compared with the bonds between atoms within molecules; the scope of their interactions are only several hundred pm; they do not have direction and saturation properties; dispersion force is the main force in most molecules. Only molecule of large polarity has remarkable orientation force. Induction force is usually small.

Physical properties of substance (such as melting point, boiling point, and so on) are related to intermolecular interactions, generally speaking, if van der Waals forces of substance are small, its melting point and boiling point would be low. We can learn from Table 10-6 that within the series HCl, HBr, HI, van der Waals forces increase, and boiling points and melting points increase in the order HCl < HBr < HI.

10.7.3 Hydrogen Bond

Normally, boiling points of a series of similar compounds containing elements in the same Group increase with their increasing molar masses. But some exceptions were noticed for hydrides of the elements in Group VA, VIA, and VIIA. Among these hydrides, the lighter compound (NH_3, H_2O, HF) has the higher boiling point, those are contrary to our expectations. This study and other related observations led chemists to postulate the existence of hydrogen bond. When an H atom is combined with an atom X such as F, O, N, etc. which has large electronegativity and small radius to form a covalent bond, the electron cloud between the two nuclei is strongly leaned to the X atom. The H atom is almost a bare proton and with large positive charge field. Thus the H atom can be directionally attracted by another atom Y such as F, O, N, etc., which also has large electronegativity, small radius and lone pair electrons in the outer, to form the structure of X—H⋯Y. The electrostatic attraction between the H atom and the Y atom (shown in the dashed line) is called **hydrogen bond**. X and Y represent O, N, or F; X—H is a molecule or the part of a molecule, and Y is a part of another molecule. The three atoms usually lie along a straight line. As shown in Figure 10-20.

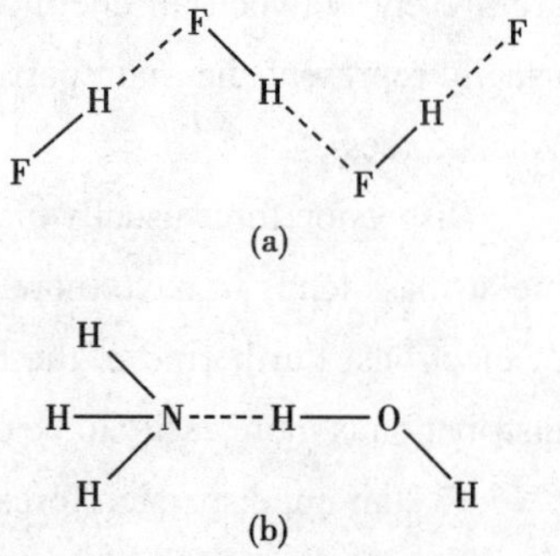

The strength of hydrogen bond is determined by the electronegativity of atom X and Y and their atomic radii. The higher electronegativity and smaller atomic radius of X and Y is, the stronger of hydrogen bond is. The electronegativity value of Cl is higher than that of N, but the atomic radius of Cl is bigger than that of N. So only the very weak hydrogen bond exits in HCl molecule. The strength order of hydrogen bond is:

Figure 10-20 Hydrogen Bonds in Hydrogen Fluoride and Ammonia
(a) HF (b) NH_3 and H_2O

$$F—H \cdots F > O—H \cdots O > O—H \cdots N > N—H \cdots N > O—H \cdots Cl$$

The average energy of hydrogen bond is stronger than that of dipole-dipole interaction (up to 42kJ·mol^{-1}). Thus, hydrogen bond is a powerful force in determining the structure and properties of many molecules.

Hydrogen bond forms not only between molecules such as HF, NH_3 (Figure 10-20) but also in molecule, such as nitric acid HNO_3, o-nitrophenol (Figure 10-21). Intramolecular hydrogen bond is not in a straight line, but a ring structure with more stability.

Hydrogen bond exists in many compounds, and it affects the nature of substance. Because the destruction of hydrogen bond needs energy, in similar compounds they can form intermolecular hydrogen bonds whose boiling point and melting point is higher than that of compounds which have no hydrogen bond. Such as the hydride of Group VA to VIIA elements, the boiling points of NH_3, H_2O and HF are higher than that of other hydrides of elements in the same group who have a larger relative atomic mass, which are due to the unusual behaviors of inter-molecular hydrogen bonds formed. Intramolecular hydrogen bond generally reduces the melting

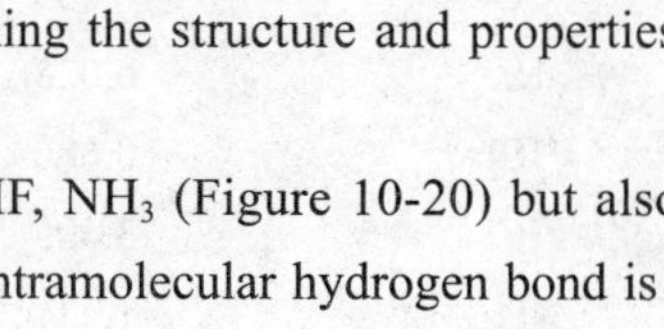

Figure 10-21 Intramolecular Hydrogen Bond in Nitric Acid and o-Nitrophenol

point and boiling point of compound. Hydrogen bond also affects solubility of material. If hydrogen bond forms between solute and solvent, the solubility would increase. If intramolecular hydrogen bond forms in solute molecule, the solubility in polar solvent would be small, but increase in non-polar solvent. For example, *o*-nitrophenol molecule can form intramolecular hydrogen bond, *p*-nitrophenol molecule does not form intramolecular hydrogen bond because of long distance between the function groups, but it can form hydrogen bond with water molecule, so, solubility of *o*-nitrophenol in water is lower than that of *p*-nitrophenol.

> **Question and thinking 10-5** For HX acid, it is known that HCl, HBr and HI are strong acids. However, HF is the only weak acid. Why?

Some biological materials, such as protein molecules, nucleic acid molecules, have hydrogen bonds. DNA (deoxyribonucleic acid) molecule, the chain by more than 2 nucleotide bases (C=O ----- H-N and C=N ----- H-N) to form hydrogen bonds between the pair and connected, that is, adenine (A) and thymine (T) form 2 hydrogen bonds, guanine (G) and cytosine (C) form 3 hydrogen bonds. They are twisting double helix structures as shown in Figure 10-22, the circle also depends on hydrogen bond to maintain and enhance its stability. Once the hydrogen bond is damaged, the molecular geometry changed, its physiological functions will lose. Therefore, to understand the hydrogen bond has a very important significance for medical students.

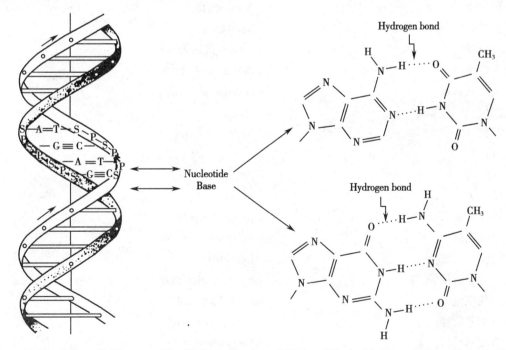

Figure 10-22 Schematic Diagram of Double Helix Structure of DNA and Nucleotide Bases to Form Hydrogen Bond

Key Terms

| 成键分子轨道 | bonding molecular orbital |
| 等性杂化 | equivalent hybridization |

不等性杂化	nonequivalent hybridization
电偶极矩	electric dipole moment
定域键	localized bond
反键分子轨道	antibonding molecular orbital
非极性分子	non-polar molecule
非键轨道	nonbonding orbital
分子轨道理论	molecular orbital theory
分子间作用力	intermolecular force
非极性共价键	non-polar covalent bond
基态	ground state
极性分子	polar molecule
极化率/极化性	polarizability
极性共价键	polar covalent bond
价层电子对互斥理论	valence shell electron pair repulsion theory
价键理论	valence bond theory
键参数	bond parameter
键长	bond length
键级	bond order
键角	bond angle
键能	bond energy
解离能	dissociation energy
离域键	delocalized bond
配位共价键	coordinate covalent bond
配位键	coordination bond
氢键	hydrogen bond
取向力	orientation force
色散力	dispersion force
诱导力	induction force
诱导偶极	induced dipole
永久偶极	permanent dipole
杂化	hybridization
杂化轨道理论	hybridization theory
杂化轨道	hybrid orbitals
正常共价键	normal covalent bond
自由基或游离基	free radical

Summary

Valence Bond Theory (VB) Covalent bond is formed by pairing of two electrons from two bonded atoms respectively and they are opposite spins, and the atomic orbitals hold bonding electrons overlap in maximum extent as far as possible by valence bond theory (VB).

Sigma bond (σ-bond) is formed by overlapping the orbitals "end-to-end". pi bond (π-bond) is formed by overlapping the orbitals "side-by-side". Generally, σ-bond is more stable than a π-bond. The single bond is built by σ-bonds. The double bond consists of one σ-bond and one π-bond. The triple bond consists of one σ-bond and two π-bonds. Coordinate bond is formed when one species is the donor of electron lone pairs and other species has the outer orbitals without electron to accept the lone pairs. The nature of coordinate bond is still the covalent bond.

Valence Shell Electron Pair Repulsion Model (VSEPR) For molecule AX_mE_n, electron pair around the central atom A tend to be located as far as possible in the space. The valence shell electron pair is composed of electron pair for σ-bond and lone electron pair of atom A. The shapes of molecules depend on not only the positions occupied by surrounding atom X attached to the A atom, but also lone pairs of electrons E by VSEPR model.

Hybridization Theory In many molecules the bonding is best described in terms of hybrid orbitals that result from the mixing or combining the atomic orbitals (s, p, d, etc.) of central atom in a molecule. In the mixing process, atomic orbital number is equal before and after hybrid. In addition, the molecular geometry depends on the character of hybridization. For example, two sp hybrid orbitals have their major lobes pointed in opposite directions, 180° apart; a set of three sp^2 hybrid orbitals has major lobes at 120° in a plane; and four sp^3 hybrid orbitals show tetrahedral geometry, having lobes at 109°28′.

Molecular Orbital Theory (MO) In MO theory, electrons belong to the entire molecule. A set of molecular orbitals formed from a linear combination of atomic orbitals is called an LCAO-MO. Three rules (same symmetry, approximate energy and maximum extent of orbital overlapping) must be satisfied for effective combination. The electron configuration in a molecule is obtained by feeding the appropriate number of electrons into these molecule orbitals and following the same rules to fill the electrons into atomic orbitals. The MO can be used to indicate the nature of bonds, forming magnetism and stability of the simple molecules.

Intermolecular forces (van der Waals forces and hydrogen bonds) Intermolecular forces are much weaker than bonding forces. They are orientation forces, induction forces and dispersion forces. Orientation forces are only in polar molecules. Induced forces exist between polar molecules and non-polar molecules, also in polar molecules. Dispersion forces present in all molecules and it is the most significant among van der Waals forces.

Hydrogen bonding Hydrogen bonding is an attractive interaction between two atoms that arises from a link of X−H⋯Y or X−H⋯X, where A and B are highly electronegative elements such as O, N, or F, and possess a lone pair of electrons. Hydrogen bond is much stronger than van der Waals forces.

Intermolecular forces affect the state of substance and some properties such as melting point, boiling point, solubility, diffusion, surface tension, etc.

Exercises

1. Distinguish the following terms:
  (1) σ bond and π bond
  (2) normal covalent bond and coordinate covalent bond

(3) polar bond and non-polar bond

(4) localized π bond and delocalized π bond

(5) equivalent hybridization and nonequivalent hybridization

(6) bonding molecular orbital and anti-bonding molecular orbital

(7) permanent dipole and instantaneous dipole

(8) van der Waals forces and hydrogen bond

2. Why is covalent bond with the nature of saturation and direction?

3. Use hybridization theory to describe the type of hybrid orbitals of central atom and predict the geometry of the following molecules and ions.

(1) CO_3^{2-} (2) SO_2 (3) NH_4^+ (4) H_2S

(5) PCl_5 (6) SF_4 (7) SF_6 (8) BrF_5

4. Use hybridization theory to describe how mixing the atomic orbitals of the central atoms leads to the hybrid orbitals in ethane C_2H_6, ethylene C_2H_4, acetylene C_2H_2 and determine the formed bond types.

5. The geometry of BF_3 is trigonal planar while NF_3 is trigonal pyramidal, use hybridization theory to explain them.

6. Explain the following changes in the hybrid types of center atom and geometry.

(1) $BF_3 \rightarrow BF_4^-$ (2) $H_2O \rightarrow H_3O^+$ (3) $NH_3 \rightarrow NH_4^+$

7. What's the difference between valence electron pair geometry of the center atom and molecule geometry? Describe NH_3 molecule as an example.

8. What is the hybridization of central atom in each of the following substances? Determine the geometry of the following molecules or ions.

(1) PH_3 (2) $HgCl_2$ (3) $SnCl_4$ (4) $SeBr_2$ (5) H_3O^+

9. A compound has a serious carcinogenicity and its composition is as follows: H 2.1%, N 29.8%, and O 68.1%, its molar mass is about $50 g \cdot mol^{-1}$. Answer the following questions:

(1) Write the chemical formula of the compound.

(2) If H and O are bonding, draw the structural formula.

(3) Determine the hybridization type of N atom and describe the types of orbitals and bonds in the molecule.

10. Write molecular orbital formula of the following two-atom molecules or ions, point out the chemical bond, and calculate bond order to determine what is the most stable? Which is the most unstable? Which of paramagnetic? Which is diamagnetic?

(1) B_2 (2) F_2 (3) F_2^+ (4) He_2^+

11. "Aged pigment" is caused by lipofuscin depositing on the skin of the elders, the formation of lipofuscin is relative with superoxide ion O_2^-. Use molecular orbital theory to predict whether O_2^- exists? If it exists, write its electron configuration and find the bond order. Compared with O_2, how about the stability and magnetic properties of O_2^-?

12. Use VB and MO to explain why stable H_2 can exist and He_2 cannot.

13. Determine types of Π-bond in the following molecules or ions.

(1) NO_2 (2) CO_2 (3) SO_3 (4) C_4H_6 (5) CO_3^{2-}

14. What are free radicals? What is active oxygen free radical?

15. Predict the geometry of the following molecules, determine whether the electric dipole moment

is zero and determine their molecular polarities.

(1) SiF_4 (2) NF_3 (3) BCl_3 (4) H_2S (5) $CHCl_3$

16. Compared the following pairs of molecules with each other, which one is more polar? Explain it simply.

(1) HCl and HI   (2) H_2O and H_2S (3) NH_3 and PH_3

(4) CH_4 and SiH_4 (5) CH_4 and $CHCl_3$ (6) BF_3 and NF_3

17. Boiling points of rare gases are listed below, from the data explain the ranking.

Name	He	Ne	Ar	Kr	Xe
Boiling Point /K	4.26	27.26	87.46	120.26	166.06

18. Rank the following in order of increasing boiling points, and explain your ranking.

(1) H_2 CO   Ne   HF   (2) CI_4 CF_4 CBr_4 CCl_4

19. Cl_2 and F_2 are gases at room temperature, Br_2 is liquid, I_2 is solid, why?

20. Ethanol (C_2H_5OH) and dimethyl ether (CH_3OCH_3) have the same molecular formula, but the boiling point of ethanol is higher than that of dimethyl ether, why?

21. Determine the intermolecular forces of the following groups.

(1) benzene and carbon tetrachloride   (2) ethanol and water

(3) benzene and ethanol   (4) liquid ammonia

22. Rank the following in order of decreasing hydrogen bonding strength, and explain your ranking.

(1) HF and HF   (2) H_2O and H_2O (3) NH_3 and NH_3

23. One organic solvent which is harmful to health, its molecular formula is AB_4, A is in Group ⅣA, B is in Group ⅦA, the electronegativity values of A and B are 2.55 and 3.16, respectively. Answer the following questions

(1) Molecular geometry of AB_4 is tetrahedral, predict the type of hybridization when atom A and B are bonding.

(2) What is bond polarity of A-B? What is molecular polarity of AB_4?

(3) AB_4 is liquid at room temperature, what molecular forces exist?

(4) If the melting point and boiling point of AB_4 are compared with that of $SiCl_4$, which is higher?

Supplementary Exercises

1. Determine the hybridization state of central atom in each of the following molecules: (a) $HgCl_2$, (b) AlI_3, and(c) PF_3. Describe the hybridization process and determine the molecule geometry in each case.

2. Use the VSEPR model to predict geometries of the following molecules and ions:

(a) AsH_3 (b) OF_2 (c) $AlCl_4^-$ (d) I_3^-

3. N_2^+ ion can be prepared by bombarding N_2 molecule with fast-moving electrons. Predict the following properties of N_2^+: (a) electron configuration, (b) bond order, (c) magnetic character, and (d) bond length relative to the bond length of N_2 (is it longer or shorter?)

4. Is the π bond in NO_2^- localized or delocalized? How can you determine whether a molecule or ion will exhibit delocalized?

5. Which of the followings can form hydrogen bond with water? CH_4, F^-, HCOOH, Na^+.

Answers to Some Exercises

[Exercises]

3.

Molecules or ions	Molecular geometry	Valence electron-pair geometry
CO_3^{2-}	Trigonal planar	Trigonal planar
SO_2	Bent	Trigonal planar
NH_4^+	Tetrahedral	Tetrahedral
H_2S	Bent	Tetrahedral
PCl_5	Trigonal bipyramidal	Trigonal bipyramidal
SF_4	Distortion tetrahedral	Trigonal bipyramidal
SF_6	Octahedral	Octahedral
BrF_5	Square pyramidal	Octahedral

4. C_2H_6: sp^3 C_2H_4: sp^2 C_2H_2: sp

5. The type of hybrid orbitals of central atom B is sp^2 equivalent hybridization and that of N is sp^3 nonequivalent hybridization.

6. (1) sp^2 equivalent hybridization → sp^3 equivalent hybridization, trigonal planar → tetrahedral

(2) no changes in the hybrid types of center atom, bent → triangular pyramid

(3) sp^3 nonequivalent hybridization → sp^3 equivalent hybridization, triangular pyramid → tetrahedral

8. (1) sp^3 nonequivalent hybridization   triangular pyramid

(2) sp hybridization   linear

(3) sp^3 equivalent hybridization   tetrahedral

(4) sp^3 nonequivalent hybridization   bent

(5) sp^3 nonequivalent hybridization   triangular pyramid

9. (1) HNO_2 (2) :Ö—N—Ö—H with σ, σ, σ bonds and lone pair on N   (3) Π_3^4

10. (1) $[(\sigma_{1s})^2(\sigma_{1s}^*)^2(\sigma_{2s})^2(\sigma_{2s}^*)^2(\pi_{2p_y})^1(\pi_{2p_z})^1]$, two single-electron π bonds, bond order is 1, paramagnetic.

(2) $[KK(\sigma_{2s})^2(\sigma_{2s}^*)^2(\sigma_{2p_x})^2(\pi_{2p_y})^2(\pi_{2p_z})^2(\pi_{2p_y}^*)^2(\pi_{2p_z}^*)^2]$, one σ bond, bond order is 1, diamagnetic.

(3) $[KK(\sigma_{2s})^2(\sigma_{2s}^*)^2(\sigma_{2p_x})^2(\pi_{2p_y})^2(\pi_{2p_z})^2(\pi_{2p_y}^*)^2(\pi_{2p_z}^*)^1]$, one σ bond and one three-electron π bond, bond order is 1.5, paramagnetic.

(4) $[(\sigma_{1s})^2(\sigma_{1s}^*)^1]$, one three-electron σ bond, bond order is 0.5, paramagnetic.

F_2^+ is the most stable, He_2^+ is the most unstable.

11. The electron configuration of oxygen molecular is

$O_2[KK(\sigma_{2s})^2(\sigma_{2s}^*)^2(\sigma_{2p_x})^2(\pi_{2p_y})^2 = (\pi_{2p_z})^2(\pi_{2p_y}^*)^1 = (\pi_{2p_z}^*)^1]$, bond order is 2. There is one electron in two three-electron π bonds respectively, therefore oxygen is paramagnetic.

The electron configuration of O_2^- is

$(\sigma_{1s})^2(\sigma_{1s}^*)^2(\sigma_{2s})^2(\sigma_{2s}^*)^2(\sigma_{2p_x})^2(\pi_{2p_y})^2(\pi_{2p_z})^2(\pi_{2p_y}^*)^2(\pi_{2p_z}^*)^1$, bond order is 1.5. There is one electron in the oxygen molecular ion O_2^-, therefore it is paramagnetic.

The magnetic property of O_2^- is weaker than that of O_2 and the stability: $O_2 > O_2^-$.

12. The electron configuration of hydrogen atom is $1s^1$ and that of He is $1s^2$. According to VB theory, there is one single-electron in hydrogen atom, after two single-electrons with opposite spin are paired, they can form a stable covalent bond. However, there is no single-electron in helium atom, it cannot form covalent bond, which means that He_2 cannot exist.

According to MO theory, the electron configuration of hydrogen molecule is $(\sigma_{1s})^2$. The bond order is 1, therefore the H_2 can exist stable. While the electron configuration of helium molecule is $(\sigma_{1s})^2(\sigma_{1s}^*)^2$. The bond order is zero, which means that He_2 cannot exist.

13. (1) Π_3^3 (2) two Π_3^4 (3) Π_4^6 (4) Π_4^4 (5) Π_4^6

15.

Molecule	Molecular geometry	Electric dipole moment	Molecular polarity
SiF_4	Tetrahedral	$=0$	Non-polar molecule
NF_3	Trigonal pyramidal	$\neq 0$	Polar molecule
BCl_3	Trigonal planar	$=0$	Non-polar molecule
H_2S	Bent	$\neq 0$	Polar molecule
$CHCl_3$	Distortion tetrahedral	$\neq 0$	Polar molecule

16. (1) HCl   (2) H_2O (3) NH_3 (4) nonpolar molecule   (5) $CHCl_3$ (6) NF_3

18. (1) $H_2 < Ne < CO < HF$ (2) $CF_4 < CCl_4 < CBr_4 < CI_4$

21. (1) Dispersion force

　　(2) Orientation force, induction force, dispersion force and hydrogen bond

　　(3) Dispersion force and induction force

　　(4) Orientation force, induction force, dispersion force and hydrogen bond

22. (1) > (2) > (3)

23. (1) sp^3 equivalent hybridization

　　(2) A-B is polar covalent bond, AB_4 is nonpolar molecule.

　　(3) Dispersion force

　　(4) The melting point and boiling point of $SiCl_4$ is higher.

[Supplementary Exercises]

1. (a) sp hybridization, linear

　　(b) sp^2 hybridization, trigonal planar

　　(c) sp^3 hybridization, trigonal pyramidal

2. (a) trigonal pyramidal   (b) bent   (c) tetrahedral   (d) linear

3. (a) $(\sigma_{1s})^2(\sigma_{1s}^*)^2(\sigma_{2s})^2(\sigma_{2s}^*)^2(\pi_{2p_y})^2(\pi_{2p_z})^2(\sigma_{2p_x})^1$

　　(b) 2.5

　　(c) Paramagnetic

　　(d) Longer bond than N_2

4. It is delocalized.

5. F^- and HCOOH

Chapter 11
Coordination Compounds

Coordination compound or complex is a kind of compounds with complex ion composition generally. They are widely applied into many fields. The coordination compound is more complex than the common compounds.

Complexes are closely related to organisms and medicine. With the in-depth study of the trace elements in the life, It is found that many essential trace elements in organism exist in the form of complexes and they are closely related to the physiological activities. Many biological catalysts — enzymes are also metal complexes, playing dominant roles in biochemical reactions in the body. Some drugs play their roles in the body by forming complexes. In addition, the analysis methods based on coordination reaction are also widely used in biochemical test, environmental monitoring and drug analysis. To know the properties of the drugs related to coordination compounds will help us to reveal the pathogenesis of some diseases in the human body and synthesize new drugs — metal complexes. Bioinorganic chemistry is one of the most active forefront sciences with many growth points in modern natural science. Therefore, it is necessary to understand nature of the complexes for medical students.

11.1 Basic Concept of Coordination Compound

11.1.1 Formation of Coordination Compound

Dissolve $CuSO_4$ in H_2O to form a blue solution in a tube, then add $6mol \cdot L^{-1}$ of ammonia drop by drop to form a dark blue solution. Add several drops of NaOH solution into the solution, neither NH_3 gas nor blue $Cu(OH)_2$ precipitate was produced. But adding several drops of $BaCl_2$ solution, the white $BaSO_4$ precipitate would be produced. It shows that there are large amount of SO_4^{2-} ions in the solution, but no Cu^{2+} ions and NH_3 can be detected almost. By the specific detecting way, we know that $[Cu(NH_3)_4]^{2+}$ called complex ion was formed in the solution. The behavior of $[Cu(NH_3)_4]^{2+}$ in solution is similar to that of weak electrolytes, which can slightly dissociate into Cu^{2+} and NH_3, and most of them still exist in the form of $[Cu(NH_3)_4]^{2+}$. that is evident for $[Cu(NH_3)_4]^{2+}$ to be more stable. The complex ions (or complex molecules) like $[Cu(NH_3)_4]^{2+}$ are formed by the combination of metal cation (or atom) such as Cu^{2+} with the anions (or neutral molecules) such as NH_3 by the coordination bond. So it could be acids, bases and salts, such as $H_2[PtCl_6]$, $[Cu(NH_3)_4](OH)_2$ and $[Cu(NH_3)_4]SO_4$ or complex molecules, such as $[Ni(CO)_4]$ etc.

11.1.2 Constitutes of Coordination Compound

Most of the complexes are composed by complex ions and counter ions with opposite charges, such as [Cu(NH$_3$)$_4$]SO$_4$. Its composition can be expressed as

$$\underbrace{\underbrace{[\underset{\downarrow}{Cu(NH_3)_4}]}_{\text{Central atom Ligands}}}_{\text{Inner sphere}} \quad \underbrace{\underset{\downarrow}{SO_4}}_{\text{Outer sphere}}$$

$$\text{Coordination compound}$$

1. Inner sphere and outer sphere

Complex ion is characteristic part of complex and consists of **central atom** and **ligands**. The complex ion is also called inner sphere, and it is usually written in square brackets. The counter ions with opposite charges are called outer sphere. The inner and outer spheres connected by ionic bond. In the aqueous solution, the coordination compounds are easily dissociated into the outer ions and complex ions which nevertheless are difficult to be dissociated. So, the complex ion and outer ions separate, but the complex ion behaves like a polyatomic ion.

2. Central atom

In complex ions (or complex molecules), a cation or atom accepting lone pairs of electrons from ligands is called the central atom which is the core part of the complex ion. Generally, it is metal ion. Most of them are transition elements, such as Ag$^+$ in [Ag(NH$_3$)$_2$]$^+$, Ni(0) in [Ni(CO)$_4$]. In addition, the atoms of some subgroup elements adjacent to them and the nonmetallic elements with high oxidation value are also the most common central atoms, such as Si(Ⅳ) in [SiF$_6$]$^{2-}$.

3. Ligand and ligating atom

In the complexes, the anion or neutral molecules that bind to the central atom with the coordination bonds are ligands, such as NH$_3$ in [Ag(NH$_3$)$_2$]$^+$, CO in [Ni(CO)$_4$], F$^-$ in [SiF$_6$]$^{2-}$ and so on. The atom donating lone pair electrons in ligands is called ligating atom, such as N atom in NH$_3$, C atom in CO, F atom in F$^-$, etc. The outermost shell of ligating atom have lone pair electron, which are non-metallic atoms with large electronegativity usually, such as N, O, C, S, F, Cl, Br, I, etc.

Ligands are divided into monodentate ligand and multidentate ligand, according to the number of ligating atom in the ligands. Monodentate ligand only has a single ligating atom. Such as pyridine, NH$_3$, H$_2$O, CN$^-$, F$^-$, Cl$^-$, etc., corresponding ligating atoms are N, N, O, C, F, Cl separately. Ligands containing two or more ligating atoms are called multidentate ligands such as ethylenediamine H$_2$N-CH$_2$-CH$_2$-NH$_2$ (or en), diethylenetriamine H$_2$NCH$_2$CH$_2$NHCH$_2$CH$_2$NH$_2$ (DEN) and ethylenediamineteraacetate ion (labeled Y^{4-}). They are bidentate, tridentate and hexadentate ligands separately. For example, labeled Y^{4-} is as following

$$\begin{array}{c} ^-\ddot{O}OCH_2C \\ ^-\ddot{O}OCH_2C \end{array} \!\!\diagdown\!\! \ddot{N}CH_2CH_2\ddot{N} \!\!\diagup\!\! \begin{array}{c} CH_2CO\ddot{O}^- \\ CH_2CO\ddot{O}^- \end{array}$$

There are four O and two N ligating atoms in Y^{4-}.

In some cases, although a few ligands have two ligating atoms, but they can only take one of them to bond with the central atom like a monodentate, because the two ligating atoms are too close in ligand molecule. These ligands are called ambidentate ligand or heterosexual biradical ligands, such as only N, O, S, N act as ligating atoms in nitro NO_2^-, nitrite ONO^-, thiocyanate SCN^-, isothiocyanate NCS^- separately.

4. Coordination number

The number of ligating atoms in complex to bond directly with central atom by coordination bonds is called coordination number. Generally, the coordination number is the number of coordination bonds between the central atom and the ligand. If ligands are monodentate ligands, the coordination number of the central atom is equal to the number of ligands. For instance, the coordination number of Cu^{2+} is 4 in $[Cu(NH_3)_4]^{2+}$. If the ligands are multidentate ligand with n of ligating atoms, the coordination number of the central atom is equal to n times the number of ligands. For example, the coordination number of Cu^{2+} is 4 instead of 2 in $[Cu(en)_2]^{2+}$ because en molecule is bidentate ligand. The coordination number of Co^{3+} is 6 instead of 4 in $[Co(en)_2(NH_3)Cl]^{2+}$. In the complexes, the common coordination numbers of central atoms are 2, 4 and 6. The common coordination numbers of some metal ions are listed in Table 11-1.

Table 11-1 Common Coordination Numbers of Some Metal Ions

Coordination number	Metal ions	Examples
2	Ag^+, Cu^+, Au^+	$[Ag(NH_3)_2]^+$, $[Cu(CN)_2]^-$
4	Cu^{2+}, Zn^{2+}, Cd^{2+}, Hg^{2+}, Al^{3+}, Sn^{2+} Pb^{2+}, Co^{2+}, Ni^{2+}, Pt^{2+}, Fe^{3+}, Fe^{2+}	$[HgI_4]^{2-}$, $[Zn(CN)_4]^{2-}$ $[Pt(NH_3)_2Cl_2]$
6	Cr^{3+}, Al^{3+}, Pt^{4+}, Fe^{3+}, Fe^{2+}, Co^{3+} Co^{2+}, Ni^{2+}, Pb^{4+}	$[PtCl_6]^{2-}$, $[Co(NH_3)_3(H_2O)Cl_2]$ $[Fe(CN)_6]^{3-}$, $[Ni(NH_3)_6]^{2+}$ $[Cr(NH_3)_4Cl_2]^+$

Three factors affects on the coordination number mainly, they are electron configurations of central atom, space effect and electrostatic effect of ligands.

The outer vacant valance orbitals of the second periodic elements are 2s and 2p sublevel, the total of four orbitals. They can accommodate four pairs of electrons at most. The maximum coordination number is 4, such as $[BeCl_4]^{2-}$, $[BF_4]^-$, etc. For the elements after the second periodic their outer vacant valance orbitals are $(n-1)d$, ns, np, nd. The coordination number can be more than 4, Such as $[AlF_6]^{3-}$, $[SiF_6]^{2-}$, etc.

The larger the size of central atom is, the smaller the shape of ligand is, the more conducive to form the complex ion with large coordination number is, such as the size of F^- is smaller than that of Cl^-, then, Al^{3+} could form $[AlF_6]^{3-}$ with coordination number 6 while $[AlCl_4]^-$ with coordination number 4. The radius of the central atom B (Ⅲ) is smaller than that of Al^{3+}, so B (Ⅲ) could only form $[BF_4]^-$ with coordination number 4.

Considering the electrostatic effect, the higher the positive charge of the center atom is, the more conducive to form the complex ion with large coordination number is, such as Pt^{2+} and Cl^- can form $[PtCl_4]^{2-}$, but Pt^{4+} can form $[PtCl_6]^{2-}$. When the central atoms are the same, the more negative charges the ligand has, the greater the repulsion between the ligands is, and the smaller the coordination number is. For example, Ni^{2+} and NH_3 can form $[Ni(NH_3)_6]^{2+}$ with coordination number 6, while Ni^{2+} and CN^- only form $[Ni(CN)_4]^{2-}$ with coordination number 4.

5. The charge of complex ion

The charge of the complex ion is equal to the algebraic sum of the total charge of the central atom and the ligand. For example, NH_3 is a neutral molecule, so the charge of $[Cu(NH_3)_4]^{2+}$ ion is 2, while the charge of $[HgI_4]^{2-}$ ion is $[1 \times (+2) + 4 \times (-1)] = -2$. Because the complex is electrically neutral, the charge of complex ion and the oxidation number of central atom is easy to know by the charge of ligands and the outer sphere ions.

> **Question and thinking 11-1** What is the difference between coordination compound and double salt?

11.1.3 Formulas and Nomenclature of Coordination Compounds

The rules of nomenclature for coordination compounds are the same as that for general inorganic compounds.

(1) Cation is written before anion. This is usual practice when naming a salt. The ending for anionic coordination is -ate, alternatively, -ic if named as an acid. For cationic and neutral coordination compound, the name of the metal is used without any characteristic ending.

(2) Nonionic or molecular coordination compound is given one word name.

(3) The inner sphere is enclosed in square brackets in the molecular formula. Within the inner sphere, the ligand's name precedes the metal's name, but in the formula the metal ion precedes the ligands. For example,

$[Cu(NH_3)_4]SO_4$, Tetraamminecopper(Ⅱ) sulfate

$[Co(NH_3)_6]Cl_3$, Hexaamminecobalt(Ⅲ) chloride

(4) Neutral ligands are named as molecule, negative ligands end in -o, and positive ligands end in -ium. Neutral ligands retain their usual name. Coordinated water is called aqua and coordinated ammonia is called ammine.

(5) The prefixes di-, tri-, tetra-, etc., are used before such simple expressions as bromo, nitro, and oxalate. Prefixes bis-, tris-, tetrakis-, etc., are used before the names of the complexes, such as ethylenediamine and trialkylphosphine. For example,

$K_3[Al(C_2O_4)_3]$ Potassium trioxalatoaluminate(Ⅲ)

$[Co(en)_2Cl_2]_2SO_4$ Dichlorobis(ethylenediamine) cobalt(Ⅲ) sulfate

(6) Oxidation state of the central atom is designated by Roman numeral in parentheses at the end of the name of coordination compound, for a negative oxidation state, a minus sign is used before Roman numeral, and 0 is used for zero. The following are examples of complexes.

$[Ni(CO)_4]$ Tetracarbonylnickel(0)

$[Pt(NH_2)(NO_2)(NH_3)_2]$ Minodiamminenitroplatinum(Ⅱ)

(7) The ligands in coordination compound are named in the sequence of "negative; neutral;

positive", without separation by hyphens. Within these categories the groups are listed in order of increasing complexity.

Formula	Chemical name
$[Cu(NH_3)_4]^{2+}$	Tetraamminecopper(II) ion
$[CoCl_2(NH_3)_4]^+$	Tetraamminedichlorocobalt(III) ion
$[Fe(en)_3]Cl_3$	Tris (ethylenediamine)iron (III) chloride
$[Ag(NH_3)_2]OH$	Diamminesilver(I) hydroxide
$H_2[PtCl_6]$	Hydrogen hexaammineplatinate(IV)
$[Co(ONO)(NH_3)_5]SO_4$	Pentaamminenitritecobalt(III) sulphate
$[Co(NH_3)_5(H_2O)]_2(SO_4)_3$	Pentaammineaquacobalt(III) sulphate
$[Co(NH_3)_2(en)_2]Cl_3$	Diamminebis(ethylenediamine)cobalt(III) chloride
$NH_4[Co(NO_2)_4(NH_3)_2]$	Ammonia diamminetetranitrocobaltate(III)
$NH_4[Cr(NCS)_4(NH_3)_2]$	Ammonia diamminetetra-isothiocyanatochromate(III)

11.2 Chemical Bond Theory of Coordination Compound

Some physical and chemical properties of complexes depend on their structures, especially the bonding natures between ligands and central atom. Chemical bond theory is used to clarify the bond natures and to indicate some properties of complex, such as the coordination number, geometry, magnetic and so on. This section focuses on the valence bond theory and the crystal field theory briefly.

11.2.1 Valence Bond Theory of Coordination Compound

1. Main points of valence bond theory

In 1931, the American Chemist, L. Pauling applied the hybrid orbital theory to the coordination compounds and proposed the **valence bond theory of coordination compound.** Basic points of valence bond theory are as follows:

a. The chemical bond between ligating atom and central atom is coordination bond, in which the lone electron pairs of ligating atom are donated to the empty orbitals of central atom, and then they share the lone pairs. The nature of coordination bond is covalent bond.

Ligand is the electron-donor, Lewis base; central atom is the electron receptor, Lewis acid; and their combination is complex ion or molecule which is also acid-base complex.

b. The empty valance orbitals of central atom will be hybridized to produce hybrid orbitals with equal number, equal energy and different orientations. The hybrid orbital can overlap with the orbital of ligands containing the lone pair electrons along the bond axis to form a coordination bond.

c. The geometry of the complex is depending on coordination number and hybridization type of central atom. Table 11-2 lists them.

In many cases, valence bond theory can't be used to predict the geometry of complex and hybridization type of central atom, but it could be used to explain the geometry and magnetic data of the complex.

Table 11-2 Hybridization Types of Central Atom and Corresponding Geometry of Coordination Compounds

Coordination number	Hybridization type	Geometry	Examples
2	sp	Linear	$[Ag(NH_3)_2]^+$, $[AgCl_2]^-$, $[Au(CN)_2]^-$
4	sp^3	Tetrahedral	$[Ni(CO)_4]$, $[Cd(CN)_4]^{2-}$, $[ZnCl_4]^{2-}$, $[Ni(NH_3)_4]^{2+}$
	dsp^2	Square planar	$[Ni(CN)_4]^{2-}$, $[PtCl_4]^{2-}$, $[Pt(NH_3)_2Cl_2]$
6	sp^3d^2	Octahedral	$[FeF_6]^{3-}$, $[Fe(NCS)_6]^{3-}$, $[Co(NH_3)_6]^{2+}$, $[Ni(NH_3)_6]^{2+}$
	d^2sp^3	Octahedral	$[Fe(CN)_6]^{3-}$, $[Co(NH_3)_6]^{3+}$, $[Fe(CN)_6]^{4-}$, $[PtCl_6]^{2-}$

2. Outer orbital complex ions and inner orbital complex ions

If the central atom is transition element, its valence orbitals often include $(n-1)$d orbitals. According to the hybridization types of central atoms, the complexes are divided into outer orbital and inner orbital types. Outer orbital coordination compound is the complex which the valence orbitals, ns, np and nd of central atom are hybridized. For examples, the sp, sp^3 and sp^3d^2 hybridizations of central atom will result in the formations of outer orbital coordination compounds with the coordination number of 2, 4 and 6. While inner orbital coordination compound is the complex which the valence orbitals, $(n-1)$d, ns and np of central atom are hybridized. Such as dsp^2 and d^2sp^3 hybridizations of central atom, it will result in inner orbital coordination compounds with the coordination number of 4 and 6.

3. Application of valence bond theory

Take the complex ion $[Fe(H_2O)_6]^{3+}$ as example, the valence electronic configuration of Fe^{3+} ion is $[Ar]3d^5$. Five electrons occupy 3d orbitals respectively. When Fe^{3+} combines with H_2O to form $[Fe(H_2O)_6]^{3+}$, the six equivalent sp^3d^2 hybrid orbitals are formed by hybridization of one 4s, three 4p and two 4d orbitals. The O atoms in H_2O molecules will donate six lone electron pairs to six sp^3d^2 hybrid orbitals of Fe^{3+} respectively to form six coordination bonds, and the outer orbital complex $[Fe(H_2O)_6]^{3+}$ ion is formed.

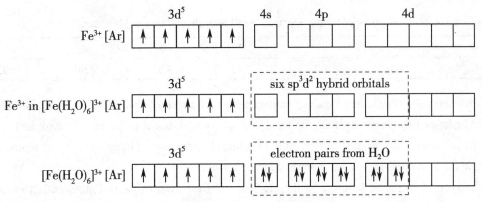

But in $[Fe(CN)_6]^{3-}$ ion, the valence electrons in 3d orbitals of Fe^{3+} would be affected by ligand CN^- greatly. Because the ligating atom in CN^- is C atom, it has the lower electronegativity than O atom. C atom has the larger tends to donate the lone electron pair to Fe^{3+} compared with O atom. Thus, CN^- ions will make Fe^{3+} rearrange the valence electrons in 3d orbitals again. After rearranging, the two empty 3d orbitals would be hybridized with one 4s orbital and three 4p orbitals to form six equivalent d^2sp^3 hybrid orbitals. Similarly, the C atoms in CN^- ions will donate six lone electron pairs to six d^2sp^3 hybrid orbitals of Fe^{3+} respectively. The inner orbital complex $[Fe(CN)_6]^{3-}$ ion is formed.

The nature of sp^3d^2 and d^2sp^3 hybridization is same in fact, but the energy level of hybrid orbitals is different. Because sp^3d^2 hybridization takes 4s, 4p and 4d orbitals to hybrid, but d^2sp^3 hybridization takes 3d, 4s and 4p orbitals to hybrid. The orbital energy level of 3d is less than 4d. So, the sp^3d^2 hybrid orbitals have the higher energy level than that of d^2sp^3. In other words, the stability of inner orbital complex is better than the outer orbital complex.

The geometry of both $[Fe(H_2O)_6]^{3+}$ and $[Fe(CN)_6]^{3-}$ is octahedron because of six bonding pairs near the central atom. In fact, sp^3d^2 and d^2sp^3 hybridization model of central atom possess the same geometry in the space.

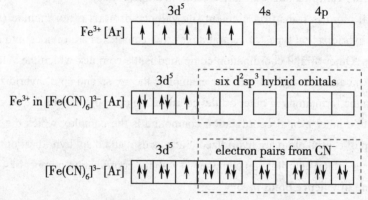

4. Magnetic moment of coordination compound

The outer or inner orbital model is usually determined by measuring the magnetic moment of the complex. Table 11-3 lists the relationship between the single electron numbers n and the theoretical value of magnetic moment μ by the approximate formula.

$$\mu \approx \sqrt{n(n+2)}\mu_B$$

Here μ_B is Bohr magnetion, $\mu_B = 9.27 \times 10^{-24} A \cdot m^2$

Table 11-3 Amount of Single Electron and Theoretical Value of Magnetic Moment (μ)

n	0	1	2	3	4	5
μ/μ_B	0.00	1.73	2.83	3.87	4.90	5.92

If we know the experimental data of magnetic moment about the complexes, hybridization type of it could be known by above formula, the inner or outer orbital model could be indicated, and then the geometry of complex ion is known also. Table 11-4 lists the experimental data of magnetic moments and the single electron numbers of some coordination compounds.

The type of complex depends on the electron configuration of central atom and the nature of ligands.

The complex must be outer orbital model when $(n-1)d$ orbital of central atom is full filled, in $(n-1)d^{10}$, such as $[Ag(CN)_2]^-$, $[Zn(CN)_4]^{2-}$, $[CdI_4]^{2-}$ and $[Hg(CN)_4]^{2-}$.

Table 11-4 Experimental Data of Magnetic Moment and Single Electron Numbers of Some Coordination Compounds

Complex	d electron numbers of central atom	μ/μ_B	Numbers of single electron	Type of coordination compound
$[Fe(H_2O)_6]SO_4$	6	4.91	4	Outer orbital
$K_3[FeF_6]$	5	5.45	5	Outer orbital
$Na_4[Mn(CN)_6]$	5	1.57	1	Inner orbital
$K_3[Fe(CN)_6]$	5	2.13	1	Inner orbital
$[Co(NH_3)_6]Cl_3$	6	0	0	Inner orbital

When the electron numbers in $(n-1)d$ orbital of central atom is less than three, there are two empty $(n-1)d$ orbitals at least, the complex must be inner orbital complex, such as $[Cr(H_2O)_6]^{3+}$ and $[Ti(H_2O)_6]^{3+}$.

The electron configuration of central atoms with $d^4 \sim d^7$ can form either inner or outer orbital complexes. The ligands become the main factor to determine the type of complexes. If the ligating atom has the larger electronegativity such as N, O and halogen atoms, etc., it is hard to donate the lone pair, then it tends to occupy the outer nd orbit of central atom to form the outer-orbit complex, such as $[FeF_6]^{3-}$ and $[Fe(H_2O)_6]^{3+}$. On the contrary, if the ligating atom has lower electronegativity such as C atom in CN^-, it is easy to donate the lone pair, and ligand has the great affect on the valence electrons in $(n-1)d$ orbital of central atoms. Then d electrons of central atom will rearrange and make out empty $(n-1)d$ orbit to form inner orbital complex, such as $[Fe(CN)_6]^{3-}$.

In some cases, the inner orbital complexes containing empty $(n-1)d$ orbitals is still unstable. For example, the valence electron configuration of V^{3+} in $[V(NH_3)_6]^{3+}$ is $[Ar]3d^2$. Two empty 3d orbitals hybridize with one 4s and three 4p orbitals to form six d^2sp^3 hybrid orbitals which have been used to form six coordination bonds with six NH_3 molecules. After that, there is still one empty 3d orbital in V^{3+} ion. Hence, $[V(NH_3)_6]^{3+}$ ion is unstable although it is an inner orbital complex ion.

Valence bond theory successfully indicates the formation of complexes, coordination number, chemical properties, geometry, magnetic property and stability of complexes. But it neglected energy variations of d orbitals of central atom affected by electric field of ligands, thus, it only could be used to explain the properties of complex in the ground state, and could not explain the other properties like color, absorption spectrum of complex and so on. These problems will be solved by the crystal field theory.

> **Question and thinking 11-2** Some metal complexes of platinum Pt are known as active *anticancer* drugs, such as $[PtCl_4(NH_3)_2]$, $[PtCl_2(NH_3)_2]$ and $[PtCl_2(en)]$. The experimental results indicate that all of them are all antimagnetic materials. Applying Valence Bond Theory to predict hybridization type of central atom and point out they are inner or outer orbital complexes.

11.2.2 Crystal Field Theory

The **Crystal field theory** (CFT) was first proposed by H. Bethe in 1929, and was not developed rapidly until it was successfully used to explain absorption spectra of metal complexes in the 1950s.

1. Main points of crystal field theory

(1) The center atom and ligands are combined by electrostatic forces. Central atom is a point with positive charge, and ligands are the points with negative charge. The interactions between them are electrostatic attraction and repulsion, other than the covalent bond.

(2) Five equivalent d orbitals would be split into different energy levels, affected by the negative electric fields of ligands. Some d orbitals raise their energy and some lower.

(3) Due to the splitting of d orbital energy level, the electrons in d orbitals will rearrange. That would make the total energy of whole system changing. After rearranging the electrons in the splitting d orbitals, if the total energy is lower than that before splitting, the complex will be more stable.

2. d Orbital splitting of the central atom in octahedral field

Take an octahedral complex as an example to understand the crystal field theory.

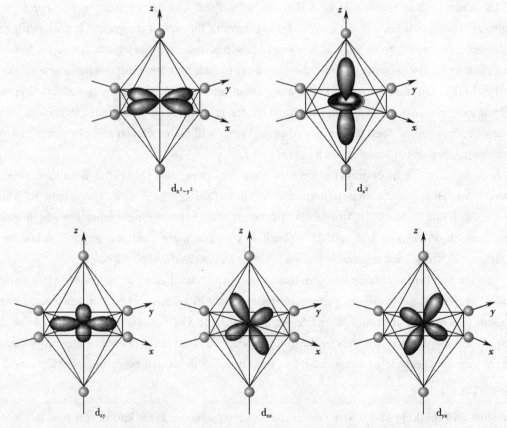

Figure 11-1 d Orbitals of Octahedral Coordination Compound and Relative Positions of Ligands

Five d orbitals have different orientations in the space. They will be influenced indifferent degree by the electrostatic fields of ligands. Assuming that six ligands are spherical distribution symmetrically, here the sphere center is central atom. The repulsion that the electrons in d orbitals are subjected to come from ligands is equal, they are still in the same energy level although their energy levels are increased because of the influence of surrounding ligands. In fact, the six ligands approach the central atom from

"+" and "−" directions along x, y, z axes (± x, ± y, ± z) as shown in Figure 11-1. Two among the six ligands directly approach facing the lobes end of d_{z^2} and $d_{x^2-y^2}$ orbitals, while other three ligangs just right approach to the lobes of d_{xy}, d_{yz} and d_{xz} orbitals. Due to the different orientations of d orbitals, electrons in d_{z^2} and $d_{x^2-y^2}$ orbitals suffer the stronger repulsion than those in d_{xy}, d_{yz} and d_{xz} orbitals. Thus, d_{z^2} and $d_{x^2-y^2}$ orbitals are higher in energy, and d_{xy}, d_{yz} and d_{xz} orbitalsare lower comparing with the original equivalent orbitals. In octahedral geometry of complex, five equivalent d orbitals of central atom split into d_γ and d_ε groups, or d_γ and d_ε orbital. d_γ group includes d_{z^2} and $d_{x^2-y^2}$ orbitals, and d_ε group includes d_{xy}, d_{yz} and d_{xz} orbitals. The orbitals in every groups are still equivalent. Thus, d_ε group has three equivalent orbitals and d_γ group has two as shown in Figure 11-2. E_0 is d orbital energy level of free central atom; E_s is d orbital energy level of central atom in spherical electrostatic field.

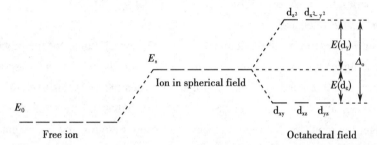

Figure 11-2 Splitting Energies of d Orbital of Central Atom in an Octahedral Field of Ligands

3. Splitting energy and its influencing factors

The splitting models and degrees of d orbitals of central atom with different geometries of complexes are different. We still take the octahedral coordination compound as the example to understand the splitting energy in crystal field theory.

The energy difference between the highest and the lowest energy levels of splitted d orbitals of the central atom is called the splitting energy notated by Δ. The energy difference between d_γ and d_ε is the splitting energy of octahedral field notated by Δ_o.

According to crystal field theory, the splitting energy between d_γ and d_ε can be calculated in an octahedral complex. Now, the average energy of the five d orbitals (E_s) is set as 0, which is taken as a comparative standard for calculation of relative energy. Furthermore, the total energy of d orbitals is equal before and after splitting in octahedral field. That is

$$2E(d_\gamma) + 3E(d_\varepsilon) = 5E_s = 0$$
$$E(d_\gamma) - E(d_\varepsilon) = \Delta_o$$

Then we get the solutions, $E(d_\gamma) = +0.6\Delta_o$ $E(d_\varepsilon) = -0.4\Delta_o$

Therefore, the energy level of d_γ orbital increases $0.6\Delta_o$ and d_ε orbital decreases $0.4\Delta_o$.

For the same types of complexes, the main influencing factors on their splitting energies are the nature of their ligands, the oxidation number and the atomic radius of the central atoms. That is a. Intensity of ligand field. For a given central atom, the splitting energy relates to the intensity of ligand field. The greater the intensity of the field is, the higher the splitting energy is. Intensity order of ligand field can be attained by spectroscopic experiments of octahedral complexes. It is

$$I^- < Br^- < Cl^- < SCN^- < F^- < S_2O_3^{2-} < OH^- \approx ONO^- < C_2O_4^{2-} < H_2O < NCS^-$$

$\approx$ EDTA < NH$_3$ < en < SO$_3^{2-}$ < NO$_2^-$ < CN$^-$ < CO

This order is called spectro-chemical series. According to this series, the splitting ability of d-orbitals under ligand I$^-$ ion into d$_\gamma$ and d$_\varepsilon$ energy levels is the weakest (the gap between energy levels is the smallest), but those under CN$^-$ and CO are the highest. So, I$^-$ is called weak or low field ligand; CN$^-$ and CO are called strong-field ligands. The intensity of other ligands depends on their central atoms. In general, ligands before H$_2$O in the series belong to weak-field ligands. The intensity of the ligands between H$_2$O and CN$^-$ in the series depends on central atoms and magnetic moments of the complexes. As narrated above, in spectro-chemical series study, that ligands with the same ligating atoms are often gotten together. Such as OH$^-$, C$_2$O$_4^{2-}$ and H$_2$O whose ligating atoms are O atom, NH$_3$ and en whose ligating atoms are N atoms. We also draw a rough conclusion for the following ligating atoms by spectro-chemical series, that is, Δ_o order generally is I < Br < Cl < F < O < N < C.

b. Oxidation number of central atom

Splitting energies depend on the oxidation number of central atoms for the complexes with the same ligands. The higher the oxidation number of central atom is, the larger the splitting energy is. Because the higher oxidation number of the central atom means the more positive charge. Such central atom makes a stronger attraction to ligands. So the distance between central atom and the ligands is closer. The stronger the repulsion between d electrons of central atom and the lone pair in ligands, the higher the splitting energy is. For example,

[Co(H$_2$O)$_6$]$^{2+}$ Δ_o = 111.3 kJ·mol^{-1};   [Co(H$_2$O)$_6$]$^{3+}$ Δ_o = 222.5 kJ·mol^{-1}

[Fe(H$_2$O)$_6$]$^{2+}$ Δ_o = 124.4 kJ·mol^{-1};   [Fe(H$_2$O)$_6$]$^{3+}$ Δ_o = 163.9 kJ·mol^{-1}

c. Atomic radius of central atom

For the complexes possessing the same ligands, when the oxidation numbers of central atoms are the same, their splitting energies will increase as the increasing of atomic radius. The bigger the atomic radius of central atom is, the farther away the d orbital of the central atom is from atomic nuclear, and the closer to the ligands the central atom is. So, the stronger the repulsion between the central atom and the ligand field is, the higher the splitting energy is.

3d^6 [Co(NH$_3$)$_6$]$^{3+}$ Δ_o = 275.1 kJ·mol^{-1}

4d^6 [Rh(NH$_3$)$_6$]$^{3+}$ Δ_o = 405.4 kJ·mol^{-1}

5d^6 [Ir(NH$_3$)$_6$]$^{3+}$ Δ_o = 478.4 kJ·mol^{-1}

The geometry of the complex is also an important influencing factor for splitting energy. The tetrahedral complexes are very different from the octahedral complexes in splitting energy significantly.

4. d-Electrons configuration of the central atom in octahedral field

In octahedral geometry of complex, d electrons configuration of central atom tends to reduce the system energy.

When the electronic configurations of central atoms are d$^1 \sim$ d^3, the electrons occupy the orbitals of d$_\varepsilon$ solely with the same spin, according to Aufbau principle and Hund's rules.

When the electronic configurations of central atoms are d$^4 \sim$ d^7, for an octahedral complex, the d electrons configuration will have two types. One is that the electrons occupy the orbitals of the lower energy levels as much as possible according to Aufbau principle, the other is that d electrons will occupy different d-orbitals as much as possible to keep the same spin according to Hund's rules. The configuration type of d electrons is determined by splitting energy Δ_o and relative values of electron pairing energy P. Electron pairing energy must be required to overcome the repulsion between two electrons if an electron occupies an orbital in which has been accommodated by another electron.

Configuration types of d electrons of central atoms in some octahedral complexes are given in Table 11-5. From it, we can find that there is only configuration case of d electrons for $d^1 \sim d^3$ and $d^8 \sim d^{10}$ in the complex according to Aufbau principle and Hund's rules, no matter its ligands are strong-field or weak-field. If d electrons is $d^4 \sim d^7$, the d electrons will occupy the orbitals in d_ε as much as possible when the central atom interacts with strong-field ligands ($\Delta_o > P$). On the contrary, the d electrons will occupy the orbitals in d_ε and d_γ as much as possible to make them in parallel spin when the central atom interacts with weak-field ligands ($\Delta_o < P$). So the number of unpaired d electrons of the latter is more than that of the former. For the complexes whose central atom possessing the same number of d electrons, the high-spin complexes will have the more unpaired d electrons, while the low-spin complexes will have less unpaired d electrons. As for these complexes whose central atoms are $d^4 \sim d^7$ state, the electron configurations would be low-spin complexes under strong-field ligands, such as NO_2^-, CN^- and CO, and be high-spin complex with weak-field ligands, such as X^-, H_2O, etc.

Table 11-5 Configuration of d Electrons of Central Atoms in Octahedral Complexes

Electrons	Weak field ($P > \Delta_o$)		Unpaired electron numbers	Strong field ($P < \Delta_o$)		Unpaired electron numbers
	d_ε	d_γ		d_ε	d_γ	
1	↑		1	↑		1
2	↑ ↑		2	↑ ↑		2
3	↑ ↑ ↑		3	↑ ↑ ↑		3
4	↑ ↑ ↑	↑	4	↑↓ ↑ ↑		2
5	↑ ↑ ↑	↑ ↑	5	↑↓ ↑↓ ↑		1
6	↑↓ ↑ ↑	↑ ↑	4	↑↓ ↑↓ ↑↓		0
7	↑↓ ↑↓ ↑	↑ ↑	3	↑↓ ↑↓ ↑↓	↑	1
8	↑↓ ↑↓ ↑↓	↑ ↑	2	↑↓ ↑↓ ↑↓	↑ ↑	2
9	↑↓ ↑↓ ↑↓	↑↓ ↑	1	↑↓ ↑↓ ↑↓	↑↓ ↑	1
10	↑↓ ↑↓ ↑↓	↑↓ ↑↓	0	↑↓ ↑↓ ↑↓	↑↓ ↑↓	0

(Rows 4–8 weak field: high-spin; rows 4–8 strong field: low-spin)

5. Crystal field stabilization energy

Electron gives priority to occupy d orbitals with lower energy level after splitting of d orbitals under negative charge field of ligands. Systematic energy will decrease when d electrons fill to splitted orbitals comparing with the case of d electrons configuration before splitting. The reduced energy of the system is **crystal field stabilization energy** (CFSE). The greater the absolute value of CFSE is, the lower the systematic energy is, and the more stable the complex is.

CFSE is related to the number of d-electrons in central atom and the intensity of ligand field. It is also associated with the complex geometry. The CFSE of an octahedral complex can be calculated as following

$$\text{CFSE} = x\text{E}(d_\varepsilon) + y\text{E}(d_\gamma) + (n_2 - n_1)P \tag{11.1}$$

Here, x and y are the numbers of electrons in d_ε and d_γ, respectively. n_1 is the number of the electron pair in d orbitals before splitting; n_2 is the number of electron pair in d_ε and d_γ orbitals after splitting. The results are showed in Table 11-6.

Table 11-6 CFSE of Complex with d^n Electrons in Octahedral Field

d electron	Weak field			Strong field		
	electron configuration		CFSE	electron configuration		CFSE
	d_ε	d_γ		d_ε	d_γ	
0	0	0	0	0	0	0
1	1	0	$-0.4\Delta_o$	1	0	$-0.4\Delta_o$
2	2	0	$-0.8\Delta_o$	2	0	$-0.8\Delta_o$
3	3	0	$-1.2\Delta_o$	3	0	$-1.2\Delta_o$
4	3	1	$-0.6\Delta_o$	4	0	$-1.6\Delta_o+P$
5	3	2	0	5	0	$-2.0\Delta_o+2P$
6	4	2	$-0.4\Delta_o$	6	0	$-2.4\Delta_o+2P$
7	5	2	$-0.8\Delta_o$	6	1	$-1.8\Delta_o+P$
8	6	2	$-1.2\Delta_o$	6	2	$-1.2\Delta_o$
9	6	3	$-0.6\Delta_o$	6	3	$-0.6\Delta_o$
10	6	4	0	6	4	0

Sample Problem 11-1 Calculate the CFSE of the octahedral complexes if Co^{3+} ion is central ion in a strong-field and in a weak-field, respectively, and compare their stability.

Solution Co^{3+} has six valence d electrons, $[Ar]3d^6$, its electron configuration is

Spherical field   Weak octahedral field $(\Delta_o<P)$ Strong octahedral field $(\Delta_o>P)$

Spherical-field, $E_s=0$

Weak-field, $CFSE = 4E(d_\varepsilon) + 2E(d_\gamma) + (1-1)P = 4\times(-0.4\Delta_o) + 2\times(+0.6\Delta_o) = -0.4\Delta_o$

Strong-field, $CFSE = 6E(d_\varepsilon) + 0E(d_\gamma) + (3-1)P = 6\times(-0.4\Delta_o) + 0 + 2P$

$= 6\times(-0.4\Delta_o) + 2P = -0.4\Delta_o - 2\times(\Delta_o-P)$

Here, $\Delta_o>P$,   so, $CFSE < -0.4\Delta_o$

According to the calculations above, both CFSEs are lower than zero, especially in a strong-field. Thus, the complex of central Co^{3+} ion in a strong-field is more stable.

6. d-d transition and the color of coordination compounds

Visible light is a mixture containing various wavelengths of light. When a substance absorbs some wavelength of light, the colour of substance what we see is the just the colour of complementary light absorbed. For example, if absorbed light is red, substance appears green-blue; if it absorbs green-blue light, the substance appears red; green-blue and red light is complementary light in colour each other. Colour relationships between complementary lights are listed in Table 11-7.

Table 11-7 Relationship between the Color of Substance and Light Absorbed

Substance colour	Colour of light absorbed	Range of wavelength absorbed/nm
Yellow green	Purple	400~425
Yellow	Deep blue	425~450
Orange Yellow	blue	450~480
Orange	Green blue	480~490
Red	Blue green	490~500
Purplish red	green	500~530
Purple	Yellow green	530~560
Deep blue	Orange Yellow	560~600
Green blue	Orange	600~640
Blue green	red	640~750

Experimental facts show that the splitting energy is approaching to the energy of visible light. Most of transition-metal complexes are colorful because most of transition metal ions have unfilled splitting d orbitals. If d electrons with lower energy level selectively absorb a certain wave-length of light, they would transit to the higher energy level of d orbitals known as d-d transition. Thus the complexes appear the color of complementary light absorbed.

For example as shown in Figure 11-3, $[Ti(H_2O)_6]^{3+}$ is red, the electronic configuration of Ti^{3+} is $[Ar]3d^1$, the d electron fills d_ε orbital whose energy level is lower in octahedral field. There is only an electron in d orbitals. If we consider absorption spectrum of complex ion $[Ti(H_2O)_6]^{3+}$, When the complex ion is illuminated by a beam of light, it can absorb light of 492.7nm, blue green region of the visible spectrum. This absorption promotes the electron from d_ε orbital to d_γ orbital. The energy of photon with wave-length of 492.7nm or 20 300cm^{-1} corresponds to the energy change of 243kJ·mol^{-1} (1cm^{-1} = 11.96J·mol^{-1}), it just equals to the splitting energy Δ_o of $[Ti(H_2O)_6]^{3+}$. Because the sample absorbs strongly the green light, it appears red. Figure 11-4 is the absorption spectrum of $[Ti(H_2O)_6]^{3+}$ ion to the different wavelength of visible light.

Because the most of complex ions absorb selectively the different wavelengths color of visible light under the different ligands with the very different splitting energies, the complexes show the different color. The stronger the ligand field intensity is, the higher the splitting energy is, the more energy is absorbed by d-d transition of d electrons, the shorter the wavelength of absorbed light is.

But some metal ions such as Zn^{2+}, Ag^+ ions, their valence electron are d^{10}, the d_γ orbitals are full filled by electrons. Thus, their complexes are colorless because no d-d transitions in them.

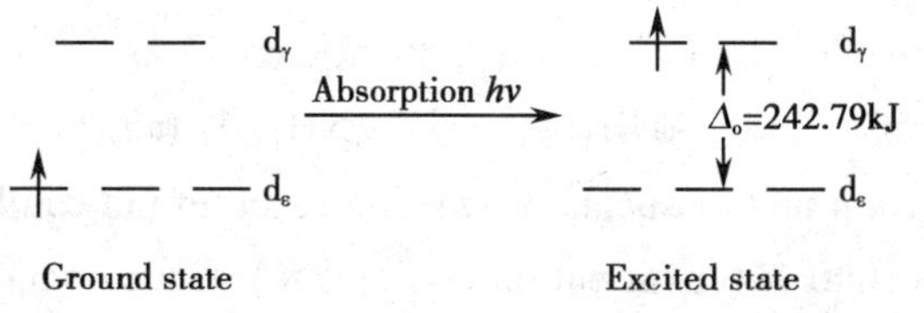

Figure 11-3 d-d Transition in $[Ti(H_2O)_6]^{3+}$ Ion

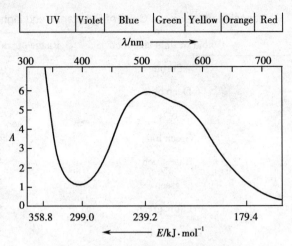

Figure 11-4 Absorption Spectrum of $[Ti(H_2O)_6]^{3+}$ Ion

In conclusion, the complex is colorful because of the d-d transitions. So, the following two requirements should be satisfied for the colorful complex:

a) The outer d-orbitals of central atoms are not full filled.

b) The splitting energy must be within the energy ranges of visible lights.

CFT can explain the color and magnetism of the complex, but it can't reasonably explain the spectro-chemical series of ligands and high splitting energy of central atoms when CO molecule is ligand. The main reason is that CFT only focuses on electrostatic interactions between central atom and ligands, and influences produced by ligands on d-orbitals of central atom, while neglecting the overlap of d-orbitals of central atom with the orbitals of ligand.

11.3 Coordination Reaction Equilibrium

The reaction between central atom and ligands is called coordination reaction, and the release of central atom and ligands from a complex ion is called dissociation. In aqueous solution, the equilibrium between formation and dissociation of the complex is known as coordination equilibrium. The difference between coordination equilibrium and the equilibrium of common reactions is that the formation reaction is the much stronger than the dissociation reaction. The general principles of chemical equilibrium are fully applicable to coordination equilibrium.

11.3.1 Coordination Equilibrium Constant

Adding excessive ammonia to $CuSO_4$ solution will generate dark blue solution of $[Cu(NH_3)_4]^{2+}$ ions, and very small amount of $[Cu(NH_3)_4]^{2+}$ ions will dissociate.

$$Cu^{2+}(aq) + 4NH_3(aq) \rightleftharpoons [Cu(NH_3)_4]^{2+}(aq)$$

When coordination reaction and dissociation reaction reach to the equilibrium, the equilibrium constant called **stability constant K_s** (or formation constant K_f) of the complex can be expressed as follows.

$$K_s = \frac{[Cu(NH_3)_4^{2+}]}{[Cu^{2+}][NH_3]^4}$$

Where $[Cu^{2+}]$, $[NH_3]$ and $[Cu(NH_3)_4^{2+}]$ are equilibrium concentrations of Cu^{2+} ions, NH_3 and $[Cu(NH_3)_4]^{2+}$ ions, respectively. K_s is stability measurement of the complex in aqueous solution for the complexes with the equal coordination numbers. The larger the K_s is, the more easily the complex ion forms, and the more stable the complex ion is. For example, at 298.15K, K_s of $[Ag(CN)_2]^-$ and $[Ag(NH_3)_2]^+$ ions are 1.3×10^{21} and 1.1×10^7, respectively, so $[Ag(CN)_2]^-$ ion is more stable than $[Ag(NH_3)_2]^+$ ion. For the complexes containing different numbers of ligands, their K_s cannot be used to compare their relative stability. K_s for the most of the complex is a large number generally, for the sake of convenience, $\lg K_s$ is often be used. The stability constants of some complexes are given in Appendix 3.

The complexes are formed and dissociated in steps. For example,

$$Cu^{2+}(aq) + NH_3(aq) \rightleftharpoons [Cu(NH_3)]^{2+}(aq) \quad K_{s1} = \frac{[Cu(NH_3)^{2+}]}{[Cu^{2+}][NH_3]}$$

$$[Cu(NH_3)]^{2+}(aq) + NH_3(aq) \rightleftharpoons [Cu(NH_3)_2]^{2+}(aq) \quad K_{s2} = \frac{[Cu(NH_3)_2^{2+}]}{[Cu(NH_3)^{2+}][NH_3]}$$

$$[Cu(NH_3)_2]^{2+}(aq) + NH_3(aq) \rightleftharpoons [Cu(NH_3)_3]^{2+}(aq) \quad K_{s3} = \frac{[Cu(NH_3)_3^{2+}]}{[Cu(NH_3)_2^{2+}][NH_3]}$$

$$[Cu(NH_3)_3]^{2+}(aq) + NH_3(aq) \rightleftharpoons [Cu(NH_3)_4]^{2+}(aq) \quad K_{s4} = \frac{[Cu(NH_3)_4^{2+}]}{[Cu(NH_3)_3^{2+}][NH_3]}$$

To combine first and second steps of the equilibrium, we obtain

$$Cu^{2+}(aq) + 2NH_3(aq) \rightleftharpoons [Cu(NH_3)_2]^{2+}(aq)$$

Its equilibrium constant is notated by β_2

$$\beta_2 = \frac{[Cu(NH_3)_2^{2+}]}{[Cu^{2+}][NH_3]^2} = \frac{[Cu(NH_3)^{2+}]}{[Cu^{2+}][NH_3]} \times \frac{[Cu(NH_3)_2^{2+}]}{[Cu(NH_3)^{2+}][NH_3]} = K_{s1}K_{s2}$$

Similarly,

$$\beta_3 = K_{s1}K_{s2}K_{s3} \quad \beta_4 = K_{s1}K_{s2}K_{s3}K_{s4}$$

Overall stability constant β_n that includes all steps of stability constants equals K_s.

Question and thinking 11-3 Which one of the following complex ion has the largest K_s and which one has the smallest K_s?

(1) $[Cr(NH_3)_6]^{3+}$ (2) $[Cr(SCN)_6]^{3-}$ (3) $[Cr(CN)_6]^{3-}$

11.3.2 Equilibrium Shift of Coordination Reaction

Like the other chemical equilibria, coordination equilibrium is also a relative and conditional dynamic equilibrium. If a system at complex ion equilibrium is disturbed by a change of acidity, precipitant or other ligands in solution, the system will shift its position to counteract the effect of the disturbance.

1. Influence of solution acidity on complex ion

According to Brönsted theory, many ligands such as F^-, CN^-, SCN^-, OH^- and NH_3 are proton acceptor, so they are bases. If ligand's basicity is strong enough and the concentration of H^+ ions in solution is high enough also, the ligands will combine with the protons, and that would make the complex ion to dissociate. For example,

$$[Cu(NH_3)_4]^{2+}(aq) \rightleftharpoons Cu^{2+}(aq) + 4NH_3(aq)$$

$$+ 4H^+$$

$$\Updownarrow$$

$$4NH_4^+(aq)$$

equilibrium shifts

$$[Cu(NH_3)_4]^{2+}(aq) + 4H^+ \rightleftharpoons Cu^{2+}(aq) + 4NH_4^+(aq)$$

$$K = \frac{[Cu^{2+}][NH_4^+]^4}{[Cu(NH_3)_4^{2+}][H^+]^4} = \frac{1}{K_s\{[Cu(NH_3)_4^{2+}]\} \times K_a^4(NH_4^+)} = \frac{K_b^4(NH_3)}{K_s\{[Cu(NH_3)_4^{2+}]\} \times K_w}$$

$$= \frac{(1.8 \times 10^{-5})^4}{2.09 \times 10^{13} \times (1.0 \times 10^{-14})^4}$$

$$= 5.0 \times 10^{23}$$

Dissociation of the complex ion caused by increasing solution acidity is called acid effect. The stronger the solution acidity is, the more unstable the complex ion is. When the solution acidity is given, the stronger the ligands' basicity is, the more unstable the complex ion is. The antiacid ability of the complex ion is relevant with its K_s. The bigger the K_s is, the stronger the antiacid ability the complex ion is. Such as the K_s of $[Ag(CN)_2]^-$ is bigger (1.3×10^{21}), so its antiacid ability is stronger. Thus, $[Ag(CN)_2]^-$ can s exist in acidic solution well.

On the other hand, the most central atoms of the complexes are transition metals ions which are easy to hydrolyze to lower their concentrations in aqueous solutions. It will promote the dissociation of the complexes. The stronger the basicity of the solution, the more beneficially for the hydrolysis of central atom is. For example,

$$[FeF_6]^{3-}(aq) \rightleftharpoons 6F^-(aq) + Fe^{3+}(aq)$$

$$+ 3OH^-(aq)$$

$$\Updownarrow$$

$$Fe(OH)_3 \downarrow$$

equilibrium shifts

$$[FeF_6]^{3-}(aq) + 3OH^-(aq) \rightleftharpoons 6F^-(aq) + Fe(OH)_3 \downarrow$$

$$K = \frac{[F^-]^6}{[FeF_6^{3-}][OH^-]^3} = \frac{1}{K_s\{[FeF_6]^{3-}\}K_{sp}(Fe(OH)_3)}$$

$$= \frac{1}{1.0 \times 10^{16} \times 2.79 \times 10^{-39}}$$

$$= 3.6 \times 10^{22}$$

The effect on the complex ion dissociation caused by combination of central metal atom with OH⁻ ions in aqueous solution is called hydrolysis. To avoid hydrolysis of central atom, a lower pH value of the solution will be better for the complex ion stability; to increase antiacid ability of the complex ion, a higher pH value of the solution will be helpful. Generally, under the premise of no hydroxide precipitation, a higher pH value of the solution will ensure the stability of complex ion.

2. Influence of precipitation reagent on complex ion

White AgCl precipitation will dissolve and generate colorless and transparent $[Ag(NH_3)_2]^+$ ions by adding a great amount of ammonia into its solution. In the above solution, a yellow precipitation will be formed immediately with the addition of NaBr solution. The reaction is as follows

$$AgCl(s) \rightleftharpoons Cl^-(aq) + Ag^+(aq) \qquad [Ag(NH_3)_2]^+(aq) \rightleftharpoons 2NH_3(aq) + Ag^+(aq)$$

equilibrium shifts $\downarrow 2NH_3(aq) \updownarrow [Ag(NH_3)_2]^+(aq)$

equilibrium shifts $\downarrow Br^-(aq) \updownarrow AgBr(s)$

In the first step, equilibrium shifts from AgCl to $[Ag(NH_3)_2]^+$ with adding NH_3 solution. But in the second step, the equilibrium shifts from $[Ag(NH_3)_2]^+$ to AgBr with adding NaBr solution. Here, the precipitation AgBr is more stable than the complex ion $[Ag(NH_3)_2]^+$. In the above example, the K_{sp} (5.35×10^{-13}) of AgBr is far less than the K_{sp} (1.77×10^{-10}) of AgCl, so Br^- can disturb the coordination equilibrium of $[Ag(NH_3)_2]^+$ ion and ammonia only can dissolve AgCl but not AgBr.

Sample Problem 11-2 Calculate AgCl solubility in the solution of $6 mol \cdot L^{-1}$ NH_3 at 298.15K. Add solid NaBr into the above solution and make $[Br^-] = 0.1 mol \cdot L^{-1}$ (ignore the volume change of solution), does any AgBr precipitation appear?

Solution The reaction of AgCl in NH_3 solution is

$$AgCl(s) + 2NH_3(aq) \rightleftharpoons [Ag(NH_3)_2]^+(aq) + Cl^-(aq)$$

Equilibrium constant of this reaction is K.

$$K = \frac{[Ag(NH_3)_2^+][Cl^-]}{[NH_3]^2} = \frac{[Ag(NH_3)_2^+][Cl^-]}{[NH_3]^2} \cdot \frac{[Ag^+]}{[Ag^+]}$$

$$= K_s\{[Ag(NH_3)_2]^+\} \cdot K_{sp}(AgCl)$$

$$= 1.1 \times 10^7 \times 1.77 \times 10^{-10} = 1.95 \times 10^{-3}$$

If the solubility of AgCl in $6.0 mol \cdot L^{-1}$ NH_3 solution is S $mol \cdot L^{-1}$, then, $[Ag(NH_3)_2^+] = [Cl^-] = S$ $mol \times L^{-1}$, $[NH_3] = (6.0 - 2S)$ $mol \cdot L^{-1}$, bring the equilibrium concentration into the formula of the equilibrium constant,

$$K = \frac{(S mol \cdot L^{-1})^2}{(6.0 mol \cdot L^{-1} - 2S mol \cdot L^{-1})^2} = 1.95 \times 10^{-3}$$

So, at 298.15K, the solubility of AgCl in $6.0 mol \cdot L^{-1}$ NH_3 solution is $0.26 mol \cdot L^{-1}$.

In the above solution, if there is AgBr precipitation, then the reaction should be

$$[Ag(NH_3)_2]^+(aq) + Br^-(aq) \rightleftharpoons 2NH_3(aq) + AgBr(s)$$

So, the equilibrium constant of the reaction is

$$K = \frac{[NH_3]^2}{[Br^-][Ag(NH_3)_2^+]} = \frac{1}{K_s\{[Ag(NH_3)_2]^+\} K_{sp}(AgBr)}$$

$$= \frac{1}{1.1 \times 10^7 \times 5.38 \times 10^{-13}} = 1.7 \times 10^5$$

The reaction quotient is

$$J = \frac{c^2(NH_3)}{c\{[Ag(NH_3)_2]^+\} \cdot c(Br^-)} = \frac{(6.0 mol \cdot L^{-1} - 2 \times 0.26 mol \cdot L^{-1})^2}{0.26 mol \cdot L^{-1} \times 0.10 mol \cdot L^{-1}} = 1155$$

on account of $J<K$, $\Delta_r G_m < 0$, AgBr precipitation appears.

Question and thinking 11-4 In aqueous solution, $Ni^{2+} + 6NH_3 \rightleftharpoons [Ni(NH_3)_6]^{2+}$, $K_s = 1.1 \times 10^8$; $Ni^{2+} + 3en \rightleftharpoons [Ni(en)_3]^{2+}$, $K_s = 3.9 \times 10^{18}$. Which substance, NH_3 or en is the better ligand to dissolve some insoluble salt of Ni(II)?

3. Influence of redox reaction on complex ion

Redox reaction of some central ions in solution can affect coordination equilibrium and make it shift to dissociate the complex ion. So, some redox reactions which originally do not happen can perform with the presence of ligands.

Sample Problem 11-3 At 298.15K and standard conditions,

(1) $Au^+(aq) + e^- \rightleftharpoons Au(s)$, $\qquad \varphi^\ominus(Au^+/Au) = +1.692V$

(2) $Au^+(aq) + 2CN^-(aq) \rightleftharpoons [Au(CN)_2]^-(aq)$, $\quad K_s = 2 \times 10^{38}$

Calculate standard electrode potentials of the following reactions.

(3) $[Au(CN)_2]^-(aq) + e^- \rightleftharpoons Au(s) + 2CN^-(aq)$, $\quad \varphi^\ominus\{[Au(CN)_2]^-/Au\}$

Solution Reaction (2) is the reverse reaction of reaction (3), according to the relationship between standard electrode potential and equilibrium constant at 298.15K, we get

$$\lg K = \frac{nE^\ominus}{0.05916V} = \frac{n\{\varphi^\ominus(Au^+/Au) - \varphi^\ominus\{[Au(CN)_2]^-/Au\}\}}{0.05916V}$$

Here, $n = 1$, so

$$\varphi^\ominus\{[Au(CN)_2]^-/Au\} = \varphi^\ominus(Au^+/Au) - \frac{0.05916V}{n}\lg K_s$$

$$= 1.692V - 0.05916V \times \lg(2 \times 10^{38})$$

$$= -0.574V$$

The above calculation has a practical application in gold mining because

$$\varphi^\ominus(O_2/OH^-) = +0.401V < \varphi^\ominus\{[Au^+/Au\} = +1.692V,$$

So the following reaction can't proceed forwardly.

$$4Au(s) + O_2(g) + 2H_2O(l) \not\longrightarrow 4OH^-(aq) + 4Au^+(aq)$$

But because

$$\varphi^\ominus(O_2/OH^-) > \varphi^\ominus\{[Au(CN)_2]^-/Au\}$$

The reaction can happen forwardly.

$$4Au(s) + 8CN^-(aq) + O_2(g) + 2H_2O(l) \rightleftharpoons 4[Au(CN)_2]^-(aq) + 4OH^-(aq)$$

Au can be oxidized into $[Au(CN)_2]^-$ by O_2 with the presence of CN^- ions. Furthermore, if Zn as reducing agent is added into the solution, Au is gotten. The reaction is

$$2[Au(CN)_2]^-(aq) + Zn(s) \rightleftharpoons [Zn(CN)_4]^{2-}(aq) + 2Au(s)$$

Question and thinking 11-5
1. Point out the electrode with the largest $\varphi^\ominus$ and the electrode with the smallest $\varphi^\ominus$.
(1) $\varphi^\ominus(Ag^+/Ag)$ 　　　(2) $\varphi^\ominus\{[Ag(NH_3)_2]^+/Ag\}$
(3) $\varphi^\ominus\{[Ag(S_2O_3)_2]^{3-}/Ag\}$ 　(4) $\varphi^\ominus\{[Ag(CN)_2]^-/Ag\}$
2. Please prove that $\varphi^\ominus\{[Ag(S_2O_3)_2]^{3-}/Ag\}$ and $\varphi^\ominus(AgBr/Ag)$ can be deduced by the $\varphi^\ominus(Ag^+/Ag)$ as well as K_s of $[Ag(S_2O_3)_2]^{3-}$ and K_{sp} of AgBr.

4. Influence of other ligands on complex ion

In the coordination equilibrium system, if another complexant is added into the system, the new complex ion may be formed. We can predict the probability by their K_s values.

Sample Problem 11-4 Can the following reaction happen at 298.15K?

$$[Zn(NH_3)_4]^{2+}(aq) + 4OH^-(aq) \rightleftharpoons [Zn(OH)_4]^{2-}(aq) + 4NH_3(aq)$$

And how many is the ratio of $[Zn(NH_3)_4^{2+}]/[Zn(OH)_4^{2-}]$ in $1 mol \cdot L^{-1}$ NH_3 solution? Which complex ion with Zn^{2+} is major form in solution?

Solution At 298.15K, the stability constant for $[Zn(NH_3)_4]^{2+}$, $K_s = 2.88 \times 10^9$; for $[Zn(OH)_4]^{2-}$, $K_s = 3.16 \times 10^{15}$.

$$[Zn(NH_3)_4]^{2+}(aq) + 4OH^-(aq) \rightleftharpoons [Zn(OH)_4]^{2-}(aq) + 4NH_3(aq)$$

Its equilibrium constant is:

$$K = \frac{[Zn(OH)_4^{2-}][NH_3]^4}{[Zn(NH_3)_4^{2+}][OH^-]^4} \times \frac{[Zn^{2+}]}{[Zn^{2+}]} = \frac{K_s[Zn(OH)_4^{2-}]}{K_s[Zn(NH_3)_4^{2+}]} = \frac{3.16 \times 10^{15}}{2.88 \times 10^9} = 1.10 \times 10^6$$

K is a big number, it means that the reaction can go forward in aqueous solution, that is, $[Zn(NH_3)_4]^{2+}$ ion can transform into $[Zn(OH)_4]^{2-}$ ion.

In $1 mol \cdot L^{-1}$ NH_3 solution, because $c_b K_b > 20 K_w$, $c_b / K_b > 500$, so

$$[OH^-] = \sqrt{c_b K_b} = \sqrt{1 \times 1.8 \times 10^{-5}} \ mol \cdot L^{-1}$$

$$[NH_3] = 1 - [OH^-] \approx 1 mol \cdot L^{-1}$$

$$\frac{[Zn(NH_3)_4^{2+}]}{[Zn(OH)_4^{2-}]} = \frac{[NH_3]^4 K_s[Zn(NH_3)_4^{2+}]}{[OH^-]^4 K_s[Zn(OH)_4^{2-}]} \approx \frac{(1 mol \cdot L^{-1})^4 \times 2.88 \times 10^9}{(\sqrt{1.8 \times 10-5} mol \cdot L^{-1})^4 \times 3.16 \times 10^{15}}$$

$$= 2.84 \times 10^3$$

So, Zn^{2+} ions are exists in $[Zn(NH_3)_4]^{2+}$ ions mainly in $1 mol \cdot L^{-1}$ NH_3 solution.

11.4 Chelate and Biological Ligands

11.4.1 Chelating Effect

Cd^{2+} can form the complexes with methylamine (CH_3NH_2) and ethylene diamine in the same coordination number as shown in Figure 11-5.

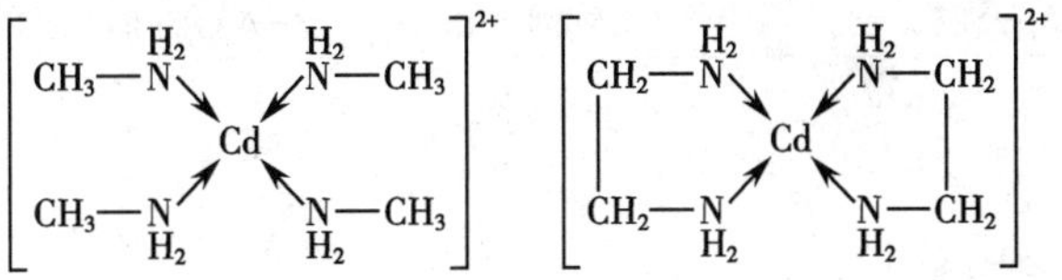

$K_s = 3.55 \times 10^6$ $K_s = 1.66 \times 10^{10}$

Figure 11-5 Structures of $[Cd(CH_3NH_2)_4]^{2+}$ and $[Cd(en)_2]^{2+}$

The difference between them is that ethylene diamine (or en) is bidentate ligand. In en molecule, two N atoms donate two lone-pairs respectively to Cd^{2+} and form a complex ion with ring structure.

This ring-like complex formed by the central atom bonding with multidentate ligand is called **chelate**. It seems as a crab tightly hold the substance with a pair of crab claws. The effect of the complex stability increased by forming chelate is known as **chelating effect**. Multidentate ligand which can form a chelate with central atom is called **chelating agent**.

The most common chelating agents are organic compounds containing amino-N and carboxyl O groups especially, such as ethylenediaminetetraacetic acid, EDTA for short, or its salts with Y^{4-} ion. Its negative ions can form chelate of five-member rings with the metal ions at most as shown in Figure 11-6. The chelate of EDTA has the very high stability. A few chelating agents are of inorganic compounds, such as sodium tripolyphosphate, whose structure chelated with Ca^{2+} ion is shown in Figure 11-7.

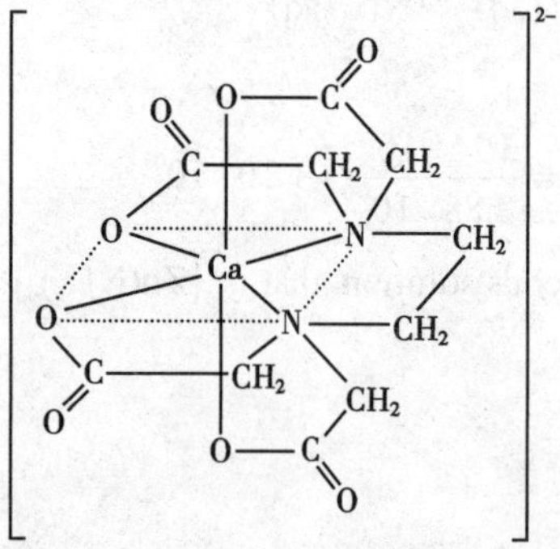

Figure 11-6 Structure of CaY^{2-}

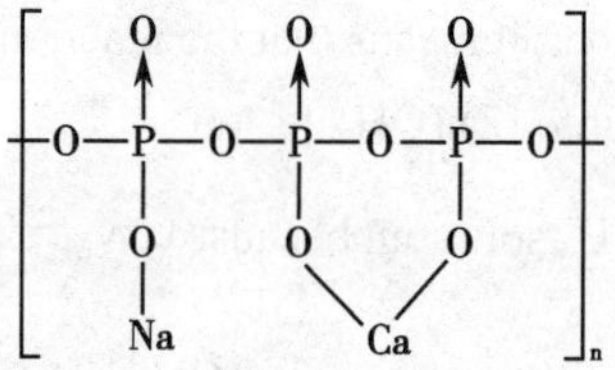

Figure 11-7 Chelate of Ca^{2+} and Tripolyphosphate

Because Ca^{2+} and Mg^{2+} can form stable chelates with sodium tripolyphosphate. So, the sodium tripolyphosphate is often added into the water boiler to avoid the formation of precipitations of Ca^{2+} and Mg^{2+} ions.

The chelates possess the very good stability generally. From thermodynamics point of view, $\lg K_s^{\ominus}$ of chelate ion and the relationship between $\lg K_s^{\ominus}$ and thermodynamic function is as follows

$$\Delta_r G_m^{\ominus} = -2.303RT \lg K_s^{\ominus} = \Delta_r H_m^{\ominus} - T\Delta_r S_m^{\ominus}$$

$\lg K_s^{\ominus}$ is determined by $\Delta_r H_m^{\ominus}$ and $\Delta_r S_m^{\ominus}$. Comparing $[Cd(en)_2]^{2+}$ with $[Cd(NH_2CH_3)_4]^{2+}$, $\Delta_r H_m^{\ominus}$ are almost equal because both of them has four coordination bonds (N → Cd), respectively, but the value of $\Delta_r G_m^{\ominus}$ of $[Cd(en)_2]^{2+}$ is smaller, and the values of its $\Delta_r S_m^{\ominus}$ and $\lg K_s^{\ominus}$ are bigger. So the stability of $[Cd(en)_2]^{2+}$ is higher than that of $[Cd(NH_2CH_3)_4]^{2+}$.

Table 11-8 Thermodynamic Functions of Several Complex Ions

Complex ion	$\Delta_r H_m^{\ominus}$ / kJ·mol^{-1}	$T\Delta_r S_m^{\ominus}$ / kJ·mol^{-1}	$\Delta_r G_m^{\ominus}$ / kJ·mol^{-1}	$\lg K_s^{\ominus}$
$[Cd(NH_2CH_3)_2]^{2+}$	−29.37	−1.92	−27.45	4.81
$[Cd(en)]^{2+}$	−29.41	3.89	−33.30	5.84
$[Cd(NH_2CH_3)_4]^{2+}$	−57.30	−20.1	−37.2	6.52
$[Cd(en)_2]^{2+}$	−56.5	4.2	−60.7	10.6
$[Zn(NH_3)_2]^{2+}$	−28.0	0.42	−28.62	5.01
$[Zn(en)]^{2+}$	−27.0	7.53	−35.10	6.15

In aqueous solution, metal ions are actually in forms of hydrated ions. Such as $[Cd(H_2O)_4]^{2+}$, when it coordinates with CH_3NH_2 to form a common complex ion, each CH_3NH_2 will replace one H_2O

molecule. The number of particles before and after the reaction are equal, $\Delta_r S_m^\ominus$ is of little change. For example,

$$[Cd(H_2O)_4]^{2+}(aq) + 4CH_3NH_2(aq) \rightleftharpoons [Cd(NH_2CH_3)_4]^{2+}(aq) + 4H_2O$$

But Cd^{2+} combines multidentate en to form chelate$[Cd(en)_2]^{2+}$, every en molecule can replace two H_2O, molecules,

$$[Cd(H_2O)_4]^{2+}(aq) + 2en(aq) \rightleftharpoons [Cd(en)_2]^{2+}(aq) + 4H_2O$$

After reaction, the number of particles in solution is more obviously, the degree of disorder will be increased, so $\Delta_r S_m^\ominus$ became larger. The driving force of chelating reaction is mainly from entropy change before and after reaction. The stability of some chelating reaction is not only related with entropy increase, but also enthalpy change.

11.4.2 Impact Factors on the Stability of Chelate

Wu can understand the stability of chelate from the structure of it also.

1. Size of chelate ring

In the vast majority of chelates, the chelates with five-or six-member ring are the most stable. The bond angles of two kinds of rings are 108° or 120°, respectively. For example, the stability constants of chelates formed by Ca^{2+} with $(-OOCCH_2)_2N(CH_2)_nN(CH_2COO-)_2$ will decrease with the increasing of n in Table 11-9. This is because the bond angle of a five-member ring is closer to that of sp^3 hybrid orbitals of C (109°28′). The smaller the tension is, the more stable is the ring is. Some ligands possessing conjugated double bonds in their structures can coordinate with central atom to form a stable six-member ring of chelates. E.g. acetylacetone, the bond angle of sp^2 hybrid orbitals of C atom in conjugated double bond is 120° which is just the bond angle of a six-member ring as shown in Figure 11-8.

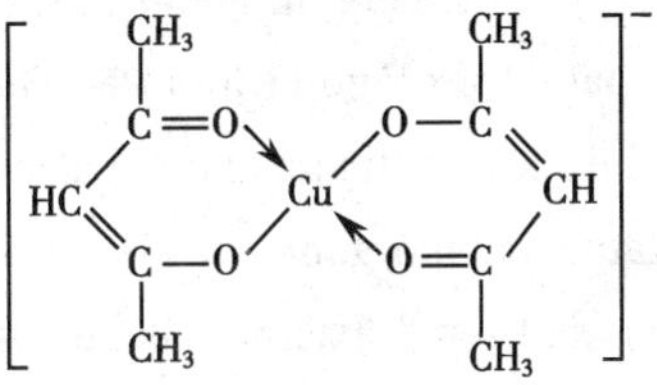

Figure 11-8 Structure of Copper(II)-bis(Acetylacetone)

Table 11-9 lgK_s of Complexes of Ca^{2+} Coordinating with EDTA Homologues

Ligand name	Numbers of ring	lgK_s
EDTA ion	Five pentagon-rings	11.0
PDTA ion	Four pentagon-rings, one six-member ring	7.1
Butanediamine EDTA ion	Four pentagon-rings, one seven-member ring	5.1
Pentamethylene diamine, EDTA ion	Four pentagon-rings, one eight-member ring	4.6

Three-member ring and four-member ring are unstable because of their high tension forces. Therefore, there are generally 2 or 3 other atoms between the neighbor ligating atoms in multidentate molecule to form a stable chelate of five- or six-member ring.

2. Number of chelate ring

When one ligating atom in multidentate coordinates with central atom, the distance between the remaining ligating atoms and the central atom would reduces, thus the probability that the remaining ligating atoms combine with the central atom increases. If one of the coordination bonds is destroyed, it would be easier to restore because other ligating atoms are still bonding with the central atom, so the chelate is very stable.

The more the ligating atoms in multidentate ligand is, the more the chelating rings will form, the more the coordination bonds with central atom will form, the smaller probability the ligands departing from the central atom is, the more stable the chelate is. Figure 11-9 indicates these.

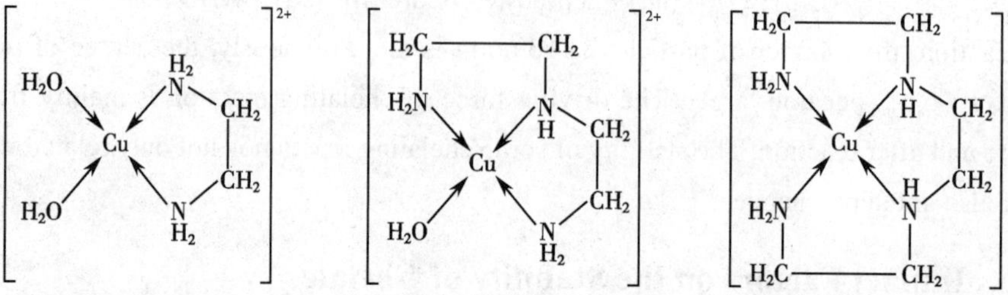

One ring $\lg\beta_1=10.67$ two rings $\lg\beta_1=15.9$ three rings $\lg\beta_1=20.5$

Figure 11-9   Relationship between the Numbers of Chelate Rings and the Stability of the Chelates

11.4.3 Biological Ligands

Essential trace metal elements in human body are mainly in the form of complexes to play their roles, respectively. In organisms, the ions or molecules coordinating with metals to form the complexes are called **biological ligands**. Biological ligands include porphyry in compounds, proteins, peptide nucleic acids, sugars and sugar proteins, lipoproteins, and other macromolecules ligands beside some organic and inorganic ions (such as amino acids, nucleotides, organic acids radicals, Cl^-, HCO_3^-, HPO_4^{2-}, etc.), and some macromolecules ligands, such as vitamins and hormones. Broadly, O_2 and CO are also biological ligands. Different ligands with various coordination groups decide their coordination abilities and coordination mode when they coordinate with central atom, and then, their biological functions are decided. The ligating atoms in the biological ligands generally are N, O and S atoms containing lone pair. Table 11-10 shows some typical metal-enzyme complexes and their functions in organisms.

Table 11-10   Some Typical Metal-Enzyme Complexes

Metal	Enzyme	Function	Metal	Enzyme	Function
Fe	ferredoxin	photosynthesis	Mn	arginase	formation of urea
	succinic dehydrogenase	dehydrogenation of butanedioic acid		pyruvate carboxylase	pyruvic acid **metabolism**
Cu	cytochrome oxidase	electron transfer	Co	ribonucleotide reductase	DNA biosynthesis
	tyrosinase	skin pigment formation		glutamic acid mutase	amino acid **metabolism**
Zn	carbonic anhydrase	hydration of CO_2	Mo	xanthine oxidase	purine **metabolism**
	carboxymethyl-phthalocyanine	Protein hydrolyze		nitrate reductase	nitrate **metabolism**
	alkaline ospholipase	ospholipase hydrolyze			

Iron is one of the most abundant metals in organisms. In mammals, the most of iron exists in form of porphyry in complexes and heme, such as hemoglobin (Hb), cytochrome C, peroxidase, and catalase.

In the hemoglobin, Fe(II) bonds with four pyrrole N atoms in porphyrin IX ring and a N atom in the side chain of histidine amino acid outside the ring to form chelate of Fe (II) which the coordination number is five. Fe(II) is not in the original plane of porphyry in IX ring, but it is above the plane about 75pm. Magnetic measurement indicates that Fe(II) in hemoglobin is in a high-spin state. When hemoglobin combines with O_2 to form oxyhemoglobin (HbO_2), Fe(II) forms a strong coordination bond with O_2 to pull Fe(II) into the plane of porphyrin ring, and then Fe(II) is hexa-coordinate. But magnetic measurement shows that Fe(II) in oxyhemoglobin is in low-spin state. For the human being, O_2 combines with hemoglobin to form oxyhemoglobin in lung, and oxyhemoglobin releases O_2 into blood as shown in Figure 11-10. This process in the body keeps going day and night to meet the needs of O_2 in the body.

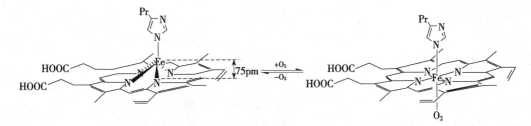

Figure 11-10　Effect of Hemoglobin on O_2 Carrier Reversibly

CO can compete the sixth ligand of Fe(II) with O_2 in heme. The bonding capacity of CO is about 200 times that of O_2. In CO poisoning, most hemoglobin is in the forms of CO hemoglobin which has lost the capability of transporting O_2 around the body and results in anoxia. Clinically, CO poisoning can be treated by hyperbaric oxygen. Hyperbaric oxygen therapy can increase the physical dissolved O_2 in plasma significantly, and significantly improve the O_2 content in the body. So the body can not only break away from hypoxia, but also speed up the dissociation of CO hemoglobin, and remove CO from the body. Hyperbaric oxygen therapy is very effective for CO poisoning. It has a very important significance on improving the cure rate, reducing the mortality and reducing the sequelae.

The large amount of CO_2 in capillary cycle of mammals must be transformed and removed as soon as possible. In the absence of enzyme catalyst, the hydration rate of CO_2 is only $7.0 \times 10^{-4} s^{-1}$, it can't meet the needs of the body. Carbonic anhydrase will increase the reaction rate by 10^9 times, it makes CO_2 will transform to HCO_3^- and dissolve into blood rapidly. In lung, HCO_3^- will be rapidly dehydrated under the catalysis of carbonic anhydrase to and quickly generate CO_2 which is exhaled. The catalysis of carbonic anhydrase greatly accelerated the transportation of CO_2, it is a kind of important zinc enzyme (Figure 11-11).

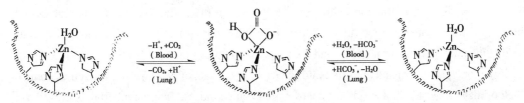

Figure 11-11　Regulation Process of Carbonic Anhydrase in Blood and Lungs

In fact, the ways of complexes to maintain the normal activities of our body are various, and they are closely related to the body's metabolism.

Key Terms

配位化合物	coordination compound
配离子	complex ion
中心原子	central atom
配体	ligand
配位原子	ligating atom
单齿配体	monodentate ligand
多齿配体	multidentate ligand
价键理论	valence bond theory
外轨配合物	outer orbital complex ions
内轨配合物	inner orbital complex ions
晶体场理论	crystal field theory
分裂能	splitting energy
电子成对能	electron pairing energy
晶体场稳定化能	crystal field stabilization energy
配位平衡常数	coordination equilibrium constant
螯合物	chelate
螯合效应	chelating effect
螯合剂	chelating agent
生物配体	biological ligands

Summary

A coordination compound or complex is the product of a Lewis acid-base reaction in which neutral molecules or anions (called ligands) bond to a central metal atom (or ion) by coordination bonds.

Most coordination compounds consist of complex ion and the ions with opposite charges of complex ion. Complex ion formed by central atom and ligands is called inner sphere. The other part with opposite charges is called outer sphere. Because coordination compounds are neutral, the total charges inner sphere is equal to the total charges of outer sphere. The inner sphere and outer sphere are combined by ionic bond.

The main points of Valence Bond Theory (VBT) are, (a) Central atom is the acceptor of lone pairs; the ligands are the donor of lone pairs. Complexes are formed through a interaction between central atom and ligands by coordination bonds; (b) To strengthen the bonding capacity, the vacant valence orbitals of central atom must hybridize to form hybrid orbitals; (c) The hybridization types decide the coordination numbers of central atoms and the shape of complex ion.

If the lone pairs of ligand are filled into the $(n-1)d$ orbitals of central atom as well as some of the n shell orbitals, it is called inner orbitals coordination compound, but into the nd orbitals as well as some of the n shell orbitals, it is called outer orbitals coordination compound.

In most of cases, we can predict the model of complex, inner orbital or outer orbital complex, by determining the magnetic moment (μ) of it. The relationship between μ and n, the number of unpaired electrons, is $\mu \approx \sqrt{n(n+2)}\mu_B$. By comparing calculated and observed values of μ to known, Types of hybrid orbitals and the geometry of complex can be known. It is possible to determine n in a complex from its paramagnetism. The compounds with no unpaired electrons are diamagnetic.

The main points of Crystal Field Theory (CFT) are, (a) The interaction between central atom and ligands in the complex is considered to be purely electrostatic force between central ion and anion ligands or attraction if the ligand is a neutral molecule; (b) The d orbitals energy level of central atom are split in the electrostatic field of the ligands; (c) The d electrons of central atom rearrange to lead to increase the stability of the system because of the lower energy of system.

The splitting energy Δ_o is the energy difference between the highest energy level and the lowest energy level of splitted d orbitals. The splitting energy of complexes in the octahedral field is as follows:

$$\Delta_o = E(d_\gamma) - E(d_\varepsilon)$$
$$E(d_\gamma) = 0.6\Delta_o$$
$$E(d_\varepsilon) = -0.4\Delta_o$$

The factors affecting splitting energy are: (a) The splitting energy is different in the different geometry of complexes; (b) For a given central atom, the more the field intensity of ligand is, the more the splitting energy is; (c) For a given ligand, the more the charge of the central atom is, the larger the splitting energy value is; (d) If the ligands and their charge number are same, the larger the radius of central atom and the further the d orbitals from nucleus is, the larger the splitting energy is.

Because of the splitting of d orbitals in the octahedral field, the algebraic sum of $-0.4\Delta_o$ per electron in d_ε level and $0.6\Delta_o$ in d_γ level for each of electrons is defined as crystal field stabilization energy (CFSE). The more the CFSE decreases, the more stable the complex is.

In the octahedral complexes, the total CFSE is given by

$$\text{CFSE} = xE(d_\varepsilon) + yE(d_\gamma) + (n_2 - n_1)P$$

The unfilled d orbitals of most transition metal ions split under the action of ligands field. After the d electrons at lower level selectively absorb a wave-length of light, the electrons migrate from d orbitals of the low energy level to that of high level, this transition is called d-d transition. Thus, the complexes appear the complementary color of light absorbed.

Complexes are formed in steps, each step has its stability constant, $K_{s1}, K_{s2}, K_{s3}, \cdots K_{sn}$. The stability constant K_s of a complex is the product of the stepwise constants characterizing the separate stages of complexing process

$$K_s = K_{s1} \cdot K_{s2} \cdot K_{s3} \cdot K_{s4}$$

K_s of complex can express their stability in aqueous solution. When the coordination number of complex ions is same, we can determine directly the stability of complex ions according to the values of K_s.

Chelate is a special complex formed between the central atom and multidentate ligand or chelating agent with two or more ligating atoms. They are very good in stability because they have five- or six-number ring structure generally.

Exercises

1. Distinguish the following concepts

(1) Inner sphere and outer sphere (2) Monodentate ligand and multidentate ligand (3) d^2sp^3 hybridization and sp^3d^2 hybridization (4) Outer orbital complex and inner orbital complex (5) Strong field ligand and weak field ligand (6) Low-spin complex and high-spin complex

2. Name the following complexes, and point out the central atom, ligands, ligating atom, and coordination number. Write out the expression of K_s.

(1) $Na_3[Ag(S_2O_3)_2]$ (2) $[Co(en)_3]_2(SO_4)_3$
(3) $H[Al(OH)_4]$ (4) $Na_2[SiF_6]$
(5) $[PtCl_5(NH_3)]^-$ (6) $[Pt(NH_3)_4(NO_2)Cl]$
(7) $[CoCl_2(NH_3)_3H_2O]Cl$ (8) $NH_4[Cr(NCS)_4(NH_3)_2]$

3. What is the chelate? What are the characteristics of chelate? List the factors which have the effects on the stability of chelate. To form a chelate of five- or six-member ring, what are the requirements for the ligands?

4. True or false

(1) Coordination compound consists of complex ions and outer layer ions.

(2) The central atoms of all coordination compounds are metal elements.

(3) Ligand numbers is equal to coordination number of central atom.

(4) The charge number of the complex ion is equal to that of the central atom.

(5) The stronger the intensity of ligand field is, the higher the splitting energy of the central atom will be in an octahedron field.

(6) The magnetic moment of outer orbital complex must be larger than that of inner orbital complex.

(7) The low-spin complexes are more stable than the high-spin complexes with the same central atom

5. It is known that the geometry of $[PdCl_4]^{2-}$ is square planar, and the geometry of $[Cd(CN)_4]^{2-}$ is tetrahedral, please analysis the hybridization of their bonding orbitals according to valence bond theory and point out whether they are paramagnetism ($\mu \neq 0$) or diamagnetism ($\mu = 0$).

6. According to the measured magnetic moments, please indicate the geometry of the following chelates, and point out whether they are inner or outer orbital complexes.

(1) $[Co(en)_3]^{2+}$, 3.82 μ_B; (2) $[Fe(C_2O_4)_3]^{3-}$, 5.75 μ_B; (3) $[Co(en)_2Cl_2]Cl$, 0 μ_B

7. An octahedral iron complex has been prepared in lab, but the oxidation number of iron ion is unknown. Its magnetic moment is $5.10\mu_B$ determined by magnetic balance, please estimate the oxidation number of iron, and indicate whether the complex is high spin or low spin.

8. According to the Valence Bond Theory and the Crystal Field Theory, please explain why the low spin $[Co(CN)_6]^{4-}$ is easily oxidized to low spin $[Co(CN)_6]^{3-}$ in the air.

9. The splitting energy Δ_o of the following complexes and the electron pairing energy P of the central ion are listed below, please estimate their magnetic moments by the distribution of the d electrons on d_ε and d_γ energy levels of the central ions. Point out which complexes are high-spin and which complexes are low-spin.

	$[Co(NH_3)_6]^{2+}$	$[Fe(H_2O)_6]^{2+}$	$[Co(NH_3)_6]^{3+}$
P/cm^{-1}	22 500	17 600	21 000
Δ_o/cm^{-1}	11 000	10 400	22 900

10. It is known that the wavelength of visible light absorbed by $[Mn(H_2O)_6]^{2+}$ is shorter than that by $[Cr(H_2O)_6]^{2+}$, point out which splitting energy is higher, and write out the electron configuration of d electrons in d_ε and d_γ energy levels of the central atom.

11. It is known that the Δ_o values of the high spin complex ion $[Fe(H_2O)_6]^{2+}$ and the low spin complex ion $[Fe(CN)_6]^{4-}$ are $124.38 kJ \cdot mol^{-1}$ and $394.68 kJ \cdot mol^{-1}$, respectively. Both of the electron pairing energies P are $179.40 kJ \cdot mol^{-1}$. Calculate their CFSE values, respectively.

12. Calculate the equilibrium constants for the following coordination reactions, respectively. And predict the reaction direction and point out which reaction is going to complete much more.

(1) $[Hg(NH_3)_4]^{2+} + Y^{4-} \rightleftharpoons HgY^{2-} + 4NH_3$

(2) $[Cu(NH_3)_4]^{2+} + Zn^{2+} \rightleftharpoons [Zn(NH_3)_4]^{2+} + Cu^{2+}$

(3) $[Fe(C_2O_4)_3]^{3-} + 6CN^- \rightleftharpoons [Fe(CN)_6]^{3-} + 3C_2O_4^{2-}$

13. Does the precipitate of $Cu(OH)_2$ come into being under the following conditions? (1) The mixture of $0.10 mol \cdot L^{-1}$ $CuSO_4$ and $0.10 mol \cdot L^{-1}$ NaOH with equal volume. (2) The mixture of $0.10 mol \cdot L^{-1}$ $[Cu(NH_3)_4]SO_4$ and $0.10 mol \cdot L^{-1}$ NaOH with equal volume. (3) The mixture of 100mL $0.10 mol \cdot L^{-1}$ $CuSO_4$, 50mL $0.10 mol \cdot L^{-1}$ NaOH, and 50mL $0.10 mol \cdot L^{-1}$ NH_3.

14. At 298.15K, the concentrations of $[Ni(NH_3)_6]^{2+}$ and NH_3 are $0.10 mol \cdot L^{-1}$ and $1.0 mol \cdot L^{-1}$ respectively in $[Ni(NH_3)_6]^{2+}$ solution. Add the ethylenediamine (en) into the solution and make the initial concentration of en is $2.30 mol \cdot L^{-1}$. Calculate the concentrations of $[Ni(NH_3)_6]^{2+}$, NH_3, $[Ni(en)_3]^{2+}$ and en at equilibrium, respectively.

15. Add 30.0mL of ammonia which mass fraction is 18.3% ($\rho = 0.929 kg \cdot L^{-1}$) into 50mL of $0.10 mol \cdot L^{-1}$ $AgNO_3$ solution and dilute the solution to 100mL with distilled water. Calculate:

(1) The concentrations of Ag^+, $[Ag(NH_3)_2]^+$ and NH_3 in the solution, respectively.

(2) Will AgCl precipitate or not with the addition of 10.0mL of $0.100 mol \cdot L^{-1}$ KCl into the solution? Calculate the minimum equilibrium concentration of NH_3 when the AgCl precipitation does not appear in the solution.

16. At 298.15K, a solution is prepared by mixing 35.0mL of $0.250 mol \cdot L^{-1}$ NaCN and 30.0mL of $0.100 mol \cdot L^{-1}$ $AgNO_3$. Calculate the concentrations of Ag^+, CN^- and $[Ag(CN)_2]^-$ in the solution.

17. It is known that, equilibrium constant of the following reaction, $K^\ominus = 4.786$.

$$Zn(OH)_2(s) + 2OH^-(aq) \rightleftharpoons [Zn(OH)_4]^{2-}(aq)$$

Calculate $\varphi^\ominus \{[Zn(OH)_4]^{2-}/Zn\}$ according to the data in Appendix 3.

18. At 298.15K, add solid KCl into the 1L of $0.05 mol \cdot L^{-1}$ $AgNO_3$ with excessive ammonia to make the initial concentration of Cl^- to be $9 \times 10^{-3} mol \cdot L^{-1}$ (neglected the volume change), answer the following questions:

(1) What is the minimum concentration of NH_3 in solution to prevent AgCl precipitation?

(2) What is the equilibrium concentration of each component in the solution, respectively?

(3) What is $\varphi\{[Ag(NH_3)_2]^+/Ag\}$ in the solution?

19. At 298.15K, the electrode potential of Ag^+ / Ag is decreased from its standard electrode potential to $-0.505V$ when $6mol \cdot L^{-1}$ $Na_2S_2O_3$ is added into Ag^+ / Ag standard electrode solution with equal volume. Answer the following questions:

(1) What is the concentration of Ag^+ in the electrode solution after adding $Na_2S_2O_3$ solution?

(2) What is the value of $K_s\{[Ag(S_2O_3)_2]^{3-}\}$?

(3) What is the concentration of each component in electrode solution, respectively, with the addition of the solid KCN into this electrode solution to make its initial concentration is $2mol \cdot L^{-1}$?

20. The equilibrium constant of the following reaction is $K^{\ominus} = 1.66 \times 10^{-3}$ at 298.15K.

$$Cu(OH)_2(s) + 2OH^-(aq) \rightleftharpoons [Cu(OH)_4]^{2-}(aq)$$

Calculate K_s of $[Cu(OH)_4]^{2-}$. What is the minimum concentration of the NaOH solution when we want to dissolve 0.1mol of $Cu(OH)_2$ in 1L of NaOH solution.

21. $\varphi^{\ominus}(Fe^{3+}/Fe^{2+})$, $K_s\{[Fe(bipy)_3]^{3+}\}$ and $K_s\{[Fe(bipy)_3]^{2+}\}$ are known. Find the formula of calculating $\varphi^{\ominus}\{[Fe(bipy)_3]^{3+}/[Fe(bipy)_3]^{2+}\}$.

22. It is known that $K_s = 2.818 \times 10^{17}$ for $[Fe(bipy)_3]^{2+}$ and $\varphi^{\ominus} = 0.96V$ in the reaction $[Fe(bipy)_3]^{3+}(aq) + e^- \rightleftharpoons [Fe(bipy)_3]^{2+}(aq)$, Calculate K_s of $[Fe(bipy)_3]^{3+}$ and point out the more stable complexes.

23. At 298.15K, how much mole of $Na_2S_2O_3$ should be added into 1L of $0.10 mol \cdot L^{-1}$ $[Ag(NH_3)_2]^+$ solution to transform $[Ag(NH_3)_2]^+$ into $Ag(S_2O_3)_2^{3-}$ completely and no $[Ag(NH_3)_2]^+$ in solution? Calculate the concentrations of $S_2O_3^{2-}$, NH_3 and $Ag(S_2O_3)_2^{3-}$ in the solution at 298.15K. $K_s\{[Ag(NH_3)_2]^+\} = 1.1 \times 10^7$, $K_s\{[Ag(S_2O_3)_2]^{3-}\} = 2.9 \times 10^{13}$.

Supplementary Exercises

1. The compound $CoCl_3 \cdot 2H_2O \cdot 4NH_3$ may be one of the hydrate isomers $[Co(NH_3)_4(H_2O)Cl_2]Cl \cdot H_2O$ or $[Co(NH_3)_4(H_2O)_2Cl]Cl_2$. A $0.10 mol \cdot L^{-1}$ aqueous solution of the compound is found to have a freezing point of $-0.56°C$. Determine the correct formula of the compound. The freezing point depression constant for water is $1.86K \cdot mol^{-1} \cdot kg^{-1}$, and for aqueous solution, molarity and molality can be taken as approximately equal.

2. Given that K_s values of $[CuY]^{2-}$ and $[Cu(en)_2]^{2+}$ are 5×10^{18} and 1.0×10^{21} respectively, try to determine which one of the two complexes is more stable.

3. Add 100mL of $0.100 0 mol \cdot L^{-1}$ NaCl solution into 100mL of $0.100 0 mol \cdot L^{-1}$ $AgNO_3$ containing excessive ammonia to produce no precipitation of AgCl. Calculate the concentration of ammonia at least in mixture solution.

4. Whether AgI precipitation will form or not when $c([Ag(CN)_2]^-) = c(CN^-) = 0.1 mol \cdot L^{-1}$ in a solution and the solid KI is added to the solution to make $c(I^-)$ equal to $0.1 mol \cdot L^{-1}$, explain the reason by calculation.

5. Given that K_s of $[Ag(NH_3)_2]^+$ and $[Ag(CN)_2]^-$ are 1.1×10^7 and 1.3×10^{21} respectively, try to determine the direction of the following reaction.

$$[Ag(CN)_2]^-(aq) + 2NH_3(aq) \rightleftharpoons [Ag(NH_3)_2]^+(aq) + 2CN^-(aq)$$

6. A Cu electrode is immersed in a solution which is composed of $1.00 mol \cdot L^{-1}$ NH_3 and $1.00 mol \cdot L^{-1}$

$[Cu(NH_3)_4]^{2+}$. When a standard hydrogen electrode is used as cathode, E_{cell} is $+0.052V$. Calculate the formation constant K_f of $[Cu(NH_3)_4]^{2+}$?

7. The following concentration cell is constructed.

$$Ag \mid [Ag(CN)_2]^- (0.1 mol \cdot L^{-1}), CN^- (0.1 mol \cdot L^{-1}) \parallel Ag^+ (0.1 mol \cdot L^{-1}) \mid Ag$$

K_s of $[Ag(CN)_2]^-$ is 1.26×10^{21}, calculate E_{cell} of the concentration cell?

Answers to some Exercises

[Exercises]

4. (1) × (2) × (3) × (4) × (5) √ (6) × (7) √

5. Pd: dsp^2, diamagnetism; Cd: sp^3, diamagnetism

6. (1) $[Co(en)_3]^{2+}$, octahedron, outer orbital complex
   (2) $[Fe(C_2O_4)_3]^{3-}$, octahedron, outer orbital complex
   (3) $[Co(en)_2Cl_2]Cl$, octahedron, inner orbital complex

7. +2, high spin

8. According to the Valence Bond Theory, the low spin $[Co(CN)_6]^{4-}$ is easy to lose one electron to form low spin $[Co(CN)_6]^{3-}$.

   According to the Crystal Field Theory, one electron of $[Co(CN)_6]^{4-}$ at d_γ orbital is easy to lose and form lower energy-level $[Co(CN)_6]^{3-}$.

9.

$[Co(NH_3)_6]^{2+}$	$d_\varepsilon^5 d_\gamma^2$	$\mu = 3.87$	high-spin
$[Fe(H_2O)_6]^{2+}$	$d_\varepsilon^4 d_\gamma^2$	$\mu = 4.90$	high-spin
$[Co(NH_3)_6]^{3+}$	$d_\varepsilon^6 d_\gamma^0$	$\mu = 0$	low-spin

10. The splitting energy of $[Mn(H_2O)_6]^{2+}$ is larger than that of $[Cr(H_2O)_6]^{2+}$.

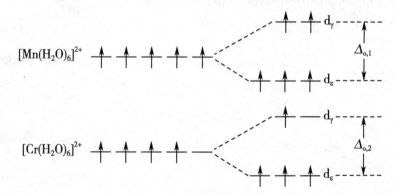

11. $[Fe(H_2O)_6]^{2+}$, CFSE $= -49.75 kJ \cdot mol^{-1}$; $[Fe(CN)_6]^{4-}$, CFSE $= -558.43 kJ \cdot mol^{-1}$

12. (1) forward (2) backward (3) forward

13. (1) $Cu(OH)_2$ precipitate (2) $Cu(OH)_2$ precipitate (3) $Cu(OH)_2$ precipitate

14. $[Ni(NH_3)_6^{2+}] = 5.5 \times 10^{-11} mol \cdot L^{-1}$ $[NH_3] = 1.60 mol \cdot L^{-1}$ $[Ni(en)_3^{2+}] \approx 0.10 mol \cdot L^{-1}$
    $[en] = 2.00 mol \cdot L^{-1}$

15. (1) $[Ag^+] = 5.4 \times 10^{-10} mol \cdot L^{-1}$, $[Ag(NH_3)_2^+] = 0.05 mol \cdot L^{-1}$, $[NH_3] = 2.90 mol \cdot L^{-1}$
    (2) no AgCl precipitation; $c(NH_3) \geqslant 0.46 mol \cdot L^{-1}$

16. $[Ag^+] = 1.91\times10^{-20}\,mol\cdot L^{-1}$, $[CN^-] = 0.043\,mol\cdot L^{-1}$, $[Ag(CN)_2^-] = 0.046\,mol\cdot L^{-1}$

17. $\varphi^\ominus\{[Zn(OH)_4]^{2-}/Zn\} = -1.28\,V$

18. (1) $c(NH_3) = 0.58\,mol\cdot L^{-1}$

    (2) $[Ag(NH_3)_2^+] = 0.05\,mol\cdot L^{-1}$, $[Ag^+] = 1.97\times10^{-8}\,mol\cdot L^{-1}$, $[NH_3] = 0.48\,mol\cdot L^{-1}$
    $[Cl^-] = [K^+] = 9\times10^{-3}\,mol\cdot L^{-1}$

    (3) $\varphi\{[Ag(NH_3)_2]^+/Ag\} = 0.343\,7\,V$

19. (1) $[Ag^+] = 4.27\times10^{-15}\,mol\cdot L^{-1}$ (2) $K_s\{[Ag(S_2O_3)_2]^{3-}\} = 2.93\times10^{13}$

    (3) $[Ag(S_2O_3)_2^{3-}] = 1.01\times10^{-7}\,mol\cdot L^{-1}$
    $[CN^-] \approx 1\,mol\cdot L^{-1}$
    $[Ag(CN)_2^-] \approx 0.5\,mol\cdot L^{-1}$
    $[S_2O_3^{2-}] \approx 3.0\,mol\cdot L^{-1}$

20. $K_s = 3.5\times10^{16}$, $c(OH^-) = 7.3\,mol\cdot L^{-1}$

21. $\varphi^\ominus\{[Fe(bipy)_3]^{3+}/[Fe(bipy)_3]^{2+}\} = \varphi^\ominus(Fe^{3+}/Fe^{2+}) - 0.059\,16\lg\dfrac{K_{s,2}}{K_{s,1}}$

22. $K_s = 3.5\times10^{16}$, $[Fe(bipy)_3]^{2+}$ is more stable.

23. $n(Na_2S_2O_3) = 0.214\,mol$, $[Ag(S_2O_3)_2^{3-}] \approx 0.10\,mol\cdot L^{-1}$, $[Ag(S_2O_3)_2^{3-}] \approx 0.10\,mol\cdot L^{-1}$,
$[NH_3] = 0.20\,mol\cdot L^{-1}$, $[Ag(NH_3)_2^+] \approx 1.0\times10^{-5}\,mol\cdot L^{-1}$

[Supplementary Exercises]

1. $[Co(NH_3)_4(H_2O)Cl]Cl_2$

2. $[CuY]^{2-}$ is the more stable than $[Cu(en)_2]^{2+}$.

3. $c(NH_3) \geqslant 1.13\,mol\cdot L^{-1}$

4. Not AgI precipitate forms in the solution.

5. The reaction is in reverse direction.
$[Ag(NH_3)_2]^+(aq) + 2CN^-(aq) \rightleftharpoons [Ag(CN)_2]^-(aq) + 2NH_3(aq)$

6. $K_s\{[Cu(NH_3)_4]^{2+}\} = 2.09\times10^{13}$

7. $E_{cell} = 1.130\,V$

（刘国杰　申小爱）

Chapter 12
Colloids

The term **colloid** was introduced in 1861 by Thomas Graham (1805—1869) to describe the highly dispersed mixture in which particles diffuse slowly and cannot pass through a semipermeable membrane (e.g. a sheepskin). After the evaporation of solvents, colloids form amorphous jellies instead of crystals. Four decades later, веймарн studied more than 200 substances and proved that anything that could form crystal was also able to form colloid in a proper way. With the development of macromolecular researches in 1930s, it was revealed that the size of macromolecule particles was similar to that of colloidal particles. Thus macromolecular solutions are considered colloidal dispersions. In liquid medium, colloidal particles can also form by the assembly of amphiphiles (molecules with at least a hydrophilic polar group and a hydrophobic group).

Colloids are significant in medical science. Many crucial materials in tissues and cells can form colloids, including proteins, nucleic acid, starch, glycogen and cellulose. Typical colloidal dispersions involve blood, body fluid, cells and cartilage. Many physiological phenomena and pathological changes of organisms are closely related to colloidal properties. Knowledge of colloids can also be widely applied to areas such as food, painting, petroleum and catalyst and so on.

Emulsions, though not included in colloids, possess huge specific surface area and have properties similar to those of colloids. Thus it is also discussed in this chapter.

12.1 Colloidal Dispersions

12.1.1 Introduction

A colloidal dispersion is a mixture in which at least one dimension of particle is in the range of 1 to 100nm. Particles in colloidal dispersions are large clusters of atoms, ions and small molecules (e.g. in ferric hydroxide sols and gold sols), or very large ions or molecules (such as in protein solutions). In colloids, the combination of colloidal particles is called dispersed phase, while the bulk substance in which colloidal particles are scattered is called dispersion medium. Both the dispersion medium and the dispersed phase can be liquids, gases or solids. Generally, the substance of relatively smaller amount is regarded as the dispersed phase, while that of larger amount is the dispersion medium. Examples of colloidal dispersion systems are listed in Table 12-1.

Table 12-1 Examples of Typical Colloidal Dispersion Systems

Dispersion medium	Dispersed phase	Type	Examples
Gas	Liquid	Liquid aerosol	Fog
Gas	Solid	Solid aerosol	Smoke
Liquid	Gas	Foam	Whipped cream
Liquid	Liquid	Emulsion	Milk
Liquid	Solid	Sol	Paints
Solid	Gas	Solid foam	Pumice
Solid	Liquid	Gel	Butter, cheese
Solid	Solid	Solid sol	Ruby glasses

12.1.2 Classification and Characteristics of Colloidal Dispersions

Colloidal dispersions can be generally classified into three types: sols, macromolecular solutions and association colloids. The types and characteristics of dispersed systems are listed in Table 12-2.

Table 12-2 Characteristics of Dispersed Systems

Size of dispersed particles	Type		Component of dispersed particles	General properties	Examples
<1nm	True solutions		Small molecules or ions	Homogeneous; thermodynamically stable; cannot be separated by filtration	Aqueous solutions of NaCl, NaOH, $C_6H_{12}O_6$
1~100nm	Colloids	Sols	Colloidal particles (aggregates of atoms, molecules or ions)	Heterogeneous; thermodynamically unstable; cannot be separated by filtration	Sols of ferric hydroxide, arsenic sulfide, silver iodide, gold, silver and sulfur
		Macromolecular solutions	Macromolecules	Homogeneous; thermodynamically stable; cannot be separated by filtration; form solutions	Aqueous solutions of proteins and nucleic acids, benzene solution of rubber
		Association colloids	Micelles	Homogeneous; thermodynamically stable; cannot be separated by filtration; form micellar solution	Sodium lauryl sulfate solution of over a certain concentration
>100nm	Coarse dispersion systems (emulsions and suspensions)		Coarse particles	Heterogeneous; thermodynamically unstable; can be separated by filtration	Milk, mud

The highly distribution of dispersed particles usually results in the rapid increase of surface area. For example, a cube whose edge length is 1cm has a surface area of 6cm². On its division into uniform tiny cubes with the edge length being 1nm (1cm = 10^7nm), the surface area would become $6 \times 10^7 \text{cm}^2$, ten million times of that before the division.

Dispersion degree is used to describe the extent that a certain dispersed phase disperses in a medium. It is usually represented by the surface area of per volume, called **specific surface area**.

$$S_0 = S / V \tag{12.1}$$

Here S_0 is the specific surface area, S is the total surface area and V is the volume of a substance, respectively. Equation 12.1 shows that specific surface area increases with the total surface area of colloidal dispersed particles, so does dispersion degree.

When materials form highly dispersed systems, their surface properties become extremely prominent due to the rapid increase of their surface area. Generally, the contacting surface between two phases is called **interface**, such as liquid-gas interface, solid-gas interface, liquid-liquid interface and solid-liquid interface. Meanwhile, surface usually refers to the interface between solid and gas or between liquid and gas.

Both the circumstance and the energy of molecules at the interface between two phases are different from those of molecules in a phase. Take the liquid-gas interface as an example as shown in Figure 12-1. Molecule A inside the liquid is attracted equally by its surrounding molecules, resulting in a total force of zero. However, for molecule B and C which stay near the surface of the liquid, the uneven attractive forces by their surrounding molecules result in the attraction towards the inside of the liquid. If surface area of liquid is increased, the work must be done to overcome the attraction by molecules in the liquid phase. Subsequently, the work is stored in the surface molecules as potential energy. Thus it is clear that the surface molecules have more energy than the internal molecules. The increased energy is termed as **surface energy**.

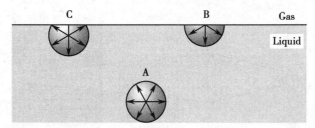

Figure 12-1 The Forces on Molecules in the Liquid or at the Interface

Both liquids and solids tend to spontaneously diminish their surface area. Droplets gather together to reduce their surface area and surface energy. Small particles in highly dispersed colloidal dispersions tend to spontaneously aggregate into large particles to reduce both the surface area and the surface energy. This property is termed as coagulation instability, indicating that colloidal dispersions are thermodynamically unstable systems.

Macromolecular solutions are homogeneous solutions in which single macromolecules disperse in the medium and there is no interface between them. Their properties are related to the macromolecular flexibility and the affinity of the medium. However, since the size of dispersed particles in macromolecular solutions is within that of colloidal particles, macromolecular solutions are also colloidal dispersions.

Dispersed particles in association colloids are micelles formed by the assembly of surface-active molecules (amphiphilic molecules with hydrophilic polar groups and hydrophobic nonpolar groups) if its concentration exceeds a certain value. The assembly of surface-active molecules is spontaneous and reversible, so association colloids are thermodynamically stable systems.

12.2 Sols

12.2.1 Properties of Sols

Sols are thermodynamically unstable dispersions formed by dispersion medium and colloidal particles with a diameter of 1 to 100nm. Colloidal particles of sols are aggregates of huge number of atoms, molecules or ions. Heterogeneity, highly distribution and aggregation instability are the characteristics of sols. The optical, kinetic and electrical properties of sols are all based on these characteristics.

1. Optical properties of sols —— Tyndall effect

When sols are irradiated with a focused strong beam of light in a dark room, a cone beam can be seen in the vertical direction to the visible light (Figure 12-2). This phenomenon can be attributed to the fact that dispersed colloidal particles whose diameter is comparable to or shorter than the wavelength of light are large enough to scatter light. Meanwhile, ions or small molecules do not significantly scatter light. Therefore, this phenomenon would not be observed when light is focused on a solution. The scattering of light by dispersed colloidal particles is called **Tyndall effect**. Thus Tyndall effect can be used to distinguish sols from the normal solutions.

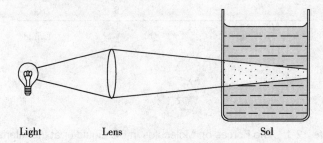

Figure 12-2 Tyndall Effect of a Sol

In 1871, Rayleigh investigated light scattering and got the following conclusions. (1) The intensity of the scattered light increases with the number of colloidal particles in unit volume. (2) For colloidal particles whose diameter is smaller than light wavelength, the larger the particle volume is, the greater the scattered light intensity is. (3) The shorter the light wavelength is, the stronger the scattered light is. Thus violet light is the easiest to be scattered among visible lights, causing the scattered light and transmitted light of a colorless sol being blue and red. (4) The larger difference in the index of refraction between the dispersion phase and the dispersion medium is, the stronger the scattered light is. In a homogeneous molecular solution, there is little difference in the index of refraction of the dispersion phase and the dispersion medium, so the scattered light is relatively weak. Hence the relative molar mass and the molecular shape of macromolecules can be measured based on the above theory.

2. Kinetic properties of sols

(1) Brownian motion. The irregular motion of colloidal particles observed with a microscope is called **Brownian motion**. This motion is caused by irregular buffeting received by colloidal particles from the kinetic motions of the molecules in the surrounding medium. The smaller the colloidal particle mass and the higher temperature is, the faster the particle moves and the stronger the Brownian motion is. Since the motion of colloidal particles can prevent them from coagulating and separating from the solvent, Brownian motion is one of the stabilizing factors to sols. Thus sols are kinetically stabilized.

(2) Diffusion and sedimentation equilibrium. The phenomenon of colloidal particles in sols migrate from the part of higher concentration to those of lower concentration is called **diffusion**. Diffusion would be easier at the higher temperature. Both the Brownian motion and the concentration difference of colloidal particles are the main reasons for diffusion.

The phenomenon of colloidal particles precipitate in gravitational field is called **sedimentation**. If the dispersion particles were big and heavy, there would be very weak Brownian motion. Thus the diffusion force approaches zero and the colloidal particles would precipitate very soon in gravitational field, just like a coarse system does. On the contrary, if colloidal particles are very small, the diffusion and the sedimentation would exist simultaneously. When the sedimentation rate equals to the spreading rate, the colloidal dispersion is in the equilibrium state. At this moment, the concentration of colloidal particles increases gradually from the top to the bottom of vessel and it forms a stable concentration gradient (Figure 12-3). This state is termed as **sedimentation equilibrium**.

Figure 12-3 Diffusion and Sedimentation Equilibrium

Since the sedimentation rate of colloidal particles is very slow in the gravity owning to their tiny size, it would take extremely long time to reach sedimentation equilibrium. Svedberg, an outstanding Sweden physicist, invented the first ultracentrifuge. With the centrifugal force being 100 000 times of the gravity, sedimentation equilibrium of colloidal particles in sols or macromolecules in macromolecular solutions can be rapidly achieved. Ultracentrifugation is one of the fundamental separation means in modern biomedical fields.

> **Question 12-1**   What is the significance of the dynamic properties of sols in biomedical?

3. Electrical properties of sols —— Brownian motion

Pour a colorful sol into a U-tube, carefully cover the two ends of the sol with colorless electrolytic solution and leave clear surfaces between the sol and the electrolytic solution at the same level. If electrodes are inserted in the electrolytic solution and direct current is applied for a while, one side of liquid surface goes up and the other goes down as shown in Figure 12-4. This experiment proves that colloidal particles carry the same type of charge. The movement of charged colloidal particles in electrical field is called **electrophoresis**. Colloidal particles of the most metal sulfide, silicic acid, gold and silver sols are negatively charged, and they will move to anode. Thus the sols are called negatively charged sols. While in the most of hydroxide sols, colloidal particles are positively charged and will move to cathode. These sols are called positively charged sols.

The dispersion medium must carry the opposite charges toward the colloidal particles because the whole system is neutral. As shown in Figure 12-5, if a sol separated by a porous membrane (e.g. activated carbon, burned magnetic disk, etc.) is applied with direct current, colloidal particles will be adsorbed by the porous membrane. Components in the medium would move to the oppositely charged electrode and pass through the porous membrane. The movement can be easily observed from the surface position variation with a capillary electro-osmosis apparatus. The directional movement of the dispersion medium in an applied electric field is called **electro-osmosis**.

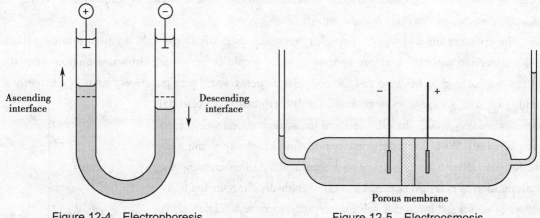

Figure 12-4   Electrophoresis   Figure 12-5   Electroosmosis

Both electrophoresis and electro-osmosis are electrical phenomena caused by relative movement of the dispersed phase and the dispersion medium. They have not only theoretical significance but also practical application values. Electrophoresis is widely used in the separation and identification of amino acids, polypeptides, proteins, nucleic acid, etc.

12.2.2   Micellar Structures and Stabilization of Sols

Although sols are thermodynamically unstable systems, many of them are relatively stable during a long period of time. This can be attributed to their kinetic and electrical properties, especially the latter.

1. Origins of surface charges on colloidal particles

Solid colloidal particles are charged because of their dissociation or selective adsorption.

(1) Selective adsorption on the interface of colloidal nuclei. Colloidal nuclei (aggregates of atoms, ions or molecules) tend to adsorb the substances so as to reduce their interface energy. They usually selectively adsorb the ions similar to their own components in the dispersions as stabilizers. Thus their interfaces would carry the charges. For example, dropping $FeCl_3$ solution into the boiling water to prepare $Fe(OH)_3$ sol. The reaction is

$$FeCl_3 \text{ (aq)} + 3H_2O \text{ (l)} \rightarrow Fe(OH)_3 \text{ (aq)} + 3HCl \text{ (aq)}$$

Many $Fe(OH)_3$ molecules in the solution aggregate to form colloid nuclei and some others react with HCl as

$$\text{Fe(OH)}_3 \text{ (s)} + \text{HCl (aq)} \rightarrow \text{FeOCl (aq)} + 2\text{H}_2\text{O (l)}$$
$$\text{FeOCl (aq)} \rightarrow \text{FeO}^+ \text{(aq)} + \text{Cl}^- \text{(aq)}$$

Thus Fe(OH)_3 colloidal nuclei would be positively charged by adsorbing FeO^+ ions with positive charge and similar to the components of Fe(OH)_3, while Cl^- ion called counter ions with negative charges of the colloidal nuclei would distribute in dispersion medium.

Another example is in the preparation of AgI sols with AgNO_3 solution and KI solution, the reaction is

$$\text{AgNO}_3\text{(aq)} + \text{KI(aq)} \rightarrow \text{AgI(sol)} + \text{KNO}_3\text{(aq)}$$

Colloidal nuclei with different charges can be prepared by vary the amount of reactants. AgI colloidal particles will adsorb I^- if excess KI is used, whereas they will adsorb Ag^+ if excess AgNO_3 is used. Hence AgI colloidal nuclei can be negatively and positively charged in these two cases, respectively.

(2) Dissociation of surface molecules on colloidal nuclei. Take silica nuclei as an example. The colloidal nuclei of silica sol consists of many $x\text{SiO}_2 \cdot y\text{H}_2\text{O}$ molecules. H_2SiO_3 molecules on their interface can dissociate into SiO_3^{2-} and H^+ as

$$\text{H}_2\text{SiO}_3 \text{ (sol)} \rightleftharpoons \text{HSiO}_3^-\text{(aq)} + \text{H}^+ \text{(aq)}$$
$$\text{HSiO}_3^-\text{(aq)} \rightleftharpoons \text{SiO}_3^{2-}\text{(aq)} + \text{H}^+ \text{(aq)}$$

Thus SiO_3^{2-} ions remain on the surface of H_2SiO_3 nuclei and make the nuclei negatively charged, while H^+ ions diffuse in the dispersion medium.

2. Electrical double layer of colloidal particles

The structure of a sol is relatively complex. If the surface of a colloidal nucleus is charged, it would attract the counter ions. At the same time, the counter ions tend to diffuse into the whole medium due to thermal motion. So, the nearer the counter ions is to surface of the colloidal nucleus, the more the number of counter ions is.

Charged colloidal nuclear surfaces always bound to a large amount of water molecules and absorb hydrolyzed counter ions. Thus a colloidal particle is wrapped by a layer of hydration membrane. If a colloidal particle moves, both the hydration membrane close to the nucleus and the counterions in this hydration membrane would move together. The hydration membrane including ions on the colloidal particle surface and the trapped counter ions is called **adsorption layer**. Other hydrated counter ions diffusely distribute around the adsorption layer, forming **diffusion layer**. The double-layered structure that is made of the adsorption layer and the diffusion layer with the opposite charges is called **diffused electrical double layer**. Colloidal particle refers to the nucleus and the adsorption layer, while micelle refers to the colloidal particle and the diffusion layer. The medium between micelles is called intermicellar liquid. Thus a sol is the integration of micelles and intermicellar liquid. For instance, the structure of a micelle of Fe(OH)_3 sol can be illustrated as follows.

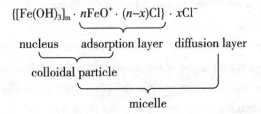

Figure 12-6 shows a micelle structure of $Fe(OH)_3$ sol. If colloidal particles move, the diffusion layer will split from the adsorption layer. The interface between them is called sliding interface. When a certain amount of electrolyte is added into a sol, some counter ions are forced to move from the diffusion layer into the adsorption layer, causing the diffusion layer become thinner. The more the electrolytes added is, the more the counter ions forced into the adsorption layer is, the thicker the diffusion layer is. When the charges on colloidal nucleus surfaces are almost neutralized by counter ions moved into adsorption layers, colloidal particles are in isoelectric state, which means colloidal particles are uncharged and will not move in electrical field. Sometimes, too much of electrolyte is added, charges on colloidal particles can sometimes even change to the opposite charges, called recharged phenomenon.

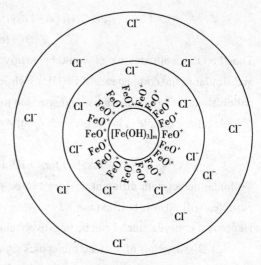

Figure 12-6   Micelle Structure

3. Stabilization of sols

The stability of a colloidal dispersion is attributed to several factors.

(1) Characteristics of charged colloidal particles. The electrostatic repulsion between two colloidal particles with the same charge prevents them from getting close, merging and becoming larger. That's why colloidal particles in a sol are rather stable. However, when the kinetic energy of colloidal particles becomes strong enough to overcome the electrostatic repulsion, colloidal particles will collide with each other and coagulate. Thus heating may cause coagulation of sols. Usually, kinetic energy of colloidal particles is not strong enough to overcome the electrostatic repulsion, so sols are stable.

(2) Protecting action of hydration membranes on the surface of colloidal particles. The hydration membrane of a micelle seems like a layer of elastic membrane that prevents the colloidal particles from colliding each other and coagulating. Thus the stability of a sol is closely related to the thickness of the hydration membrane on the colloidal particles. The thicker the hydration membrane is, the more stable the sol is.

(3) Thermodynamic stability. Brownian motion of the highly dispersed colloidal particles is also one important reason for keeping a sol to be stable in the gravity field.

4. Coagulation of sols

Generally, if the stabilization factors are influenced, colloidal particles will collide, stick to each other even to precipitate from the dispersion medium, this phenomenon is called **coagulation**. There are many factors that may cause a sol to coagulate, such as heating, radiation and the addition of electrolytes, the last on is the main reason.

(1) Effect of electrolytes on coagulation. Coagulation caused by electrolytes is mainly because the added electrolytes can change the structure of the adsorption layer. With the addition of electrolyte, more counter ions will move from the diffusion layer to the adsorption layer. Thus the charges on surface of colloidal particles will be neutralized and the adsorption layer will become thinner little by little. As a

result, the stability of sols decreases and coagulation occurs finally.

Sols are very sensitive to electrolytes. Although electrolyte of extremely small amount may make a sol more stable during preparation, electrolyte of a very slightly overdose will cause coagulation. **Critical coagulation concentration** (c.c.c.), the lowest concentration (mmol·L^{-1}) of an electrolyte that is needed to make a certain amount of sol to coagulate in a certain period of time, is often used to evaluate the coagulation ability of an electrolyte. Different electrolyte has the different coagulation ability. The lower the critical coagulation concentration is, the stronger the electrolytic coagulation ability is. Critical coagulation concentrations of several electrolytes to three kinds of sols are listed in Table 12-3.

Table 12-3 Critical Coagulation Concentration (mmol·L^{-1}) of Electrolytes to Several Sols

As_2S_2 (Negatively charged sols)		AgI (Negatively charged sols)		Al_2O_3 (Positively charged sols)	
LiCl	58	$LiNO_3$	165	NaCl	43.5
NaCl	51	$NaNO_3$	140	KCl	46
KCl	49.5	KNO_3	136	KNO_3	60
KNO_3	50	$RbNO_3$	126		
$CaCl_2$	0.65	$Ca(NO_3)_2$	2.40	K_2SO_4	0.30
$MgCl_2$	0.72	$Mg(NO_3)_2$	2.60	$K_2Cr_2O_7$	0.63
$MgSO_4$	0.81	$Pb(NO_3)_2$	2.43	$K_2C_2O_4$	0.69
$AlCl_3$	0.093	$Al(NO_3)_3$	0.067	$K_3[Fe(CN)_6]$	0.08
$\frac{1}{2}Al_2(SO_4)_3$	0.096	$La(NO_3)_3$	0.069		
$Al(NO_3)_3$	0.095	$Ce(NO_3)_3$	0.069		

Coagulation abilities of electrolytes follow the rules below.

a. The critical coagulation concentration of a sol is extremely sensitive to the charge numbers of counter ions. The larger the counter ions charge is, the stronger the coagulation ability is. For the counter ions with one, two or three charges, the ratio of their critical coagulation concentration approximates

$$(1/1)^6 : (1/2)^6 : (1/3)^6 = 100 : 1.8 : 0.14$$

That is, the critical coagulation concentration is in inverse ratio to the sixth power of the counter ions charges, namely **Schulze-Hardy rule**.

b. Different counter ions with the same charge may be near but different in coagulation abilities actually. For example, when coagulating a negatively charged sol with monovalent cations, the sequence of coagulation ability is

$$H^+ > Cs^+ > Rb^+ > NH_4^+ > K^+ > Na^+ > Li^+$$

Similarly, the coagulation ability sequence of the monovalent anions to coagulate a positively charged sol is

$$F^- > Cl^- > Br^- > I^- > CNS^-$$

The above sequences are called **lyotropic series**.

c. Coagulation abilities of some organic ions are very strong, especially for the ions of surfactant (such as soap) and polyamide. They can effectively destroy a sol and rapidly cause coagulation. This may be attributed to the very strong adsorption between colloidal nuclei and organic ions.

(2) Intercoagulation of sols. Intercoagulation will happen when two oppositely charged sols are mixed. If a positively charged sol is mixed with a negatively charged sol in a proper proportion, all charges on the surface of colloidal particles are neutralized, and resulting in completely intercoagulation. Otherwise, intercoagulation will be incomplete or even no intercoagulation will happen. For example, since colloidal suspensions in foul water are usually negatively charged, as addition of alum [$KAl(SO_4)_2 \cdot 12H_2O$] to foul water, the positively charged $Al(OH)_3$ sol formed by Al^{3+} hydrolyzation will coagulate the colloidal suspensions and make water clean.

(3) Protection and sensibilization of macromolecules. On the addition of a macromolecular solution to a sol, macromolecules may be adsorbed to and encapsulate the colloidal particles as shown in Figure 12-7(a). Thus the affinity between colloidal particles and dispersion medium increases. The stability of the sol increases as a result. However, in some cases, a small amount of macromolecular solution cannot increase the stability of a sol. On the contrary, it decreases the stability of a sol or even results in coagulation, called sensibilization. Sensibilization may be attributed to the fact that the amount of the added macromolecule is too small to cover the surfaces of the colloidal particles fully. Thus the colloidal particles attach to the added macromolecules as in Figure 12-7(b), gradually grow and coagulation happens finally.

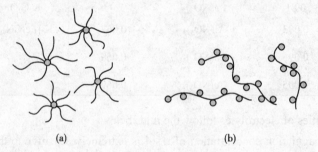

Figure 12-7 Scheme of Protection (a) and Sensitization (b) of Macromolecules to Sols

12.2.3 Aerosols

1. Formation of aerosols

Aerosol is formed by dispersing tiny solid or liquid particles into gases. For example, smoke and dust are aerosols formed by dispersing solid particles in air, whereas fog is an aerosol formed by dispersing tiny droplets in air. Figure 12-8 shows the diameter scopes of dispersed particles in different aerosols. Since the dispersion degree of smoke or fog (particle diameter 0.01~1nm) is relatively higher than that of dusts (particle diameter 1~1 000nm), the stability of the former is relatively higher than that of the latter.

2. Relationship between aerosol and environment

Aerosols have been paid extensive attention in preventative medicine. Dusts produced in industrial and agricultural production float in atmosphere for a long time, hence pollute the environment and endanger human health. The dynamic and electrical properties of dust aerosols are similar to that of sols. Chances of dusts being inhaled in by living bodies and their stability in atmosphere are directly related to the size and the charge of the dispersed particles. Thus charged dust particles are easily held by human bodies. Generally, particles with a diameter of larger than 10μm are accessible to the nasal cavity and will mainly deposit in upper respiratory tract, while particles with a diameter of 2~10μm mainly deposit in the bronchial and will directly penetrate into alveoli or even be trapped on alveolar walls. Particulate matter whose particle diameter is less than or equals to 2.5μm (PM2.5) can reach the non-ciliated zones of lungs. Due to their large specific surface area and strong adsorption ability, they can bring heavy metals, sulfates, organic toxicants or even virions with themselves, and bringing more severe influence to human bodies. Pneumonoconiosis is the disease whose major symptom is the fibration of heart and lung tissues caused by the inhalation of industrial dusts for a long time. Silicosis, caused by the inhalation of crystal silica dusts, is one of the most dangerous pneumonoconiosis.

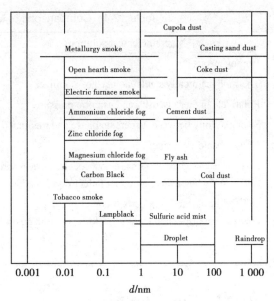

Figure 12-8 Ranges of Aerosol Particle Diameters

In the view of special effects of particulate matters on human health and environment, the content of the airborne particulate matter begins to be regarded as an important air quality standard. The World Health Organization (WHO) recommended that the annual and the daily averaged PM2.5 should not exceed 10μg/m^3 and 25μg/m^3, respectively. In 1990s, the US and the UK began to monitor PM2.5. Major cities in China began to monitor PM2.5 since 2012.

12.3 Macromolecular Solutions

Macromolecular compounds have the huge relative molar mass of $10^4 \sim 10^6$ degree. Proteins, nucleic acids and glycogens are the typical life-related bio-macromolecules. Macromolecular compounds also include polymers (e.g. natural rubber and polystyrene) and non-polymers (e.g. natural lignin). Many properties of macromolecular compounds, such as difficulty in dissolving, swelling and large viscosities, are all related to their huge relative molar mass. By dissolving macromolecular compound in a proper medium, **macromolecular solutions** can form automatically, with the size of single molecule reaches that of colloidal particles. Thus some properties of macromolecular solutions are similar to those of sols, such as low diffusion rate and incapable of passing through a semipermeable membrane. However, since the macromolecular solutions are solutions in nature, they also have the different properties compared with sols. In Table 12-4, the different properties of macromolecular solutions and sols are compared.

Table 12-4 Comparisons of Macromolecular Solution and Sol

Macromolecular solutions	Sols
Homogeneous	Heterogeneous
Thermodynamically stable, no need of stabilizer	Thermodynamically unstable, need stabilizer
Relatively higher viscosity and osmotic pressure	Relatively lower viscosity and osmotic pressure
Strong affinity between dispersion phase and medium	Weak affinity between dispersion phase and medium
Unnoticeable Tyndall effect	Noticeable Tyndall effect
No obvious change when a small amount of electrolyte is added; salting out when a large amount of electrolyte is added	Coagulation when a small amount of electrolyte is added
Gel can be formed in a certain condition; dry gel can be dispersed in dispersion medium to form solution again	Coagulation after removal of dispersion medium; sediment by coagulation is very difficult to be dispersed again in dispersion medium only in some special ways

12.3.1 Introduction

1. Characteristics of macromolecular structures

A macromolecular compound usually contains carbochain that is built up by a large number of units called monomers. The repeating frequency of the monomer is called the degree of polymerization denoted by n. For example, natural rubber is made of the combination of thousands of isoprene units ($-C_5H_8-$), so its chemical formula can be expressed as $(C_5H_8)_n$. Glycans such as cellulose, starch, glucogen and glucan are made of huge amount of glucose units ($-C_6H_{10}O_5-$), so their general chemical formula can be expressed as $(C_6H_{10}O_5)_n$. Amino acids are the structural units of proteins. A macromolecular compound is a mixture of homologues with different degree of polymerization. Thus both its degree of polymerization and the relative molar mass are averaged values.

Different macromolecular compounds have the different lengths and different linking ways in their units, such as linear structure and dendritic structure. Many polymers can also crosslink to form dendritic structures.

There are many single bonds in the macromolecular chain. Each single bond can rotate around the bond axis of the neighbored single bond, called internal side rotation as shown in Figure 12-9. Internal side rotations may lead to the structural variation of carbochain and the change in distance subsequently between two ends of the chain. **Flexibility** is used to describe the variable macromolecular chains.

Alkane chains that are made of C and H atoms are usually relatively flexible because the interaction between these atoms is relatively too small to prevent the internal side rotation of carbon-carbon bonds. If polar groups such as -Cl, -OH, -CN and -COOH are bonded to alkane chains, both the interactions between atoms and the resistance to the internal

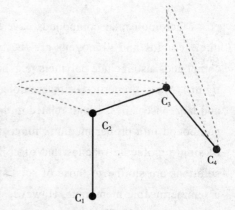

Figure 12-9 Internal side Rotations of Macromolecules

side rotations increases, hence the macromolecular flexibility decreases and the rigidity will increase. Variation will become more prominent with more polar substituent groups and shorter distance between the polar substituent groups. Flexible macromolecules are easy to curl or even form coils. On the contrary, rigid macromolecules are difficult to curl, forming rod shapes at the extreme.

The properties of a macromolecular solution mainly depend on the structural features and the state of the macromolecules in dispersion medium. Besides flexibility, the affinity between macromolecules and the medium also plays a key role. For example, a solvent of strong affinity eliminates the cohesion and makes macromolecules spread, called a fine solvent of the macromolecule. Similarly, a bad solvent of a macromolecule refers to the solvent that has weak affinity to the macromolecule, in which the macromolecular long chain would curl spontaneously.

2. Formation of macromolecular solutions

The apparent difference between macromolecular solution and the solution of normal molecules with low molar mass is **swelling**. It means that the solvent molecules slowly get into the space of the curled macromolecular chains, inducing the whole macromolecule to spread out with its volume to even ten times of its original size. Swelling is a prelude when the macromolecular compounds dissolve. A macromolecular compound can first swell, and then dissolve completely in the solvent that has strong affinity with it. Many macromolecular compounds have very strong affinities with H_2O molecules that hydration layers form around them. This is the main reason why the aqueous macromolecular solutions are stable. A macromolecular solution is a thermodynamically stable system since macromolecular compounds can dissolve automatically in their fine solvents.

12.3.2 Properties of Protein Solutions

Many macromolecular compounds like proteins exist in the form of ions in solution, so they are generally called **polyelectrolytes**. Those polyelectrolytes with many charged groups bonded to the molecular chains have the high charge density and the strong affinity to polar solvents. Polyelectrolytes can be classified into cations, anions and zwitterions according to the charge type carried in their aqueous solutions.

The composition and the structure of a protein determine the properties of its solution. It is well known that proteins are polypeptide chains made up of 20 kinds of amino acids. Amino acids are linked by peptide bonds. There are many varied kinds of dissociable groups in polypeptide chains, such as proton donors and proton receptors. Proteins acquire charges mainly through the ionization of carboxyl groups $-COOH$ that denote protons to amino groups $-NH_2$. Phenolic hydroxyl groups and guanidinium groups may also contribute the charge to proteins.

Both the net charge and the charge distribution of proteins vary with the pH of solution. For example, charges on proteins will change by merely adjusting the pH. The pH of a solution in which protein molecules carry the same quantity of positive charges and negative charges (the net charge is zero) is known as **isoelectric point** noted in pI.

If pH is more than pI, carboxyl groups would denote more protons, and protein molecules would carry negative charges to be $R{\diagup}^{COO^-}_{\diagdown NH_2}$. If pH is less than pI, amino groups would receive more protons and protein molecules would carry positive charges to be $R\begin{smallmatrix}\diagup COOH\\ \diagdown NH_3^+\end{smallmatrix}$. If pH equals pI, protein molecules are in the isoelectric state as $R\begin{smallmatrix}\diagup COO^-\\ \diagdown NH_3^+\end{smallmatrix}$. The equilibrium shift of a protein molecule can be illustrated as follows.

$$R\begin{smallmatrix}\diagup COO^-\\ \diagdown NH_2\end{smallmatrix} \underset{OH^-}{\overset{H^+}{\rightleftharpoons}} R\begin{smallmatrix}\diagup COO^-\\ \diagdown NH_3^+\end{smallmatrix} \underset{OH^-}{\overset{H^+}{\rightleftharpoons}} R\begin{smallmatrix}\diagup COOH\\ \diagdown NH_3^+\end{smallmatrix}$$

$$\text{pH>pI} \qquad\qquad \text{pI} \qquad\qquad \text{pH<pI}$$

$$\Updownarrow$$

$$(R\begin{smallmatrix}\diagup COOH\\ \diagdown NH_2\end{smallmatrix})$$

For example, pI of human serum albumin (HSA) is 4.64, if it is dissolved in a buffer of pH 6.0, HSA exists in the form of anions; while in that of pH 4.0, it exists in the form of cations. By adjusting the pH to 4.64, HSA molecules are in the isoelectric state. Proteins in their isoelectric state cannot move in an applied electrical field, and they are easy to coagulate. Different proteins have the different compositions and conformations, so they have varied pI (Table 12-5).

Table 12-5 Isoelectric Point (pI) of Proteins

Proteins	Source	pI	Proteins	Source	pI
Protamine	Sperm of salmon	12.0~12.4	Lactalbumin	Cow's milk	5.1~5.2
Cytochrome C	Horse heart	9.8~10.3	Gelatin	Animal skin	4.7~4.9
Myoglobin	Muscle	7.0	Ovalbumin	Chicken egg	4.6~4.9
Hemoglobin	Rabbit blood	6.7~7.1	Pepsin	Cow's milk	4.6
Myosin	Muscle	6.2~6.6	Casein	Pig stomach	2.7~3.0
Insulin	Cattle	5.3~5.35	Fibroin	Natural silk	2.0~2.4

Not only electrophoretic property, but also other properties such as solubility, viscosity and osmotic pressure and so on are closely related to the charge state and the protein charge quantities in solution. The electrostatic attraction and repulsion between charged groups and the net charges on proteins affect the hydration of protein and the molecular chain flexibility, leading to some property changes of protein solutions. At pI, the affinity of proteins to water and the hydration degree decrease dramatically, hence protein molecular chains stick to each other and the solubility reduced. At a pH below or above pI, proteins carry a net positive or negative charge. Thus their affinity to water will increase, and lead to the protein molecular chains spread and the hydration degree increases. The protein solubility increases as a result.

Proteins and nucleic acids are macromolecules of great biological significance. Different proteins and nucleic acids carry the different net charges, so they have different mobility in an electrical field. Based on this reason, a series of electrophoretic techniques such as polyacrylamide gel electrophoresis (PAGE), sodium dodecyl sulfate-polyacrylamide gel electrophoresis (SDS-PAGE) and isoelectric focusing electrophoresis (IEF) are used to separate and identify bio-macromolecules widely.

12.3.3 Destabilization of Protein Solutions

Hydration is the key factor to stabilize a protein solution. If the large amount of inorganic salt like ammonium sulfate $(NH_4)_2SO_4$ or sodium sulfate Na_2SO_4 is added to a protein solution, the hydration degree of protein molecules would decrease dramatically due to the intense hydration of inorganic ions, often resulting in the precipitation of protein. The effect of making protein precipitate from solution by large amount of inorganic salt is called **salting out**; whose essence is the protein dehydration. $(NH_4)_2SO_4$ is the best inorganic salt for protein salting out because of its high and temperature-independent solubility (the concentration of its saturated solution reaches $4.1 \text{mol} \cdot L^{-1}$ at 25℃). Furthermore, it is a mild chemical that proteins keep their biological activities even at a very high concentration.

The salting out ability of an electrolyte is mainly related to the ion type, but not the charge of ions. Typically, the salting out ability of anions is much stronger than that of cations, the normal ions are sequenced as follows.

$$SO_4^{2-} > C_6H_5O_7^{3-} > C_4H_4O_6^{2-} > CH_3COO^- > Cl^- > NO_3^- > Br^- > I^- > CNS^-$$

The salting out ability sequence of cations also known as the lyotropic series is as follows.

$$NH_4^+ > K^+ > Na^+ > Li^+$$

Besides inorganic salts, organic solvents that have strong affinity to water such as ethanol, methanol and acetone also decrease the hydration degree of proteins, resulting in the precipitation of proteins from water.

Variations in temperature and the pH also destabilize a macromolecular solution. Since most proteins have the minimum solubility at pI, the precipitation of proteins can be realized by merely adjusting pH.

12.3.4 Osmotic Pressure and Membrane Equilibrium of Macromolecular Solutions

1. Osmotic pressure of macromolecular solutions

Osmosis occurs when a macromolecular solution is separated from pure solvent by a semipermeable membrane, just like that of small molecular solutions. However, the osmotic pressure of a linear macromolecular solution usually doesn't obey van't Hoff Equation. That is, the osmotic pressure changes are much larger but not the proportional relationship between concentration and osmotic pressure. One reason for this phenomenon is that a large amount of solvent molecules are trapped in the space of the long curled macromolecular chains. As a result, the number of the effective solvent molecules in unit volume decreases rapidly when the concentration of macromolecular compound increases. Moreover, a macromolecule can form different structural domains because of its flexibility. Each of these domains is equivalent to a structural unit of small molecules. These structural domains of a macromolecule are relatively independent, and it may produce osmotic effect that is equivalent to that produced by many small molecules. As a result, a macromolecular solution at low concentration is not an ideal solution. The relationship between its osmotic pressure π and the mass concentration (ρ_B, $g \cdot L^{-1}$) approximately follows the corrected formula below.

$$\frac{\pi}{\rho_B} = RT\left(\frac{1}{M_r} + \frac{B\rho_B}{M_r}\right) \tag{12.2}$$

Here, M_r is the relative molar mass of the macromolecular compound; B is a constant. By measuring the osmotic pressures and plotting $\frac{\pi}{\rho_B}$ against ρ_B, a straight line can be obtained (Figure 12-10). Then the relative molar mass of the macromolecular compound can be calculated since the intercept of this line is $\frac{RT}{M_r}$.

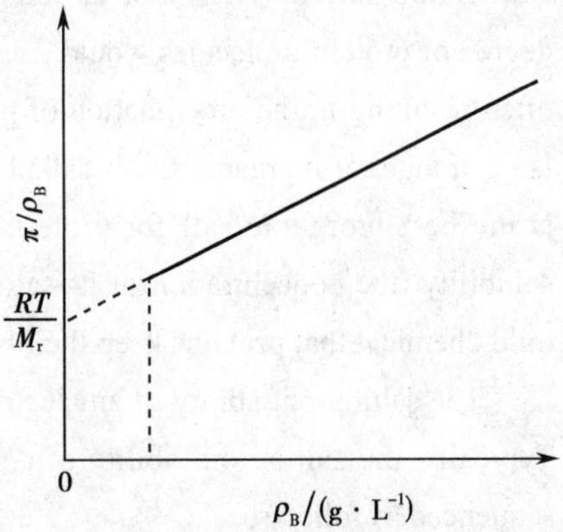

The osmotic pressure caused by macromolecular compounds like proteins in living bodies is termed as **colloidal osmotic pressure**, which plays an important role in maintaining the blood volume and the balance of water or electrolytes between the exterior and the interior of blood vessels.

Figure 12-10 Calculation of Relative Molecular Mass of Macromolecular Compounds by Extrapolation Method

2. Membrane equilibrium

If a protein solution is separated from a solution of small ions by a semipermeable membrane, the small ions can permeate the semipermeable membrane while the protein ions cannot. At the meantime, due to the electrostatic attraction between protein ions and small ions, small ions fail to distribute evenly on both side of the membrane at equilibrium so that the neutralization of the solution can be kept. This phenomenon is called **membrane equilibrium** or **Donnan equilibrium**. Membrane equilibrium is significant to the investigation of the electrolytic small ion distribution between the exterior and the interior of cells.

Figure 12-11 illustrates the separation of a protein (NaP) solution from NaCl solution by a semipermeable membrane. Suppose the initial concentration of Na^+ and P^- in the internal side and the external side the membrane is c_1 and c_2, respectively as shown in Figure 12-11(a). Since small ions can permeate through the semipermeable membrane but P^- cannot, Cl^- ions permeate from the external side to the internal side of the membrane. Thus Na^+ ions of the same amount have to permeate to the external side of the membrane also, so that the neutralization of the solution can be kept. Suppose there are $x\ mol\cdot L^{-1}$ of Cl^- ions and $x\ mol\cdot L^{-1}$ of Na^+ ions permeate from the external side to the internal side, the distribution of ions are showed in Figure 12-11(b) on equilibrium. Since the ion diffusion rate v is proportional to the concentration of ions,

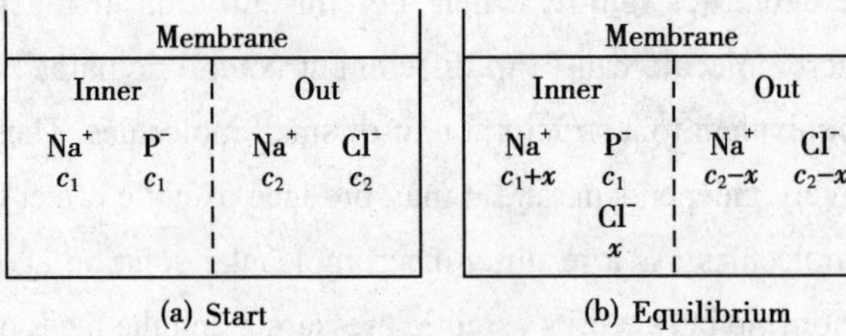

Membrane					Membrane			
Inner		Out			Inner		Out	
Na^+	P^-	Na^+	Cl^-		Na^+	P^-	Na^+	Cl^-
c_1	c_1	c_2	c_2		c_1+x	c_1	c_2-x	c_2-x
					Cl^-			
					x			
(a) Start					(b) Equilibrium			

Figure 12-11 Diagram of Membrane Equilibrium

$$v = kc(\text{Na}^+) \times c(\text{Cl}^-)$$

at the equilibrium. $v_{in} = v_{out}$. Hence

$$[\text{Na}^+]_{out} \times [\text{Cl}^-]_{out} = [\text{Na}^+]_{in} \times [\text{Cl}^-]_{in} \tag{12.3}$$

where $[\text{Na}^+]_{out}$, $[\text{Cl}^-]_{out}$, $[\text{Na}^+]_{in}$ and $[\text{Cl}^-]_{in}$ are the equilibrium concentrations of ions in the external or internal side of the membrane, respectively.

A prerequisite to establish Donnan equilibrium is that the product of the electrolyte ion concentrations should be equal on both sides of the *membrane* at equilibrium. Substituting the equilibrium concentrations of ions into Equation 12.4, we can get

$$(c_1 + x)x = (c_2 - x)^2$$

$$x = \frac{c_2^2}{c_1 + 2c_2} \text{ or } \frac{x}{c_2} = \frac{c_2}{c_1 + 2c_2} \tag{12.4}$$

Equation 12.4 shows that, if Donnan equilibrium is reached, x or $\dfrac{x}{c_2}$ (also called diffusion fraction) is determined by the initial concentration c_1 of NaP in the internal side of the membrane and the initial concentration c_2 of NaCl in the external side of the membrane.

If $c_1 \gg c_2$, $x = \dfrac{c_2^2}{c_1 + 2c_2} \approx 0$, there will be no permeation from the external side to the internal side.

If $c_2 \gg c_1$, $x = \dfrac{c_2^2}{c_1 + 2c_2} \approx \dfrac{c_2}{2}$, about half of NaCl in the external side will permeate to the internal side. Thus the concentration of NaCl on both sides of the membrane will approximately equals to each other.

If $c_2 = c_1$, $x = \dfrac{c_2^2}{c_1 + 2c_2} \approx \dfrac{c_2}{3}$, about $\dfrac{1}{3}$ of NaCl will permeate to the internal side.

Membrane equilibrium is a very common physiological phenomenon. The ionic permeability of cellular membrane is determined not only by the size of membrane pores, but also by the protein concentration in cells which has a certain effect on the ionic permeability outside a cell and the distribution of electrolyte ions on both sides of cellular membranes. For example, Cl^- ions can freely pass through the membranes of red blood cells, while the concentration of Cl^- in cells is 70% of that in extracellular plasma. One of main reasons is the Donnan effect caused by the relatively high concentration of protein anions in red blood cells. Of course cellular membranes are not ordinary semipermeable membranes since they have complicated structures and functions. There are many factors affecting the distribution of electrolytic ions inside and outside a cell membrane. The membrane equilibrium is merely one of them.

12.3.5 Gels

Under circumstances such as temperature dropping or solubility decreasing, the viscosity of many macromolecular solutions would become larger and larger. Finally, these macromolecular solutions would lose their mobility and form semi-solid mesh-structured materials called **gels**. The process of forming a gel is called **gelation**. You can get gels by solving agar, gelatin or animal gum in warm water and allow them to cool down without vibration.

When we discuss the gels, dispersion medium is considered to be solvent. The normal solvent is water generally.

In gelation, the linear macromolecules in solution get close to each other and crosslink on many sites to form network-shaped skeleton. Dispersion medium is involved in the networks. The amount of solvent comprised in the networks may be very large. For instance, the water content of solid agar is roughly 0.2%, but that of an agar gel can reach about 99.8%. Blood clots contain a large amount of water. Other substances such as human muscles and tissues can also be regarded as gels. They have the certain strength of network structure to maintain their shapes, while exchanges of metabolites can also be realized in the network structure.

Gels such as dextrane gel and polyacrylamide gel are produced by the crosslinking of one or several kinds of compounds. These artificial polymeric gels are commonly used in column chromatography and electrophoresis in molecular biology and biochemistry researches.

Gels can be classified into two types, rigid gels and elastic gels. Rigid gel such as silica gel and iron hydroxide has rigid network structure because of the strong crosslinking. Thus, if it is dried, molecules in dispersion medium will leave the network, but no obvious changes in their volumes and shapes can be observed. On the contrary, gels formed by flexible macromolecular compounds are usually elastic gels, such as gelatin, agar and polyacrylamide. This kind of gels will shrink and become elastic after being dried. If a dried elastic gel is put back into a proper solvent, it will swell or even dissolve in the solvent completely again.

Main properties of gels are as below, which are closely related to their network structure.

(1) Swelling. If a dried elastic gel is put into a proper solvent, it will absorb solvent molecules and its volume will expand by swelling. If swelling spontaneously stops at a certain degree, it is called finite swelling. On the contrary, some gels will not stop swelling until their networks disappear in solvent, called infinite swelling.

(2) Bound water. Gels absorb water in swelling. Water molecules tightly bound to gels are called bound water. Permittivity, vapor pressure, freezing point and boiling point of bound water are all lower than those of pure water.

The study of bound water in gels is very important in biology. For example, the anti-drought or anti-cold ability of vegetables may relate to the characteristics mentioned above. The amount of bound water in muscular tissues of human body decreases with the age increasing. Older people have much lower amount of bound water in their muscular tissue than younger people have.

(3) Syneresis. If an elastic gel is exposed to air for a period of time, the part of water will automatically separate from the gel to causing its volume diminish gradually. This phenomenon is called **syneresis**. Syneresis can be regarded as the subsequent gelation of a macromolecular solution, in which the junctions between macromolecule chains continuously increase and the solvent molecules are pushed out of the gel network, resulting in a smaller gel volume. Human serum used in clinical assays is prepared in the syneresis of blood clots.

Products based on gels are widely used in medical science. For example, donkey-hide gelation, a traditional Chinese medicine, is a gel; dried silica gel is a drying agent commonly used in laboratory; the other substances such as artificial semipermeable membranes and leathers are xerogel. The gel networks are rather flexible and movable. Bio-macromolecules and proteins are movable in gels applied in electrical field. Based on these characteristic, gel-based electrophoresis and chromatography have been developed and widely used in the separation of proteins.

12.4 Surfactants and Emulsions

12.4.1 Surfactants

The surface molecules of liquid are always pulled inward of liquid by inside of other molecules of liquid, and they are not attracted as intensely by the molecules in the neighbouring medium. Therefore, all of surface molecules are subject to an inward force. Solution surface tends to adsorb solutes to reduce the surface tension. At a constant temperature and pressure, force on the surface in the unit length is defined as **surface tension**. It is represented by σ with the unit of $N \cdot m^{-1}$. There are three kinds of water surface tension change on the addition of different solutes. (1) Water surface tension increases on the addition of some inorganic salts such as NaCl, NH_4Cl, Na_2SO_4 and KNO_3, and polyhydroxyl organic compounds such as cane sugar and mannitol. The plot(1) in Figure 12-12 indicates it. (2) Water surface tension decreases gradually on the addition of most of organic compounds such as alcohol, aldehyde, carboxylic acid and ester, the plot (2) in Figure 12-12 indicate it. (3) With the increasing concentration of soap and synthetic detergents like the metal salts, sulfates and benzenesulphonates containing organic acid more than 8 carbon atoms, water surface tension decreases sharply at the beginning, then remains roughly unchanged. The plot (3) in Figure 12-12 shows it. Substance that can dramatically decrease water surface tension are called **surface-active substances** (SAS) or **surfactant**, while that make water surface tension increase or just decrease a little is called surface inactive substance.

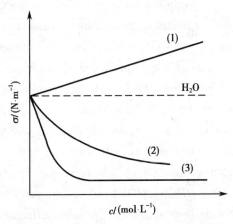

Figure 12-12 Variations of Surface Tension with Concentration

If the adsorption of a solute on the surface reduces the solvent surface tension, more solute would be maintained on the surface. Therefore, the solute concentration on the surface is higher than that in the solution, this is called positive adsorption. On the contrary, if the adsorption of a solute on the surface increases solvent surface tension, more solute would be maintained in the solution. In that case, the solute concentration on the surface is lower than that in the solution, it is called negative adsorption. In a word, surface-active substances and surface inactive substances cause the positive adsorption and the negative adsorption, respectively.

There are generally two types of group in the molecules of surfactants. One is the non-polar hydrophobic or lipophilic group which are straight or branched hydrocarbon; another is the polar hydrophilic group such as $-OH$, $-COOH$, $-NH_2$, $-SH$ or $-SO_2OH$. A characteristic of surfactant is that its molecules simultaneously contain both a non-polar hydrophobic group and a polar hydrophilic group as shown in Figure 12-13.

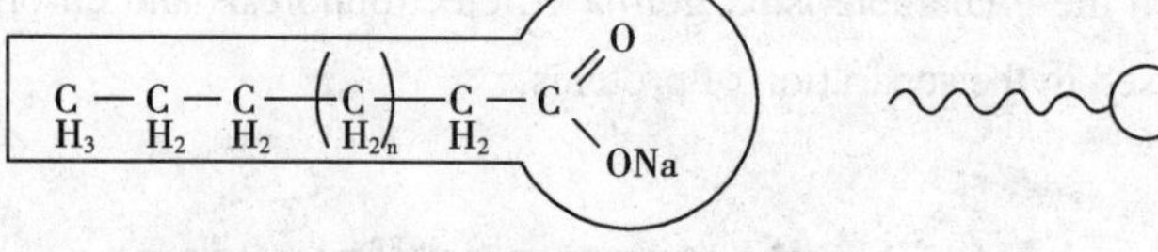

Lyophobic group　　Hydrophilic group　　　　　Diagram

Figure 12-13　Diagram of Surfactant (Soap)

Take soap called sodium fatty acid as an example, when it is dissolved in water, the hydrophilic carboxyl terminal groups get into water, while the long lipophilic hydrocarbon chains tend to leave water. If amount of the soap is small, the soap molecules will directionally concentrate on the water surface as in Figure 12-14. Surfactants tend to gather on the surface of a solution, on the interface between two immiscible liquids, and on the contacting interface between liquid and solid, thus the surface tension of liquid will be reduced.

Surfactants play the important role in life sciences. For instance, lipids forming cellular membranes like phospholipids and glycolipids and so on as well as bile salts secreted by gallbladder are surfactants.

Now, many kinds of surfactants are used in medicine. These surfactants can be classified into ionic surfactants and nonionic surfactants. The former can be further classified into anionic surfactants, cationic surfactants and zwitterionic surfactants according to the charges carried. Anionic surfactants have anionic hydrophilic groups, for example, the hydrocarbon chain in fatty acid salt usually contain 11~17 carbon atoms, with its general formula being RCOO-M. Here M represents base metal, alkali metal or NH_4^+. Sodium dodecyl sulfate (R-OSO_3Na) is another example for anionic surfactants. Cationic surfactants have cationic hydrophilic groups. Quaternary amine salts such as benzalkonium bromide, a common external antiseptic, are important cationic surfactants used in medicine. A zwitterionic surfactant molecule has both anionic and cationic hydrophilic groups. Amino acid surfactants (RNH_2CHCH_2COOH) and betaine surfactants [$RN^+H_2(CH_2)_2CH_2COO^-$] are the two major types of zwitterionic surfactants. Anionic surfactants have the better decontamination ability. That's why the main ingredients of detergents are anionic surfactants. On the other hand, cationic surfactants have the better bactericidal effect. Zwitterionic surfactants have both of these abilities. Nonionic surfactants are surfactants whose hydrophilic groups are linked by non-dissociable hydroxyl group-OH or ether bond-O-.

12.4.2　Association colloids

When a very small amount of surfactant is added to pure water, the surfactant molecules will be adsorbed orderly on water surface to form a thin layer of membrane. However, if the surfactant reaches a certain amount, the surfactant molecules would gather together by stacking their hydrophobic groups, forming assemblies in which the hydrophilic groups are towards the water while the hydrophobic groups are towards the interior, it is called **micelle** as shown in Figure 12-14. The formation of micelles decreases the contact interface between water and the hydrophobic groups, this making the whole system in more stable. Solutions formed by micelles are called **association colloids**.

The lowest concentration of a surfactant needed to form micelles is called **critical micelle**

concentration (CMC). CMC is affected by temperature, the structure and amount of surfactant, molecular association degree, pH and the existence of electrolytes.

In an association colloid in which surfactant concentration is close to the CMC, micelles are in spherical shape and display similar association degree. With the concentration increasing, the micelles varied to circular cylinder shape or even layered structure because the micelles are getting bigger and the association number is increasing. Figure 12-15 indicates these cases.

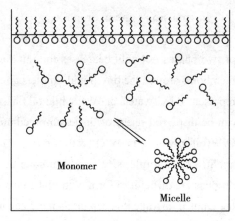

Figure 12-14 Diagram of Micelle Formation

Insoluble organics such as oil from animals or plants can be solubilized in water with micelles formed by surfactants. This effect is known as solubilization. For example, soaps and synthetic detergents can be used to wash the oil on clothes because of solubilization. Solubilization of surfactants is also applied widely in laboratory and pharmaceutical industry.

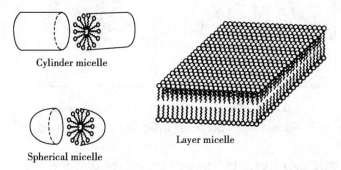

Figure 12-15 Diagram of Different Micelle Shapes

12.4.3 Emulsions

An **emulsion** is a coarse dispersion system in which a liquid dispersed phase is dispersed into another immiscible liquid dispersion medium. Generally, one phase in an emulsion is water and another phase is oil including organic solvent with weak polarity. The dispersed particles in an emulsion are relatively big so that they can be seen by an ordinary optical microscope.

An emulsion is an unstable system. For example, the immiscible mixture of oil and water is rock violently, they can disperse each other, but, after a period of standing, the two liquids would separate from each other and the emulsion disappears, because the surface energy would increase sharply to form a thermodynamically unstable system after dispersing the liquid droplets. As a result, the tiny liquid droplets will gather together spontaneously during their collision each other to decrease the energy of the system. Accordingly, the addition of **emulsifying agent**, the third substance to the system, is a prerequisite for getting a stable emulsion. The function of an emulsifying agent is called emulsification. Emulsifying agents are surfactants usually. For example, fats in food have to be emulsified into tiny droplets before they are absorbed in the intestine walls. Here bile salts act as the emulsifying agent.

In emulsions, the hydrophilic groups in surfactant molecules point to the water phase and the hydrophobic groups point to the oil phase. The orientation arrangement of surfactants on the interface of two phases not only decreases the interfacial tense, but also results in the formation of a protective membrane around the tiny liquid droplets and stabilizes the emulsion. Water phase in an emulsion can be represented by 'water' or 'W'; while oil phase can be represented by 'oil' or 'O'. Both 'Oil' and 'Water' can be dispersed phase or dispersion medium. Therefore, emulsions can be classified into two types of O/W(oil-in-water) or W/O (water-in-oil). The emulsion type is mainly determined by the emulsifying agent. For example, since sodium soap dramatically decreases the surface tension of water, water droplets are difficult to form with this emulsifying agent, forming O/W emulsions. On the other hand, if the emulsifying agent is calcium soap that can be dissolved in oil and dramatically decreases the surface tension of oil, W/O emulsions would form. Figure 12-16 illustrates the two different types of emulsions.

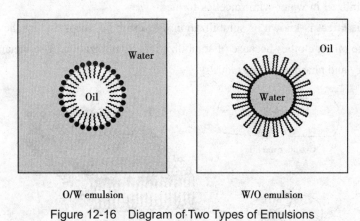

Figure 12-16 Diagram of Two Types of Emulsions

Emulsions are commonly used in both medical science and daily life. Milk, fish liver oil emulsion and clinical fat emulsion injection are emulsions. In order to increase the medicine dosage, medicines for injection are usually made as W/O emulsions that can be absorbed slowly by human body after their degradation. For example, Salk and others found that the antibody level of patients treated with emulsified influenza vaccine is about 10 times higher than that of patients treated with routine methods, and to be well more than two years. The type of an emulsion can be determined by two methods named staining method and dilution method. With the staining method, a small amount of dye that can only dissolve in oil is added to a sample of the emulsion. Then the stained emulsion is observed with a microscope. If the whole emulsion displays the color of the dye, it would be a W/O emulsion. On the contrary, if only the droplets of the dispersed phase display the color of the dye, it would be an O/W emulsion.

> **Question 12-2**   If a hydrophilic solid surface is treated by a suitable surfactant such as the water repellent, what will happen?

Dilution method is developed based on the principle that emulsions are easy to be diluted by their dispersion medium. By subsequently adding a drop of emulsion and a drop of water on a clean and dry glass slide, the emulsion would be the type of O/W if the emulsion distributes well in water; Otherwise, it would be the type of W/O.

Key Terms

吸附层	adsorption layer
气溶胶	aerosol
结合胶体	association colloid
Brown 运动	Brownian motion
聚沉	coagulation
胶体	colloid
临界聚沉浓度	critical coagulation concentration
临界胶束浓度	critical micelle concentration
扩散双电层	diffused electrical double layer
扩散	diffusion
扩散层	diffusion layer
分散度	dispersion degree
Donnan 平衡	Donnan equilibrium
电渗	electroosmosis
电泳	electrophoresis
乳化剂	emulsifying agent
乳状液	emulsion
柔韧性	flexibility
凝胶	gel
胶凝	gelation
界面	interface
等电点	isoelectric point
感胶离子序	lyotropic series
高分子溶液	macromolecular solution
胶束	micelle
沉降	sedimentation
沉降平衡	sedimentation equilibrium
溶胶	sol
比表面	specific surface area
表面能	surface energy
表面张力	surface tension
表面活性物质	surface-active substance
表面活性剂	surfactant
脱液收缩	syneresis
Tyndall 效应	Tyndall effect

Summary

Colloids include sols, macromolecular solutions and association colloids. Their mutual features are the diameter of dispersed particles in the range of 1~100nm, the slow diffusion rate, the permeation through the filter paper and the disability in passing semipermeable membranes.

The basic features of sols are heterogeneity, high dispersity and coagulation instability. It has optical, kinetic and electrical behaviors, such as Tyndall effect, Brownian motion or sedimentation equilibrium and electrophoresis or electro-osmosis, respectively. The colloidal particles in sols can be charged because of the preferential adsorption of resemble ions onto the surface of the colloidal nucleus or the dissociation of molecules on the surface. Charged colloidal particles adsorb opposite ions to form the electrical double layer. Colloidal particle refers to the nucleus and the adsorption layer, while micelle refers to the colloidal particle and the diffusion layer. Key factors that stabilize sols are the charged colloidal particles and the protection of hydration membranes. Colloids are sensitive to electrolytes. The amount of electrolyte needed for the coagulation is critical coagulation concentration of the electrolyte. Coagulation also occurs when oppositely charged sols are mixed.

Aerosol, a colloidal dispersion system with gas as dispersion media, is important in preventive medicine.

Macromolecular solutions are stable and homogeneous colloidal systems. Differs from sols, the size of macromolecules in 1~100nm and also somewhat differing from the ordinary solution. The properties of a macromolecular solution depend on the macromolecular. The carbochain of polymer is connected with chain unit, while both the degree of polymerization and the relatively molecular weight of polymer are average values. The hydration membranes around macromolecules make the solution stable. The relative molar mass is obtained by osmotic pressure measurement with extrapolation. Polyelectrolytic solution, such as a protein solution, is characterized by the isoelectric point, the Donnan equilibrium, and salting out effect.

A gel forms when a macromolecular solution is coagulated under certain circumstances. It may be elastic or non-elastic. The essential distinctions between two types of gels are their behavior on swelling, hydration and syneresis.

Surfactants tend to form positive adsorption at the surface of solutions to remarkably decrease surface tension. Their molecules simultaneously have at least a hydrophobic group and a hydrophilic group. Micelles are formed when the amount of surfactant accessing to water reached a degree. Stable association colloids are formed by micelles. Critical micelle concentration is the lowest concentration of surfactant when it begins to form micelles. Micelles have sphere, cylindrical and plate structures. Though emulsions belong to suspensions, surfactant makes them stable. Emulsions can be "oil-in-water" (O/W) type and "water-in-oil" (W/O) type.

Exercises

1. Mercury Hg vapor is poisonous. Which one is the most dangerous if liquid mercury is (1) in a beaker; (2) in a beaker with a layer of water on Hg surface; (3) distributed into mercury drops with a diameter of 2×10^{-4} cm? Why?

2. How much work should be done to distribute a drop of water with a radius of 1.00mm to small drops with a radius of 1.00×10^{-3} mm at 20℃ under a pressure of 100kPa? The value of σ for water is 72.8mN·m^{-1} at 20℃. (1N = 10^{-2} J·cm^{-1})

3. What is the surface energy and surface tension? What's the relationship between them?

4. Why sols can usually exist relatively stable while they are regarded as unstable system?

5. Why deltas form at the entrance of the Yangtze River and the Peral River to the sea?

6. Why sols show Tyndall effects? Please explain the essential reason.

7. Mix 12mL 0.02mol·L^{-1} of KCl solution with 100mL 0.05mol·L^{-1} of $AgNO_3$ solution to prepare AgCl sol. Please write down the structural formula of the micelles.

8. Suppose the same volume of 0.008mol·L^{-1} KI and 0.01mol·L^{-1} $AgNO_3$ solution are mixed to prepare AgI sol. Sequence the coagulation ability of the following electrolytic solutions of $MgSO_4$, $K_3[Fe(CN)_6]$ and $AlCl_3$ with the same concentration and the same volume. Suppose the same volume of 0.01mol·L^{-1} KI and 0.008mol·L^{-1} $AgNO_3$ solution are mixed to prepare AgI sol. Sequence the coagulation ability of above three electrolytic solutions with the same concentration and the same volume.

9. In order to prepare negatively charged AgI sol, how many liters of 0.005mol·L^{-1} $AgNO_3$ solution should be added to 25mL of 0.016mol·L^{-1} KI solution at most?

10. Both sol A and sol B carry unknown charges. If a small amount of $BaCl_2$ or more amount of NaCl have the same coagulation ability to sol A, while a small amount of Na_2SO_4 or more amount of NaCl have the same coagulation ability to sol B. Try to determine what kind of charge do sol A and sol B carry, respectively?

11. Why are the sols and macromolecular solutions stable? What methods can be used to damage their stability, respectively?

12. What are the identical and different characteristics for macromolecular solutions and small molecular solutions?

13. What is the relationship between protein electrophoresis and solution pH? What is the electrophoresis direction if pI of a protein is 6.5 and protein ions go in a solution of pH 8.6 under an applied electronic field?

14. What is gel? What are the main characteristics of gel? What is the prerequisite to form a gel?

15. What is surfactant? Explain the reason why surfactant can reduce the surface tension of the solvent according to its characteristics in molecule structure.

16. What is critical micelle concentration (CMC)? What are the different behaviors of surface-active agents if the concentration of it is more than or less then CMC, respectively?

17. How many types of emulsions are there? What are the meanings of each type?

Supplementary Exercises

1. Indicate the fundamental difference between a colloidal dispersion and a true solution.

2. Explain the difference between lyophilic and lyophobic sols, with reference to some of the properties in which differ.

3. Explain the Tyndall effect and how it may be used to distinguish between a colloidal dispersion and a true solution.

4. Why do particles in a specific colloid remain dispersed?

5. Suppose water droplets with a radius of 1mm are dispersed into droplets with a radius of 1nm at 298.15K. How many times is the specific surface area increased?

6. According to the basic properties, stabilization and coagulation of sols, list main methods that can be used for the detection and removal of PM2.5.

Answers to Some Exercises

[Exercises]

1. (3)

2. 9.13×10^{-4} J

7. $[(AgCl)_m \cdot nAg^+ \cdot (n-x)NO_3^-]^{x+} \cdot x(NO_3^-)$

8. $K_3[Fe(CN)_6] > MgSO_4 > AlCl_3$, $AlCl_3 > MgSO_4 > K_3[Fe(CN)_6]$

9. 80 mL

10. Sol A is negatively charged, while sol B is positively charged

13. Omitted; the protein ions would move towards the anode

[Supplementary Exercises]

5. 10^6 times

（林　毅）

Chapter 13
Titrimetric Analysis

The primary tasks of analytical chemistry are to detect the chemical composition of substances, to identify the chemical structure of compounds, and to determine their contents. The analytical methods can be classified into qualitative analysis, quantitative analysis and morphological analysis. The quantitative analysis can be further sorted into chemical analysis and instrumental analysis according to the adopted methods. Wherein, chemical quantitative analysis are the methods based on the chemical reactions of the compound, and instrumental analysis are the methods based on the specific instruments by the physical or physicochemical properties of the substances.

Titrimetric analysis, also known as volumetric analysis, is one of the most common chemical analysis methods among quantitative analysis in analytical chemistry. Titration consists of acid-base titration, redox titration, coordination titration and precipitation titration. It is mainly used to determine the content of a sample whose content is greater than 1%, and widely used in many fields due to its advantages such as rapidity, simplicity and high accuracy (relative error $<0.2\%$). Combined with all kinds of modern titration analyzers, titration can be developed to 400 different types which refer to a variety of applications in product quality control, detection and administration, product monitoring in petroleum chemical industry, metallurgy, coal power, food, tobacco, pharmaceutical industries, and so on.

13.1 Error and Deviation

13.1.1 Causes and Classification of Error

The measured values of physical quantities are impossible to be absolutely consistent with the true values, and can only approach to them with the development of human ability to understand the objective world and the improvement in instrumental precision. Since a sample analysis usually involves multiple steps and different physical quantities, together with various limited factors such as cost, time and environment, the analysis results always refer to tiny unreliability and uncertainty that is called **error**. There are plenty of causes that lead to errors in quantitative analysis. Thus, errors can be sorted into **systematic error** and **random error** according to their characteristics and sources.

1. Systematic error

Systematic errors arise from invariable sources and their values will be relatively constant if repeating the determination under identical conditions, therefore it is called **determinate error**. According to the sources, systematic errors can be categorized as follows:

(1) Method error. It is originated from defects of analytical methods, for example, an uncompleted reaction, an interfering substance, considerable gap between the end point and the stoichiometric point for the inappropriate indicators, etc.

(2) Instrumental error and reagent error. The instrumental errors arise from the inherent limitations of equipments, for example, the poor sensitivity of analytical balance, the low accuracy of volumetric glass wares and so on. The reagent errors are derived from minor impurities or interfering substances in distilled water or reagents.

(3) Operational error. It is the error resulting from the subjective factors or the habits of individual operators, for example, the ways to read data and the vision sensibility of **operators** to distinguish colors.

2. Random error

Random error caused by some changing factors is hard to control, thus also called undeterminable error. For instance, the slight fluctuations in temperature, humidity, pressure, voltage or instrumental performance, all can affect the obtained data or results.

13.1.2 Evaluation of Analytical Result

1. Error and accuracy

Accuracy is defined as proximity between a measured value x and the true value T. **Error** is the difference between x and the T. It is a measure of accuracy. There are two ways to express errors

Absolute Error $\qquad\qquad\qquad E = x - T \qquad\qquad\qquad$ (13.1)

Relative Error $\qquad\qquad\qquad E_r = \dfrac{E}{T} \times 100\% \qquad\qquad\qquad$ (13.2)

Where E is the absolute error, and Er is the relative error. The lower the error is, the higher the accuracy is.

Relative error reflects the percentage of the error in the true value, thus it is a more rational way to express accuracy. The error may be positive or negative.

Sample Problem 13-1 Analytical balance was used to weigh two Na_2CO_3 samples, and their readings were 1.638 0g and 0.163 8g, respectively. If their true weights were 1.638 1g and 0.163 9g, calculate their absolute errors and relative errors, respectively.

Solution The absolute errors are

$$E_1 = 1.638\ 0g - 1.638\ 1g = -0.000\ 1g$$
$$E_2 = 0.163\ 8g - 0.163\ 9g = -0.000\ 1g$$

The relative errors are

$$E_{r1} = \dfrac{-0.000\ 1g}{1.638\ 1g} \times 100\% = -0.006\%$$

$$E_{r2} = \dfrac{-0.000\ 1g}{0.163\ 9g} \times 100\% = -0.06\%$$

The above results showed that the relative errors are different although their absolute errors are equal. The greater the mass of analyte is, the lower the relative error is, and the higher the accuracy is under the same weighing condition.

2. Deviation and precision

The accuracy can be evaluated only if the true value is known. Therefore, **precision** is usually used to estimate the reliability of the measured values. The precision, defined as proximity among the measured values of some parallel detections, can be evaluated by **deviation** that refers to reproducibility among the measured values. The differences between any measuring value x_i and their arithmetical mean $\bar{x}$ is called absolute deviation d.

$$d = x_i - \bar{x} \tag{13.3}$$

Obviously, the closer to zero the deviation is, the better the precision of the measurements is.

Practically, mean absolute deviation $\bar{d}$, mean relative deviation d_r, and standard deviation s are commonly used to express the precision of analysis results. $\bar{d}$, d_r and s are defined as follows

$$\bar{d} = \frac{|d_1| + |d_2| + |d_3| + \cdots + |d_n|}{n} \tag{13.4}$$

In Equation 13.4, $|d|$ is the absolute value of each absolute deviation, n is the sample number

$$d_r = \frac{\bar{d}}{\bar{x}} \times 100\% \tag{13.5}$$

The most common measure of data dispersion around the mean is the standard deviation

$$s = \sqrt{\frac{d_1^2 + d_2^2 + d_3^2 + \cdots + d_n^2}{n-1}} \tag{13.6}$$

In titrimetric analysis, the relative deviation of the analytical results commonly shall be less than 0.2%.

Accuracy refers to veracity while precision refers to reproducibility. Note that even if the measurements are very precise, their accuracy might be poor. For example, an incorrectly calibrated instrument cannot give accurate readings although its precision is very likely not affected. On the other hand, an accurate value is rarely obtained if precision is poor. In Figure 13-1, the bull's-eye was compared with the true value and accuracy refers to how close to the bull's-eye the bullet strikes were when five bullets were shot. When the measured values are discrete and far from the true value, both accuracy and precision are poor, as shown in Figure 13-1A. When the measured values are concentrated but far from the true value, the precision is high but the accuracy is poor, as shown in Figure 13-1B. If the measuring values are concentrated on the "bull's-eye", the true value, both accuracy and precision are good, as shown in Figure 13-1C. This shows that high precision does not necessarily mean the good accuracy but the good accuracy of the analysis results must be based on the high precision.

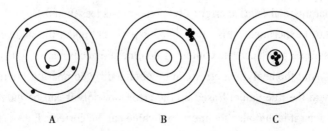

Figure 13-1 Accuracy versus Precision

What should be noted is that the error calculated using the relative true value is still the deviation in a sense, because the true value of the content of substance is actually unknown. Therefore, people do not discriminate them in practice.

13.1.3 Methods to Improve the Accuracy of Measurement

The accuracy would be undoubtedly affected by the factors that produce errors while the precision only depends on the random errors. As a result, it will be due to the presence of systematic errors if having high precision but poor accuracy, and they can be eliminated according to the sources.

1. Reducing Systematic error

Systematic error is the main source of inaccuracy of measurement and the following methods are usually used to improve the accuracy of the results.

(1) Instrument calibration. The systematic errors caused by inaccurate instruments can be eliminated or reduced by calibrating the the used instrument. Instruments like analytical balance, burettes and pipettes should be calibrated, and then using the corrected value to calculate.

(2) Choice of analytical methods. It is crucial to select proper analytical methods because of the applicability of different quantitative analytical methods. For example, titrimetric method has high accuracy and low sensitivity, so it can be applied into macro-analysis (mass fraction $\omega > 1\%$) but not in micro-analysis (mass fraction $\omega < 1\%$). The instrumental analysis methods having high sensitivity are required to perform the micro-analysis despite of low accuracy. Moreover, the method errors derived from the limitation of analysis method itself are usually important sources of systematic errors, thus the related causes should be found and eliminated. For example, choosing an appropriate indicator can reduce end point error while an appropriate masking reagent can remove the influence of interfering ions in titration analysis.

In the process of analysis, the total errors derived from each step of the measurement should be controlled within available limits. For example, the limit of an analytical balance is 0.000 1g, the absolute error is ± 0.000 1g. If a sample is weighed by indirect method, it will need weighing twice and produce an absolute error of ± 0.000 1 × 2g. In order to keep the relative error less than 0.1%, the mass of sample should be more than 0.2g ($d / d_r = 0.000\,2g / 0.1\%$). Therefore, the mass of sample should be large enough to decrease the relative error. The weighing of the reagent with the greater molar mass will have a smaller relative error than the one with the smaller molar mass if both had the same amounts. Another example is the titration in which the absolute error of each reading is ± 0.01mL. Two readings (initial volume and final volume) are required to determine the total consumed volume, and this cause a total absolute error of ± 0.02mL. Therefore, the consumed volume in each titration should be more than 20.00mL to ensure that the total relative error of two readings is less than 0.1%.

(3) Standard control experiment. It is a paralleled experiment under the same experiment conditions in which a standard sample with accurate content is determined accompanying as the test of the sample in the same methods. And the analytical error can be obtained by comparing the measured value of the control experiment with the true value. Based on the error, not only it can be judged whether there is a systematic error and what it is, but also the measured value can be corrected, ensuring that the measured value approaches the true value.

(4) Blank control experiment. It is the detection carried out without any testing samples by the same methods, procedures and under analysis conditions. The result of this determination is called blank value. The accurate analysis result can be obtained by eliminating the blank value from the measured value. The blank experiment can be used to remove or reduce the error introduced by the vessels and the impurities in the reagents and even the distilled water. The blank value should not be too great, otherwise the reagents have to be purified or the vessels have to be changed.

2. Reducing random error

Random errors are arbitrary. The probabilities to make positive errors and negative errors are equivalent. So they can be removed by increasing the numbers of repetitive measurements but not by calibrating instruments. A sample should be measured for 3~5 times to calculate the mean value and get the more accurate analysis result based on the higher precision.

During the determination, what should be paid attention to is the errors caused by neglectful or incorrect operations such as wrong reagents, inaccurate reading and miscalculation, etc. In this case, the results must be abandoned.

13.2 Significant Figures

13.2.1 Concepts of Significant Figures

In expressing the experimental results, the data not only reflect the measured values but also the accuracy of measurement. **Significant figures**, also known as **significant digits**, are digits that carry both numerical and precision meaning, including all the certain digits and an uncertain digit. The uncertain digit is the trailing digit whose error is ±1.

The uncertain digit of significant figures is usually estimated based on the scale of the used instruments reflecting their precision. For example, the reading of a burette is 24.02mL where "24.0" is certain and accurate but the trailing digit "2" is estimated and uncertain. This indicates that the accuracy of the burette is 0.01mL and the potential absolute error is ±0.01mL. The actual volume of the solution should be 24.02 ± 0.01mL. Vice versa, proper instrument can be chosen according to the accuracy of the digits. For instance, the transfer pipettes or burettes should be used to transfer 20.00mL of solution, while the measuring pipettes and measuring cylinders should be used to transfer 2.00 and 2.0mL of solutions, respectively. And 2mL is an approximate volume that can be measured by general measuring vessels.

Among the numbers of 0 to 9, "0" is not a significant figure when it is used as the leading digits for a number. Taking 20.50mL volume of a solution as example, the unit is "milliliter" and it is 0.020 50L if the unit is "liter", and the "0" in front of the "2" is only a leading digit but not a significant figure. Therefore, the leading zero is never significant digit, and the zeros appearing anywhere between two non-zero digits and the trailing zeros in a number containing a decimal point are significant. However, the significance of trailing zeros in a number without a decimal point can be ambiguous.

The number 4 200mL is an integral number with trailing zeros, so it is impossible to determine its significant digits. In this case, scientific notation should be used to express the significant digits. In the scientific notation all numbers are written in the form of ($m \times 10^n$), where the exponent n is an integer, and the number m is any real number between "1" to "9". Thus, 4 200mL can be expressed as 4.2×10^3 (two significant digits), 4.20×10^3 (three significant digits) and 4.200×10^3 (four significant digits) by different instruments with different accuracy.

The numbers used to represent multiples and submultiples do not have the restriction in number of significant figures. The defined constant and the international agreement values, etc, having no error and called accurate values, do not have restriction in number of significant figures, either. Taking Na_2CO_3 as example, the molar mass $MM.(\frac{1}{2}Na_2CO_3) = (106.0 / 2) g \cdot mol^{-1}$, here, 2 is natural number that representing submultiple when the elementary unit of the substance is $\frac{1}{2}Na_2CO_3$. Another examples are the definition of pound, 1b = 0.453 592 37kg, and the velocity of light in the air $c = 299\ 792\ 458 m \cdot s^{-1}$. Besides, the constants such as the relative atomic mass and equilibrium constants, etc have no restriction in the number of significant figures either in general calculation.

The significant figures in logarithm operation such as pH, pK and lg$c$, etc, depend only on their decimal digits. Their integer parts are only the exponent of "10" in their antilogarithm but not real significant digits. For instance, pH 10.20 has two significant digits, and it is identical with the expression $[H_3O^+] = 6.3 \times 10^{-11} mol \cdot L^{-1}$.

13.2.2 Arithmetic Rules for Significant Figures

1. Rounding

In practical measurement, a variety of measuring instruments are used and multiple determinations are carried out commonly. Thus, keeping the significant digits based on the propagation of error law are necessary. This process is termed as **rounding**. The rounding rules are as follows:

(1) If abandoned digit is less than 5, drop it and all digits after it.

(2) If abandoned digit is larger than 5, or it is 5 followed by other non-zero digits, add 1 to the last significant digit. For example, rounding 13.025 1 to four significant digits, it is 13.03.

(3) If abandoned digit is just 5 without any other followed numbers or followed only by zeros, drop it; if the last significant digit is odd, add 1 to be even after rounding. For example, four significant digits after rounding off 13.015 and 13.025 are both 13.02.

The original data can be only rounded for one time. For example, 2.749 5 can only be rounded to a two significant digits 2.7 directly, while it is not permitted to round to 2.75, and then followed take the second rounding to 2.8. Rounding should be conducted after computation if the results need computation.

2. Addition and subtraction

The significant figures of the results after addition and subtraction depend on the value with the lowest precision, namely the least decimal digits, among the ones involved in the arithmetic.

3. Multiplication and division

The significant figures of the results after multiplication and division depend on the value with the greatest error, namely the least significant digits, among the ones involved in the arithmetic.

Sample Problem 13-2 Determining the relative atomic mass of Mg element using displacement method, a set of obtained data are as follows: $m_{Mg}=0.035\,2$g, $t=26.0℃$, $V_{H2}=37.40$mL, $p=100.15$kPa, $p_{H_2O}=3.362\,9$kPa. Calculate the atomic mass of Mg based on the above data.

Solution
$$A_{rMg} = \frac{m_{Mg}RT}{(p-p_{H_2O})V_{H_2}}$$

$$A_{rMg} = \frac{0.035\,2\text{g} \times 8.314\text{kPa}\cdot\text{L}\cdot\text{K}^{-1}\cdot\text{mol}^{-1} \times (273.15+26.0)\text{K}}{(100.15\text{kPa}-3.362\,9\text{kPa}) \times 37.40 \times 10^{-3}\text{L}} = 24.2\text{g}\cdot\text{mol}^{-1}$$

Generally, using the calculator to get the result, we only round the final arithmetic results but not the result of each arithmetic step. And the significant figures can be preset in some scientific calculators.

> **Question and thinking 13-1** Why should the rounding of significant figures obey the rounding rule?

13.3 Principle of Titrimetric Analysis

13.3.1 Introduction to Titrimetric Analysis

1. Concepts and features of titrimetric analysis

Titrimetric analysis is the quantitative measurement of concentration of a given **analyte** (or sample) in solution by completely reacting it with a stoichiometric equivalent amount of a **standard solution**. The standard solution is called **titrant** whose accurate concentration has been known. The stoichiometric mixture is named **equivalence point** or **stoichiometric point.** The content of analyte can be calculated with the consumed volume and the concentration of titrant. The process, in which the standard solution is titrated by burette to the analyte solution until the reaction is just completed, is termed as **titration.**

Unfortunately, there is usually no obvious indication to signal the equivalence point for most chemical reactions. The **end point** is often indicated by an appropriate **indicator** which changes its color or generates precipitate near the equivalence point. The end point of the titration is a point at which the titration is completed, and it is signaled by the indicator. As the end point and the equivalence point do not just coincide equally, the difference between them will generate the **titration error**, the main error sources of titrimetric analysis. The accuracy of titrimetric analysis depends on the degree of the titration reaction and the choice of indicator.

The characteristics of the titrimetric analysis are rapid, simple reaction and high accuracy.

2. The basic requirements for chemical reactions in titrimetric analysis

A reaction in titrimetric analysis must satisfy the following conditions:

(1) Titration reaction should react completely in stoichiometric or equivalent proportions according to the chemical equation, generally more than 99.9%.

(2) It must be quick, or it can be accelerated by heating or catalyst.

(3) There is no side reaction in titration. Impurities in the analyte do not react with the standard solution, or they can be masked by the proper methods.

(4) The method to determine the end point of titration must be reliable and convenient.

3. Steps of titrimetric analysis

Titrimetric analysis generally includes three steps: (1) Preparation of standard solution, (2) Standardization of the standard solution, and (3) Analyte determination.

13.3.2 Classification of Titrimetric Analysis

1. Classification according to titrimetric methods

(1) Direct titration. All requirements of titration reaction are satisfied and the content of analyte can be directly titrated with the standard solution. For example, the NaOH solution can be directly titrated by the HCl standard solution.

(2) Back titration. Some reaction cannot meet all requirements of titration reaction. For instance, when the reaction is slow, the analyte is solid or there is not any suitable indicator to show the end point of the titration, then excess standard solution will be added into the analyte solution to ensure the reaction to be completed. Then, the remaining standard solution is titrated with another standard solution, thus determining the content of analyte. That is **back titration**. For example, solid $CaCO_3$ cannot be titrated directly by HCl standard solution because of the small solubility of it. But if $CaCO_3$ is completely dissolved by adding an excessive HCl standard solution, then the remaining HCl standard solution is titrated by the NaOH standard solution, and a better result can be obtained easily.

(3) Indirect titration. In some cases, analytes can not react with titrants directly. Then, the content of the analyte can be only determined by choosing the other chemical reactions. This method is indirect titration. For example, Ca^{2+} react with $C_2O_4^{2-}$ to CaC_2O_4 precipitate quantitatively which can be isolated by filtering. CaC_2O_4 is dissolved in H_2SO_4 solution to produce $H_2C_2O_4$ stoichiometrically. Then, the concentration of $H_2C_2O_4$ can be titrated by $KMnO_4$ standard solution, and the content of Ca^{2+} in the solution can be known by indirect titration.

(4) Replacement titration. The replacement titration is performed by titrating the product derived from the reaction between analyte and an appropriate reagent in advance.

2. Classification according to chemical reaction types

The titrimetric analysis can be divided into four types, acid-base titration, redox titration, complexometric titration and precipitation titration.

13.3.3 Preparation of Standard Solution

There are two methods to prepare the standard solution, in a direct manner or an indirect manner.

1. Direct manner

If a reagent is available in pure and stable state, the standard solution can be prepared by accurately weighing and dissolving it with an appropriate solvent in a volumetric flask with a defined volume. The substance which can be directly used to prepare a standard solution is called **primary standard substance**. A primary standard substance must comply with the following requirements:

(1) The composition of the primary standard substance must agree with its formula. For hydrated substances, such as $H_2C_2O_4 \cdot 2H_2O$, the number of water molecules in it should be the same with the formula.

(2) The substance must be highly pure (main composition is more than 99.9%). Impurities in the substance do not affect the accuracy of the titration.

(3) The substance must be stable, hard to absorb water and carbon dioxide, as well as not easy to be oxidized in the air.

(4) The reaction between the analyte and the primary standard substance must be in the stoichiometric coefficients but no side reactions.

(5) Its molar mass should be heavier enough to reduce the weighing errors.

2. Indirect manner

If the reagent cannot act as the standard primary substance, the standard solution of it with approximate concentration can be prepared firstly, and then the accurate concentration will be standardized by the primary standard substance or another standard solution. This process is **standardization**.

13.4 Acid-Base Titration

Acid-base titration is based on the proton transfer reaction for the quantitative titration analysis. It is widely used to determine the content of samples which can directly or indirectly react with acidic or basic substances. In most acid-base titrations, it usually doesn't have obvious indicator to signal the approaching of the equivalence point. A suitable indicator which changes color of mixture solution at a specific pH range can be used to determine the end point. Therefore, we need to know the pH change condition in titration and the pH range corresponding to the color change of indicator, to choose a appropriate indicator to indicate the end point.

13.4.1 Acid-Base Indicator

1. Principle of color change of an acid-base indicator

In general terms, acid-base indicator is commonly organic weak acid such as phenolphthalein and litmus, or organic weak base such as methyl orange and methyl red which changes its color as the chemical structure of indicator changes. For example, methyl orange in solution has the equilibrium as follows

$$(CH_3)_2\overset{+}{N}=\!\!\!\!\bigcirc\!\!\!\!=N-\underset{H}{N}-\!\!\bigcirc\!\!-SO_3^- \rightleftharpoons (CH_3)_2N-\!\!\bigcirc\!\!-N=N-\!\!\bigcirc\!\!-SO_3^- + H^+$$

red(quinoid, acid type color)   yellow(azo, alkali type color)

If the concentration of H_3O^+ increases, the equilibrium will shift to the left side. Methyl orange is red in the solution which presents acid type color by changing the weak base into its conjugate acid. On the other hand, if the concentration of H_3O^+ decreases, the equilibrium will shift to the right side, methyl orange is yellow in the solution which presents alkali type color.

Now, take an indicator of weak acid HIn and its conjugate base In$^-$ as an example to interpret the principle of color change of an acid-base indicator. HIn has the dissociation equilibrium in aqueous solution as follows

$$HIn(aq) + H_2O(l) \rightleftharpoons H_3O^+(aq) + In^-(aq)$$

$$K_{HIn} = \frac{[H_3O^+][In^-]}{[HIn]} \tag{13.7}$$

Where K_{HIn} is the acid dissociation constant of HIn; [HIn] and [In$^-$] are the equilibrium concentrations of HIn and In$^-$, respectively.

or

$$[H_3O^+] = K_{HIn} \times \frac{[HIn]}{[In^-]}$$

Take the negative logarithm to both sides of above equation, we get

$$pH = pK_{HIn} + \lg\frac{[In^-]}{[HIn]} \tag{13.8}$$

In Equation 13.8, pK_{HIn} is a constant at a certain temperature. The color of HIn in solution will depend on the mixing color of HIn and In$^-$. If the pH of solution changes, the ratio of $\frac{[In^-]}{[HIn]}$ and the solution color will changes. This is the principle of color change of acid-base indicator by changing the pH of solution in acid-base titration.

2. Range of color change of acid-base indicator

Due to the sensibility of human eyes, when $\frac{[In^-]}{[HIn]} \geq 10$, the color of HIn is covered by In$^-$, our eyes only can see the color of In$^-$; in contrast, when $\frac{[In^-]}{[HIn]} \leq 0.1$, we only can see the color of HIn; when $0.1 < \frac{[In^-]}{[HIn]} < 10$, what we can see is the mixing color of HIn and In$^-$.

If the pH value of solution changes from $(pK_{HIn} - 1)$ to $(pK_{HIn} + 1)$, the indicator will turn its color from acid type to alkali type. For example, methyl orange turns its color from ted to yellow. On the contrary, if the pH value of solution changes from $(pK_{HIn} + 1)$ to $(pK_{HIn} - 1)$, the indicator will turn its color from yellow to red. Therefore, $(pK_{HIn} \pm 1)$ is called the color change range of acid-base indicator. The indicators with different pK_{HIn} have different color range. When $pH = pK_{HIn}$, or $[In^-] = [HIn]$, the indicator shows just mixing colors of HIn and In$^-$ which is called **color change point**. For example, the pK_{HIn} of methyl orange is 3.7 and it will change color over the pH range of 2.7~4.7.

In fact, the vision sensibility of human eyes to different colors is very dissimilar, and the actual indicator color range is not calculated by pK_{HIn}, but by vision sensibility. Most of the actual color change ranges of indicators are less than two pH unit, for example, methyl orange is pH 3.2~4.4. A list of several common acid-base indicators, along with their pK_{HIn}, color changes, and corresponding pH ranges, is provided in Table 13-1.

Table 13-1 Acid-base Indicator in Common Use

Indicator	pK_{HIn}	pH Range	Acid color	Transition color	Base color
Thymol blue (first transition)	1.7	1.2~2.8	red	orange	yellow
Methyl orange	3.7	3.2~4.4	red	orange	yellow
Bromophenol blue	4.1	3.1~4.6	yellow	blue-purple	purple

					Continue
Indicator	pK_{HIn}	pH Range	Acid color	Transition color	Base color
Bromocresol green	4.9	3.8~5.4	yellow	green	blue-green
Methyl red	5.0	4.8~6.0	red	orange	yellow
Bromothymol blue	7.3	6.0~7.6	yellow	green	blue
Neutral red	7.4	6.8~8.0	red	orange	yellow
Phenolphthalein	9.1	8.2~10.0	colorless	pink	fuchsia
Thymol blue (second transition)	8.9	8.0~9.6	yellow	green	blue
Thymolphthalein	10.0	9.4~10.6	colorless	light blue	blue

13.4.2 Titration Curves and Choices of Indicator

In acid-base titration, the proper indicator must be used to reduce the titration error. The end point should approaches to the equivalence point as far as possible. Consequently, it is important to understand the pH change in titration process. The pH change caused by adding a small amount of acid or base standard solutions near the stoichiometric point is important especially.

The graphs drawn using the volume of the added acid or base standard solution as abscissa against the pH of the mixture as ordinate are termed as **acid-base titration curves**. The curve helps us to choose the proper indicator by the pH change in the titration.

1. Titration between strong acids and strong bases

(1) Titration curve. Take titrating 20.00mL of 0.100 0mol·L^{-1} HCl with 0.200 0mol·L^{-1} NaOH as example to describe the pH change in the process of titration.

$$H_3O^+ (aq) + OH^- (aq) \rightarrow H_2O(l)$$

a) Before titration, the concentration $c(H_3O^+)$ is equal to the initial concentration of the HCl solution, $c(H_3O^+) = 0.100\ 0\text{mol} \cdot \text{L}^{-1}$.

$$pH = -\log c(H_3O^+) = -\log c(HCl) = -\log 0.100\ 0 = 1.00$$

b) Before the equivalence point, the pH of the mixture depends on the concentration of remaining HCl in solution. After adding 19.98mL of NaOH standard solution, $c(H_3O^+)$ is

$$c(H_3O^+) = \frac{0.100\ 0\text{mol} \cdot \text{L}^{-1} \times 0.02\text{mL}}{20.00\text{mL} + 19.98\text{mL}} = 5 \times 10^{-5} \text{mol} \cdot \text{L}^{-1}$$

$$pH = 4.30$$

c) At the equivalence point, HCl is completely neutralized by NaOH, and the solution is neutral.

$$K_w = 1.00 \times 10^{-14} = [H_3O^+][OH^-] = [H_3O^+]^2$$

$$[H_3O^+] = 1.000 \times 10^{-7} \text{mol} \cdot \text{L}^{-1}$$

Thus, the pH at the equivalence point is 7.00.

d) After the equivalence point, the pH is determined by the concentration of excess OH$^-$ in solution. For example, after adding 20.02mL (titration error is -0.1%) of NaOH solution, the concentration of OH$^-$ is

$$c(OH^-) = \frac{0.100\ 0\text{mol} \cdot \text{L}^{-1} \times (20.02 - 20.00)\text{mL}}{20.00\text{mL} + 20.02\text{mL}} = 5 \times 10^{-5} \text{mol} \cdot \text{L}^{-1}$$

$c(H_3O^+)$ is

$$c(H_3O^+) = \frac{K_w}{c(OH^-)} = \frac{1\times10^{-14}}{5\times10^{-5}} = 2\times10^{-10}\,mol\cdot L^{-1}$$

$$pH = 9.70$$

In the above way, the pH values in different stages of titration are calculated and listed in Table 13-2. The titration curves referring to titrate strong acid with strong base is shown in Figure 13-2. The abscissa is the volume of added NaOH solution, and the pH of the mixture is the ordinate. Figure 13-2 is just based on Table 13-2 in the different stages of titration.

a) From the beginning of the titration to that 19.98mL of NaOH solution was added, the pH changed from 1.00 to 4.30, thus showing a smooth curve.

b) Around the stoichiometric point pH 7.00, the pH bursts from 4.30 to 9.70 with 5.40 units increment when the situation changes from 0.02mL remaining HCl solution to 0.02mL excess NaOH. The corresponding titration errors is from −0.1% to +0.1%. The bursting change of pH is termed as **titration jump** or **inflection point**. The pH range of the inflection point is called the range of titration jump. And the vertical part of the curve is titration jump pH 4.30∼9.70 whose midpoint has pH 7.00.

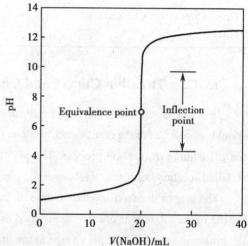

Figure 13-2 Titration Curve of 20.00mL of 0.100 0mol·L^{-1} HCl with 0.100 0mol·L^{-1} NaOH

Table 13-2 Data of Titration for 20.00mL of 0.100 0mol·L^{-1} HCl by 0.100 0mol·L^{-1} NaOH

Volume (mL) of Titrant, NaOH / mL	Volume (mL) of rest HCl / mL	Volume (mL) of excess NaOH / mL	pH
0.00	20.00		1.00
18.00	2.00		2.28
19.80	0.20		3.30
19.98	0.02		4.30
20.00	0.00		7.00
20.02		0.02	9.70
20.20		0.20	10.70
22.00		2.00	11.70
40.00		20.00	12.50

c) After the titration jump, the pH change tends to be slow if NaOH solution is added continuously, thus the last part of the curve turns to be flat.

If titrating 20.00mL of 0.100 0mol·L^{-1} NaOH solution with 0.100 0mol·L^{-1} HCl standard solution, an inverted titration curve with identical shape can be obtained similarly.

(2) **Choices of indicator.** Titration error derived from the indicator is less than ±0.1%, and it meets the requirement of titration analysis as long as the end point signaled by indicator is in the pH range of titration jump, even though the end point is not identical with the stoichiometric point. So the principle

for choosing indicator can be concluded into that the pH range of the color change partially or entirely lies in the range of titration jump, the indicator can be used to signal the specific titration end point. According to this principle, the pH range of the titration jump in titrating strong acid with strong base is 4.30~9.70, therefore, methyl orange (pH=3.2~4.4), phenolphthalein (pH=8.2~10.0) and methyl red (pH=4.8~6.0) can be used as the indicators, as shown in Figure 13-3.

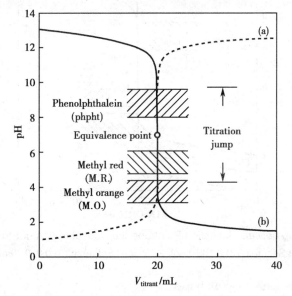

Figure 13-3 Strong Acid-Strong Base Titration Curve for
(a) 20.00mL 0.100 0mol·L^{-1} HCl Solution with 0.100 0mol·L^{-1} NaOH Solution;
(b) 20.00mL 0.100 0mol·L^{-1} NaOH Solution with 0.100 0mol·L^{-1} HCl Solution

However, the sensibility of human vision to color change should be considered when choosing the indicator in actual titration. The color changes are easy to identify when the color of phenolphthalein turns from colorless to pink and the one of methyl orange turns from yellow to orange. Peoples are sensible to the color changing from pale to dark. Therefore, phenolphthalein is usually used as indicator when titrating strong acid with strong base, and methyl orange is usually chosen when titrating strong base with strong acid.

(3) **The relationship between inflection point and the concentrations of acid and base**. The narrow or wide pH range of titration jump is associated with the concentrations of titrant and sample. For instance, take 1.000, 0.100 0 and 0.010 00mol·L^{-1} NaOH solutions to titrate identical concentration of HCl solutions respectively, the obtained titration curves are shown in Figure 13-4. Their titration jump are pH 3.30~10.70, pH 4.30~9.70 and pH 5.30~8.70, respectively. It can be seen that the titration jump will decrease by 2 pH units when the concentrations of acid or base decrease by 2 times. It indicates that the smaller the concentration of NaOH is, the narrower the pH range of titration jump is, so the less the choices of the indicators. When the titrant is very dilute, e.g. $c < 10^{-4}$ mol·L^{-1}, the inflection point is not distinct or even miss, so it is impossible to use an indicator to show the end point. In contrast, the greater the concentration is, the wider the pH range of titration jump is, and the more the choices of indicators is. However, the larger amount the titrant in one drop of concentrated solution is,

and the greater the error is. Therefore, the concentration of acid or base titrant should be within 0.1 to $0.5 \text{mol} \cdot \text{L}^{-1}$ generally.

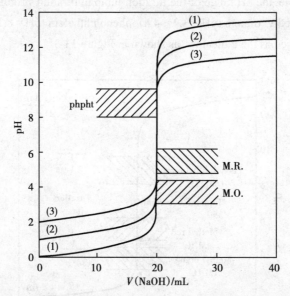

Figure 13-4   Relationship between Inflection Point and the Concentration of Strong Acid or Strong Base
(1)$1.000 \text{mol} \cdot \text{L}^{-1}$ NaOH Solution; (2) $0.100 0 \text{mol} \cdot \text{L}^{-1}$ NaOH Solution; (3) $0.010 00 \text{mol} \cdot \text{L}^{-1}$ NaOH Solution

2. Titration of weak acids or weak bases

(1) **Titration curves.** Take the titration of 20.00mL of $0.100 0 \text{mol} \cdot \text{L}^{-1}$ HAc solution with $0.100 0 \text{mol} \cdot \text{L}^{-1}$ NaOH solution as example. The pH changes during the titration were computed and listed in Table 13-3. The titration curve was shown in Figure 13-5.

Table 13-3   Data for Titration of 20.00mL of $0.100 0 \text{mol} \cdot \text{L}^{-1}$ HAc with $0.100 0 \text{mol} \cdot \text{L}^{-1}$ NaOH

NaOH added/mL	Composition of solution	Calculation for [H$_3$O$^+$]	pH
0.00	HAc	$[H_3O^+] = \sqrt{K_a c}$	2.88
10.00	HAc + Ac$^-$		4.75
18.00	HAc + Ac$^-$	$[H_3O^+] = K_a \times \dfrac{[HAc]}{[Ac^-]}$	5.70
19.80	HAc + Ac$^-$		6.74
19.98	HAc + Ac$^-$		7.80
20.00	Ac$^-$	$[OH^-] = \sqrt{K_b c_b} = \sqrt{\dfrac{K_w}{K_a} c_{\text{salt}}}$	8.73
20.02	OH$^-$ + Ac$^-$		9.70
20.20	OH$^-$ + Ac$^-$	$[OH^-] = \dfrac{c(\text{NaOH})V(\text{NaOH})}{V_{\text{total}}}$	10.70
22.00	OH$^-$ + Ac$^-$		11.68
40.00	OH$^-$ + Ac$^-$		12.50

(Inflection point: 7.80 to 9.70)

(2) **Choices of indicators**. Comparing Figure 13-5 with Figure 13-2, the titration characteristics of weak monoprotic acid with strong base can be concluded as follows:

a) The initial pH 2.88 is higher (Figure 13-5) compared with pH 1.00 (Figure 13-2), because HAc is a weak acid.

b) The pH change at the beginning part of titration is not flat, and then turnning flat on as addition of NaOH solution. This is caused by a small amount of NaAc formed at the beginning of the titration, which has common ion effect to the dissociation of the HAc. Adding NaOH decreases the concentration of H_3O^+ fastly, producing a sharp change in pH. However, a buffer solution composed of NaAc and HAc is formed when more NaOH solution is added, and resists the pH change on addition of NaOH, causing a flatter part in the curve. Approaching to the stoichiometric point, the concentration of the HAc becomes smaller and the buffer capacity of the solutions decreases, and continuing titration causes greater pH change, thus the titration curve turns sharp.

c) The solution is not neutral at stoichiometric point. When the stoichiometric point is reached, HAc is completely neutralized by NaOH to produce NaAc which is a weak base, and the pH is 8.73 but not 7.00.

d) The pH range of titration jump (pH 7.80~9.70) is narrower than the one of the titration between strong acid and strong base with equivalent concentrations. Based on the titration jump, the indicators that change color in basic pH range should be used. Thus phenolphthalein can be used.

(3) **The relationship between the inflection point and strength of acid and base**. Titrating weak acid with strong acid, the titration jump is associated with not only the concentration of the weak acid but also its strength. As shown in Figure 13-6, it is the titration of $0.100\,0\,mol \cdot L^{-1}$ weak acids with different strength. The smaller the K_a value is, the narrower the titration jump is. When $K_a \leqslant 10^{-7}$, the titration jump of $0.100\,0\,mol \cdot L^{-1}$ weak acid is too indistinctive to determine using common indicators. It has been demonstrated that weak acid can be accurately titrated by strong base only when $K_a \geqslant 10^{-8}$.

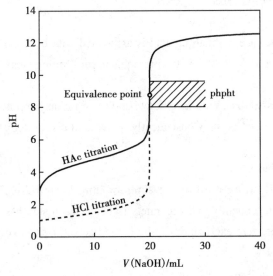

Figure 13-5 Titration Curve of 20.00mL $0.100\,0\,mol \cdot L^{-1}$ HAc Solution Being Titrated with $0.100\,0\,mol \cdot L^{-1}$ NaOH Solution

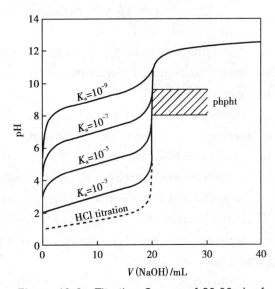

Figure 13-6 Titration Curves of 20.00mL of $0.100\,0\,mol \cdot L^{-1}$ Weak Acids and HCl Solution Titrated with $0.100\,0\,mol \cdot L^{-1}$ NaOH. Ka of Weak Acids is 10^{-3}, 10^{-5}, 10^{-7} and 10^{-9} respectively

The titration of weak monobasic base with strong acid is similar with that of weak monoprotic acid. For example, the titration curve of 20.00mL of $0.100\,0\,mol \cdot L^{-1}$ $NH_3 \cdot H_2O$ with $0.100\,0\,mol \cdot L^{-1}$ HCl solution is shown as Figure 13-7.

It can be seen that the curve of titrating weak monobasic base with strong acid is inverted to the one of titrating weak monoprotic acid with strong base. The titration jump in Figure 13-7 is pH 4.30~6.30, which lies in the acidic region. And the stoichiometric point is pH 5.28, so the mixture is weakly acidic and the indicators such as methyl orange and methyl red that change their color in acidic region should be chosen.

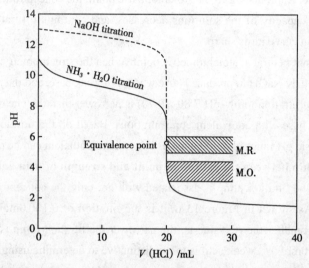

Figure 13-7 Titration Curves for 20.00mL of 0.100 0mol·L^{-1} NH$_3$·H$_2$O Solution Being Titrated with 0.100 0mol·L^{-1} HCl Solution

The pH range of titration jump of titrating weak base with strong acid is associated with both the strength and concentration of weak bases. Therefore, $c(B)K_b \geqslant 10^{-8}$ is the prerequisite for titrating weak base with strong acid.

Because the reactions between weak acids and weak bases are incomplete and the titration jump is indistinctive, those titrations are difficult to signal the end point. Consequently, weak acid can be only titrated by strong base, vice versa.

3. Titration of polyprotic acids or multibasic bases

The reactions involve multiple steps when titrating multiprotic acids or multibasic bases. Whether the reaction of each step can be titrated independently and accurately (whether each step has titration jump) and whether two consecutive steps interfere each other can be judged according to two prerequisites below:

(1) To ensure obvious titration jump of each step, $c(A)K_a \geqslant 10^{-8}$ or $c(B)K_b \geqslant 10^{-8}$.

(2) The ratio between the dissociation constants of two consecutive steps should be greater than 10^4, that is $K_{ai} / K_{ai+1} > 10^4$ (or $K_{bi} / K_{bi+1} > 10^4$) to ensure the titration step by step. Otherwise, the titration of the next step will disturb that of the previous step, because the H$_3$O$^+$ ions produced by the next step begin to react with the titrant although the H$_3$O$^+$ ions from the previous step have not been neutralized completely yet.

Taking the titration of 0.100 0mol·L^{-1} H$_3$PO$_4$ ($K_{a1} = 6.9 \times 10^{-3}$, $K_{a2} = 6.1 \times 10^{-8}$, and $K_{a3} = 4.8 \times 10^{-13}$)

with $0.1000\text{mol}\cdot\text{L}^{-1}$ NaOH solution as example, there are two but not three inflection points in the titration curve in Figure 13-8. Because $cK_{a1} \geq 10^{-8}$ and $K_{a1}/K_{a2} = 1.13 \times 10^5 > 10^4$, there is obvious titration jump on the first stoichiometric point, and the H_3O^+ produced by the first-step dissociation can be titrated accurately. Referring to the second step, because $cK_{a2} \approx 10^{-8}$ and $K_{a2}/K_{a3} = 1.27 \times 10^5 > 10^4$, there is obvious titration jump on the second stoichiometric point. However, because $cK_{a3} \ll 10^{-8}$, the third step has no obvious titration jump and it cannot be accurately titrated by strong base.

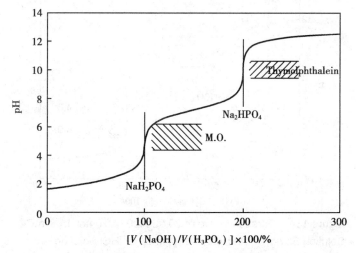

Figure 13-8 Titration Curve for 20.00mL $0.1000\text{mol}\cdot\text{L}^{-1}$ H_3PO_4 Solution Titrated with $0.1000\text{mol}\cdot\text{L}^{-1}$ NaOH Solution

At the first stoichiometric point, the reaction product is NaH_2PO_4 whose pH is approximately

$$pH = \frac{1}{2}(pK_{a1} + pK_{a2}) = \frac{1}{2} \times (2.16 + 7.21) = 4.68$$

So, methyl red can be used as indicator to signal the first end point of the titration.

At the second stoichiometric point, the reaction product is Na_2HPO_4 whose pH is approximately

$$pH = \frac{1}{2}(pK_{a2} + K_{a3}) = \frac{1}{2} \times (7.21 + 12.32) = 9.76$$

So, thymolphthalein can be used as indicator to show the second end point of the titration.

When titrating multibasic bases, for example, titrating $0.0500\text{mol}\cdot\text{L}^{-1}$ Na_2CO_3 with $0.1000\text{mol}\cdot\text{L}^{-1}$ HCl solution, there are two infection points and the titration curve is shown as Figure 13-9.

$$pH = \frac{1}{2}(pK_{a1} + pK_{a2}) = \frac{1}{2} \times (6.35 + 10.33) = 8.34$$

At the first stoichiometric point, the product is ampholyte $NaHCO_3$ whose pH is about 8.34. The two successive steps of the titration overlap with each other because $K_{b1}/K_{b2} = 9.54 \times 10^3 < 10^4$. Combined with the buffer action of the HCO_3^-, the titration jump is not obvious, causing a great titration error. Using $NaHCO_3$ with identical concentration as the reference solution or using the mixed indicators (the pH of color change is 8.3) that consists of methyl red and thymol blue, the accuracy of the titration can be improved.

At the second stoichiometric point, the product is H_2CO_3 mainly existing as the dissolving CO_2 in the solution, and its saturated concentration is 0.04mol·L^{-1}. Thus the concentration of H$_3$O$^+$ is

$$[H_3O^+] = \sqrt{K_{a1}c(A)} = \sqrt{4.5 \times 10^{-7} \times 0.040} = 1.3 \times 10^{-4} \text{mol·L}^{-1}$$

pH = 3.87, $c(B)K_{b2} \approx 10^{-8}$, the titration jump is obvious, so methyl orange can be used as indicator. However, the end point might shift to lower pH because of forming a saturated CO$_2$ solution to increase the concentration of H$_3$O$^+$.

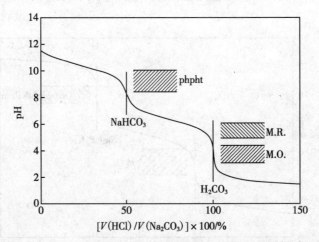

Figure 13-9   Titration Curve of 20.00mL 0.050 0mol·L^{-1} Na$_2$CO$_3$ Solution Being Titrated with 0.100 0mol·L^{-1} HCl Solution

13.4.3   Preparation and Standardization of Acid and Base Standard Solutions

1. Preparation and standardization of acid standard solution

HCl and H$_2$SO$_4$ are widely employed in the preparation of the acid standard solutions. HCl is generally preferred because HCl is more stable. However, HCl cannot be used to prepare standard solution of accurate concentration directly because of its volatility. Usually, concentration of HCl solution closed to the required one is prepared firstly, and then it is standardized by primary standard substance.

The primary standard substances used the most to standardize HCl are anhydrous sodium carbonate Na$_2$CO$_3$ or sodium tetraborate Na$_2$B$_4$O$_7$·10H$_2$O. Pure Na$_2$CO$_3$ is readily prepared and inexpensive. However, Na$_2$CO$_3$ is hygroscopic to absorb CO$_2$ in the air, thus it should be heated to 270~300℃ for 1h, then cool in a desiccator before use. The compositions of borax contain crystal water, thus borax should be stored in a humidistat with a relative humidity of 60%.

Methyl orange can be used as the indicator to standardize HCl solution using Na$_2$CO$_3$.

Standardizing HCl solution with borax, the reaction equation is as follows

$$Na_2B_4O_7 \text{(aq)} + 2HCl \text{(aq)} + 5H_2O \text{(l)} = 4H_3BO_3 \text{(aq)} + 2NaCl \text{(aq)}$$

The equivalence point is at pH 5.1 and methyl red can be used as indicator.

2. Preparation and standardization of base standard solution

NaOH and KOH are often used in the preparation of the base standard solutions. NaOH is relatively cheap and common, but it has strong moisture absorption and easy to absorb CO$_2$ in the air. Therefore, it can only be prepared near to the required concentration (about 0.1mol·L^{-1}). The common primary standard substances used to standardize NaOH solution include oxalic acid $H_2C_2O_4 \cdot 2H_2O$ and potassium acid phthalate $KHC_8H_4O_4$. The latter has high stability because of the large molecular mass and readily available in high purity form. Therefore, it is the ideal primary standard substance to standardize NaOH solution. The reaction is

$$KHC_8H_4O_4 \text{ (aq)} + NaOH \text{ (aq)} = KNaC_8H_4O_4 \text{ (aq)} + H_2O \text{ (l)}$$

The phenolphthalein could be employed to indicate the end point of the titration because the equivalence point of the reaction is pH 9.1.

$H_2C_2O_4 \cdot 2H_2O$ is quite stable and it has the small molar mass. It is a diprotic acid, and the ratio of K_{a1} to K_{a2} is less than 10^4. $H_2C_2O_4$ has just one inflection point producing $Na_2C_2O_4$ when used to standardize the solution of NaOH. Phenolphthalein is employed to signal the endpoint of titration whose equivalence point is at pH 8.4.

13.4.4 Applications of Acid-Base Titrations

Some application examples are listed below to illustrate the practical applications of acid-base titration.

1. Determination of nitrogen in analyte

The nitrogen in organic sample such as proteins can be determined by the Kjeldahl method. First of all, the concentrated H_2SO_4 and the catalyst $CuSO_4$ are added into the sample, and heated to decompose, converting the nitrogen to NH_4^+. Then the concentrated NaOH solution is added into the obtained solution to produce NH_3. Finally, the NH_3 is distilled and the content of nitrogen is determined. This method is called Kjeldahl method and commonly used in biochemistry and food analysis.

There exists the weak acid NH_4^+ ion in the solution of $(NH_4)_2SO_4$ and NH_4Cl after digestion. The acidity of NH_4^+ is too weak to be titrated directly by the alkali standard solution. So we use distillation method to determine the content of ammonia nitrogen. Excess concentrated NaOH is added into the sample and heated to distill NH_3 which is absorbed by H_3BO_3 solution to produce $(NH_4)_2B_4O_7$. Then it can be titrated with the HCl standard solution, producing NH_4Cl and H_3BO_3. The equivalence point of this reaction is approximately at pH 5, methyl red or the mixed indicators consisting of methyl red and bromocresol green can be used as the indicator.

The reaction is as follows,

$$NH_3 \text{ (g)} + H_3BO_3 \text{ (aq)} = NH_4BO_2 \text{ (aq)} + H_2O \text{ (l)}$$
$$NH_4BO_2 \text{ (aq)} + HCl \text{ (aq)} + H_2O \text{ (l)} = NH_4Cl \text{ (aq)} + H_3BO_3 \text{ (aq)}$$

Since 1mol of HCl is equivalent to 1mol nitrogen based on the reaction, this method is accurate and reliable but more time-consuming.

2. Determination of acetylsalicylic acid (aspirin) content

Acetylsalicylic acid is commonly used as antipyretic analgesics. Its molecular structure contains carboxyl group and it can be directly titrated by standard solution of NaOH. Phenolphthalein as the indicator, the reaction of the titration is

Determination of acetylsalicylic acid can be performed by dissolving the sample in quantitative volume of ethanol and controlling the temperature under 10℃ to prevent the ester group in acetylsalicylic acid from hydrolysis, which may make a higher result than the true value.

Question and thinking 13-2 There are many methods to titrate a extremely weak acid or weak base, such as a non-aqueous titration, increasing concentration titration (determination of NaAc), reinforce titration (determination of boric acid), potentiometric titration(instrument analysis) and indirect titration. Try to refer them by the related information and clarify their principles.

13.5 Oxidation-Reduction Titrations

Oxidation-reduction titration is a type of titration based on a redox reaction. It has a wide range of applications. It is not only used to determine the content of the oxidizing substances or the reducing substances directly, but also to indirectly determine the content of the substances that can quantitatively react with the oxidizing agents or the reducing agent. The redox reactions used in the oxidation-reduction titration must need the following requirements:

(1) The analyte must initially be present in a single oxidation state or reduction state well suitable for titration.

(2) Redox titration reaction must be stoichiometric and the equilibrium constant K should be large enough, generally $K > 10^6$.

(3) Redox reactions used in titration must be rapid enough, or can be improved by heating or adding the catalyst.

(4) There must be appropriate indicators to indicate the endpoint of titration.

Redox titrations can be categorized by the types of standard solutions employed. The redox titrations include potassium permanganate method, iodimetry, potassium dichromate method, cerium sulphate method and potassium bromate method.

13.5.1 Potassium Permanganate Method

1. Basic principle of potassium permanganate method

Potassium permanganate method is a redox titration based on the strong oxidizing agent $KMnO_4$ as the standard solution. In an acidic solution, $KMnO_4$ is reduced to Mn^{2+}, and the reduction is represented by the following equation

$$MnO_4^-(aq) + 8H_3O^+(aq) + 5e^- = Mn^{2+}(aq) + 12H_2O\ (l) \qquad \varphi^\ominus = 1.507V$$

In the weak acid, neutral or weak alkaline solutions, MnO_4^- is reduced to brown MnO_2 precipitation. Therefore, if $KMnO_4$ is used in a redox titration, the titration must be performed in strong acidic solution. H_2SO_4 is the most suitable acid to keep the acidic medium, rather than HNO_3 and HCl, because HNO_3 has oxidizing property, meanwhile HCl can be oxidized by $KMnO_4$ to lead the side reaction. The suitable concentration of H_2SO_4 is within $0.5 \sim 1.0$ mol·L^{-1} generally, otherwise, higher concentration can decompose $KMnO_4$ into Mn^{2+} and O_2.

KMnO$_4$ is purple if the concentration of MnO$_4^-$ is more than 2×10^{-6} mol·L^{-1}, while reductive product Mn^{2+} is nearly colorless, so KMnO$_4$ itself can be a good indicator. The titration end point is reached if the reddish color does not disappear in 30 second. We should know that the reducing substances in the air will react with KMnO$_4$, the reddish color cannot persistent.

The reducing substances such as Fe^{2+}, H$_2$O$_2$, C$_2$O$_4^{2-}$, etc, can be directly titrated in acidic solution by KMnO$_4$. Back titration can be used to determine the oxidizing substances such as MnO$_2$ which can not be titrated by KMnO$_4$. And indirect titration can be employed to determine the content of the non-oxidizing or non-reducing substance such as Ca^{2+}.

2. Preparation and standardization of potassium permanganate solution

(1) Preparation of potassium permanganate solution. The purity of KMnO$_4$ is not high and it often contains a few of impurities such as MnO$_2$, sulfates and nitrates. Besides, the distilled water usually contains a minority of organic impurities that may reduce KMnO$_4$. To obtain the KMnO$_4$ solution with the stable concentration, the prepared KMnO$_4$ solution should be heated to mild boiling for one hour to make the organic impurities react KMnO$_4$ completely, followed by stewing for 2 to 3 days and filtrating to remove the impurity MnO$_2$ using a sintered glass filter. Commonly, the concentration of the KMnO$_4$ solution is about 0.02 mol·L^{-1} and it is usually stored in a brown glass bottle.

(2) Standardization of KMnO$_4$ solution. Standardization may be accomplished by using the primary standard substances such as sodium oxalate Na$_2$C$_2$O$_4$, (NH$_4$)$_2$SO$_4$·FeSO$_4$·6H$_2$O and pure iron wire. The Na$_2$C$_2$O$_4$ is often used to standardize the KMnO$_4$ solution because this reagent is readily obtained anhydrous and hygroscopic. In H$_2$SO$_4$ solution, the ionic reaction equation between KMnO$_4$ and Na$_2$C$_2$O$_4$ is

$$MnO_4^-(aq) + 5C_2O_4^{2-}(aq) + 8H^+(aq) = Mn^{2+}(aq) + 10CO_2(g) + 4H_2O(l)$$

It is very slow at the beginning of reaction, but it can be accelerated by heating or the self-catalyst Mn^{2+}. If drops of MnSO$_4$ solution was added before titration and the analyt mixture was heated to 40℃ but not over 90℃, the titration become fast, H$_2$C$_2$O$_4$ will partially decompose.

3. Application of potassium permanganate method

To determine the content of H$_2$O$_2$ in the commercial hydrogen peroxide, the reductive H$_2$O$_2$ can be titrated directly by the KMnO$_4$ standard solution in the acidic solution. The reaction between KMnO$_4$ and H$_2$O$_2$ is

$$2MnO_4^-(aq) + 5H_2O_2(aq) + 6H^+(aq) = 2Mn^{2+}(aq) + 5O_2(g) + 8H_2O(l)$$

The content of H$_2$O$_2$ in the commercial hydrogen peroxide is about 30%. It should be diluted before titration. Furthermore, the titration should be carried out at room temperature because H$_2$O$_2$ is easy to decompose under heating.

13.5.2 Iodometric Titration

1. Basic principle of iodimetric titration

Iodometric titration is the titrimetric analysis based on the oxidative property of I$_2$ and the reductive property of I$^-$. The half reaction of I$_2$ in the iodometric titration is

$$I_2 + 2e^- = 2I^- \qquad \varphi^\ominus = 0.535\,5\text{V}$$

I$_2$ is a moderate oxidizing agent that can react with the strong reducing agents, while I$^-$ is a moderate reducing agent that can react with lots of oxidizing agents.

Iodometry, also known as iodometric titration, can be sorted into two types: **direct iodometry** and **indirect iodometry**. The direct iodometry is used in the titration of the reducing substances such as S^{2-}, SO_3^{2-}, Sn^{2+} and $S_2O_3^{2-}$, which have the lower standard electrode potentials than $\varphi^{\ominus}(I_2/I^-)$, and can be directly titrated by I_2 standard solution. The indirect iodometry is used in the titration of the oxidizing sustances such as ClO_3^-, CrO_4^{2-}, MnO_4^-, MnO_2 and Cu^{2+}, which have the higher standard electrode potentials than $\varphi^{\ominus}(I_2/I^-)$. That is, the oxidizing substances are reduced by excess I^- firstly and then the formed I_2 is quantitatively titrated by the $Na_2S_2O_3$ standard solution.

The starch solution is used as indicator in the iodometric titration. The starch reacts sensitively with I_2 to form a blue complex. However, it will be affected by the temperature and pH of the solution.

2. Preparation and standardization of the iodometric standard solutions

(1) Preparation and standardization of the I_2 standard solution. I_2 is volatile, corrosive and slightly soluble. It is usually prepared into the aqueous solution of approximate concentration, and the KI solution is added to form the complex ion I_3^- so as to improve the solubility of I_2 and reduce its volatility. The prepared I_2 solution is usually standardized by the $Na_2S_2O_3$ standard solution or the primary standard substance As_2O_3.

Standardizing reaction I_2 is

$$I_2(aq) + 2Na_2S_2O_3(aq) = 2NaI(aq) + Na_2S_4O_6(aq)$$

(2) Preparation and standardization of the $Na_2S_2O_3$ standard solution. $Na_2S_2O_3 \cdot 5H_2O$ may contains the impurities such as the elementary sulfur S, Na_2CO_3 and Na_2SO_4, and it is efflorescent and deliquescent easily. Moreover, it may be decomposed and oxidized by CO_2, O_2 and microorganism in aqueous solution. Consequently, the $Na_2S_2O_3$ standard solution needs to be prepared using the cooled and newly boiled distilled water, and maintaining the pH 9~10 by adding a small amount of Na_2CO_3 as stabilizing agent. Standing for 9 to 10 days, the $Na_2S_2O_3$ solution is standardized by the I_2 standard solution or the primary standard substances. The common primary standard substance is $K_2Cr_2O_7$. In the acidic solution, $K_2Cr_2O_7$ reacts with KI to I_2 is

$$Cr_2O_7^{2-}(aq) + 6I^-(aq) + 14H^+(aq) = 2Cr^{3+}(aq) + 3I_2(s) + 7H_2O(l)$$

Take standard $Na_2S_2O_3$ solution to titrate formed I_2.

3. Application of the iodometric titrations

(1) Determining the content of vitamin C by the direct iodometric titration Vitamin, also called ascorbic acid, is a strong reducing agent and can be quantitatively oxidized by iodine to dehydroascorbic acid, the reaction is

$$\underset{\substack{\| \\ O}}{CH} - \underset{\substack{\| \\ OH}}{C} = \underset{\substack{| \\ OH}}{C} - \underset{\substack{| \\ H}}{C} - \underset{\substack{| \\ OH}}{\overset{H}{C}} - \underset{\substack{| \\ H}}{\overset{OH}{C}} - H + I_2 \rightleftharpoons \underset{\substack{\| \\ O}}{CH} - \underset{\substack{\| \\ O}}{C} - \underset{\substack{\| \\ O}}{C} - \underset{\substack{| \\ H}}{C} - \underset{\substack{| \\ OH}}{\overset{H}{C}} - \underset{\substack{| \\ H}}{\overset{OH}{C}} - H + 2HI$$

During the titration, HAc was added to assure the acidic solution so as to avoid ascorbic acid being oxidized by other oxidizing substances.

> **Question and thinking 13-3** One of the important indexes to evaluate the pollution degree of water is chemical oxygen demand (COD), also called chemical oxygen consumption. The common national standard method to determine COD is $K_2Cr_2O_7$ titrimetry in which $K_2Cr_2O_7$ is used to oxidize the reductive substances such as the organic, the nitrites, the ferrites and the sulfides. COD can be calculated by the amount of the remaining oxidants. The unit of COD is ppm or $mg \cdot L^{-1}$ usually. The smaller the COD value is, the less the water is polluted. What is the COD of the environmental water around you? Consult the relevant literatures to design an experiment to determine it and conduct the experiment in lab.

(2) Determining the content of NaClO by the indirect iodometry. NaClO is a kind of bactericide, and it can be transformed into Cl_2 which can oxidize I^- into I_2. Then the content of NaClO can be determined by titrating I_2 formed with $Na_2S_2O_3$ standard solution.

13.6 Complexometric Titration

13.6.1 Basic Principle of Complexometric Titration

Complexometric titration is based on the metal-ligand complexation reaction and widely used to determine the content of metallic ion. The vast majority of titration analysis is carried out between the interaction of multidentate ligand and metallic ion, which is called **chelatometric titration**. Ethylenediamine tetraacetic acid H_4Y and its disodium salt Na_2H_2Y, both are denoted as EDTA, and they are the common chelators used in the chelatometric titration. EDTA can chelate with most metal ions to form the water-soluble chelates with five-member rings. The reaction between metallic ions and EDTA can be expressed as follow

$$M^{n+}(aq) + H_2Y^{2-}(aq) \rightleftharpoons MY^{n-4}(aq) + 2H^+(aq)$$

1. The main factors affecting EDTA titrimetry

(1) Acidity of solution. As the reaction proceeds, the acidity of solution will gradually increase and ethylenediamine tetraacetate Y^{4-} can form polyprotic acids with H_3O^+ step by step. Therefore, the stronger the acidity of the solution is, the greater the tendency of the coordination equilibrium shifting to the left is, and the quantitative reaction is not complete. This phenomenon is called the **acid effect**. If the acidity is too weak, metallic ions especially transition metallic ions are prone to hydrolysis and form $M(OH)_n$ precipitation. Therefore, conducting EDTA titrimetry, the acidity should be strictly controlled using buffer solution.

(2) Effects of the coexisting metallic ions on the EDTA titrimetry. Since EDTA can react with most metal ions, when the other ions are simultaneously present in the solution, they often interfere with each other and affect the accuracy of the measurement results. Therefore, adjusting the pH value of the solution or adding the masking agents before titration can avoid the interference.

(3) Effects of the coexisting ligands on EDTA titrimetry. If there are other ligands L in the solution which can react with the metal ion M^{n+}, the chance of M^{n+} to react with EDTA will be less. This

phenomenon is called **coordination effect**. The more stable the ML is, the greater the coordination effect is, and the less complete the titration reaction is. So the coordination effect should be considered during the complexometric titration. And it can be eliminated by adjusting the pH of solution.

2. Metallochromic indicator

The end point of the coordination titration can be signalled by the **metallochromic indicator**. The metallochromic indicator, denoted as HIn, are the water-soluble organic dyes which can form color complexes MIn^{n-1} with the metallic ions. EDTA competes with HIn to complex M^{n+}, and the reaction between MIn^{n-1} and EDTA is

$$MIn^{n-1}(aq) + H_2Y^{2-}(aq) \rightleftharpoons MY^{n-4}(aq) + In^-(aq) + 2H^+(aq)$$

The metallochromic indicators must satisfy the following requirements:

(1) The color difference between free HIn and complex MIn^{n-1} should be remarkably different.

(2) MIn^{n-1} must be stable enough, $K_s(MIn^{n-1}) > 10^4$, but less stable than MY^{n-4}, $K_s(MY^{n-4})/K_s(MIn^{n-1}) > 10^2$. If the stability of MIn^{n-1} is too poor, it will cause the end point appearing ahead of time. If the stability is too high, EDTA cannot compete to seize M^{n+} from the MIn^{n-1} to release the free HIn, so the color may not change at the end point.

The most commonly used indicators include eriochrome black T (EBT), calcium red, xylenol orange and so on. Here take EBT as an example to explain the mechanism of color change of the metallochromic indicator. EBT, denoted as NaH_2In, is a kind of weak acidic azo dye. EBT and metal ions can form a wine-red complex, which has the following dissociation equilibrium in aqueous solution.

$$H_2In^- \underset{}{\overset{pK_{a1}=6.3}{\rightleftharpoons}} HIn^{2-} \underset{}{\overset{pK_{a2}=11.6}{\rightleftharpoons}} In^{3-}$$

(purple)　　　　(sky-blue)　　　　(orange)

pH < 6.3　　　　pH 6.3～11.6　　　pH > 11.6

In the above formula, if the pH is lower than 6.3, the free EBT shows purple color which is similar with that of the MIn^{n-1}. Thus, the pH must be maintained in the range of 9.0～10.5 by NH_4Cl-NH_3 buffer solution usually. EBT is employed as indicator for determination of Zn^{2+}, Ca^{2+} and Mg^{2+} ions. For example, when the Mg^{2+} ions are titrated with EDTA, several drops of EBT indicator will be added firstly to react with the Mg^{2+} ions, forming a wine-red complex of $MgIn^-$. After adding EDTA, the free Mg^{2+} ions will react with EDTA to form a colorless metal complex ion MgY^{2-}, until the all of free Mg^{2+} ions in solution are consumed completely. And then the excess EDTA will seize the Mg^{2+} ions from the $MgIn^-$ complexes to release the free EBT to solution, showing its own blue color and indicating the end point.

13.6.2　Preparation and Standardization of EDTA Standard Solution

EDTA standard solution is usually indirectly prepared. First, EDTA solution of approximate concentration is prepared using EDTA disodium salt. And then it is standardized by the primary standard substances such as Zn, ZnO, $ZnSO_4$, $CaCO_3$ and $MgCO_3 \cdot 7H_2O$, EBT is indicator. The common concentration of EDTA standard solution is about 0.01～0.05mol·L^{-1}.

13.6.3　Applications of Complexometric Titration

EDTA is a versatile titrant that can be used for direct analysis of virtually all metal ions. In inorganic drugs analysis, it is commonly employed as titrant for determination of the contents of Ca^{2+}, Mg^{2+}, Al^{3+} and Bi^{3+} etc.

1. Determination of the content of Ca^{2+} ions in drugs containing calcium

There are many drugs containing calcium, such as $CaCl_2$, calcium gluconate and calcium lactate. These drugs can be determined by EDTA titration according to Chinese Pharmacopoeia. When calcium gluconate was determined, weigh exactly a certain amount of calcium gluconate sample and dissolve it in distilled water, then adjust the pH of the solution to 10 with NH_3-NH_4Cl buffer solution, and finally titrate it using EDTA and with EBT as the indicator until the color of the solution changes from wine red to blue. That reaches the end point.

2. Determination of water hardness

Water hardness is the total concentration of Ca^{2+} ions and Mg^{2+} ions in water sample. Similarly, the water hardness can be determined by titrating water sample with EDTA as the titrant and EBT as the indicator at pH 10 adjusted by the NH_3-NH_4Cl buffer. When the color of the mixture changes from wine red into blue, the end point is reached. The stability order of the complexes in the solution is that, $CaY^{2-} > MgY^{2-} > MgIn^- > CaIn^-$.

13.7 Precipitation Titration

Precipitation titrations are quantitative titration methods based on the precipitation reaction. The most important precipitating reagent is silver nitrate $AgNO_3$. At present, **argentometric methods** based on $AgNO_3$ are mainly used to determine Cl^-, Br^-, I^- and SCN^- ions, etc, which can react with Ag^+ to insoluble silver salt, or to determine organic substance or sample which can release the above ions after pretreatment. For example, the precipitation titrations are employed to determine Cl^- in blood serum, to monitor Cl^- in table-water and to measure haloid in certain drugs. According to the indicator used, the argentometric methods can be classified into different types.

1. The argentometric method that $AgNO_3$ is the standard solution and K_2CrO_4 is the indicator is called **Mohr method**. When titrating Cl^- by Mohr method, since the K_{sp} of $AgCl$ and Ag_2CrO_4 at 25 ℃ are 1.8×10^{-10} and 1.1×10^{-12}, respectively. So $AgCl$ has smaller solubility than Ag_2CrO_4. When the titration starts, Ag^+ reacts firstly with Cl^- to $AgCl$ precipitate; and all Cl^- ions is completely precipitated at the stoichiometric point. Continuing titration, excess Ag^+ combines CrO_4^{2-} to form a brick-red color of precipitation, indicating the end point.

2. The argentometric method that $AgNO_3$ is the standard solution and $NH_4Fe(SO_4)_2$ is the indicator is called **Volhard method**. Volhard method is used in the precipitation titration under strongly acidic condition maintained by $0.3 mol \cdot L^{-1}$ HNO_3 solution which prevents the hydrolysis of the Fe^{3+} and the precipitation formed with the ions such as PO_4^{3-}, CO_3^{2-}, S^{2-}, etc. Volhard method is sorted into direct method and indirect method.

Direct method is usually used to determine the content of Ag^+. KSCN or NH_4SCN is the standard solution to titrate Ag^+ in HNO_3 medium. Excess of SCN^- ions react with Fe^{3+} in indicator to a red complex and to signal the end point of the titration when all the Ag^+ ions have been quantitatively precipitated by SCN^- ions.

Indirect method is used to determine the content of halide ion X^-. Firstly, excess $AgNO_3$ standard solution is added into the HNO_3-acidified samples containing X^-, and $NH_4Fe(SO_4)_2$ is indicator, the remaining Ag^+ is back-titrated with NH_4SCN standard solution. After forming AgSCN precipitating completely, Excess of SCN^- ions react with Fe^{3+} in indicator to a red complex and to signal the end point of the titration.

K_{sp}(AgSCN) is smaller than that of AgCl at the same temperature. Therefore, when Cl^- is determined by back titration, the AgCl precipitate should be filtered off or isolated using organic solvent such as nitrobenzene before converting into AgSCN. In determining I^- by the indirect method, $AgNO_3$ standard solution should be added into solution to form AgI precipitation fully, and then, indicator $NH_4Fe(SO_4)_2$ is added into the mixture to prevent the side-reaction between Fe^{3+} and I^-.

3. The argentometric method using $AgNO_3$ as the standard solution and adsorption indicator to determine the end point is called **Fajans method**. Adsorption indicators are a class of organic dyes and can be categorized into two types. 1) Organic weak acids such as fluorescent yellow HFIn and its derivatives, they can dissociate into anions in the solution. 2) Organic weak base, it can dissociate into cations in the solution. Here, the titration of Cl^- with $AgNO_3$ standard solution is used to illustrate the principle of Fajans titration. HFIn dissociates into yellow-green FIn^- ion, so it can be used as the indicator in titration. Before the stoichiometric point, there is excessive Cl^- in the solution, AgCl precipitate absorbs Cl^- to carry the negative charge, so FIn^- is not absorbed by the precipitate and indicates the yellow-green color. After the stoichiometric point, there is excessive Ag^+ and AgCl precipitate absorbs the Ag^+ to carry positive charge, thus the precipitate strongly absorbs the negative FIn^-. The structure of FIn^- is changed and its color changes from yellow green to pink color, that is the end point of titration.

Key Terms

滴定分析	titrimetric analysis
误差	error
偏差	deviation
系统误差	systematic error
偶然误差	random error
准确度	accuracy
精密度	precision
绝对误差	absolute error
相对误差	relative error
绝对偏差	absolute deviation
相对偏差	relative deviation
有效数字	significant figure
运算规则	arithmetic rules
修约	round/rounding
滴定	titrate
滴定剂	titrant
标准溶液	standard solution

滴定终点	end point
化学计量点	stoichiometric point
等当点	equivalence point
滴定误差	titration error
指示剂	indicator
直接滴定法	direct titration
返滴定法	back titration
标定	standardization
一级基准物质	primary standard substance
酸碱滴定	acid-base titration
酸碱指示剂	acid-base indicator
pH 变色范围	pH range of color change
滴定曲线	titration curve
滴定突跃	titration jump
拐点	inflection point
氧化还原滴定	oxidation-reduction titrations
高锰酸钾法	potassium permanganate method
碘量法	iodometric titration
配位滴定法	complexometric titration

Summary

The test measurements need the significant figures after rounding the data obtained by experiment devices and rounding rule.

Titration is the regular method that a standard solution is used to titrate the sample solution in order to know the concentration or moles of the sample based on the chemical reaction and exact calculation formula. According to the characteristics of the chemical reaction between the standard solution and the tested sample, titration is classified into acid-base titration, redox titration, complexometric titration, precipitation titration and so on.

Indicators such as acid-base indicators and metallochromic indicators are used to properly show the titration end points which coincide closely with the stoichiometric points. A proper indicator may show sharp end point and reduce the titration error to give an accurate results.

Back titration is a process in which the excess reagent A is added to react with the sample solution, and then another reagent B is used to titrate the excess of A left, so as to know the amount of initial A and left A in the mixture, we can easily know how much of reagent A was used and the concentration of sample furthermore.

Exercises

1. How many significant figures are there in each of following numbers?
   (1) 2.032 1   (2) 0.021 5   (3) $pK_{HIn} = 6.30$ (4) 0.01%   (5) 1.0×10^{-5}

2. As the weighing error of analytical balance is ± 0.000 1g, the maximum error for twice weighing

will reach ± 0.000 2g if the sample is weighed by indirect method. How much is the sample weight required at least to guarantee the relative weighing error less than 0.1%?

3. Two students independently measured the content of copper in the same sample, their data are as follows:

	Content in sample (mass fraction, ω%)		
Student A	0.361 0	0.361 2	0.360 3
Student B	0.364 1	0.364 2	0.364 3

If the true value of the mass fraction was 0.360 6, whose data is more accurate? Whose data is more precise? Why?

4. If NaOH solution was standardized by potassium hydrogen phthalate, what will be the measured concentration of NaOH solution in the following cases?

(1) The initial burette reading of the NaOH solution was 1.00mL, but it was recorded as 0.10mL by mistake.

(2) The mass of potassium hydrogen phthalate was 0.351 8g, but it was recorded as 0.357 8g by mistake.

5. If K_{In} of weak basic indicator is 1.3×10^{-5}, what is the pH interval of color change of it?

6. Weigh 1.335 0g of analytical pure reagent Na_2CO_3 to prepare 250.00mL of the primary standard substance solution, and use it to titrate HCl solution whose concentration is approximate $0.1 mol \cdot L^{-1}$. If 25.00mL of Na_2CO_3 solution is required to react with 24.50mL HCl solution completely, calculate the accurate concentration of HCl solution.

7. How many grams of primary standard substance of oxalic acid $H_2C_2O_4 \cdot 2H_2O$ are required to standardize about 25mL of $0.1 mol \cdot L^{-1}$ NaOH solution? Can the relative error be controlled within the range of ± 0.05% while $H_2C_2O_4 \cdot 2H_2O$ is used as primary standard substance? If $H_2C_2O_4 \cdot 2H_2O$ is replaced by potassium hydrogen phthalate $KHC_8H_4O_4$, what is the result?

8. What are the differences between equivalence point and end-point in titrimetry? And what are the relationships among the equivalence points, end-point and neutralization point in all kinds of acid-base titrations?

9. A weak monoprotic acid is titrated by NaOH standard solution. The equivalence point is reached with the addition of 40.00mL of NaOH standard solution, the pH of solution becomes 6.20 with the addition of 16.00mL of NaOH solution. Calculate the dissociation constant of this weak acid?

10. In the following acids (assuming they are all $0.100 0 mol \cdot L^{-1}$ and 20.00mL), which can be titrated by NaOH standard solution directly? And which cannot be? What kinds of indicators should be used to signal each endpoint, and how many inflection points do they have respectively, whether they can be titrated directly or not?

(1) Formic acid (HCOOH) $pK_a = 3.75$
(2) Succinic acid ($H_2C_4H_4O_4$) $pK_{a1} = 4.16$, $pK_{a2} = 5.61$
(3) Maleic acid ($H_2C_4H_2O_4$) $pK_{a1} = 1.83$, $pK_{a2} = 6.07$
(4) Phthalic acid ($H_2C_8H_4O_4$) $pK_{a1} = 2.89$, $pK_{a2} = 5.51$
(5) Boric acid (H_3BO_3) $pK_{a1} = 9.27$

11. If $0.100 0 mol \cdot L^{-1}$ citric acid H_3Cit ($H_3OHC_6H_4O_6$) is titrated with $0.100 0 mol \cdot L^{-1}$ NaOH standard solution, how many inflection points will it have? And what kinds of indicators should be used to each of the inflection points? (Three successive acid-dissociation constants value of H_3Cit: $pK_{a1} = 3.13$,

$pK_{a2}=4.76$, $pK_{a3}=6.40$)

12. Weigh 0.683 9g sample containing Na_2CO_3, $NaHCO_3$ and other impurities which don't react with acid, dissolve it in the water. The prepared solution is titrated by 23.10mL of $0.200\,0\text{mol} \cdot L^{-1}$ HCl with phenolphthalein as indicator until the color of the solution changes from pink to colorless. After that, the resulting solution is continuously titrated by 26.81mL of HCl solution with methyl orange as the second inflection point indicator until the color of the solution changes from yellow to orange. Calculate the mass fractions of two major components of Na_2CO_3 and $NaHCO_3$ in the sample, respectively.

13. Accurately weigh 1.000g sample of crude ammonium salt and add it into the excess NaOH solution to produce the NH_3 which is absorbed by 50.00mL of $0.250\,0\text{mol} \cdot L^{-1}$ H_2SO_4 solution. The excess of H_2SO_4 is then back titrated by 1.56mL of $0.500\,0\text{mol} \cdot L^{-1}$ NaOH to reach the end point. Calculate the mass fraction of ammonia in the sample.

14. After 0.475 0g sample to be digest by concentrated H_2SO_4 and a catalyst, adding excess NaOH. Then the distilled NH_3 is absorbed into 25.00mL of HCl solution. The remaining HCl is titrated with 13.12mL of $0.078\,93\text{mol} \cdot L^{-1}$ NaOH solution. If 25.00mL of HCl solution reacts exactly with 15.83mL of NaOH solution, try to calculate the nitrogen content of the sample.

15. Solve 0.412 2g of acetylsalicylic acid $C_9H_8O_4$ sample into 20mL of ethanol, add 2 drops of phenolphthalein indicator in the solution, take $0.103\,2\text{mol} \cdot L^{-1}$ NaOH standard solution to titrate it at the temperature lower than 10℃. At the drop endpoint, 21.08mL of NaOH solution was consumed. Calculate the mass concentration of acetylsalicylic acid in this sample.

16. Accurately transfer 25.00mL of hydrogen peroxide H_2O_2 sample solution into a 250mL volumetric flask and diluted it to mark with the distilled water. Then, 25.00mL of the prepared solution which has been acidified by adding H_2SO_4 was titrated by 35.86mL of $0.027\,32\text{mol} \cdot L^{-1}$ $KMnO_4$ standard solution to reach the end point. Calculate the mass volume fraction of H_2O_2 in the original sample solution.

17. In order to measure the content of Ca^{2+} in blood, it is usually precipitated to form CaC_2O_4, and then CaC_2O_4 is re-dissolved in H_2SO_4 and titrated by $KMnO_4$ standard solution. Here, 5.00mL of blood serum sample was diluted to 50.00mL, then take 10.00mL of the diluted sample is titrated by 1.15mL of $0.002\,00\text{mol} \cdot L^{-1}$ $KMnO_4$ to reach the end point. How many milligrams of Ca^{2+} are in 100mL of blood?

18. Add the excess KI and appropriate amount of H_2SO_4 into a water solution containing 2.622g bleaching powder, then, 35.58mL of $0.110\,9\text{mol} \cdot L^{-1}$ $Na_2S_2O_3$ standard solution has been used to titrate the liberated I_2 immediately, Calculate the mass fraction of available chlorine $Ca(OCl)_2$ in the bleaching powder. $MM.[Ca(OCl)_2]=143.0\text{g} \cdot \text{mol}^{-1}$.

19. Accurately weigh 0.198 8g vitamin C (ascorbic acid) and dissolve it in 100mL of fresh boiled distilled water with the addition of 10mL of diluted acetic acid and appropriate amount of starch indicator. After that, the solution is titrated by 22.14mL of $0.050\,00\text{mol} \cdot L^{-1}$ I_2 standard solution to reach the end point. Calculate the mass fraction of ascorbic acid in the sample.

20. A certain amount of ZnO could react with 20.00mL of $0.100\,0\text{mol} \cdot L^{-1}$ HCl solution completely. If the same amount of ZnO was titrated by $0.050\,00\text{mol} \cdot L^{-1}$ EDTA standard solution, what milliliter of EDTA standard solution would be required to reach the end point?

21. 0.210 0g primary standard substance of $CaCO_3$ is dissolved in HCl solution, then dilute it to

250.00mL solution. A portion of 25.00mL solution is pipetted and titrated by EDTA standard solution buffering to pH 10, eriochrome black T(EBT) is the indicator. After adding 20.15mL of EDTA, the end point is reached. Calculate the concentration of EDTA standard solution.

22. 100mL of a water sample is titrated by $0.01048\mathrm{mol\cdot L^{-1}}$ EDTA after the pH is adjusted to 10, EBT is indicator, adding 14.20mL to reach the end point. In another 100mL of same water sample, Mg^{2+} has been precipitated to $Mg(OH)_2$ by adjusting pH to 13, it is titrated by above EDTA solution using murexide as indicator, adding 10.54mL of EDTA to reach the end point. Calculate the concentration of Ca^{2+} and Mg^{2+} respectively in the sample of water.

23. Acidify 10.00mL of waste solution containing silver by HNO_3 from radiology department of the hospital, titrate it by 23.48mL of $0.04382\mathrm{mol\cdot L^{-1}}$ NH_4SCN standard solution with $NH_4Fe(SO_4)_2$ as the indicator to reach the end point. Calculate the content of Ag^+ in the waste solution ($\mathrm{g\cdot L^{-1}}$).

Supplementary Exercises

1. If you had to do the calculation of $(29.837 - 29.24) / 32.065$, what is the correct significant figure?

2. Two monoprotic acids, both $0.1000\mathrm{mol\cdot L^{-1}}$ in concentration, are titrated with $0.1000\mathrm{mol\cdot L^{-1}}$ NaOH solution. The pH at the equivalence point for HX is 8.8 and that for HY is 7.9. Answer the following questions:

(a) Which is the weaker acid?

(b) Which indicators in followings could be used to titrate each of these acids?

	Methyl orange	Methyl red	Bromthymol blue	Phenolphthalein
pH ranger for color change	3.2~4.4	4.8~6.0	6.0~7.5	8.2~10.0

3. A sample of 0.1276g of an unknown monoprotic acid, HA, was dissolved in 25.00mL of water and titrated with $0.06330\mathrm{mol\cdot L^{-1}}$ NaOH solution. The volume of base required to reach the stoichiometric point was 18.40mL.

(a) What is the molar mass of the acid?

(b) After 10.00mL of base had been added in the titration, the pH was determined to be 5.87. What is K_a for the unknown acid?

4. How many grams of $H_2C_2O_4\cdot 2H_2O$ are needed to prepare 250.00mL of $0.1000\mathrm{mol\cdot L^{-1}}$ $C_2O_4^{2-}$ standard solution and $0.1000\mathrm{mol\cdot L^{-1}}$ H_3O^+ standard solution?

5. A student titrated 25.00mL of NaOH solution with standard sulfuric acid solution. It took 13.40mL of $0.05550\mathrm{mol\cdot L^{-1}}$ H_2SO_4 to neutralize NaOH in the solution. What was the molarity of NaOH solution?

6. A freshly prepared solution of NaOH was standardized with $0.1024\mathrm{mol\cdot L^{-1}}$ H_2SO_4.

(a) If 19.46mL of NaOH was neutralized by 21.28mL of H_2SO_4, what was the molarity of the base?

(b) How many grams of NaOH were there in each liter of this solution?

7. 0.5720g of sample in a mixture containing Na_2CO_3, $NaHCO_3$ and inert impurities is titrated by $0.1090\mathrm{mol\cdot L^{-1}}$ HCl, after adding 15.70mL to reach the phenolphthalein end point and a total of 43.80mL to reach the methyl orange end point. What is the percent each of Na_2CO_3 and $NaHCO_3$ in the mixture? $MM.(NaHCO_3) = 84.0\mathrm{g\cdot mol^{-1}}$, $MM.(Na_2CO_3) = 106.0\mathrm{g\cdot mol^{-1}}$.

Answers to Some Exercises

[Exercises]

1. (1) 5   (2) 3   (3) 2   (4) 1   (5) 2

2. 0.2g

3. The accuracy of measured values from student A is higher than the one of student B, the measured values of student B have higher precision than student A as $d_r(A) > d_r(B)$.

4. (1) Lower   (2) Higher

5. pH 8.11～10.11

6. 0.102 8 mol·L^{-1}

7. 0.15g; No, Yes

9. $pK_a = 6.38$

10. (1) Yes, phenolphthalein; inflection point 6.75～9.70; one inflection point

    (2) Yes, thymolphthalein; pH 9.6; one inflection point

    (3) Yes, methyl orange, pH 3.95 and thymolphthalein, pH 9.30; two inflection points

    (4) Yes, thymolphthalein or phenolphthalein; pH 9.0; one inflection point

    (5) No

11. One inflection point; thymolphthalein

12. 0.716 0; 0.091 14

13. 0.412 5

14. $\rho(H_2O_2) = 33.30$ g·L^{-1}

15. 23.0 mg

16. 0.053 80

17. 0.980 7

18. 20.00 mL

19. 0.010 42 mol·L^{-1}

20. $c(Ca^{2+}) = 0.001\ 105$ mol·L^{-1}; $c(Mg^{2+}) = 0.000\ 383\ 6$ mol·L^{-1}

21. 11.09 g·L^{-1}

[Supplementary Exercises]

1. 0.019.

2. (a) HX is weaker.   (b) Phenolphthalein

3. (a) 365 g·mol^{-1} (b) $K_a = 1.6 \times 10^{-6}$

4. 3.150g, 4.5g

5. 0.005 950 mol·L^{-1}

6. (a) 0.224 0 mol·L^{-1} (b) 8.958g

7. $\omega(Na_2CO_3) = 31.71\%$ $\omega(NaHCO_3) = 44.98\%$

（李雪华　赖泽锋）

Chapter 14
Ultraviolet-Visible Spectrophotometry

Spectrophotometry is an important qualitative and quantitative analytical method based on the light absorption characteristic of a given substance which is measured by a spectrophotometer. Usually, spectrophotometry is divided into three categories as ultra-violet spectrophotometry when the wavelength coverage is from 200nm to 380nm, visible spectrophotometry when the wavelength coverage is from 380nm to 780nm, and infrared spectrophotometry when the wavelength coverage is from 780nm to 3×10^5nm.

Spectrophotometry is characterized by high sensitivity, good accuracy, simple operation, and rapid determination. The lowest concentration of detection is 10^{-5}mol·L^{-1}~10^{-6}mol·L^{-1}, and the relative deviation of detection is 2%~5%, so it is specially applied to the micro- and trace- analysis. Spectrophotometry, as one of the conventional analytical methods, has been extensively applied in the medicine, sanitation, environmental protection, chemical industry, etc. In this chapter, visible spectrophotometry will be described in detail and ultraviolet spectrophotometry will be introduced briefly.

14.1 Absorption Spectrum of Substance

14.1.1 Selective Absorption of Light by Substance

When a substance or a solution is irradiated by a beam of light, the collision between the photons and the molecules, atoms or ions of the substance occurs. Owing to the energy levels of a particle are quantized, the light absorption occurs when the photon's energy (hv) equals to the energy difference between the ground state and the excited state of the particles. The photon's energy is transferred to their molecules, atoms or ions, which makes these particles transit to a higher energy level (the excited state) from the lowest energy level (the ground state).

$$M \text{ (the ground state)} + hv \rightarrow M^* \text{ (the excited state)}$$

The above process is called as the light absorption by a substance. The photon's energy is within $(1.6 \times 10^{-19}$~$3.2 \times 10^{-18})$J, so the wavelength of the absorption light is within the visible or ultraviolet region. Different substance has different energy, so the absorption energy of the photon is different and the absorption wavelength is also different. That is, the different substance absorbs light selectively.

The light can be classified as monochromatic light and polychromatic light according to the number of wavelength. Monochromatic light is of single wavelength. The light which is composed of more than one wavelength is called polychromatic light. The white light (such as sunlight, incandescent lamp light) is polychromatic light whose wavelength ranges are from 400nm to 760nm. The relationship between the colors of the substance and the light absorbed is listed in Table 11-7.

In visible light region, the violet light has the highest energy because of its shortest wavelength, while the red light has the lowest energy due to its longest wavelength. The experiments show that if two kinds of different color light are mixed in a certain proportion, the white light is formed. These two colors are called complementary color each other, such as yellow and blue, purple and green, etc. which are located at two opposites and shown in Figure 14-1.

A color of the solution is different from that of the others because it absorbs light selectively. When a beam of white light passes through a solution, the colorless solution absorbs none of the light, but a colored solution can absorb the specific light of its complementary color. For example, the aqueous solution of KMnO₄ is purple because of its absorption of green light; CuSO₄ solution is blue because of its absorption of yellow light. So the color of a solution is the complementary color of the light that it absorbs. The darker the color of the solution is, the greater the amount of light it would absorb. The color we perceive depends not only on the wavelength of light, but on its intensity.

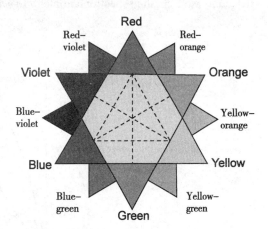

Figure 14-1 Diagrammatical Drawing for the Complementary Colors

> **Question and thinking 14-1** What's the practical significance of absorption spectra? How to draw an absorption spectrum?

14.1.2 Absorption Spectrum of Substance

The **absorbance** of a solution is a word used to define the absorption of light by the substance. Different solution has different absorption of light, which can be displayed by an absorption spectrum. An ultraviolet-visible **absorption spectrum** or **absorption curve** is essentially a graph of absorbance represented by the symbol A versus the wavelength represented by the symbol λ in a range of ultraviolet or visible regions. The absorption spectrum shows how A varies with λ, which is very important for a qualitative or quantitative analysis to select an optimal absorption wavelength.

The curves in Figure 14-2 represent the absorption spectra with various concentrations. They have similar shapes and the same wavelength of maximum absorbance. The wavelength of maximum absorbance is called as λ_{max}, the λ_{max} of tri-orthophenanthroline ferro complex ion is 508nm (the absorption light is green), and this is the reason why its color is violet-red.

Different substance may show different absorption spectra and the different wavelengths of the maximum response. So it can be applied to the qualitative analysis. Additionally, for the same substance with different concentration, the absorbance increases with the concentration around the absorption peak. So we can determine the content of a given substance according to its absorbance at a specific wavelength. Because of the maximum absorption response at λ_{max}, the λ_{max} is usually selected as the detection wavelength in spectrophotometry to achieve the high sensitivity.

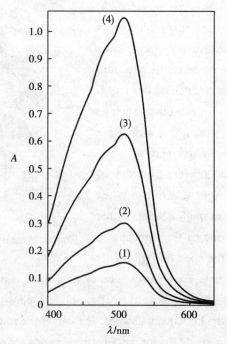

Figure 14-2　The Absorption Spectrum for tri-Orthophenanthroline Ferro Coordination ion with Different Concentrations

14.2　Fundamental Principle of Spectrophotometry

14.2.1　Transmittance and Absorbance

The attenuation of light as it passes through a sample is described quantitatively by two separate, but related terms are transmittance and absorbance.

When a beam of parallel monochromatic light passes through a homogeneous, non-scattering, colored solution, some of the light is absorbed, some is transmitted and some is reflected from the vessel surface just as Figure 14-3.

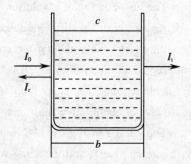

Figure 14-3　The Schematic Diagram of Light Passing through the Solution

If the intensity of the incident light is I_0; the intensity of the absorption light is I_a; the intensity of the transmittance light is I_t; and the intensity of the reflection light is I_r, then

$$I_0 = I_a + I_t + I_r \qquad (14.1)$$

The sample and the reference solution are put into the absorption cells, respectively, each cell has the same thickness and is made of the same material. A monochromatic light (the intensity is I_0) passes through these solutions, respectively. Because their intensity of the reflection light I_r is similar, so Equation 14.1 can be rewritten as

$$I_0 = I_a + I_t \tag{14.2}$$

the more I_a is, the less I_t is, the stronger the absorption ability of a given compound is.

Transmittance T is defined as the fraction of the original light that passes through the sample, the ratio of the intensity of the transmittance light I_t to the intensity of the incident light I_0.

$$T = \frac{I_t}{I_0} \tag{14.3}$$

The negative logarithm of light transmittance is called absorbance, which is indicated by the symbol A. The larger A is, the stronger the absorption of light by the solution.

$$A = -\lg T = -\lg \frac{I_t}{I_0} = \lg \frac{I_0}{I_t} \tag{14.4}$$

Absorbance is the more common way for expressing the attenuation of light because, as shown in the next part, it is a linear function of the analyte's concentration.

14.2.2 Beer-Lambert Law

The light absorption is related not only with the characteristics of the solution, but also with other factors such as the wavelength of the incident light, concentration of the solution, the thickness of the absorption cell and the temperature, etc. The quantitative relationship between the absorbance and the concentration or the quantitative relationship between the absorbance and the length of light path through the cuvette in cm were observed by Beer in 1760 and Lambert in 1852 respectively.

From Lambert's studies, absorbance A is proportional to the pathlength of a sample solution when a beam of appropriate wavelength light passes through a solution of a fixed concentration. This observation is called as Lambert's law which can be expressed as

$$A = k_1 b \tag{14.5}$$

where b is the path length of the solution (or the thickness of the absorption cell, the cell path length); k_1 is the proportionality constant. Lambert's law is applied into the homogeneous medium.

From Beer's studies, absorbance A is proportional to the concentration of the solution when a beam of light of appropriate wavelength passes through a solution in a fixed cuvette. The relationship between the absorbance and the concentration is called as Beer's law which can be expressed as

$$A = k_2 c \tag{14.6}$$

where c is the concentration of the solution; k_2 is the proportionality constant. The Beer's law can be applied when the light is a monochromatic light.

Combining Equation 14.5 and 14.6, a new equation is obtained as

$$A = \varepsilon b c \tag{14.7}$$

where ε is the analyte's **molar absorptivity** with unit of $L \cdot mol^{-1} \cdot cm^{-1}$; b is the pathlength of the solution (or the thickness of the absorption cell) with unit of cm; c is the molarity of the solution with unit of $mol \cdot L^{-1}$.

When concentration is expressed in mass concentration ρ with unit of $g \cdot L^{-1}$, the molar absorptivity is replaced by **mass absorptivity** a with unit of $L \cdot g^{-1} \cdot cm^{-1}$.

$$A = ab\rho \tag{14.8}$$

Equation 14.7 and 14.8 are called as **Beer-Lambert law**, the fundamental law for the absorption of non-scattering light by homogeneous non-scattering medium and the quantitative analysis of spectrophotometry. The absorptivity ε and a are related with the analyte's properties, the wavelength of the incident light, the solvent, and the temperature.

The mass absorptivity a and molar absorptivity ε can be converted to each other as

$$\varepsilon = aM_B \tag{14.9}$$

where M_B is the molar mass of the analyte with unit of $g \cdot mol^{-1}$.

The molar absorptivity ε is usually replaced by **specific extinction coefficient** in medical study. Specific extinction coefficient equals to the absorbance of a solution prepared by dissolving 1g analyte in 100mL solvent, and measured in an absorption cell of 1cm thickness. It is symbolized as $E_{1cm}^{1\%}$. The relationship between $E_{1cm}^{1\%}$ and ε or a can be inferred as

$$E_{1cm}^{1\%} = \frac{10 \times \varepsilon}{M} \tag{14.10}$$

$$a = 0.1 E_{1cm}^{1\%} \tag{14.11}$$

From Beer-Lambert law, absorbance A is proportional to the concentration c or the pathlength b of solution, but the relationship between the transmittance T and the concentration of the solution c as well as the thickness b is exponential.

$$-\lg T = \varepsilon bc$$
$$T = 10^{-\varepsilon bc} \tag{14.12}$$

Pay attention to the following points when applying the Beer-Lambert law.

(1) The Beer-Lambert law can be applied only for the monochromatic light. The determination will deviate more from the Beer-Lambert law when the wavelength coverage gets wider. The law can be applied not only in visible spectrophotometry but also in ultraviolet one.

(2) If there were two or more than two solutes in the solution simultaneously and there was no interaction between the coexisting substances at the same wavelength and without changing the absorption coefficient of the coexistence, the absorbance is equal to the sum of the absorbance of each coexisting solutes. That is

$$A = A_a + A_b + A_c + \cdots \tag{14.13}$$

where A is the total absorbance, $A_a, A_b, A_c, \cdots$ is the absorbance of absorbing species a, b, c, $\cdots$, respectively.

When the wavelength of incident light is different, the absorptivity ε or a are different too. If the value of absorptivity is larger, the intensity of absorption is stronger and the sensitivity of determination is higher.

(3) The spectrophotometry is only applied to the determination of micro-content components. If the solution is concentrated, its absorption characteristics will deviate from the Beer-Lambert law.

> **Question and thinking 14-2** What's the important meaning of molar absorptivity? How to obtain it? Which factors will produce an effect on the molar absorptivity?

Sample Problem 14-1 The relative molar mass of a known compound is $25\text{g}\cdot\text{mol}^{-1}$. It is dissolved in the ethanol to get a $0.150\text{mmol}\cdot\text{L}^{-1}$ solution. The percent transmittance is 39.8% at 480nm while the thickness of the absorption cell is 2.00cm. Calculate the molar absorptivity and mass absorptivity (ε and a) of this compound under the above conditions.

Solution The molar absorptivity and mass absorptivity (ε and a) of this compound under the above conditions can be figured out from Beer-Lambert law.

$$A = -\lg T = \varepsilon bc \quad \text{or} \quad A = -\lg T = ab\rho$$

Because, $c = 0.150 \times 10^{-3}\text{mol}\cdot\text{L}^{-1}$, $b = 2.00\text{cm}$, $T = 0.398$

$$\varepsilon_{480\text{nm}} = \frac{-\lg T}{bc} = \frac{-\lg(0.398)}{2.00\text{cm} \times 0.150 \times 10^{-3}\text{mol}\cdot\text{L}^{-1}} = 1.33 \times 10^3 \text{L}\cdot\text{mol}^{-1}\cdot\text{cm}^{-1}$$

$$a_{480\text{nm}} = \frac{-\lg T}{b\rho} = \frac{-\lg T}{bcM} = \frac{-\lg(0.398)}{2.00\text{cm} \times 0.150 \times 10^{-3}\text{mol}\cdot\text{L}^{-1} \times 251\text{g}\cdot\text{mol}^{-1}} = 5.30 \text{L}\cdot\text{g}^{-1}\cdot\text{cm}^{-1}$$

Sample Problem 14-2 The absorbance of a system consisting of enzyme and adenosine monophosphate AMP is determined as follows:

$$A_{280\text{nm}} = 0.46 \qquad A_{260\text{nm}} = 0.58$$

Molar absorptivity of the enzyme is $2.96 \times 10^4 \text{L}\cdot\text{mol}^{-1}\cdot\text{cm}^{-1}$ at 280nm and $1.52 \times 10^4 \text{L}\cdot\text{mol}^{-1}\cdot\text{cm}^{-1}$ at 260nm. The molar absorptivity of AMP is $2.4 \times 10^3 \text{L}\cdot\text{mol}^{-1}\cdot\text{cm}^{-1}$ at 280nm and $1.5 \times 10^4 \text{L}\cdot\text{mol}^{-1}\cdot\text{cm}^{-1}$ at 260nm. The thickness of the absorption cell is 1.00cm. Calculate the concentrations of enzyme and AMP in the solution.

Solution Because the absorbance A is additive. That is, $A = A_{\text{enzyme}} + A_{\text{AMP}}$

When λ is at 280nm, $A_{280\text{nm}} = A_{\text{enzyme},280\text{nm}} + A_{\text{AMP},280\text{nm}} = 0.46$

When λ is at 260nm, $A_{260\text{nm}} = A_{\text{enzyme},260\text{nm}} + A_{\text{AMP},260\text{nm}} = 0.58$

According to Beer-Lambert law,

$$A_{280\text{nm}} = \varepsilon_{\text{enzyme, 280nm}} bc_{\text{enzyme}} + \varepsilon_{\text{AMP, 280nm}} bc_{\text{AMP}}$$

$$0.46 = 2.96 \times 10^4 \text{L}\cdot\text{mol}^{-1}\cdot\text{cm}^{-1} \times 1.00\text{cm} \times c_{\text{enzyme}} + 2.4 \times 10^3 \text{L}\cdot\text{mol}^{-1}\cdot\text{cm}^{-1} \times 1.00\text{cm} \times c_{\text{AMP}}$$

$$A_{260\text{nm}} = \varepsilon_{\text{enzyme, 260nm}} bc_{\text{enzyme}} + \varepsilon_{\text{AMP, 260nm}} bc_{\text{AMP}}$$

$$0.58 = 1.52 \times 10^4 \text{L}\cdot\text{mol}^{-1}\cdot\text{cm}^{-1} \times 1.00\text{cm} \times c_{\text{enzyme}} + 1.5 \times 10^4 \text{L}\cdot\text{mol}^{-1}\cdot\text{cm}^{-1} \times 1.00\text{cm} \times c_{\text{AMP}}$$

Combining the two equations, c_{enzyme} and c_{AMP} can be got as

$$c_{\text{enzyme}} = 1.4 \times 10^{-5} \text{mol}\cdot\text{L}^{-1}$$

$$c_{\text{AMP}} = 2.5 \times 10^{-5} \text{mol}\cdot\text{L}^{-1}$$

14.3 Improving Sensitivity and Accuracy in Determination

14.3.1 Main Sources in Spectrophotometric Errors

The calibration curve is a straight line according to the Beer's law. But in practice, the curve will bend especially when the concentration of the colored solution is high, which is called the deviation of Beer's law. The more the departure is, the more the error is. There are many reasons for the deviation mainly by instruments, solution itself and subjective factors.

Question and thinking 14-3 Which factors can be selected or controlled in order to improve the sensitivity and the accuracy of spectrophotometry in actual determination?

1. Errors from instruments

Instrument errors are caused by the low sensitivity of photoelectric tube, the low accuracy of the photocurrent measurement, the instability of the light source, the inaccurate reading, and other factors. These errors cause the transmittance difference ΔT between the measured value T and the true value of it. Consequently, the determined concentration error Δc is produced. The relationship between the relative error on the concentration and the transmittance can be derived from Beer's law.

$$\frac{\Delta c}{c} = \frac{0.434 \Delta T}{T \lg T} \tag{14.14}$$

The transmittance error ΔT is generally in the range of $\pm 0.02 \sim \pm 0.01$.

If $\Delta T = 0.01$, different T values are substituted in Equation 14.14 to obtain the corresponding relative errors of concentration, $\frac{\Delta c}{c}$. Plotting $T \times 100$ and $\frac{\Delta c}{c}$ in the x/y system (x, $T \times 100$; y, $\frac{\Delta c}{c}$) to form the curve shown in Figure 14-4.

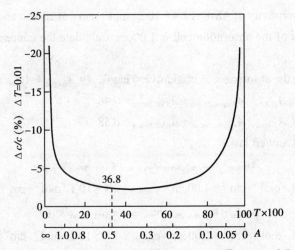

Figure 14-4 The Relationship between the Measurement Errors and the Transmittance

Figure 14-4 shows that the relative error of concentration is greater when the transmittance of the solution is tremendous or tiny. The relative error of concentration is lowest when $T\% = 36.8\%$ or $A = 0.434$. In the practical determination, the transmittance is regulated in the range of 20% to 65%, i.e., the absorbance is in the range of 0.2 to 0.7 by adjusting the concentration of the sample solution or the path length to get more accurate results.

2. Deviation from Beer's law

If the solution deviates from Beer's law, the linearity of the A-c curve becomes worse and the straight line becomes bent. The main reasons for these phenomena are listed below.

(1) Chemical factors. The concentration of the solution will change owing to the dissociation, association or solvation of the absorbing substance when it is considered unstable, which leads to the alteration of absorbance.

(2) Optical factors. Beer's law can be applied only when the light is monochromatic. However, the light from the monochromator is usually chromatic light within the narrow wavelength coverage. The deviation from Beer's law is led by different absorptive abilities of the lights with different wavelengths. The greater the absorptivity difference is, the more the deviation is (Figure 14-5).

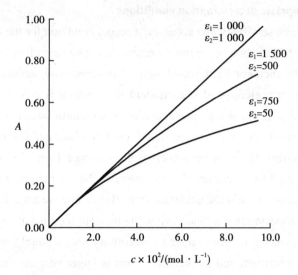

Figure 14-5 The Deviation from Beer's Law Caused by the Mixture of Two Different Absorptive Light

3. Subjective errors

Subjective errors are caused by improper operation. These errors would be caused by the different treatment for the sample and standard solutions, the different dosage of chromomeric agent, the different storage time and the different reaction temperature. So the sample solution must be manipulated carefully according to the operation procedure in order to decrease this kind of error as far as possible.

14.3.2 The Way to Improve Sensitivity and Accuracy

1. Select the appropriate colorants

The concentration of a colorless solution cannot normally be determined by visible spectrophotometry. The colorless analyte in the sample must be transformed into colored derivative by a chemical reaction which is called as a color reaction. The addition of a color reagent leads to a color reaction and the absorbance of the colored product can then be measured with a colorimeter. The reagent added in the reaction is called as a color developer. The commonly used color reaction is chelating reaction from which the product is not only stable but also of characteristic color. The most commonly used color developers are some organic reagents.

In the analysis, the appropriate color reaction should be selected, usually considering the following factors:

(1) High sensitivity. ε of the colored product must be large. The value of ε is generally about $10^4 \sim 10^5 \mathrm{L \cdot mol^{-1} \cdot cm^{-1}}$.

(2) Good selectivity. The color developer can react only with a single component or a few components. In practical analysis, the color reaction occurs only between the color developer and the

analyte component but not other components. The color of the developer should be quite different from the color of the reaction product to avoid the interference on the determination of the color developer.

(3) The color developer has no apparent absorption at the detection wavelength.

(4) The color product formed in reaction must be stable.

2. Select the appropriate determination conditions

In order to obtain high sensitivity and accuracy, a proper condition for the color reaction should be provided, the conditions for the absorbance measurement should be controlled and optimized too.

(1) Selection of the incident light wavelength. The sensitivity, accuracy, and selectivity are heavily affected by the wavelength of the incident light which is selected from the absorption spectrum. Ordinarily, λ_{max} is chosen as the detection wavelength because at which the analyte's absorptivity is maximum and the highest sensitivity could be obtained. The incident light is actually polychromatic light within the narrow wavelength coverage from the monochromator in the spectrophotometer. Figure 14-6 illustrates that the curve is flat over the wavelength range near λ_{max}, the variation of the absorptivity and the deviation from Beer's law are tiny. The relationship between the absorbance and the concentration is linear when the incident light is monochromatic light "a" with λ_{max} as the detection wavelength. However, the absorptivity varies hugely with the wavelength "b" in the steep part of the spectrum, and the negative error is larger because the deviation from Beer's law is greater. The relationship between the absorbance and the concentration is not linear when the monochromatic light "b" is selected as the incident light. If the absorption peak is sharp near λ_{max}, this detection wavelength should but not used, and the wavelength in other region where the absorbance varies little can also be selected as the detection wavelength, on the assumption that the sensitivity permits.

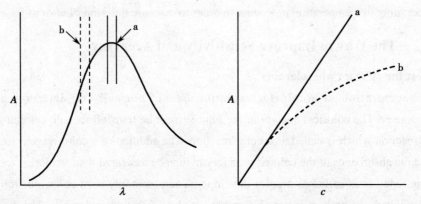

Figure 14-6 The Relationship between the Selection of Incident Light and Beer's Law

(2) The amount of color developer. Coordination reaction is frequently used as a color reaction. The color developer is usually excessive to ensure to transform the analyte to the colored compound completely. However, the more amount of color developer does not necessarily render the better results. Some side reactions accompany with the increasing amount of color developer, the structures of some products may be changed that may result in the change of the color of the solution, which is not good for determination. The amount of color developer is decided by the experiment in which the concentration

of the analyte and other reaction conditions are fixed, measuring the absorbance at different amount of color developer, and then plotting the curve by the absorbance versus the amount of color developer as shown in Figure 14-7. The absorbance is nearly a constant value when the amount of color developer is in the range from "a" part to "b" part. Thus, the appropriate amount of color developer can be determined within this range.

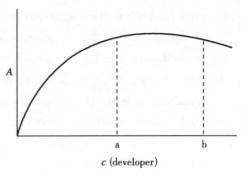

Figure 14-7 The Relationship between the Absorbance and the Usage amount of Color Developer

(3) Acidity of the solution. A color developer is usually a weak organic acid (demonstrated as HL here), the alteration of the acidity will affect the color reaction.

$$HL \rightleftharpoons H^+ + L^-$$
$$M^{n+} + nL^- \rightleftharpoons ML_n(\text{colored compound})$$

where M^{n+} is the analyte ion. The lower the acidity is, the more favorable the color reaction is. Lower acidity can increase the concentration of the colored compound. However, the metal hydroxide precipitate would form if the acidity is too low. For example, sulfosalicylic acid anion (Sal^{2-}) and ferric ion Fe^{3+} can react to form three coordination compounds with different colors under different acidity, whose coordination ratio are 1:1, 1:2 and 1:3, respectively.

pH	Coordination compound	Color
1.8~2.5	$[(Fe(Sal))]^+$	Purple red
4~8	$[(Fe(Sal)_2)]^-$	Orange
8~11	$[(Fe(Sal)_3)]^{3-}$	Yellow
>12	$Fe(OH)_3 \downarrow$	

The optimum acidity of the color reaction is determined by the experiment in which the concentration of analyte and the other reaction conditions are fixed, measuring the absorbance of the solution at different pH, and then plotting the curve by the absorbance versus the pH. The optimum acidity range can be inferred from that curve.

(4) The coloration temperature. A color reaction usually takes place at room temperature. Heating is necessary to activate reaction sometimes. But some colored compounds would decompose at the higher temperature. Therefore, the curve of absorbance versus temperature must be plotted from the experiment for different color reactions so as to find the appropriate reaction temperature.

(5) The coloration time. Some color reactions are fast but some is slow, so the proper time is needed for the color reaction to form the stable color product. On the other hand, some colored compounds are easily to be decomposed by oxidation or photochemical reaction. So the absorbance A and the reaction time t curve must be plotted to figure out the appropriate reaction time.

3. Select the correct blank solution

A proper blank solution or reference solution must be adopted in absorbance determination. In most spectrophotometric analyses, it is important to prepare a blank solution as a reference to eliminate all of the irrelevant absorption of the analyte which includes the incident light absorption by solvent and other substances, the scattering when the light passing through the solution, and the reflection by the absorption cell surface. There are three kinds of blank solution commonly used as follows.

(1) Solvent blank. We can adopt solvent as blank solution, if the color developer and other reagents used to prepare the sample solution are colorless, additionally, other color substances have no interference to the analyte. It is called as solvent blank.

(2) Reagent blank. If the color developer is colored, the sample solution has little or no absorption under determinate condition, a reagent blank is utilized to adjust absorption by color developer. A reagent blank is the solution containing all components except analyte, but it has not been subjected to all sample preparation procedures. The reagent blank is just a standard solution without the analyte itself in the standard curve method.

(3) Sample blank. If the sample itself is colored (e.g. other colored ions are mixed in the sample solution), but the color developer is colorless and does not react with the other components other than the analyte, a sample blank could be used as blank correction. A sample blank is such a solution containing all components except the color developer.

4. Interference of coexisting ions and its elimination

Sometimes, the coexisting ions themselves are colored, or sometimes they can react with the color developer to produce the colored compounds. All of them will interfere with the determination. The frequently used methods to avoid the interference are as follows.

(1) Control the acidity of the color reaction to ensure that the color developer react only with the analyte. The selectivity becomes better by the acidity regulation.

(2) Add the masking agent into the sample solution to react with the interfering ions to form colorless and stable compounds, to make them not to react with color developer.

(3) Separate the interfering ions before determination by ion exchange, forming precipitation or solvent extraction, etc. to avoid their interference.

14.4　Visible Spectrophotometry

14.4.1　Spectrophotometer

Spectrophotometry is the manner that hinges on the quantitative analysis of molecules depending on how much light is absorbed by colored compounds. Spectrophotometry uses **spectrophotometer**, that can measure a light beam's intensity as a function of its color (wavelength).The elementary components of a spectrophotometer and their correlation could be represented by the following block diagrams.

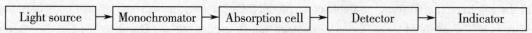

The function of each component is briefly introduced as follows.

(1) Light source. The tungsten lamp is often adopted as the light source of visible spectrophotometer. This lamp can emit continuous spectra with the wide wavelength coverage from 320nm to 2 500nm. The visible wavelength region is from 360nm to 1 000nm.

(2) Monochromator. A monochromator is an optical device that transmits a selectable narrow wavelength band of light from a wide range of wavelength available at the input. It is generally

composed of dispersive components such as a diffraction grating. The advantage of a grating monochromator is that the wavelength coverage available is wider, the dispersive power for each wavelength is almost the same, and the wavelength of the spectra has equal interval, all of which make it more applicable.

(3) Absorption cell. In spectrophotometer, sample containers are called absorption cells, or cuvettes. In visible spectrophotometry, the cells are made of optical glass. The cells used in the determination must be matched each other, that is, they must have the same thickness and transparency. The path length of the absorption cells in spectrophotometry are variable, such as 0.5cm, 1.0cm, 2.0cm, etc, all of which can be selected for analysis. Usually 1.0cm is the most popular path length in the determination.

(4) Detector. The light intensity will be changed before and after passing through the solution owing to the absorption of analyte. The function of a detector is to convert the intensity of transmittance light to an electric signal. The commonly used photoelectric transformation components are photoelectric cell or phototube and photomultiplier. A photoelectric cell is a type of vacuum tube that is sensitive to light. For example, a phototube emits electrons from a photosensitive, negatively charged surface (the cathode) when struck by visible light or ultraviolet radiation. Electrons flow through a vacuum to a positively charged collector, and generated current is proportional to the radiation intensity. The detector response depends on the wavelength of the incident photons. A Photomultiplier is more sensitive than phototube, furthermore, it has self-amplification. Therefore, photomultipliers are utilized extensively as detectors in current spectrophotometers.

(5) Indicator. The electric signals output from the photoelectric cell are input to the indicator after being amplified by the amplifier. The indicator is generally equipped with a microampere electricity meter, recorder, digital displayer and printer, etc. which can be connected with a computer. So the operation condition, absorption spectra and other data can be displayed on the screen. Meanwhile, the data can be processed and recorded, which make the determination more convenient and accurate.

14.4.2 Quantitative Analysis Methods

The frequently used quantitative analysis methods are based onthe calibration curve and standard comparison method.

(1) Calibration curve. In analytical chemistry, a **calibration curve**, also known as a standard curve, is a general method to determine the concentration of a substance in an unknown sample by comparing the unknown one to a set of standard samples of known concentration. To construct such a curve, standard solutions of various known concentrations were prepared, the maximum wavelength is selected as the detection wavelength, and cuvettes with the same path length were used to measure the resulting absorbance of the standard solutions. Then a straight line passing through the origin was plotted by Beer-Lambert law. Figure 14-8 shows a calibration curve made by the standard solutions of vitamin B_{12}. So the operator can measure the response of the unknown and, using the calibration curve, can interpolate to find the concentration of analyte.

A calibration curve is applied to the regularity determination in batches. The sample solution must be assayed under the same condition as the standards. Besides that, the concentration of the sample solution must lie within the linear range of the calibration curve.

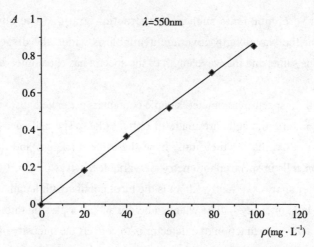

Figure 14-8　The Schematic Drawing of the Calibration Curve for Vitamin B_{12}

(2) Standard comparison method. A standard solution whose concentration symbolized as c_s is closed to that of the sample solution must be prepared and the absorbance of this standard solution A_s is measured. The absorbance of sample solution A_x is measured under the same condition. Then the concentration of the sample solution can be calculated as follows

$$c_x = \frac{A_x}{A_s} \times c_s \qquad (14.15)$$

This method is used in the unusual analytical work. It is simple and convenient, but the concentration of the analyte and the standard solution must be similar, otherwise the determination error will be greater.

14.4.3　Application in the Clinical Examination

(1) Determination of the serum iron. Iron is one of the essential trace elements for human body. Iron in serum will decrease when falling ill of anemia, blood loss, denutrition, infection, etc. On the contrary, iron in serum will increase when the hepatopathy or defective blood formation occurs. Under the acidic condition, the iron which combines with transferrin will be ionized as Fe^{3+}, then a reduction reaction of Fe^{3+} to Fe^{2+} is carried out by adding the reducing agent ascorbic acid or hydroxylamine hydrochloride, at last Fe^{2+} reacts with ferrozine disodium salt to give a purple red coordination compound with an absorption at 578nm, and its absorbance has a proportional relationship with the iron concentration in serum. Therefore, the iron amount in serum can be measured by this method.

(2) Determination of seralbumin. Seralbumin can transport the fatty acids, bile pigments, amino acids, steroid hormones in body fluids and maintain normal osmotic pressure of blood. In clinic, it can be used to cure shock and burn because it can replenish the blood loss due to the operation, unexpected accident or massive haemorrhage. It also can be used as a plasma extender. A positive charge seralbumin in pH 4.2 buffer solution will combine with a negative bromocresol green to produce blue green coordination compound which has the absorption at 630nm. The absorbance of this coordination compound is proportional to the concentration of seralbumin. The amount of seralbumin will be obtained by comparing with the standard albumin solutions which are taken through all steps of the analytical procedure.

14.5 Ultraviolet Spectrophotometry

The visible spectrophotometry is employed by a colored solution or a solution whose analytes can react with color developer to generate the colored compounds. The visible wavelength coverage is from 380nm to 760nm. Most of substances have no obvious absorption at visible light region, but they have characteristic absorption at near ultraviolet (200~380nm), so they can be analyzed by selecting ultraviolet light as light source in spectrophotometry. Generally, if a spectrophotometer can be used in ultraviolet region, it can be used in visible region too. Additionally, its precision is higher than that of the visible spectrophotometer. It can be applied to quantitative analysis, qualitative analysis, purity identification, determination of some physical and chemical constants and molecular structure elucidation of organic compounds in combination with other analytical methods.

14.5.1 Main Classifications of Ultraviolet Spectrophotometer

The ultraviolet spectrophotometer can be classified into two types according to the optical system. They are single-wavelength spectrophotometer and dual-wavelength spectrophotometer. The single-wavelength spectrophotometers include two types as single-beam spectrophotometer and dual-wavelength spectrophotometer.

1. The single-wavelength single-beam spectrophotometer

A single-wavelength single-beam spectrophotometer employs a tungsten lamp and a hydrogen lamp as the light source which wavelength coverage is usually from 200nm to 800nm. There is only one beam of monochromatic light from the light source to detector. And a grating is used to disperse the light. A single-wavelength single-beam spectrophotometer is widely applied in laboratories because of its simple construction, reasonable price and convenient operation. Its optical system is illustrated in Figure 14-9.

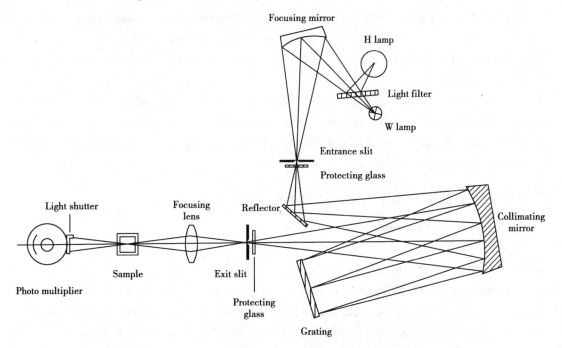

Figure 14-9　The Optical System of the Single-Beam Spectrophotometer

There are two kinds of light sources available for analysis, one is tungsten lamp applied to visible spectrophotometry; the other is hydrogen lamp applied to ultraviolet spectrophotometry (200~400nm).

The light passes through the focusing lens, and through the entrance slit to the reflector. Then it is reflected to the collimating mirror and dispersed by the grating, and then it is reflected back to the collimating mirror. The collimating mirror focuses each wavelength at a different point on exit slit. After that, the light passes through the focusing lens then through an absorption cell. The transmitted light goes into a blue sensitive phototube (200~625nm) or a red sensitive phototube (625~1 000nm), or a photomultiplier tube. Finally, transmittance and absorbance of the analyte can be displayed directly by the indicator.

Ultraviolet light can be absorbed by glass which will cause interference for measurement. So the lens, hydrogen lamp, photocell, and cuvette in ultraviolet spectrophotometer must be made of quartz which makes the ultraviolet light pass without absorption nearly.

2. The single-wavelength double-beam spectrophotometer

The optical principle of a double-beam spectrophotometer is that the monochromatic light from the monochromator is divided into two beams of the alternative monochromatic light. The light alternately passes through the sample and the reference (blank), directed by a rotating mirror (the chopper) into and out of the light path. When light passes through the sample, the detector would measures irradiance P. When the chopper diverts the beam through the reference cuvet, the detector measures P_0. The beam is chopped several times per second, and the circuitry automatically compares P and P_0 to obtain transmittance and absorbance. This procedure provides automatic correction for changes of source intensity and detector response with time and wavelength because the power emerging from the two samples is compared so frequently.

3. The dual-wavelength spectrophotometer

The optical principle of a dual-wavelength spectrophotometer is that the light from the same light source is divided into two beams, and then they pass through two monochromators to provide two beams of light with different wavelength λ_1 and λ_2. These lights alternately irradiate the same absorption cell to give absorbance difference value ΔA ($\Delta A = A_2 - A_1$). Since this method does not need the reference absorption cell, it can avoid the errors from absorption cell mismatch, and the errors of difference in refractive index and scattering between the sample solution and the reference solution. Its great advantage is that the analyte can be quantitatively assayed under the absorption interference from the background or coexisting components.

14.5.2 Application of Ultraviolet Spectrophotometry

1. Qualitative identification

Most compounds are characteristic in their spectra, such as the location, the shape, the intensity and the absorptivity of the absorption peak. Comparing the absorption spectrum of the sample with that of the standard substance or with the graph collected in literature, they maybe the same compounds if their absorption spectra are exactly the same. They are definitely not the same compounds if their absorption spectra are significantly different. The compound can be identified by comparison of λ_{max}, ε (λ_{max}) or a (λ_{max}), and the ratio of absorbance, etc.

(1) Comparison of λ_{max}, ε (λ_{max}) or a (λ_{max}). Two compounds would have similar absorption spectra if they have the same functional groups. For example, cortisone acetate and prednisone acetate cannot be identified only by λ_{max} because their λ_{max} (238±1nm) are nearly the same in anhydrous ethanol. The comparison of a (238nm) can help us distinguish them because a (238nm) of cortisone acetate is 39.0L·g^{-1}·cm^{-1}, while a (238nm) of prednisone acetate is 38.5L·g^{-1}·cm^{-1}.

(2) Comparison of absorbance (or the ratio of absorptivity). When a given substance has two or more than two absorption peaks in its ultraviolet absorption spectrum, it can be identified by the absorbance ratios from the different absorption peaks. For example, Chinese Pharmacopoeia (edition in 2015) specifies that there are three absorption peaks at 278nm, 361nm, and 550nm in the absorption spectrum of vitamin B$_{12}$, and the absorbance ratios are

$$\frac{A_{361}}{A_{278}} = 1.70 \sim 1.88 \qquad \frac{A_{361}}{A_{550}} = 3.15 \sim 3.45$$

2. Quantitative determination

The light absorption is consistent with Beer-Lambert law in the near ultraviolet region, so the quantitative determination methods are identical with the visible spectrophotometry. The ultraviolet spectrophotometer is usually more sensitive than the visible one. Comparison method is commonly used in quantitative determination. The determination can also be developed directly by comparison with mass absorptivity or molar absorptivity recorded in literature, or indirect determination the content of the drug by measuring the absorbance of the decomposition product. For example, leucogen is a drug used extensively to treat all leucocytopenia and aplastic anemia. For cancer patients, it is a perfect drug to increase leucocyte after chemoradiotherapy. In strong alkaline medium, the degradation product of leucogen, α-formacyl ethyl phenylacetate has a absorption peak at 300nm. So the determination of leucogen can be accomplished by the absorbance determination of its degradation product.

Sample Problem 14-3 a (361nm) of vitamin B$_{12}$ is 20.7L·g^{-1}·cm^{-1}. 30.0mg of the sample is weighed accurately and is dissolved in 1 000mL water. Its absorbance is 0.618 at 361nm in 1.00cm cuvette. Calculate the mass fraction of vitamin B$_{12}$ in the sample solution.

Solution According to Beer-Lambert law, the mass concentration of vitamin B$_{12}$ in the sample solution can be calculated by the following calculation:

$$\rho_{determination} = \frac{A_{361}}{ab} = \frac{0.618}{20.7L \cdot g^{-1} \cdot cm^{-1} \times 1.00cm} = 0.029\ 9g \cdot L^{-1}$$

The sample solution is prepared as 30.0mg·L^{-1}, that is, $\rho_{sample} = 0.030\ 0g \cdot L^{-1}$, so the mass fraction of vitamin B$_{12}$ in sample solution can be calculated as follows:

$$w(VB_{12}) = \frac{\rho_{determination}}{\rho_{sample}} = \frac{0.029\ 9g \cdot L^{-1}}{0.030\ 0g \cdot L^{-1}} = 0.997$$

14.5.3 Analysis of Organic Compound Structures

An ultraviolet spectrum is produced by chromophores and auxochrome in the molecule. The spectrum cannot reflect the all characteristics of a molecule, but it is of great value to analyze the conjugated structure and aromatic ring structure of the compounds. The type, the location and the number of the substituent of compounds can be deduced according to the intensity and the number of ultraviolet absorption peaks.

No absorption in the range of 200nm to 800nm will be observed in the UV spectrum of a compound with open chain or cyclic conjugated system in its structure. A compound containing two conjugated systems in its structure would give the absorption in the range of 210nm to 250nm in UV spectrum. And a compound containing a carbonyl group in its structure would have the absorption in the range of 250nm to 300nm in its UV spectrum. For example, the structure of chloral hydrate (obtained by dissolving trichloroacetic aldehyde in water) can be deduced as follows.

From the absorption spectrum of trichloroacetic aldehyde in hexane, the maximum absorption wavelength is 290nm, the molar absorptivity ε (290nm) is 33L·mol^{-1}·cm^{-1}, which agrees with the characteristic absorption of carbonyl at 250~300nm, so the structure of trichloroacetic aldehyde in hexane is still CCl_3CHO. But in the aqueous solution of trichloroacetic aldehyde (chloral hydrate), there is no absorption at 290nm which indicates that there is no carbonyl in the aqueous solution of trichloroacetic aldehyde, water molecule is added to the carbonyl group to form a new structure of chloral hydrate. That is

$$Cl_3C-\underset{\underset{OH}{|}}{\overset{\overset{OH}{|}}{C}}-H$$

The molecular structures of the constitutional isomer can be concluded by ultraviolet absorption spectrum. For example, the maximum absorption wavelength of abietic acid (Ⅰ) and levo-abietic acid (Ⅱ) are 238nm with a corresponding molar absorptivity of 15 100L·mol^{-1}·cm^{-1} and 273nm with a corresponding molar absorptivity of 7 100L·mol^{-1}·cm^{-1}, respectively.

(Ⅰ) (Ⅱ)

Key Terms

吸收光谱	absorption spectrum
吸光度	absorbance
吸收曲线	absorption curve
标准曲线	calibration curve
检测器	detector
激发态	excited state
基态	ground state
指示器	indicator
光源	light source

质量吸光系数	mass absorptivity
摩尔吸光系数	molar absorptivity
朗伯比尔定律	Beer-Lambert law
单色光	monochromatic light
单色器	monochromator
分光光度计	spectrophotometer
分光光度法	spectrophotometry
标准曲线法	calibration curve method
透光率	transmittance

Summary

Spectrophotometry is a frequently used qualitative and quantitative analytical method based on the light absorption properties of a given substance and light absorption law. When a beam of light with energy, hv, passes through a solution, substance in a solution will absorb the light because the energy difference between the ground and an excited state of substance is just equal to the energy of the light. This energy gap varies in discrete substances with their varied electronic structures. Thus, a specific substance absorbs a specific wavelength of light, which is referred to its selectivity to light. If wavelength of absorbed light is in visible region (380~780nm), substance displays its complementary color.

Beer-Lambert law is the foundation of a spectrophotometric method for analysis. The law can be expressed as $A = \varepsilon bc$, where c is the molarity of the substance and ε is the molar absorptivity. If mass concentration, ρ, is used instead of molarity, ε is replaced by mass absorption coefficient, a, then $A = ab\rho$. Coefficients a and ε are related by molar mass M_B, that is $\varepsilon = aM_B$.

Generally, there are two kinds of spectrophotometers used for qualitative and quantitative analysis. One is the visible spectrophotometer utilizing visible light source such as a tungsten lamp, the other is ultraviolet (UV) spectrophotometer using light source at wavelength 200~380nm, commonly from a hydrogen lamp. There are many ways to quantitatively determine the concentration of a solution. Calibration curve, standard comparison is commonly used.

To improve the sensitivity and precision of a measurement, it is important to take the following aspects into account:

1. To select a suitable detection wavelength (usually at λ_{max}).
2. To ensure the absorbance to be in the range of 0.2~0.7.
3. To select a proper chromogenic developer.
4. To control proper test conditions.

Exercises

1. What's the principal feature of spectrophotometry comparing with the chemical analysis?
2. What's the mass absorptivity? What's the molar absorptivity? How about their relationship? Why the wavelength of λ_{max} is usually selected as the monochromatic light used for quantitative analysis in spectrophotometry?

3. What's an absorption spectrum? What's a calibration curve? What practical applications do they have respectively?

4. What components does a spectrophotometer consist of? What's the function of each component?

5. A solution is consistent with Beer-Lambert law. When its concentration is c_1, the transmittance is T_1, what's the corresponding transmittance when the concentration is changed to $0.5c_1$ or $2c_1$ keeping the same pathlength? Which is the largest one?

6. The absorbance of a solution which is composed of ferrous ion 0.500μg per milliliter at a wavelength of 508nm in a 2.00cm cell is 0.198. Calculate the molar absorptivity of tri-orthophenanthroline ferrous coordination compound at 508nm.

7. A colored solution with the concentration of 2.0×10^{-4} mol·L^{-1} exhibited an absorbance of 0.120 in a 3.00cm absorption cell. Diluted the solution with equal volume of water and another absorbance of 0.200 was obtained in a 5.00cm absorption cell at the same wavelength. Is the solution still consistent with Beer-Lambert law under such condition?

8. The maximum absorption wavelength of cardiotonic dobutamine is at 260nm with a molar absorptivity of 703L·mol^{-1}·cm^{-1}, the relative molecular mass of dobutamine is 270. Take one tablet of this drug, dissolve it into water, and dilute to 2.00L. Put it aside for a while, then the supernatant is prepared to measure at 260nm in a 1.00cm absorption cell, and an absorbance of 0.687 is obtained. Calculate the mass of dobutamine in this tablet.

9. A compound with a relative molecular mass of 125 has a molar absorptivity of 2.5×10^5 L·mol^{-1}·cm^{-1}. Now an absorbance of 0.600 is obtained using 1.00cm absorption cell after the solution is diluted by 200 times. How many grams of this compound are weighed accurately to prepare 1 liter of above solution?

10. The intensity of transmitted light is half of that of incident light when a monochromatic light of a certain wavelength passes through a solution with pathlength of 1.00cm. How much are transmittance T and absorbance A when the pathlength is 2.00cm?

11. The human's hematological capacity can be measured as follows. Intravenation 1.00mL Evans blue, and collecting blood sample after 10-minute cyclic blending. Centrifugalized the blood sample, the plasma over whole blood is 53%. An absorbance of 0.380 is measured using a 1.00cm absorption cell. On the other hand, 1.00mL Evans blue is added to volumetric flask and diluted to 1.00L. Take 10.0mL of this solution and dilute it to 50.0mL, under the same condition, the absorbance is 0.200. Suppose all Evans blue is distributed in plasma, calculate the human's hematological capacity (L).

12. Vitamin C has a specific extinction coefficient of 560 at 245nm, $E_{1cm}^{1\%}$ (245nm) = 560. A 0.050 0g sample of Vitamin C is dissolved in 100mL sulfuric acid solution with a concentration of 5.00×10^{-3} mol·L^{-1}, then take 2.00mL of this solution and dilute it to 100mL. The resulting solution has an absorbance of 0.551 at λ_{max} (254nm) with a 1.00cm absorption cell. Calculate the mass fraction of Vitamin C in the sample.

Supplementary Exercises

1. 4.12×10^{-5} mol·L^{-1} solution of the complex $[Fe(Ophen)_3]^{2+}$ has a measured absorbance of 0.48 at 508nm in a sample cell with the path length 1.00cm. Calculate the molar absorptivity, then

the absorptivity in units of milligrams of Fe per liter (0.04mmol·L^{-1} solution of the complex is also 0.04mmol·L^{-1} in iron, and the gram atomic weight of Fe is 55.85g·mol^{-1}).

2. If monochromatic light passes through a solution of length 1.00cm. The ratio of I_t/I_0 is 0.25. Calculate the changes in transmittance and absorbance for the solution of a thickness of 2.00cm.

3. A solution containing 1.00mg iron (as the thiocyanate complex) in 100mL was observed to transmit 70.0% of the incident light compared to an appropriate blank. (1) What is the absorbance of the solution at this wavelength? (2) What fraction of light would be transmitted by a solution of iron four times as concentrated?

Answers to Some Exercises

[Exercises]

5. $T_2 = T_1^{1/2}$, $T_3 = T_1^2$, T_2 is the largest one.
6. $\varepsilon = 1.10 \times 10^4 L·mol^{-1}·cm^{-1}$
7. $\varepsilon_1 = 200 L·mol^{-1}·cm^{-1}$, $\varepsilon_2 = 400 L·mol^{-1}·cm^{-1}$, $\varepsilon_1 \neq \varepsilon_2$. The solution is not consistent with Beer-Lambert law.
8. $c = 9.77 \times 10^{-4} mol·L^{-1}$, mass is 0.528g
9. $c = 2.4 \times 10^{-6} mol·L^{-1}$, mass is 0.060g
10. $A_1 = 2A$, $T_1 = T^2$
11. $V = 4.97L$
12. $\omega = 0.986$

[Supplementary Exercises]

1. $\varepsilon = 1.17 \times 10^4 mol·L^{-1}$, $a = 0.209 L·mg^{-1}·cm^{-1}$
2. $A_1 = 0.602$, $A_2 = 1.20$, $T_2 = 0.062$
3. (1) $A = 0.155$, (2) $T_2 = 0.24$

（陈志琼）

Chapter 15
Brief Introduction to the Modern Instrumental Analysis

Instrumental analysis is an analytical method to analyze the chemical composition, amounts and chemical structures of the substances by precision instruments. Analytical instruments are generally composed of signal generator, detector, sensor, signal handler (amplifier) and readout device. Compared to other methods in chemical analysis, instrumental analysis has its special advantage in sensitivity, simplicity, quickly and automatic analysis. Now, instrumental analysis is the most important aspect of modern analytical chemistry.

Instrumental analysis plays the very important roles in scientific research, especially in biochemistry, bioengineering and medicines, etc. It is of great help in realizing the origin of life, life process, disease control and genetic information. It is prerequisite in the research on medicine content, drug action and mechanism, drug metabolism and decomposition, disease diagnosis, drug abuse, and so on.

According to its characteristics and functions, instrumental analysis can be divided into several categories as follows. ①Optical analysis: ultraviolet / visible spectrophotometry, infrared spectrophotometry, chemiluminescence analysis, fluorescence spectrometry, atomic absorption spectroscopy, nuclear magnetic resonance spectrometry, Raman spectrometry and electron diffraction spectrum. ②Electrochemical analysis: potentiometric method, conductance method, stripping voltammetry, polarographic analysis, coulometry, etc. ③Chromatography analysis: gas chromatography, liquid chromatography, ion chromatography, supercritical fluid chromatography, thin layer chromatography and capillary electrophoresis etc. ④Other instrumental analysis: such as, mass spectrometry, thermal analysis and radiochemical analysis.

In this chapter we will give a briefly introduction to several typical widely used modern instrumental methods including their theoretical basis and potential applications in biomedicine.

15.1 Atomic Absorption Spectrometry

Atomic absorption spectrometry (AAS) or atomic absorption spectroscopy is a spectroanalytical procedure for the quantitative determination of chemical elements using the absorption of light by free atoms in the gaseous state.

Atomic absorption spectrometry is developed rapidly in the 1970s and it is a sensitive, accurate and rapid technique for determining the concentration of a particular element (the analyte) in a sample to be analyzed. AAS can be used to determine over 70 different elements and is widely applied in medicine, biology, environmental protection, geology, metallurgy, agriculture, chemical industry and so on.

15.1.1 Fundamental Principles of Atomic Absorption Spectrometry

1. The formation of atomic absorption spectrometry

When a beam of light passes through the atomic vapor of the sample, the electrons in the ground state atoms will absorb energy from the electromagnetic radiation. This amount of the absorbed energy is exactly equal to the energy difference (ΔE) between the ground state and an excited state (usually, the first excited state). Each element in the periodic table has its specific electron arrangement fashion, so different atom has the different energy state, and atom will absorb a specific wavelength of light with different energy. If ΔE is the energy difference from the ground state to transit to the first excited state of atom, the relationship between the energy of electron transition and the wavelength (λ) or frequency (v) can be described by

$$\Delta E = hv = h\frac{c}{\lambda} \tag{15.1}$$

where h is Planck's constant, 6.626×10^{-34} J·s; c is the velocity of light in the unit of m·s^{-1}.

2. Measurements by atomic absorption spectrometry

If the light of the particular frequency passes through an atomic vapor layer of the sample and incident intensity is I_0, the electrons in the ground state of atom will absorb the light of the incident photons and is accompanied by attenuation of the radiation. The intensity of the transmitted radiation I_v is described quantitatively by Beer-Lambert law

$$I_v = I_0 e^{-K_v cL} \tag{15.2}$$

$$A = -\lg T = \lg\frac{I_0}{I_v} = 0.4342 K_v cL \tag{15.3}$$

In Equation 15.2 or 15.3, K_v is the corresponding absorption coefficient of frequency; c is the concentration of the ground state atom; L is the thickness of an atomic vapor (or the pathlength of a monochromatic light passing through a sample); I_v is the intensity of the transmitted radiation and A is the absorbance.

When L and λ are constant, then

$$A = Kc \tag{15.4}$$

where K is a constant and it is related to the wavelength of the radiation and the nature of the sample. The absorbance is proportional to the concentration of the absorbing species. It works very well for the dilute solutions of most substances.

15.1.2 Atomic Absorption Spectrophotometers

Atomic absorption spectrometer was also known as atomic absorption spectrophotometer. A flame atomic absorption spectrometer consists of a light source, a sample-atomizer, a monochromator, and a detection system as shown in Figure 15-1.

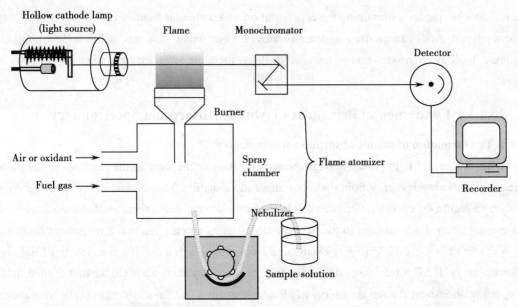

Figure 15-1 Schematic Diagram for an Atomic Absorption Spectrophotometer

> **Question and thinking 15-1** Indicate the similarities and differences between the atomic absorption spectrophotometer and the UV-Vis spectrophotometer. What are the common types of atomic absorption spectrophotometers?

15.1.3 Brief Introduction to the Experimental Techniques

1. Quantitative analysis method

(1) Standard curve. The usual procedure is to prepare a series of standard solutions, and the standards solution and the samples are then aspirated into the flame at the same condition and the absorbance is displayed by the instrument. The standard curve is a plot of absorbance A vs. the concentration c of the standard solutions. The unknown concentration is determined according to the standard curve. More details are described in Section 14.4.2.

(2) Standard addition method. Standard addition is especially appropriate when the sample compositions are unknown or more complicated and they affects the analytical signal. In standard addition method, the known quantities of analyte are added into the sample, from the increased signal, we deduce how much analyte was in the original unknown sample. This method requires a linear response to analyte.

A typical procedure involves preparing several solutions containing the same amount of sample, but different amounts of standard solution. For example, five 25mL volumetric flasks are each filled with 10mL of sample solution, then the standard solution is added in differing amounts, such as 0.00, 1.00, 2.00, 3.00, and 4.00mL, the flasks are then diluted to the mark and mixed well. Measure the absorbance A and plot the amounts or concentration of the added standard solution on the abscissa and the absorbance on the ordinate. If the curve does not pass through the origin, there is the tested element. We prolong the curve until it intersects abscissa, the distance between original to intersecting point indicate the concentration of the tested element.

Sample Problem 15-1 The amount of lithium Li in the serum can be measured by standard addition method. A standard solution of Li is prepared by dissolving 1.598 8g of lithium sulfate monohydrate $Li_2SO_4 \cdot H_2O$ (spectral purity), and transferring it to a 500.00mL volumetric flask and diluted to mark with distilled water. Four aliquots of 0.10mL each are taken from the same sample solution in blood, then it is added the lithium standard solution of 0.0, 10.0, 20.0, 30.0μL respectively and diluting to 5.00mL with distilled water. The sample is analyzed by AAS using the hollow-cathode lamp, an air acetylene flame, a wavelength of 670.8nm, lamp current of 5mA and a slit width of 0.5nm. The absorbance is found to be 0.201, 0.414, 0.622 and 0.835 in order. Calculate the amount of Li in the serum. The molar mass of $Li_2SO_4 \cdot H_2O$ is $127.9 g \cdot mol^{-1}$.

Solution (1) The concentration of lithium standard solution is

$$c_s = \frac{1.598\ 8g}{127.9g \cdot mol^{-1} \times 0.500L} \times 2 = 0.050\ 0 mol \cdot L^{-1}$$

(2) After adding 10.0, 20.0, 30.0μL lithium standard solution to 5.00mL sample solution to be tested (containing aliquots blood of 0.1mL each), the concentration of lithium added is

$$c_{s1} = \frac{0.050\ 00 mol \cdot L^{-1} \times 10.0 \times 10^{-6}L}{5.00 \times 10^{-3}L} = 1.00 \times 10^{-4} mol \cdot L^{-1}$$

$$c_{s2} = 2.00 \times 10^{-4} mol \cdot L^{-1} \qquad c_{s3} = 3.00 \times 10^{-4} mol \cdot L^{-1}$$

(3) From the experimental data of the standard addition method, a standard curve of absorbance vs. concentration of Li is plotted in Figure 15-2. Prolong the curve, the concentration c_x of Li is

$$c_x = 1.00 \times 10^{-4} mol \cdot L^{-1}$$

(4) The concentration of Li in the serum sample is

$$\frac{1.00 \times 10^{-4} mol \cdot L^{-1} \times 5.00 mL}{0.100 mL} = 5.00 \times 10^{-3} mol \cdot L^{-1}$$

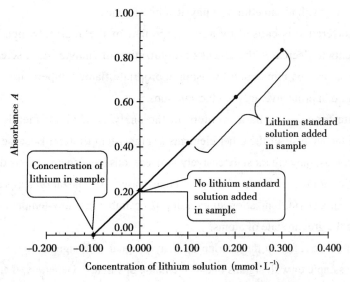

Figure 15-2 The Working Curves for the Method of Standard Additions to Assay the Amount of Lithium

2. Sensitivity and detection limit

(1) Sensitivity (S). The **sensitivity** of an analysis technique is a measure of its ability to distinguish similarly only a few species concentration at a desired confidence level. Sensitivity is equal to the slope of the calibration curve in AAS. It is reported as the absorbance A change in signal per unit change in the concentration c or mass m of analyte.

$$S = \frac{\delta A}{\delta c} \quad \text{or} \quad S = \frac{\delta A}{\delta m} \tag{15.5}$$

(2) Detection limit (D). The **detection limit** is the lowest quantity of a substance that can be distinguished from the absence of that substance (the blank value) with a stated confidence level. Typically, the ratio of signal to noise should be at least 3 for a measurement of an element.

$$D_c = \frac{c}{A} \cdot 3\sigma \ \mu g/mL \quad \text{or} \quad D_m = \frac{m}{A} \cdot 3\sigma \ \mu g \tag{15.6}$$

Where c is the concentration of the analyte; m is the mass of the analyte; A is the average absorbance of the analyte; σ is the standard deviation by determining more than 10 times for the reagent blank's signal.

Sensitivity and detection limits are important index for measuring analytical methods and instrument performance.

3. Optimization of the operative condition

The sensitivity and accuracy of AAS analysis largely depends on the optimum selection of the operative conditions, such as ①adjusting the lamp current; ②selecting the absorption wavelength; ③selecting the appropriate slit width; ④optimization of the atomization process; ⑤selecting the appropriate amount of the sample.

4. Types of interference and their elimination

The interference in AAS is generated from the physical interference, chemical interference, ionization interference, spectral interference and so on.

Physical interference occurs when the physical nature of analyte such as the solution viscosity, the surface tension and the relative density changes to reduce the absorbance. Generally, a dilution method and a standard addition method can eliminate physical interference.

Chemical interference is caused by any component of the analyte which reacts with other concomitant elements to decreases the extent of atomization of analyte. As a selective interference, chemical interference can be eliminated by using appropriate flame temperature, adding a releasing agent, a buffer agent or a protective agent to the solutions.

Ionization interference can be a problem in the analysis of alkali metals at relatively low temperature and in the analysis of the other elements at higher temperature. Because alkali metals have low ionization potentials, they are most extensively ionized. Ions have energy levels different from those of the ground state of atoms, so the desired signal is decreased. An ionization suppressor decreases the extent of ionization of analyte. Ionization suppression is desirable in a low-temperature flame in which we want to observe the ground state of atoms.

Spectral interference refers to the overlap of analyte signal with signals due to the other elements or molecules in the sample or with signals due to the flame or furnace. Generally, there are two forms of spectral interference: spectral line interference and background interference. Spectral line interference can be eliminated by minimizing the slit width or choosing another wavelength for analysis. Background interference can be eliminated by using background correction with the adjacent non-resonance line, a hydrogen or deuterium continuous light sources or on the basis of Zeeman Effect.

15.1.4 Application of Atomic Absorption Spectrometry

Atomic absorption spectrometry has the advantages of high sensitivity, good selectivity, low detection limit, little interference, simple and fast operation, etc. It plays the very important roles in many fields, especially in environmental monitoring, medical hygiene and food analysis, etc.

1. Application in environmental monitoring

Atomic absorption spectrometry has been widely applied in environmental monitoring for the detection of metals (such as zinc, copper, lead, cadmium, chromium, etc.) in water, the analysis of atmospheric environmental quality and the analysis of solids in the soil. For example, atomic absorption spectroscopy can be applied to detect the metal elements of surface water, and to evaluate their current status and development trends. AAS can be used for supervisory monitoring of heavy metals in the waste-water from production and living facilities. These are absolutely necessary for routine environmental monitoring.

2. Application in medical hygiene

Clinically, atomic absorption spectrometry is used to detect various trace elements (copper, zinc, iron etc.), and poisonous elements, for example, lead, in blood and body fluids, in hair, even in biological organs and tissues, as well as the in drugs.

3. Application in food analysis

The metal elements in food are derived from the natural elements and the external pollution elements in the process of food storage and processing, all of which can be determined by AAS. There are many types of foods that can be determined by AAS, such as grains, dairy products, eggs, meat, fish, vegetables, fruits, nuts and beverages. For example, atomic absorption spectrometry is used to determine copper in rice, cobalt in corn flour, selenium in rice, tea and garlic, and lead in pine eggs by micro-injection.

15.2 Molecular Fluorescence Spectroscopy

Emission of a photon occurs when an analyte in a higher-energy state returns to a lower-energy state. The higher-energy state can be achieved in several ways, including thermal energy, radiant energy from a photon, or by a chemical reaction. Emission following the absorption of a photon is also called photoluminescence. It is divided into two categories: fluorescence and phosphorescence. Fluorescence is the emission of light by a substance that has absorbed light or other electromagnetic radiation. It is a form of luminescence. Fluorescence can be classified in two categories: atomic fluorescence and molecular fluorescence. The present discussion mainly focuses on the molecular fluorescence analysis. It involves using a beam of light, usually ultraviolet light, that excites the electrons in molecules of certain compounds and causes them to emit light, typically, but not necessarily, visible light. **Molecular fluorescence analysis** is a method to qualitatively or quantitatively analyze the sample by the characteristics and its intensity of fluorescent spectrum of molecules.

The most important properties of fluorescence spectroscopy or fluorometry are better selectivity and higher sensitivity that is often one to three orders of magnitude higher than UV/V absorption spectrophotometry. When the analyte is fluorescent, direct determination is possible; otherwise, a variety of indirect methods using derivatization, formation of a fluorescent complex or fluorescence quenching have been developed. Fluorescence is also a powerful tool for investigating the structure and dynamics of matter or living systems at the molecular or supramolecular level. So fluorescence analysis plays a very important role in medicine and clinical analysis, especially in medicine metabolism research in vivo.

15.2.1 Fundamental Principles of Molecular Fluorescence Spectroscopy

1. Formation of molecular fluorescence

There are a series of the electron levels, paired electrons in most of molecules. Absorption of an ultraviolet or visible photon promotes a valence electron from its ground state to an excited state with conservation of the electron spin. The excited state is divided into two genres: the singlet excited states S and the triplet excited states T. The **singlet excited state** is an excited state in which all electron spins are paired (or opposite spins) and the **triplet excited state** is an excited state in which unpaired electron spins (or parallel spins) occurs.

Let's assume that the molecule initially occupies the lowest vibrational energy level of ground state electron. The ground state in Figure 15-3 is a singlet state labeled S_0. Absorption of a photon of correct energy excites the molecule to one of several vibrational energy levels in the first excited electronic state S_1 or the second electronic excited state S_2, both of S_1 and S_2 is singlet states. Fluorescence and phosphorescence are two familiar forms by emitting a photon. Once a molecule is excited by absorption of a photon, it can return to the ground state with emission of fluorescence, but many other pathways for de-excitation are also possible: internal conversion, i.e. directly return to the ground state without emission of fluorescence; intersystem crossing, possibly followed by emission of phosphorescence; vibrational relaxation, and triplet transitions, etc. These de-excitation pathways may compete with fluorescence emission if they take place on a time-scale comparable with the average time called lifetime during which the molecules stay in the excited state. Fluorescence is caused by the energy emission of photon from the lowest vibrational energy level V_0 in state S_1 to the vibrational energy levels in state S_0.

2. Excitation and emission spectra

Emission and excitation spectra are the characteristic spectra of all the fluorescent molecules. They are the basis for a quantitative or qualitative analysis in the fluorescence.

An **excitation spectrum** is measured by varying the excitation wavelength and measuring emitted light at one particular wavelength. An excitation spectrum is a graph of emission intensity F versus excitation wavelength λ_{ex}. An excitation spectrum looks very much like an absorption spectrum because, the greater the absorbance at the excitation wavelength is, the more molecules are promoted to the excited state and the more emission will be observed.

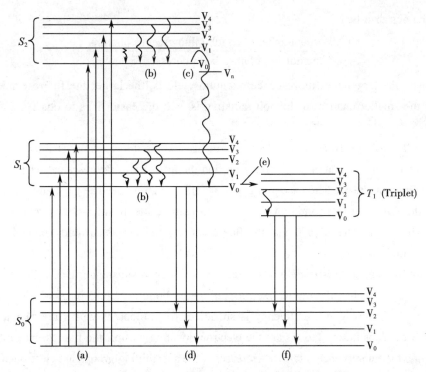

Figure 15-3 The Energy Diagram Comparing Fluorescence and Phosphorescence
(a) Absorption; (b) Vibrational Relaxation; (c) Internal Conversion;
(d) Fluorescence; (e) Intersystem Crossing; (f) Phosphorescence

If we hold the excitation wavelength fixed and scan through the emitted radiation, an **emission spectrum** or a fluorescence spectrum is produced. An emission spectrum is a graph of F versus emission wavelength λ_{em}. Usually, the emission wavelength in the most emission intensity is selected as the detection wavelength.

3. Characteristics of the fluorescence spectrum

(1) The Stokes shift. Stokes shift is the difference between positions of the band maxima of the absorption and emission spectra of the same electron transition. A molecular fluorescence spectrum is shifted to the higher wavelengths than its absorption spectrum, because the change in energy for fluorescent emission is generally less than that for absorption.

(2) No relationship between the shape of fluorescence emission spectrum and the excitation wavelength. Generally, the same shape of fluorescence emission spectrum can be observed. If the molecule is excited by the different excitation wavelengths.

(3) The "mirror image" rule. Because the differences between the vibrational levels in the ground states is similar to that in the excited states for the fluorescent substances commonly, the excitation spectrum is roughly the mirror image of the emission spectrum. This phenomenon is called the "mirror image" rule too.

4. Fluorescence efficiency and its affecting factors

(1) Fluorescence efficiency. The **fluorescence efficiency** or **fluorescence quantum yield** Φ_F is defined as the ratio of the number of photons emitted (over the whole duration of the decay) to the

number of photons absorbed.

$$\Phi_F = \frac{\text{the number of emitted fluorescence photons}}{\text{the number of absorbed fluorescence photons}} \qquad (15.7)$$

Obviously, the larger the fluorescence quantum yield is, the larger the fluorescence emission intensity is; the smaller chance of the non-radiative is, and the easier it is to observe a fluorescent compound.

(2) The relationship between the molecular structure and the fluorescence. The molecular structure is closely related to the generation of fluorescence and the intensity of fluorescence. The characteristics of the fluorescence can be predicted according to the molecular structure. The generation of fluorescence must satisfy the following two conditions: one is that the molecule must have the strong absorption by ultraviolet-visible light. The other is that the fluorescence efficiency is sufficient enough to meet the requirement of analysis.

Generally, the characteristic of the favorable fluorescence substance is as follows: ①Most of fluorescent compounds has the delocalized π conjugated bond structure in molecules, such as the aromatic rings; ②Fluorescence efficiency is sufficient for molecules with rigid planar structures generally; ③If the substituting groups are the electron-donating groups, the fluorescence intensity will increases, while the fluorescence intensity decreases if the substituting groups are electron-withdrawing groups.

(3) The effect of environmental factors on fluorescence. The environmental factors, such as temperature, solvent, pH and fluorescent quenching agent (factors that can cause the decrease of fluorescence intensity), all of them will affect the fluorescence efficiency, even affect the molecular structure and the stereochemical configuration and thus affect the shape and intensity of fluorescence spectra.

> **Question and thinking 15-2** What is fluorescence efficiency? What molecular structure of the compounds has high fluorescence efficiency?

15.2.2 Quantitative Analysis of Fluorescence Spectroscopy

1. The relation between fluorescence intensity and the concentration

When a solution in the sample cell is excited by a light source, fluorescence can be obtained from different orientation and the fluorescence intensity F can be measured. It is not appropriate that the fluorescence is observed in the orientation of transmitted light, since transmitted light intensity I_t affect fluorescence determination. In most cases, fluorescence is observed in a horizontal plane at 90° to the propagation direction of the incident beam, i.e. in direction F as shown in Figure 15-4. I_0 represents the incident light intensity, their relation is

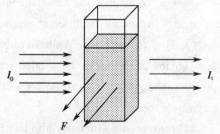

Figure 15-4 The Fluorescence of Solution in Sample Cell

$$F = K'(I_0 - I_t) \tag{15.8}$$

Where K' is the constant which is dependent on the fluorescence efficiency.

Furthermore, if c is the concentration of the sample solution, L is the thickness of the sample cell. Equation 15.8 can be expressed using Beer-Lambert law:

$$F = 2.303 K' I_0 \varepsilon c L \tag{15.9}$$

I_0 is a constant at the given wavelength and the stable light source; L is also a constant for the same sample cell. Then Equation 15.9 can be simplifies into

$$F = Kc \tag{15.10}$$

Generally, $\varepsilon c L \leq 0.05$, it should be noted that the fluorescence intensity is proportional to the concentration only for diluted solutions. The fluorescence intensity was linearly related to the fluorescence concentration in solution, which is the basis for quantitative analysis in the fluorescence.

2. Quantitative analysis methods

(1) Standard curve method. The general procedure to prepare a calibration graph or Beer's law plot can be performed by plotting the fluorescence intensity F versus a series concentrations of standard solutions. A straight line can be obtained and then unknown concentrations of sample can be interpolated from the calibration curve by comparing the fluorescence intensity of the analyte at the same conditions.

(2) The proportional method. If the calibration curve passes through the origin, the proportional method can be used by measuring in the linear range. A standard solution is prepared with pure fluorescent substance, the concentration of the standard solution is c_s in the linear range and the fluorescence intensity F_s can be determined, then the fluorescence intensity of the analyte, F_x is measured at the same conditions. If F_0 is the fluorescence intensity of blank solution, the analyte concentration c_x can be calculated by the proportion method.

$$\frac{F_s - F_0}{F_x - F_0} = \frac{Kc_s}{Kc_x} \tag{15.11}$$

K is the same number as determining the same fluorescent substance. So,

$$\frac{F_s - F_0}{F_x - F_0} = \frac{c_s}{c_x} \quad \text{and} \quad c_x = \frac{F_x - F_0}{F_s - F_0} c_s \tag{15.12}$$

The proportion method is very convenient by determining an analyte concentration according to a standard solution of fluorescent substance. But the obtained result in this way is a lower reliability than the calibration curve method.

Sample Problem 15-2 The amount of ethinylestradiol in compound norethisterone tablets (Norethisterone Co) can be assayed by fluorometric analysis. 20 tablets are ground to a fine powder and placed in a 250.0mL volumetric flask and diluted to mark with dehydrated alcohol. After filtration, the 5.00mL filtrate is diluted to 10.00mL, the fluorescence intensity of the resulting solution is found to be 61 at an emission wavelength of 185nm and an excitation wavelength of 307nm. At the same condition, the fluorescence intensity of the 1.4μg·mL^{-1} standard solution of ethinylestradiol is found to be 65. Calculate the amount of ethinylestradiol in every tablet.

Solution According to the proportion method,

$$c_x = \frac{F_x \times c_s}{F_s}$$

$$c_x = \frac{61 \times 1.4 \mu g \cdot mL^{-1}}{65} = 1.3 \mu g \cdot mL^{-1}$$

The amount of ethinylestradiol in one tablet is

$$= \frac{1.3 \mu g \cdot mL^{-1} \times 10.00 mL \times 250.0 mL \div 5.00 mL}{20} = 32 \mu g$$

15.2.3 Fluorescence Spectrophotometer

Fluorometer and spectrofluorometers are the most common instruments designed for measuring molecular fluorescence. In a fluorometer, the excitation and emission wavelengths are selected with absorption or interference filters. When a monochromator is used to select the excitation and emission wavelengths, the instrument is called a spectrofluorometer. Typically, the instrument consists of four components: light source; monochromator or filter; sample cell and detector.

Figure 15-5 depicts the simplified outline of the instrument. The light from light source passes through the excitation monochromator which allows the special wavelengths to be selected to induce fluorescence after passing through sample cell. Fluorescence is collected in the direction of 90° with the incident beam and detected in monochromator by a photomultiplier. Automatic scanning of wavelengths is achieved by the motorized monochromators, which are controlled by the electronic devices and the computer in which the data are stored.

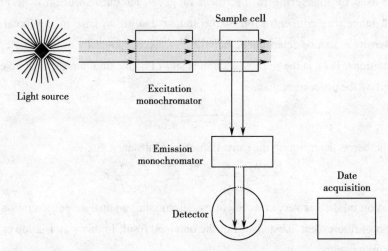

Figure 15-5 Schematic Diagram for a Conventional Spectrofluorometer

15.2.4 Applications of Fluorescence Spectroscopy

1. The fluorescence analysis of inorganic compound

Generally, most inorganic ions are not fluorescent substance except for a few metal ions such as UO_2^+. Some metal ions or most of diamagnetism ions can be determined indirectly by reacting with an organic ligand to form a fluorescent complex. The common fluorophore reagents are 8-hydroxyquinoline, 2-hydroxy-3-naphthalenecarbonic acid, 2,2'-dihydroxy-azobenzene and so on. About 70 elements, such as Ca, Mg, Zn, Al, Cd, Co, B, Si and F etc., can be determined by fluorescence analysis. And the fluorescence method can be adopted for measuring nitride, cyanide, sulfide, and peroxide etc.

2. The fluorescence analysis of organic compound

Most organic compounds that contain aromatic, heteroatomic rings and a long conjugated aliphatic functional group are fluorescent. Sometimes, a good sensitivity and selectivity can be obtained by linking the fluorescent reagent by the chemical modification or association technique. For example, a fluorophore reagent can be attached to an analyte by chemical reaction and the use of the fluorescent reagent expands the application scope in fluorescence analysis. The organic compounds which can be determined by fluorescence analysis generally include polycyclic amines, naphthols, purines, polycyclic aromatics, amino acids and proteins which contain aromatic rings or aromatic heterocycles, alkaloids in drugs, steroids, antibiotics, vitamins and so on. The most common fluorescence reagents are the following substances, fluorescent amine, o-phthalaldehyde, 1, 2-naphthoquinone-4-sulfonic acid sodium salt and 5-dimethylamino naphthalene-1-sulfonyl chloride and so on. Though deoxyribonucleic acid (DNA) has not enough fluorescence efficiency, but it can be determined by means of labeling reagents having proper molecules with fluorescent probe. The interactions between DNA and small molecules are based on the transformation of the fluorescence intensity. The fluorescent dyes can be a kind of labeling reagents in gene inspection, and fluorescent dyes are gradually used as markers to replace isotopic markers.

15.3 Chromatography

15.3.1 Brief Introduction to Chromatography

The **chromatography** was first proposed in 1906 by Michael S. Tswett, a Russian botanist when he studied to separate plant pigments. He packed a vertical glass tube with the dry calcium carbonate pellets, placed the extract of green leaves at the top of the calcium carbonate column and the plant extract was carried through the column by pouring petroleum ether from the top to down continuously. As the plant extract moved through the column, the pigments in plant extract separated into individual color bands. Tswett named the technique chromatography, the vertical glass tube is called the **packed column**, the calcium carbonate pellets are called the **stationary phase**, petroleum ether is called the **mobile phase** or **eluant**. With the development of chromatographic technology, chromatographic method is widely used to separate the colorless substances generally, but its name is still used.

Chromatography is an analytical method that is widely used for separation, determination, and identification of the components in the mixed compounds. After a century of development, the chromatographic technology and chromatograph are more perfect. As an excellent way of separation, chromatography has been used in various qualitative and quantitative analysis fields, for example, organic compounds analysis, clinical assay, medicine, environment, food and gasoline industrial.

Chromatography can be divided into categories on the basis of the different aspects.

(1) The classification according to the mobile phase. In **gas chromatography (GC)** the mobile phase is an inert gas, in **liquid chromatography (LC)** it is a liquid and in **supercritical fluid chromatography (SFC)** it is a supercritical fluid.

(2) The classification according to chromatographic bed shape. It can be column chromatography;

paper chromatography; thin layer chromatography; capillary chromatography and so on.

(3) The classification according to chromatographic separation mechanisms. It can be absorption chromatography; partition chromatography; ion exchange chromatography; gel chromatography; bioaffinity chromatography and so on.

15.3.2 General Principle of Chromatography for Separation

1. The Process of chromatographic separation

Separation of different components in chromatography is based on their different distribution coefficients in the same stationary phase at certain temperature. Figure 15-6 shows the separate processes of substance A and B. When the sample solution containing A and B is placed on the top of the column packed with solid particles, the two components in sample solution are adsorbed by the stationary phase or the adsorbent. As the mobile phase flows into the column continuously, the components is dissolved in mobile phase, this process is called "desorption". The components, moved down with the mobile phase, are adsorbed again when they meet the new adsorbent particles. Then the process of adsorption, desorption, adsorption, desorption repeats in the column (up to $10^3 \sim 10^6$ times). If the force between the components and stationary phase is stronger, the components are adsorbed more easily by the stationary phase, and they move down the column more slowly and vice versa. After two components go through the long column, supposing the component A is with weak adsorption capacity, so the substance A will go out of the column first, followed by the component B with strong adsorption capacity, thus A and B can be separated.

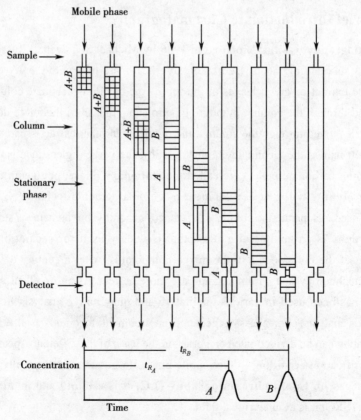

Figure 15-6 The Separate Processes of Chromatography

2. The basic terminology of chromatography

A plot of the detector response as a function of elution time or volume of eluted mobile phase is known as a "chromatogram" in Figure 15-7, basic terminologies about chromatogram were introduced.

1) **Base line**, the OD line in Figure 15-7, is traced on the chromatogram when only mobile phase is eluting from the detector. It is a line parallel to horizontal axis when the experimental conditions are stability.

2) **Peak height h**, the BA line in Figure 15-7, is the vertical dimension from the top of chromatographic peak to the baseline.

3) **peak area A** is the area inside ACD.

4) **Peak width W** known as the baseline width the IJ line in Figure 15-7 is determined by the intersection with the baseline of tangent lines drawn through the inflection points on either side of the chromatographic peak.

5) **Peak width at half height $W_{1/2}$** is the GH line as shown in Figure 15-7.

6) **Standard deviation σ**, the half of EF line in Figure 15-7, is the half of peak width at $0.607h$. So, $W = 4\sigma$, $W_{1/2} = 2.354\sigma$.

7) **Retention value**, known as retention parameter which can tell the retention state of each component in column is the main qualitative parameters in chromatography. Usually it can be measured by the time (min) or the volume of mobile phase (cm^3).

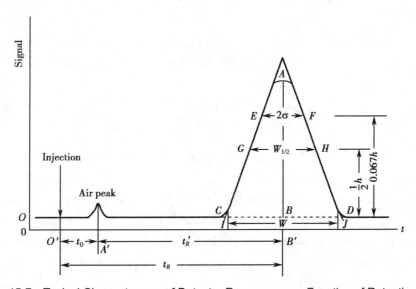

Figure 15-7 Typical Chromatogram of Detector Response as a Function of Retention Time

8) **The retention time, t_R** is the elapsed time from the introduction of the solute to the maximum peak value.

9) **Void time, t_0** is the time required for unretained components to move from the point of injection to the detector.

10) **Adjusted retention time, t'_R** is the difference between the retention time of a given component t_R and the column's void time t_0.

$$t'_R = t_R - t_0 \tag{15.13}$$

In addition, there are some volume parameters such as retention volume, void volume and adjusted retention volume and so on. These parameters have no obvious differences from the time parameters, only base on the volume of eluted mobile phase to show the chromatogram.

11) **Resolution**, R is a quantitative measurement of separation degree between two chromatographic peaks. The following expression is used to calculate R between two compounds 1 and 2.

$$R = \frac{(t_{R2} - t_{R1})}{\frac{1}{2}(W_1 + W_2)} = \frac{2\Delta t_R}{W_1 + W_2} \tag{15.14}$$

In general, the separation degree between two chromatographic peaks will improves with an increasing in R. If the resolution is 1.5, it means that the separation degree of the two components was 99.7%. So, R equals to 1.5 is often used as the sign that the two adjacent peaks are completely separated.

3. Plate theory

In 1941, Martin and Synge proposed the **plate theory** model of chromatography, and they treated the chromatographic column as though it consists of discrete sections at which partitioning of the sample between the stationary and mobile phases occur. They called each section a **theoretical plate.** After equilibria of the multiple distributions, the mixture was separated, and having more theoretical plates increases the efficiency of the separation process. For the sample emerging from a column of length L, the number of plates n in the entire column is the length L divided by the **plate height** H:

$$n = \frac{L}{H} \tag{15.15}$$

Then, the plate height is called the height equivalent to a theoretical plate. The number of theoretical plates can be determined from chromatogram according to the retention time and the width of the peak.

$$n = 16\left(\frac{t_R}{W}\right)^2 \tag{15.16}$$

Or

$$n = 5.54\left(\frac{t_R}{W_{1/2}}\right)^2 \tag{15.17}$$

A column's efficiency is defined in terms of the number of theoretical plates n or the height of a theoretical plate H. A column's efficiency improves with an increase in the number of theoretical plates or a decrease in the height of a theoretical plate and W (or $W_{1/2}$). When evaluating the column efficiency by the number of theoretical plates, the compositions, stationary phase, mobile phase, operation condition and so on must be specified.

15.3.3 Chromatographic Instruments

1. Gas chromatograph

The schematic diagram for a typical gas chromatograph is shown in Figure 15-8, and it is made of five parts: carrier gases system (carrier gas, purification and flow controllers), sample injection system (vaporizer), column, detector and data processing system.

Gas chromatograph is widely used because of its relatively low cost, easy-to-use, high-efficiency and fast analysis. And it is applied to analyze the substances which are easy to be vaporized, such as volatile oil, organic acid and ester in Chinese herbal and so on. In addition, drug molecules with high boiling point can be converted into derivates with low boiling point through derivatization reaction, and then they can be analyzed by gas chromatography.

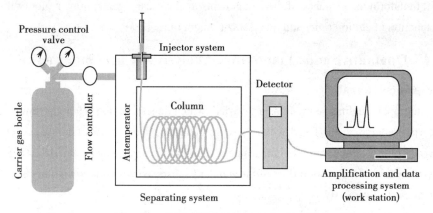

Figure 15-8　Schematic Diagram of Gas Chromatograph

2. High-performance liquid chromatograph

A high-performance liquid chromatograph HPLC mainly includes solvent reservoir, pump, column, detector, data-processing systems, as shown in Figure 15-9. Pumping pressures of several hundred atmosphere are required to obtain reasonable flow rates in HPLC. And this method has the characteristics such as high pressure, fast analysis, high performance and high sensitivity. In addition, some auxiliary devices such as gradient elution, auto-injection, degassing, can be equipped according to some requirements.

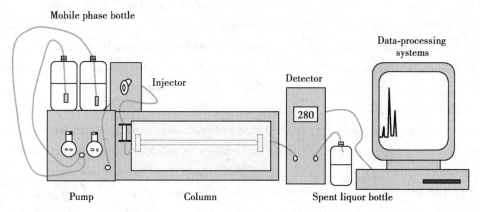

Figure 15-9　Schematic Diagram for a High-Performance Liquid Chromatograph

The main advantage of HPLC is that it can analyze non-vaporizable compositions in a wide range. For example, it can separate and purify biological macromolecules, such as, proteins, peptides, nucleic acids, and saccharides.

3. The chromatograph workflow

The sample solution has been injected and carried into the column by carrier gases system of a gas chromatograph or high-pressure infusion system of HPLC. After the different components in the sample are separated into bands there, they are carried out of the column and go into the detector. The detector such as a thermal conductivity detector (TCD), a hydrogen flame ionization detector (FID) and an electron capture detector (ECD) or an ultraviolet detector, a fluorescence detector and an electrochemical detector will transform the response of the components to electrical signals; draw a plot of the detector's signal as a function of elution time, and that is a chromatogram.

15.3.4 Qualitative and Quantitative Analysis of Chromatography

1. The qualitative analysis

Chromatography qualitative methods are varied, which depend on the retention time. The retention time of a certain component is fixed when the chromatography conditions (such as column packing, column length, column temperature, column diameter, column pressure, flow rate, detecting current etc.) are fixed by comparing the retention time of the standard substance with unknown objects.

2. The quantitative analysis

Under certain chromatographic separation conditions, the mass(m_i) or concentration of the component (i) to be determined is proportional to the response signal of detector, that is to say, it is proportional to the peak area A_i on the chromatogram.

$$m_i = f_i \times A_i \tag{15.18}$$

here f_i is correction factor. In the same chromatographic conditions, different material's f_i is different, and the same material's f_i in different conditions is also different. So f_i is the basis of quantitative analysis for chromatography. The peak area and the correction factor should be measured correctly before the quantitative calculation.

In Modern chromatographic techniques the automatic integration procedure was commonly used to integrate the peak area of the component to be measured, and then quantify the peak area.

(1) External standard method. External standard method is also known as the standard curve method. Make standard solution of the components to be determined, and then analyze it by chromatography. Under the strictly consistent condition, draw the plot of peak area versus the concentration to obtain the specification curve, from which the content of the testing component will be determined.

Without using correction factor, external standard method is more accurate, but the changes of operating conditions have a great influence on the accuracy of the results and it requires high accuracy of the injection volume, so it is suitable for the rapid analysis of large numbers of samples.

(2) Internal standard method. An internal standard is a known amount of a compound, different for analyte, that is added to the unknown. Signal from analyte is compared with signal from the internal standard to find out how much analyte is present. As a result, the relative compared method can be used for calculation. The sample to be determined (weighing m) and internal standard (weighing m_s) were added into the solution, and the percentage content $\omega_i\%$ of the component (i) is obtained based on Equation 15.19.

$$\omega_i\% = \frac{m_i}{m} \times 100\% = \frac{f_i}{f_s} \times \frac{A_i \times m_s}{A_s \times m} \times 100\% \tag{15.19}$$

where A_i and A_s are the peak areas of the component and internal standard respectively; f_i and f_s are the correction factor of the component and internal standard respectively.

When the correction factor is unknown, the "internal standard comparison method" can be used, which is an application of the internal standard method. In the same volume standard solution and sample solution containing the component i, the same amount of internal standard substances is added into the sample solution to prepare a reference solution and a test solution respectively, and then the injection is performed. By Equation 15.20, calculate the concentration of the component in the sample solution.

$$(c_i)_{sample} = \frac{(A_i / A_s)_{sample}}{(A_i / A_s)_{reference}} \times (c_i)_{reference} \tag{15.20}$$

An internal standard must meet the following requirements: ①It will not be contained in the samples; ②It has similar performance to the components to be determined; ③It will not react with the samples; ④The position of the peak in chromatogram should be nearby the components to be determined without interference.

15.3.5 Application of Chromatography

1. Application in pharmacy

Chromatography is widely used in drug analysis, including determination of drug content, analysis of compound preparations and drug metabolism research. For example, gas chromatography is commonly used for the analysis of amphetamine stimulant drugs and volatile oil of traditional Chinese medicine. And HPLC is commonly used for the determination of sulfanilamide drugs, water-soluble vitamins and antibiotics.

2. Application in medicine

In clinical examination, health surveillance, and toxicology research, it is necessary to determine the concentration of harmful substances and their metabolites in blood, urine, or other tissues. For example, the concentration of fatty acids and amino acid methyl ester in the blood, and the concentration of oxalate and the trace metabolite of chloropropanol in urine, β-chlorolactic acid, were detected by gas chromatography. The concentration of phenylacetone and bilirubin in the blood were commonly determined by HPLC.

3. Analysis of pollutants in water and agricultural products

The determination of pesticide residues in agricultural products (organic phosphorus, organochlorine, carbamates, pyrethroids, etc.), and the determination of benzene and its homologues in industrial waste water were mainly detected by gas chromatography. High performance liquid chromatography (HPLC) was used for the determination of phenolic compounds in water and Sudan red in food.

4. Combination of chromatographic techniques

With the development of life science, capillary electrophoresis and chromatography have more and more applications in the study of proteomics and metabonomics. The whole chromatography science, including capillary electrophoresis and microchip, is developing to the direction of high-efficient separation, high-flux, high sensitivity detection and multi-dimensional analysis. On the other hand, in order to compensate for the weaker qualitative analysis of chromatography, combining the chromatograph with other instruments has been developed, in particular the combination technique of liquid chromatography and mass spectrometry which has matured in recent years. It plays a great role in separation and identification of biomacromolecules. So the combination technique of chromatography and other instruments will be an important field of analytical chemistry.

15.4 Mass Spectrometry and GC-MS or LC-MS

15.4.1 Brief Introduction of Mass Spectrometry

Mass spectrometry (also mass spectroscopy, is abbreviated as MS) is a technique for studying the masses of atoms or molecules or fragments of molecules. It has long been used to measure isotopes and decipher organic structures. Mass spectrometry carried out at the beginning of last century, which was concerned with the behavior of positive ions in magnetic and electrostatic fields. The first true analytical application of mass spectrometry was described in 1940. Beginning in about 1960, interest in mass spectrometry shifted toward its use for the identification and structural analysis of complex compounds. In 2002, John B. Fenn and Koichi Tanaka were awarded the Nobel Prize for their development of soft desorption ionization methods for mass spectrometric analyses of biological macromolecules. Mass spectrometry is the most powerful detector for chromatography, offering both qualitative and quantitative information, providing high sensitivity, and distinguishing different substances with the same retention time.

A mass spectrum is a plot of ion intensity as a function of the ion's mass-to-charge ratios. A mass spectrometer consists of five components including a sampling system, an ion source, a mass analyzer, a detection system and a vacuum system as shown in Figure 15-10.

To obtain a mass spectrum, gaseous species desorbed from condensed phases are ionized, for example by bombarding them with electrons. These charged fragments ions are accelerated by an electric field and then separated according to their mass-to-charge ratio m/z by a magnetic field. The ions of the same mass-to-charge ratio will undergo the same amount of deflection and be detected by a mechanism capable of detecting charged particles by an electron multiplier. The mass spectra displayed the relative abundance of detected ions as a function of the mass-to-charge ratio. The sample can be identified by correlating known a characteristic fragmentation pattern.

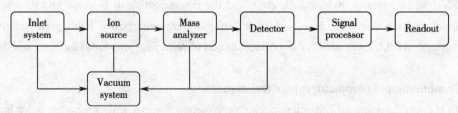

Figure 15-10 Schematic Diagram for a Mass Spectrometer

15.4.2 Mass Spectrogram

The mass spectrum generally contains an array of peaks of differing heights representing the distribution of ions by mass in a sample. The most intense peak in a mass spectrum is called the base peak. Intensities of other peaks are expressed as a percentage of the base peak intensity.

Almost all stable molecules have an even number of electrons. When one electron is lost, the resulting cation with one unpaired electron is designated, the **molecular ion**($M^{+\cdot}$). After ionization, usually has enough internal energy ($\sim$1eV) to break into fragments. The molecular ion is used to identify the molecular mass of an unknown compound. The intensity of the molecular ion peak depends upon the stability of the ionized particle. Unfortunately, with electron ionization, some compounds do not exhibit a molecular ion, because $M^{+\cdot}$ breaks apart so efficiently. However, fragments provide clues to the structure of an unknown. To find the molecular mass, we can obtain a chemical ionization mass spectrum, which usually has a strong peak for MH^+. The **isotopic peak**, beyond the molecular ion (M + 1, M + 2) provides the information on elemental composition.

Interpretation of mass spectra to elucidate molecular structure is an important field. The molecular ion can be broken apart to give the many peaks. **Fragmentation** patterns can even unravel the structures of large biological molecules. The electron beam energy and electron ionization causes bond rupture with the formation of fragments which have smaller masses. The origin and structure of the high-strength fragment ion peaks can be analyzed to speculate the structure of the compound.

15.4.3 Chromatography-Mass Spectrometry

Chromatography-Mass Spectrometry is a powerful tool for analysis of complex mixtures. Mass spectrometry is widely used as the detector in chromatography to provide both qualitative and quantitative information. Mass spectrometry requires high vacuum to prevent molecular collisions during ion separation. The problem in combining the two techniques is to remove the huge excess of matter between the chromatograph and mass spectrometer.

A common combination is **gas chromatography-mass spectrometry** (GC-MS, or GC / MS) to identify different substances within a test sample. In this technique, gas chromatograph is used to separate different compounds and felicitously evolved to employ narrow capillary columns. The capillary column is connected directly to the inlet of the mass spectrometer through a heated transfer line. The most common type of mass spectrometer (MS) associated with a gas chromatograph (GC) is a quadrupole mass spectrometer. The mass spectrometer identifies different substances by breaking each molecule into ionized fragments and detecting these fragments using their mass-to-charge ratio. It is not possible to make an accurate identification of a particular molecule by gas chromatography or mass spectrometry alone. The mass spectrometry process normally requires a very pure sample while gas chromatography using a traditional detector cannot differentiate between multiple molecules. Combining the two processes reduces the possibility of error, and it typically increases eventually detected certainty in the sample.

Similarly to GC-MS, **quid chromatography-mass spectrometry** (LC / MS or LC-MS) separates compounds chromatographically before they are introduced into the ion source and mass spectrometer. The LC-MS system contains an interface that efficiently transfers the separated components from the LC column into the MS ion source, because liquid chromatography creates a huge volume of gas when solvent vaporizes at the interface between the column and the mass spectrometer. As a requirement, the interface should not interfere with the ionizing efficiency and the vacuum conditions of MS system. Nowadays, most extensively applied LC-MS interfaces are based on atmospheric pressure ionization (API) strategies like **electrospray ionization** (ESI), atmospheric pressure chemical ionization (APCI), and atmospheric pressure photo-ionization (APPI).

Mass spectrometry provides valuable qualitative and quantitative information. These include molecular weight determination, determining of molecular formulas and identifying the unknown compounds from fragmentation patterns. Nowadays MS is very commonly used in analytical laboratories and it may be applied in a wide range of sectors including biotechnology, environment monitoring, food processing, pharmaceutical, agrochemical, and cosmetic industries.

15.5 Inductively Coupled Plasma Atomic Emission Spectrometry

Atomic emission spectroscopy (AES) is a chemical analysis method that was often used the intensity of light emitted based on electric arc and electric spark source at a particular wavelength to determine the quantity of an element in a sample. Atomic emission based on the plasma was introduced in 1964 and in the last 20 years the scope of atomic emission spectroscopy has been considerably enhanced by the application of plasma techniques.

When a valence electron in the higher-energy level returns to the lower-energy level, atomic emission occurs. Therefore, it consists of a series of discrete lines at wavelengths corresponding to the difference in energy between two atomic orbitals. The intensity I of an emission line is proportional to the number of atoms N^* populating the excited state

$$I = kN^*$$

where k is a constant which related to the efficiency of the transition. For a system in thermal equilibrium, the population of the excited state is related to the total concentration of atoms N by the Boltzmann distribution. When the light source is stable and the temperature is constant, after selecting the analysis line for an element to be tested, under the ideal state, the spectral line intensity is proportional to the sample concentration c.

$$I = ac$$

where a is a constant associated with the evaporation, atomization, and excitation process of the sample in the light source.

Inductively Coupled Plasma (ICP) is a type of atomic emission spectroscopy that produces excited atoms and ions that emit electromagnetic radiation at wavelengths characteristic of a particular element. ICP can be generated by directing the energy of a radio frequency generator into a suitable gas, usually ICP argon. The high temperature, stability, and relatively inert argon environment eliminate much of the interference encountered with flames.

ICP source comprises three concentric silica quartz tubes (Figure 15-11), each of which is open at the top. High-purity argon gas carried the sample, in the form of an aerosol, is fed through the plasma gas inlet and passes through the central tube through which flows a radio-frequency current. The quartz torch is protected from overheating by Ar coolant gas. The plasma gas flows in helical pattern which provides stability and helps to isolate thermally the outside quartz tube.

The plasma is initiated by a spark from a Tesla coil probe and is thereafter self-sustaining. The resulting ions and their associated electrons from the Tesla coil then interact with the fluctuating magnetic field. This generates enough energy to ionize more argon atoms by collision excitation. The electrons generated in the magnetic field are accelerated perpendicularly to the torch. At high speeds, cations and electrons will form closed circular eddy current, and will collide with argon atoms to produce further ionization which causes a significant temperature raise. A steady state is created with a high electron density maintaining a temperature of 6 000 to 10 000K. This torch is the spectroscopic source and it contains all the analyte atoms and ions which have been excited by the heat of the plasma.

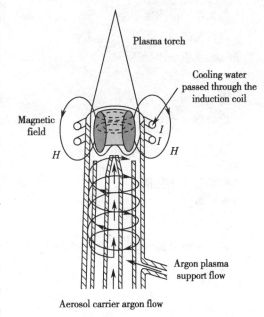

Figure 15-11 ICP Torch

Thus, the plasma source produces a greater number of excited emitted atoms, especially in the ultraviolet region. Further, the plasma source is able to reproduce atomization conditions with a far greater degree precision than that obtained by classical arc and spark spectroscopy. This feature is especially important for the multi-element determinations over a wide concentration range. The success of ICP leans on its capability to analyze large numbers of samples in a short period with very good detection limits for most elements. ICPs used today are often connected to different detection systems, such as ICP mass spectrometry and ICP atomic emission spectrometry.

Key Terms

原子吸收光谱	atomic absorption spectrometry
空心阴极灯	hollow cathode lamp (HCL)
物理干扰	physical interference
化学干扰	chemical interference
离子干扰	ionization interference
光谱干扰	spectral interference
检测限	detection limit
分子荧光	molecular fluorescence
单重态	singlet excited state
三重态	triplet excited state
分子荧光光谱	molecular fluorescence spectroscopy
激发光谱	excitation spectrum
发射光谱	emission spectrum
荧光寿命	fluorescence lifetime
荧光效率	fluorescence efficiency

荧光量子产率	fluorescence quantum yield
激发光源	excitation light source
气相色谱法	gas chromatography
色谱柱	packed column
固定相	stationary phase
流动相	mobile phase
洗脱剂	eluant
液相色谱法	liquid chromatography
超临界流体色谱	supercritical fluid chromatography
吸附色谱	absorption chromatography
分配色谱	partition chromatography
离子交换色谱	ion exchange chromatography
凝胶色谱	gel chromatography
基线	base line
峰高	peak height
标准偏差	standard deviation
保留体积	retention value
保留时间	retention time
死时间	dead time
调整保留时间	adjusted retention time
塔板理论	plate theory
速率理论	rate theory
范氏方程	van Deemter equation
分离度	resolution factor
质谱	mass spectrometry
质荷比	mass-to-charge ratio
离子源	ion source
电子轰击源	electron impact source
化学电离源	chemical ionization source
质量分析器	mass analyzer
分子离子	molecular ion
同位素峰	isotope peak
碎片	fragmentation
原子发射光谱	atomic emission spectroscopy
电感耦合等离子体	inductively coupled plasma

Summary

Briefly, atomic absorption is due to the transitions between atomic energy levels. The process that an electron absorbing the radiation energy migrates from the ground state to the excited states is called an atomic spectral transition. Atomic absorption spectroscopy is inherently a single-element method, and an atomic adsorption spectrometer consists of four major parts they are the light source, atomizer, monochromator, and transducer or detector. Atomic spectrometry has the advantages of high sensitivity, good selectivity, low detection limit and less interference. It is widely used in many fields such as environmental monitoring, medical hygiene and food analysis and so on.

Fluorescence spectrometry is a type of electromagnetic spectroscopy that analyzes fluorescence from a sample. A light source for the appropriate energy region is required in the process of fluorescence spectrometry, usually ultraviolet light and visible light. An ultraviolet light excites the electrons in the molecules and then luminescence that comes from the sample is measured. The output measurement can be described simply by $F=Kc$, where c is the concentration of the sample and K is the proportionality constant related to wavelength, intensity and pathlengths of the excitation light. Fluorescence intensity of exciting light from sample is proportional to the concentration of the sample. Fluorescence analysis is widely used in the fields of medicine analysis, clinical diagnosis, food security and environmental monitoring, and so on.

Chromatography is a powerful separation technique that is used to isolate and identify analyte from a mixture. Chromatographic separation utilizes the selective partitioning of components between a stationary phase that is immobilized within a column, and a mobile phase that passes through the column with the sample. The effectiveness of a separation is described by the resolution between the chromatographic bands for two components. Column efficiency is defined in terms of the number as well as the height of a theoretical plate. Analytical chromatography normally operates with the smaller amounts of a sample and seeks to measure the relative proportions of analytes in a mixture. A combination of chromatography and other analytical instruments such as mass spectrometry will further identify and quantify the analytes. Gas chromatography (GC) is useful to analyze of volatile samples such as volatile oil in Chinese herbal medicine. High performance liquid chromatography (HPLC) is used to measure the components which are not easily vaporized even to separate and purify biological macromolecules such as protein, peptides, nucleic acids and sugars.

A mass spectrum (MS) is a plot of the ion signal as a function of the mass-to-charge ratio. The utility of mass spectrometry arises from the fact that the ionization process generally produces a family of positive particles whose mass distribution is characteristic of the parent species. Mass spectrum is used to determine accurate measure of the molar mass of the analyst, elucidate the chemical structures of molecules, and identification of compounds from fragmentation patterns. The combination of chromatography and mass spectrometry allows the separation and identification to be carried out simultaneously, it typically increases convenient and accurate eventually in qualitative and quantitative analysis.

In an inductively coupled plasma (ICP), a radio-frequency induction coil heats Ar^+ ions to $6\,000\sim 10\,000K$. At this high temperature, emission is observed from electronically excited atoms and ions. There is little chemical interference, the temperature is very stable and little self-absorption is observed. Plasma emission spectroscopy does not require a light source and is capable of measuring about 70 elements simultaneously.

Exercises

1. What is the fundamental principle of atomic absorption spectroscopy?

2. Cadmium poisoning may be occurred as cadmium accumulation in animals. A kidney function disorder may occur if the amount of cadmium is up to $50\mu g \times g^{-1}$ in the kidney cortex. Assume a sample solution is prepared by weighing 0.256 6g of the preconditioning kidney cortex dissolving in 10.00mL of water. The cadmium concentration of the sample solution can be measured by using standard addition method. A series of sample solution are shown in the following table. After diluting to 5.00mL with distilled water, the absorbance of these solution are obtained in the following table. Calculate the amount of cadmium in the kidney cortex.

No.	Volume of sample mL	Volume of standard cadmium solution ($10\mu g \cdot mL^{-1}$) added mL	Absorption
1	2.00	0.00	0.042
2	2.00	0.10	0.080
3	2.00	0.20	0.116
4	2.00	0.40	0.190

3. The concentration of lead in a solution is determined by atomic absorption spectroscopy using the lead of hollow-cathode lamp at 283.3nm. Two aliquots of 50.0mL each are taken from the same original sample solution. Nothing is added to the first aliquot, and the absorbance of the solution is found to be 0.325. Then, $300\mu L$ of a $50.0 mg \cdot L^{-1}$ lead standard solution is added to the second aliquot, and gives an absorbance of 0.670 at 283.2nm. What is the concentration of lead in the original sample solution?

4. Generally, fluorometry is more sensitive than spectrophotometry. Please explain it.

5. Which of the following method can change the fluorescence quantum yield? Please explain it. (1) Increase the temperature; (2) Decrease the concentration of fluorescence substance; (3) Increase the viscosity of the solvent.

6. Why the calibration curve in molecular fluorescence is a nonlinear for high concentration?

7. Acetylsalicylic acid, also known by trade name Aspirin, is used as antipyretic analgesic. The concentration of acetylsalicylic acid $C_9H_8O_4$ in aspirin tablets can be determined by hydrolyzing to the salicylate ion $C_7H_5O_2^-$ and determining the concentration of the salicylate ion spectrofluorometrically. A stock standard solution is prepared by weighing 0.077 4g of salicylic acid $C_7H_6O_2$ into a 1L volumetric flask and diluted to mark with distilled water. A set of calibration standards is prepared by pipetting 0, 2.00, 4.00, 6.00, 8.00, and 10.00mL of the stock solution into separate 100mL volumetric flasks containing 2.00mL of $4 mol \cdot L^{-1}$ sodium hydroxide and diluting to mark with distilled water. The fluorescence of the calibration standards was measured at an emission wavelength of 400nm using an excitation wavelength of 310nm; results are listed in the following table.

Volume of the standard solution / mL	The fluorescence intensity (F)	Volume of the standard solution / mL	The fluorescence intensity (F)
0.00	0.00	6.00	9.18
2.00	3.02	8.00	12.13
4.00	5.98	10.00	14.96

Several aspirin tablets are ground to fine powder in a mortar and pestle. A 0.101 3g portion of the powder is placed in a 1L volumetric flask and diluted to mark with distilled water. A portion of this solution is filtered to remove insoluble binders, and a 10.00mL aliquot transferred to a 100mL volumetric flask containing 2.00mL of 4mol·L^{-1} sodium hydroxide. After diluting to mark, the fluorescence of the resulting solution is measured to be 8.69. What is the w/w% acetylsalicylic acid in the aspirin tablets?

8. What kinds of quantitative methods are commonly used in chromatographic analysis? Which method is more preferred for samples with loss of pretreatment?

9. What is the qualitative method commonly used in chromatographic analysis?

10. Please compare the differences in analytical method and application between gas chromatography and HPLC chromatography.

11. Briefly describe the structure of the mass spectrometer.

12. Briefly introduce the detection principle of ICP.

Supplementary Exercises

1. Figure shows the results of the preliminary data taken while developing a fluorometric assay for the amino acid glycine. The glycine reacts with a reagent, fluorescamine, which forms a fluorescent product with amines. The three scans show are (a) the relative intensity of emission as the excitation wavelength was scanned-an excitation spectrum at a fixed-emission wavelength; (b) the emission spectrum found at a fixed-excitation wavelength; and (c) a plot of the relative fluorescence intensity with both excitation and emission wavelengths fixed but varying the pH. Assume the spectral bandshape does not change with the pH. To optimize the assay's sensitivity to glycine, at what wavelengths should the excitation and emission monochromators be set, and what should the pH of the solution be?

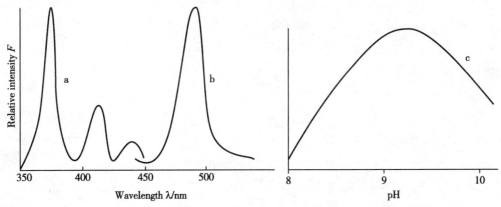

2. A solution of 5.00×10^{-5} mol·L^{-1} 1,3-dihydroxynaphthelene in 2mol·L^{-1} NaOH has a fluorescence intensity of 4.85 at a wavelength of 459nm. What is the concentration of 1,3-dihydroxynaphthelene in a solution with a fluorescence intensity of 3.74 under identical conditions?

3. The following data were obtained for three compounds separated on a 20-m capillary column.

compound	t_R / min	W / min
A	8.04	0.15
B	8.26	0.15
C	8.43	0.16

(1) Calculate the number of theoretical plates for each compound and the average number of theoretical plates for the column.

(2) Calculate the average height of a theoretical plate.

(3) Calculate the resolution for each pair of the adjacent compounds.

Answers to Some Exercises

[Exercises]

2. $22.6 \mu g \cdot g^{-1}$

3. $0.280 g \cdot L^{-1}$

7. 65.0%

[Supplementary Exercises]

1. Excitation 367nm, emission 483nm, pH = 9.35

2. $3.85 \times 10^{-5} mol \cdot L^{-1}$

3. (1) $n_A = 46\,000$ $n_B = 48\,500$ $n_C = 44\,400$ $n_{average} = 46\,300$

　(2) 0.43mm   (3) $R_{AB} = 1.5$ $R_{BC} = 1.1$

（尹计秋）

Chapter 16
Nuclear Chemistry and Its Applications

Nuclear Chemistry is the subfield of chemistry dealing with the reactions, processes, properties, structure, isolation and identification of nuclear. It mainly studies the properties of nuclei, nuclear structure, and nuclear transformation rules using the theory of chemical and physical. Radiochemistry and nuclear physics are two main subsidiary fields of nuclear chemistry, its research results have been widely used in various fields. Our concern is the application of radio chemistry and nuclear medicine in the fields of medicine, health, environment and health research.

Radiochemistry is the part of chemistry which deals with radioactive materials. It includes the production of radio nuclides and their compounds by processing irradiated materials or naturally occurring radioactive materials, the application of chemical techniques to nuclear studies, and the application of radioactivity to the investigation of chemical, biochemical or biochemical problems.

Nuclear Medicine is a medical specialty involving the application of radioactive substances in the diagnosis and treatment of diseases. It is the product of integration of medicine and modern science and technology such as nuclear technology, electronic technology, computer technology, chemistry, physics and biology. It mainly studies the application of isotopes, ray beam produced by the accelerator and nuclear radiation of radioisotopes in medicine.

16.1 Basic Concept of Nuclear Chemistry

16.1.1 Nuclear Chemistry Evolution and Its Applications

In 1895 Roentgen W.C. discovered the X rays. Curie M. and her husband, Curie P. obtained polonium after a careful purification of pitchblende in 1898. And in 1902 Curie M. announced that she had been able to isolate about 0.1g of pure radium chloride from more than one ton of pitchblende waste. The determination of the atomic weight of radium and the measurement of its emission spectrum provided the final proof that a new element had been isolated. In 1903 Rutherford E. and Soddy F. published a remarkable series of papers to formulate a theory of radioactive transformation. They posited that radioactive decay involved the spontaneous transformation of atoms (one element into those of another), accompanied by the emission of particles. Soddy F. and Fajans K. discovered the law of radioactive displacement, and then the concept of isotopes was proposed in 1910. In 1912, Hevesy G.C. created the theory of radioactive tracer technique. Though the potential of the tracer technique was clearly demonstrated in such studies, the input to chemistry and other fields remained somewhat limited as long as radioactivity was confined to the heaviest elements occurring in the natural decay series. Nuclear transmutation by α-particle bombardment specifically the nuclear reaction ($^{14}N+\alpha \rightarrow\ ^{17}O+p$) had been discovered by Rutherford in 1919, when he observed energetic protons during the bombardment of nitrogen by α particle from a natural source. The epoch-making event that led to a vast expansion of nuclear and radiochemistry was Curie's and Joliot's discovery of artificially produced radioactivity. Nuclear fission of heavy elements was discovered in 1938 by Hahn O. et al., and then the utilization of nuclear energy was accelerated.

The nuclear chemistry research of China began in 1934. The uranium extraction and purification technology were mastered in 1950. In 1955, heavy water reactor and circular accelerator were built for research in Beijing. Just August in that year, radio nuclides such as ^{25}Na, ^{32}P, ^{60}Co, were synthesized. Atomic bomb and hydrogen bomb test were success in 1964 and 1965, respectively, which indicates that China's nuclear technology has reached a high level. The main research directions for China's nuclear chemistry are preparation and application of radio nuclides, separation, analysis and application of radioactive materials, preparation and application of radiation sources, production and recycling of nuclear fuel, and radioactive waste processing and utilization.

16.1.2 Nucleon, Nuclide and Isomers

In the context of nuclear science, protons and neutrons are called **nucleons**, because they reside in the nucleus. A **nuclide** is an atomic species characterized by the specific constitution of its nucleus, by its number of protons Z, its number of neutrons N, and its nuclear energy state, denoted by $^A_Z X$, where X is the chemical symbol of the element with Z protons and A mass number. For example, the most abundant nuclide of uranium has 92 protons and 146 neutrons, so its atomic number is 92, its nucleon number is 238 (92+146), and its symbol is $^{238}_{92}U$. Nuclides with the same number of protons in the nucleus but with differing numbers of neutrons are called **isotopes**, such as $^{234}_{92}U$, $^{235}_{92}U$ and $^{238}_{92}U$. Isotopes have very similar chemical properties because they have the same electron configurations. In some cases a nucleus may exist for some time in one or more excited states and it is differentiated on this basis. Such nuclei that necessarily have the same atomic number and mass number are called isomers. ^{60m}Co and ^{60}Co are isomers; the ^{60m}Co nuclide exists in a high energy or excited state and decays spontaneously by emission of a γ-ray with a half-life of 10.5min to the lowest energy, the ground state, designated by ^{60}Co.

$$^{60m}_{27}Co \xrightarrow{\gamma 10.5 min}\ ^{60}_{27}Co$$

The symbol m stands for meta stable state, while no symbol (or g) refers to the ground state.

16.1.3 Radioelement and Radioactive Series

All the nuclides with the atomic number greater than $Z=83$ are radioactive, as we have noted. When an unstable nucleus decays, it may give off one of three common forms of radiation alpha (α) emission, beta(β) emission and gamma(γ) emission, that phenomenon was called **radioactive**.

Natural radioactive elements, such as uranium-238, give a radioactive decay series, a sequence in which one radioactive nucleus decays to a second, which then decays to a third, and so forth. Eventually, a stable nucleus, which is an isotope of lead, is reached. Three radioactive decay series are found naturally. One of these series begins with uranium-238. In the first step, uranium-238 decays by alpha emission to thorium-234.Thorium-234 in turn decays by beta emission to protactinium-234, which decays by beta emission to uranium-234.After the decay of protactinium-234 and the formation of uranium-234, there are a number of alpha-decay steps. The final product of the series is lead-206.Natural uranium is 99.28% uranium-238, which decays as we have described. However, the natural element also contains 0.72% uranium-235. This isotope starts a second radioactive decay series, which consists of a sequence of alpha and beta decays, ends with lead-207. The third naturally occurring radioactive decay series begins with thorium-232 and ends with lead-208. All three radioactive decay series found naturally end with an isotope of lead. The use of radioactive isotopes has a profound effect on the practice of medicine. Gamma radiation from cobalt-60 is more commonly used in the treatment of cancer.

16.1.4 Mass Defect and Nuclear Binding Energy

The equivalence of mass and energy explains the otherwise puzzling fact that the mass of an atom is always less than the sum of the masses of its constituent particles. For example, the helium-4 atom consists of two protons ($2 \times 1.007\,276$ amu), two neutrons ($2 \times 1.008\,665$ amu), and two electrons($2 \times 0.000\,549$ amu). Total mass of particles is $4.032\,98$ amu.

The mass of the helium-4 atom is $4.002\,60$ amu, so the mass difference is

$$\Delta m = (4.002\,60 - 4.032\,98) \text{ amu} = -0.030\,38 \text{ amu}$$

The mass difference is explained as follows. When the nucleons come together to form a nucleus, energy is released. According to Einstein's equation, there must be an equivalent decrease in mass.

$$\Delta E = mc^2$$

Where c is light speed, 3.00×10^8 m/s.

The **mass defect** of a nucleus is the difference between the nucleon mass and atomic mass. The **nuclear binding energy** of a nucleus is the energy needed to break a nucleus into its individual protons and neutrons. Thus, the binding energy of the helium-4 nucleus is the energy change for the reaction. Both the binding energy and the corresponding mass defect are reflections of the stability of the nucleus.

16.2 Radioactive Decay and Nuclear Equation

16.2.1 Radioactive Decay

Radioactive decay involves a transition from a definite quantum state of the original nuclide to a definite quantum state of the product nuclide. The energy difference between two quantums levels involved in the transition corresponds to the decay energy. This decay energy appears in the form of electromagnetic radiation and as the kinetic energy of the products.

The mode of radioactive decay is dependent upon the particular nuclide involved. Radioactive decay can be characterized by α-, β-, and γ-radiation. **Alpha-decay** is the emission of helium nucleus. **Beta-decay** is the creation and emission of either electrons or positrons, or the process of electron capture. **Gamma-decay** is the emission of electromagnetic radiation where the transition occurs between energy levels of the same nucleus. An additional mode of radioactive decay is that of internal conversion in which a nucleus loses its energy by interaction of the nuclear field with that of the orbital electrons, causing ionization of an electron instead of γ-ray emission. A mode of radioactive decay which is observed only in the heaviest nuclei is that of spontaneous fission in which the nucleus dissociates spontaneously into two roughly equal parts. This fission is accompanied by the emission of electromagnetic radiation and of neutrons. In the last decade, some unusual decay modes have been observed for nuclides very far from the stability line, namely neutron emission and proton emission. A few very rare decay modes like ^{12}C-emission have also been observed.

1. Alpha-decay(abbreviated α) emission of a $^{4}_{2}$He nucleus, or alpha particle, from an unstable nucleus. An example is the radioactive decay of radium-226.

$$^{226}_{88}Ra \rightarrow ^{222}_{86}Rn + ^{4}_{2}He$$

The product nucleus has an atomic number that is two less than that of the original nucleus, and a mass number that is four less.

2. Beta-decay(abbreviated β or β⁻) emission of a high-speed electron from an unstable nucleus. Beta emission is equivalent to the conversion of a neutron to a proton. An example of beta emission is the radioactive decay of Bi-210.

$$^{210}_{83}Bi \rightarrow ^{210}_{84}Po + ^{0}_{-1}e$$

The product nucleus has an atomic number that is one more than that of the original nucleus. The mass number remains the same.

3. Gamma-decay(abbreviated γ) emission from an excited nucleus of a gamma photon, corresponding to radiation with a wavelength of about 10^{-12}m. In many cases, radioactive decay results in a product nucleus that is in an excited state. As in the case of atoms, the excited state is unstable and goes to a lower-energy state with the emission of electromagnetic radiation. For nuclei, this radiation is in the gamma-ray region of the spectrum. An example is meta stable Co-60, which is used in medical diagnosis.

$$^{60m}_{27}Co \rightarrow ^{60}_{27}Co + \gamma$$

The product nucleus is simply a lower-energy state of the original nucleus, so there is no change of atomic number and mass number.

16.2.2 Nuclear Equations

The decay of uranium-238 can be write as an equation for a nuclear chemical reaction, and represent the uranium-238 nucleus by the nuclide symbol $^{238}_{92}U$. The radioactive decay of $^{238}_{92}U$ by alpha-particle emission (loss of a $^{4}_{2}$He nucleus) can be written as the following form.

$$^{238}_{92}U \rightarrow ^{234}_{90}Th + ^{4}_{2}He$$

This is an example of a **nuclear equation,** which is a symbolic representation of a nuclear reaction. Normally, only the nuclei are represented. It is not necessary to indicate the chemical compound or the electron charges for any ions involved, because the chemical environment has no effect on nuclear processes.

Reactant and product nuclei are represented in nuclear equations by their nuclide symbols. Other particles are given the following symbols, in which the subscript equals the charge and the superscript equals the total number of protons and neutrons in the particle (mass number).

Proton: $^{1}_{1}H$ or $^{1}_{1}p$ Neutron: $^{1}_{0}n$

Electron: $_{-1}^{0}e$ or $_{-1}^{0}\beta$ Positron: $_{1}^{0}e$ or $_{1}^{0}\beta$
Gamma photon: $_{0}^{0}\gamma$;   α: $_{2}^{4}He$ or $_{2}^{4}\alpha$

Sample Problem 16-1 Write the balanced equations for the following nuclear reactions

(1) $_{3}^{6}Li$ was bombarded with neutron particles to form $_{1}^{3}H$.

(2) $_{92}^{239}U$ was bombarded with α particle to form $_{94}^{239}Pu$.

Solution   (1) $_{3}^{6}Li + _{0}^{1}n \rightarrow _{1}^{3}H + _{2}^{4}He$

(2) $_{92}^{239}U + _{2}^{4}\alpha \rightarrow _{94}^{239}Pu + 2_{0}^{1}n$

Sample Problem 16-2 Complete the following nuclear equations.

(1) $_{53}^{122}I \rightarrow _{54}^{122}Xe +$?

(2) $_{26}^{59}Fe \rightarrow _{-1}^{0}e +$?

Solution   (1) $_{53}^{122}I \rightarrow _{54}^{122}Xe + _{-1}^{0}e$

(2) $_{26}^{59}Fe \rightarrow _{-1}^{0}e + _{27}^{59}Co$

16.2.3 Half-Life and Radioactivity

1. Half-life

The **half-life** (symbol $t_{1/2}$) of a radioactive nucleus is defined as the time it taken for half of the radionuclide's atoms to decay. In other words, the probability of a radioactive atom decaying within its half-life is 50%. Half-life of some isotopes is very short, only a fraction of a second. Half-lives of other isotopes may be considerably longer, and can be measured in minutes, days, years, and sometimes even millions of years or longer.

2. Radioactivity

Radioactive decay, also known as nuclear decay or radioactivity, is the process by which a nucleus of an unstable atom loses energy by emitting particles of ionizing radiation. A phenomenon that a material spontaneously emits kind of radiation including energetic alpha particles, beta particles, and gamma rays is considered radioactive. The rate of radioactive decay is that the number of nuclei disintegrating per unit time is found to be proportional to the number of radioactive nuclei in the sample.

The SI unit of radioactive activity is the becquerel (Bq), in honor of the scientist Henri Becquerel. One Bq is defined as one transformation (or decay or disintegration) per second. Another unit of radioactivity is the curie (Ci), which was originally defined as the amount of radium emanation (Radon-222) in equilibrium with one gram of pure radium, isotope Ra-226. At present it is equal, by definition, to the activity of any radionuclide decaying with a disintegration rate of 3.7×10^{10} Bq, so that 1 curie (Ci) = 3.7×10^{10} Bq.

3. Specific activity(Bq·kg^{-1} or Bq·g^{-1}) is defined as the amount of radioactivity or the decay rate of a particular radionuclide per unit mass of the radionuclide. The old unit of "activity" was curie, Ci, while that of "specific activity" was Ci·g^{-1}.

16.2.4 Carbon-14 Dating Method

Natural carbon is composed of three isotopes. They are 98.89% ^{12}C, 1.11% ^{13}C, and 0.000 000 000 10% ^{14}C. ^{14}C is the most important basis in radiocarbon dating. ^{14}C atoms are constantly being produced in our upper atmosphere through neutron bombardment of nitrogen atoms.

$$_{7}^{14}N + _{0}^{1}n \rightarrow _{6}^{14}C + _{1}^{1}H$$

Once formed, the ^{14}C is quickly oxidized to produce carbon dioxide, which is then converted into many different substances in plants. When animals eat the plants, they acquire the ^{14}C. For these reasons, ^{14}C is found in all living things. When the animal or plant dies, it stops exchanging carbon with its environment, and from that point onwards the amount of the ^{14}C it contains begins to decrease as the ^{14}C undergoes radioactive decay.

$$^{14}_{6}C \rightarrow {}^{14}_{7}N + {}^{0}_{-1}e$$

Measuring the amount of ^{14}C to ^{12}C in a sample from a dead plant or animal such as a piece of wood or a fragment of bone provides information that can be used to calculate when the animal or plant died.

16.3 The Introduce of Radioactive Tracer Technique — PET-CT

A **radioactive tracer**, is a chemical compound in which one or more atoms have been replaced by a radioisotope so by virtue of its radioactive decay it can be used to explore the mechanism of chemical reactions by tracing the path that the radioisotope follows from reactants to products.

Positron emission tomography (PET) is a nuclear medical imaging technique that produces a three-dimensional image or picture of functional processes in the body. The system detects pairs of gamma rays emitted indirectly by a positron-emitting radionuclide (tracer), which is introduced into the body on a biologically active molecule. Three-dimensional images of tracer concentration within the body are then constructed by computer analysis. 3D imaging is often accomplished with the aid of a **computational tomography (CT)** scan performed on the patient during the same session, in the same machine.

16.3.1 Principles of Radioactive Tracer Technique

The basic idea behind the use of radiotracers is that all the isotopes of a given element will behave the same chemically. A primary factor is whether a radioisotope of the element to be traced is available with the proper characteristics (half-life, particle energy, etc.).A second factor is whether the tagged compound desired is commercially available or can be easily synthesized. Furthermore, the available specific activity may be low for the proposed experimental use of the tagged compound.

PET is based on the detection of annihilation photons (γ) released when radio nuclides, such as ^{18}F, ^{13}N, ^{11}C, and ^{15}O, emit positrons that undergo annihilation with electrons. The photons are detected by coincidence imaging as they strike scintillation crystals made of bismuth germinate (BGO), lutetiumoxyorthosilicate (LSO), or gadolinium silicate (GSO). Bombarding target material with protons that have been accelerated in a cyclotron produces positron-emitting radio nuclides, which are then used to synthesize radiopharmaceuticals that are part of biochemical pathways in the human body, such as fluorodeoxyglucose (FDG) in glucose metabolism and ^{11}C labeled methionine and choline in protein metabolism and membrane biosynthesis, respectively.

The typical dose of FDG is 10mCi injected intravenously. Patient activity and speech are limited for 20min immediately following injection of the radioisotope to minimize physiologic up take by muscles. To our knowledge, there are no contraindications to FDG administration. Imaging is initiated approximately 60min following the injection of FDG. Patients undergo catheterization if necessary, or they void just before being positioned on the PET-CT table. They are positioned either with the arms above the head or with the arms at the side. Except for patients being studied for head and neck cancer, arms above the head is the preferred position to decrease beam-hardening artifact during the CT portion of the examination. However, not all patients can maintain this position comfortably without moving for the entire study (PET and CT), and arms by the side is an alternative. A whole-body PET study (neck through pelvis) follows an enhanced whole-body CT study. The CT study takes approximately 60~70s to complete and the PET study takes approximately 30~45min, depending on the coverage required.

16.3.2 PET-CT Imaging Process

1. The short-lived positron emitters, ^{11}C, ^{13}N, ^{15}O, and ^{19}F commonly used in positron emission to mography (PET), that are produced in cyclotrons.

2. Preparation of radiotracers and their compounds in the radiochemical laboratory.

3. Depending on the type of nuclear medicine exam you are undergoing, the dose of radiotracer is then injected intravenously, swallowed or inhaled as a gas. The CT exam will be done first, followed by the PET scan.

4. Image reconstruction using coincidence statistics, when the examination is completed.

16.3.3 The Features of PET-CT

1. The diagnostic accuracy of PET in certain clinical on ecological examinations including lung, colorectal, oesophagus cancers, melanoma, lymphomas, breast, head-and-neck cancers, sarcoma is higher than other diagnostic imaging techniques. By identifying changes in the body at the cellular level, PET imaging may detect the early onset of disease before it is evident on other imaging tests such as CT or MRI.

2. The gamma photons easily pass through human tissue, so they can be recorded by scintillation detectors placed around the body. Nuclear medicine examinations offer information that is unique — including details on both function and structure — and often unattainable using other imaging procedures. Because both scans are performed at one time without the patient having to change positions, there is less room for error.

3. Most widely used ^{18}F-labeled radiotracer is FDG. When taken into tissue, the FDG is converted in fluoroglucose-6-phosphate, which cannot be metabolized further and is trapped in the tissue. The trapped tracer is used then for imaging those organs that metabolize glucose most rapidly.

16.3.4 Radionuclide for Clinically Therapy and Diagnosis

The most rapidly expanding area of tracer use is in nuclear medicine. Nuclear medicine deals with the use of radiation and radioactivity to diagnose and treat disease. The two principal areas of endeavor, diagnosis and therapy, involve different methods and considerations for radiotracer use. (As an aside, we note that radiolabeled drugs for patients are called radiopharmaceuticals.) A list of radionuclides commonly used in diagnosis is shown in Table 16-1. Most diagnostic use of radiotracers is for imaging of specific organs, bones, or tissue. Typical administered quantities of tracer are 1~30mCi for adults. Nuclides used for imaging should emit photons with the energy between 100 and 200keV, have small decay branches for particle emission (to minimize radiation damage), have a half-life that is equal to about (1.5 × the duration of the test procedure), and be inexpensive and readily available.

Table 16-1 Radionuclide for Clinically Therapy and Diagnosis

Radionuclide	Half-life	Emitted radiation	Application
^{51}Cr	28d	γ	Diagnosis of spleen and intestinal dysfunction
^{60}Co	5y	β	Therapy for cancer
^{123}I	12h	γ	Thyroid disease diagnosis
^{131}I	8d	β and γ	Thyroid disease therapy and diagnosis
^{59}Fe	45d	β and γ	Bone and anemia therapy and diagnosis
^{81}Kr	2×10^5y	γ	Lung imaging
^{99}Tc	6h	γ	Brain, heart, lung, thyroid, gall bladder, skin, lymph node, bone, liver, spleen, and kidney imaging
^{201}Tl	7h	γ	Arteria coronaria function diagnosis
^{32}P	14.3d	β	Polycythemia vera therapy, malignant ascites control
^{90}Sr	28.8d	β	Hemangiomas
^{137}Cs	30y	γ	Measure fill level of consumer products
^{192}Ir	74.2 d	γ	Pipeline, boiler and aircraft weld radiography

16.4 Nuclear Reactions and Radioactive Emissions

16.4.1 Nuclear Reaction

A **nuclear reaction** is semantically considered to be the process in which two nuclei, or else a nucleus of an atom and a subatomic particle from outside the atom, collide to produce one or more nuclides that are different from the nuclide(s) that began the process. Thus, a nuclear reaction must cause a transformation of at least one nuclide to another. If a nucleus interacts with another nucleus or particle and then they separate without changing the nature of any nuclide, the process is simply referred to as a type of nuclear scattering, rather than a nuclear reaction.

16.4.2 Nuclear Fission Reaction and Nuclear Fusion Reactions

Natural nuclear reactions occur in the interaction between cosmic rays and matter, and nuclear reactions can be employed artificially to obtain nuclear energy, at an adjustable rate, on demand. Perhaps the most notable nuclear reactions are the nuclear chain reactions in fissionable materials that produce induced **nuclear fission**, and the various **nuclear fusion reactions** of light elements that power the energy production of the Sun and stars. Both of these types of reactions are employed in nuclear weapons.

Nuclear fission reaction is a nuclear reaction in which a heavy nucleus splits into lighter nuclei and energy is released. **Fusion reactions**: two light nuclei join to form a heavier one, with additional particles (usually protons or neutrons) thrown off to conserve momentum.

16.4.3 Nuclear Chain Reaction

When the uranium-235 nucleus splits, approximately two or three neutrons are released. If the neutrons from nuclear fission are absorbed by other uranium-235 nuclei, these nuclei split and release even more neutrons. In this way, a chain reaction can occur. A **nuclear chain reaction** is a self-sustaining series of nuclear fissions caused by the absorption of neutrons released from previous nuclear fissions. The numbers of nuclei that split quickly multiply as a result of the absorption of neutrons released from previous nuclear fissions. The chain reaction of nuclear fissions is the basis of nuclear power and nuclear weapons.

16.4.4 Nuclear Radiation and Nuclear Radiation Protection

When radiation passes through matter, it may transfer energy to atoms, molecules, or ions by colliding. These collisions may be elastic that is only kinetic energy may be transferred, raising the temperature of the material. Eventually this kinetic energy is converted to heat, which is given off to the surroundings. Frequently, interaction of radiation with matter results in inelastic collisions. Inelastic collisions are divided into excitation and ionization.

Excitation is an electron in the matter jumping to higher energy level by absorbing low energetic radiation in collision. When it drops back to the ground level, energy is given off, often as UV or visible light. Because the electron is not lost from the atom, the radiation that causes excitation is called nonionizing radiation. Nonionizing radiation is relatively harmless for human beings. The scintillation counter is used to detect and measure radiation. Ionization is an electron in the matter is removed from an atom to form a positively charged ion by absorbing energetically enough radiation in collision.

Ionization is directly dependent on the energy of the incoming radiation. When the energy is high enough in collision, cation-electron pairs are produced. The high-energy radiation that gives rise to this effect is called **ionizing radiation**. The free electron of the pair often collides with another atom and ejects a second electron.

$$\text{Atom} \xrightarrow{\text{ionizing radiation}} \text{ion}^+ + e^-$$

Background radiation is the ionizing radiation emitted from a variety of natural and artificial radiation sources. They come from two sources, natural and artificial radiation. The most radiation exposures come from natural radiation. Natural radiation levels are different in different regions. The cosmic radiation a kind of natural radiation comes from the Sun and outer space, consisted by positively charged particles and gamma radiation. It is much more intense at high elevations due to the atmosphere shielding cosmic rays decreases. The sources of most natural radiation are uranium and thorium in rocks and soil. Radioactive radon gas is produced when uranium and thorium decay. Radon has a short half-life (4 days) and decays into other solid particulate radium-series radioactive nuclides by emitting alpha particles. The largest radiation exposure to artificial sources is due to X-rays. These amounts can be higher if operator of the X-ray machine is inexperienced or if the machine itself is defective. Clearly excessive diagnostic use of X-ray should be avoided. The radiation exposures from nuclear testing and radioactive waste are slight for most people, but exposures for those living near testing sites and nuclear energy facilities may be many times higher.

The radioactive effect on human beings depends on the amount of radiation absorbed and the types of radiation. The three types of nuclear radiation have different penetrating power. In general, penetrating power is inversely related to the mass and charge of the particle. The more strongly a particle interacts with the matter, the more slightly it penetrates.

Alpha particles have larger mass and higher charge in three type emission, so they interact with the matter most strongly. As a result, they penetrate so little that they can be absorbed by tissue paper or the outer layers of human skin (about 40μm, equivalent to a few cells deep).They are not generally dangerous to life unless they are ingested or inhaled. If ingested or inhaled, an alpha emitter such as plutonium-239, which causes grave localized damage through wide ionization.

Beta particles and positrons have less charge and much less mass than alpha particles, so they interact with the matter less strongly and penetrate deeper. Beta particles are able to penetrate the matter to a certain extent and can change the structure of struck molecules. If beta emitters are ingested, the damaging effects are greatly aggravated. For example, strontium-90, can replace calcium in bones, where it does the serious damage. Specialized heavy clothing or a thick (0.5cm) piece of metal can stop these particles.

Gamma rays are neutral and massless, so they interact least with the matter and penetrate deepest. Gamma rays are the most dangerous because the energy can ionize many layers of living tissue. They can cause cancer and hereditary disease by inducing DNA alteration. It is difficult to shield them as they have no charge. A block of lead of a few inches thick is needed to stop them.

Beyond ultraviolet are higher energy kinds of radiation which are used in medicine and which we all get in low doses from space, from the air, and from the earth and rocks. Collectively we can refer to these kinds of radiation as ionizing radiation. It can cause damage to matter, particularly living tissue. At high levels it is therefore dangerous, so it is necessary to control our exposure. Furthermore, many of us owe our lives and health to such radiation produced artificially. Medical and dental X-rays discern hidden problems. Other kinds of ionizing radiation are used to diagnose ailments, and some people are treated with radiation to cure disease. We all benefit from a multitude of products and services made possible by the careful use of such radiation. People living in granite areas or on mineralized sands receive more terrestrial radiation than others, while people living or working at high altitudes receive more cosmic radiation. A lot of our natural exposure is due to radon, a gas which seeps from the Earth's crust and is present in the air we breathe.

There are four ways in which people are protected from identified radiation sources:

Limiting Time For people who are exposed to radiation in addition to natural background radiation through their work, the dose is reduced and the risk of illness essentially eliminated by limiting exposure time.

Distance In the same way that heat from a fire is less the further away, the intensity of radiation decreases with distance from its source.

Shielding Barriers of lead, concrete or water give good protection from penetrating radiation such as gamma rays. Radioactive materials are therefore often stored or handled under water, or by remote control in rooms constructed of thick concrete or even lined with lead.

Containment Radioactive materials are confined and kept out of the environment. Radioactive isotopes for medical use, for example, are dispensed in closed handling facilities, while nuclear reactors operate within closed systems with multiple barriers which keep the radioactive materials contained. Rooms have a reduced air pressure so that any leaks occur into the room and not out from the room.

Nuclear chemistry has been studied very intensively for more than a century. Compared with many factors which influence on human health, it is well understood scientifically.

Chapter 16 Nuclear Chemistry and Its Applications

Key Terms

核化学	nuclear chemistry
放射化学	radiochemistry
核医学	nuclear medicine
核子	nucleon
核素	nuclide
同质异能素	isomers
放射元素	radioelement
放射系	radioactive series
质量亏损	mass defect
核结合能	nuclear binding energy
放射性衰变	radioactive decay
核化学方程式	nuclear equation
α 衰变	alpha-decay
β 衰变	beta-decay
γ 衰变	gamma-decay
质子	proton
中子	neutron
电子	electron
正电子	positron
γ 光子	gamma photon
半衰期	half-life
放射性活度	radioactivity
放射性碳-14 测定年代法	carbon-14 dating method
放射性核素示踪	radioactive tracer
正电子发射断层扫描	positron emission tomography computational tomography(PET-CT)
核裂变反应	nuclear fission reaction
核聚变反应	nuclear fusion reactions
核链式反应	nuclear chain reaction
核辐射	nuclear radiation
核辐射防护	nuclear radiation protection
致电离辐射	ionizing radiation

Summary

Nuclear chemistry is the subject of researching nucleus (stability and radioactive) reaction, nature, structure, separation, identification and application. Nuclear chemistry used combination method of physical and chemical to research the principle of nuclear properties, nuclear structure and nuclear transformation. Research findings have been applied in many areas.

Radiochemistry is mainly researched radionuclide preparation, separation, purification, identification, chemical state in extremely low concentrations as well as nuclear transformation nature and behavior of products. Additionally, radiochemistry is also researched some substances in chemical reaction and some radioactive isotope chemical reactions.

Nuclear medicine is a new subject, which is used nuclear technology to diagnose therapy and research some diseases. Nuclear medicine is the combination product of nuclear technology, electronic technique, computer technology, chemistry, physics, biology and other modern sciences with medicine. Nuclear medicine is mainly studied isotopes, the ray beam as acceleration and radioactive isotope produced by nuclear radiation in medical application such as PET-CT instrument in clinical diagnosis.

Exercises

1. Interpretation of nouns: isotopes, isomers, nuclides.
2. What are the types of radioactive decay?
3. Write the equations of nuclear equilibrium for the following transformation processes.
(1) Bi-210 decays through beta decay
(2) Th-232 decays into Ra-228
(3) Y-84 releases a positron
(4) Ti-44 captures an electron
4. Describe the working principle of PET-CT and the medicament used in the detection process.
5. How to distinguish between ionizing radiation and non-ionizing radiation, and propose protection scheme?

Supplementary Exercises

1. Write equations for the following nuclear reactions:
(1) bombardment of $_3^6 Li$ with neutrons to produce $_1^3 H$.
(2) bombardment of $_{92}^{238} U$ with neutrons to produce $_{94}^{239} Pu$.
(3) bombardment of $_7^{14} N$ with neutrons to produce $_6^{14} C$.
(4) the reaction of two deuterium nuclei to produce a nucleus of $_2^3 He$.

2. The half-life of cobalt-60 is 5.3 a. How much of a 1.000mg sample of cobalt-60 is left after a 15.9 a period?

Answers to Some Exercises

[Exercises]

3. (1) $_{83}^{210} Bi \rightarrow {}_{84}^{210} Po + {}_{-1}^{0} e$ (2) $_{90}^{232} Th \rightarrow {}_{88}^{228} Ra + {}_{2}^{4} He$
(3) $_{39}^{84} Y \rightarrow {}_{38}^{84} Sr + {}_{1}^{0} e$ (4) $_{22}^{44} Ti + {}_{-1}^{0} e \rightarrow {}_{21}^{44} Sc$

[Supplementary Exercises]

1. (1) $_3^6 Li + {}_0^1 n \rightarrow {}_2^4 \alpha + {}_1^3 H$
(2) $_{92}^{238} U + {}_0^1 n \rightarrow {}_{92}^{239} U$ $_{92}^{239} U \rightarrow {}_{-1}^0 e + {}_{93}^{239} Np$ $_{93}^{239} Np \rightarrow {}_{-1}^0 e + {}_{94}^{239} Pu$
(3) $_7^{14} N + {}_0^1 n \rightarrow {}_6^{14} C + {}_1^1 H$
(4) $_1^1 H + {}_1^1 H \rightarrow {}_2^3 He + {}_0^1 n$

2. 0.125mg

(武世奎)

Appendix

Appendix 1 Legal Units of Measurement

Table 1 SI Units

Quantity	Name	Symbol
Length	meter	m
Mass	kilogram	kg
Time	second	s
Electric current	ampere	A
Temperature	kelvin	K
Amount of substance	mole	mol
Luminous intensity	candela	cd

Table 2 SI Derived Units with Special Names and Symbols

Physical quantity	SI Derived Unit		
	Name	Symbol	expressed in terms of SI base units and other SI derived units
plane angle	radian	rad	$1\text{rad} = 1\text{m/m} = 1$
solid angle	steradian	sr	$1\text{sr} = 1\text{m}^2/\text{m}^2 = 1$
frequency	hertz	Hz	$1\text{Hz} = 1\text{s}^{-1}$
force	newton	N	$1\text{N} = 1\text{kg} \cdot \text{m/s}^2$
pressure, stress	pascal	Pa	$1\text{Pa} = 1\text{N/m}^2$
energy, work, quantity of heat	joule	J	$1\text{J} = 1\text{N} \cdot \text{m}$
power, radiant flux	watt	W	$1\text{W} = 1\text{J/s}$
electric charge, quantity of electricity	coulomb	C	$1\text{C} = 1\text{A} \cdot \text{s}$
electric potential difference, electromotive force	volt	V	$1\text{V} = 1\text{W/A}$
capacitance	farad	F	$1\text{F} = 1\text{C/V}$
electric resistance	ohm	Ω	$1\Omega = 1\text{V/A}$
electric conductance	siemens	S	$1\text{S} = 1\Omega^{-1}$
magnetic flux	weber	Wb	$1\text{Wb} = 1\text{V} \cdot \text{S}$
magnetic flux density	tesla	T	$1\text{T} = 1\text{Wb/m}^2$
inductance	henry	H	$1\text{H} = 1\text{Wb/A}$
Celsius temperature	degree Celsius	℃	$1℃ = 1\text{K}$
luminous flux	lumen	lm	$1\text{lm} = 1\text{cd} \cdot \text{sr}$
illuminance	lux	lx	$1\text{lx} = 1\text{lm/m}^2$
activity (of a radionuclide)	becquerel	Bq	$1\text{Bq} = 1\text{s}^{-1}$
absorbed dose specific energy (imparted) kerma	gray	Gy	$1\text{Gy} = 1\text{J/kg}$
dose equivalent	sievert	Sv	$1\text{Sv} = 1\text{J/kg}$

Table 3 SI Prefix

Factor	Name	Symbol
10^{24}	yotta	Y
10^{21}	zetta	Z
10^{18}	exa	E
10^{15}	peta	P
10^{12}	tera	T
10^{9}	giga	G
10^{6}	mega	M
10^{3}	kilo	k
10^{2}	hecto	h
10^{1}	deca	da
10^{-1}	deci	d
10^{-2}	centi	c
10^{-3}	milli	m
10^{-6}	micro	μ
10^{-9}	nano	n
10^{-12}	pico	p
10^{-15}	femto	f
10^{-18}	atto	a
10^{-21}	zepto	z
10^{-24}	yocto	y

Table 4 Commonly Used SI Units

Quantity	Name	Symbol	Value in SI units
Time	minute	min	$1\text{min} = 60\text{s}$
	hour	h	$1\text{h} = 60\text{min} = 3\,600\text{s}$
	day	d	$1\text{d} = 24\text{h} = 86\,400\text{s}$
Plane angle	degree	°	$1° = (\pi/180)\text{rad}$
	minute	'	$1' = (1/60)° = (\pi/10\,800)\text{rad}$
	second	"	$1'' = (1/60)' = (\pi/648\,000)\text{rad}$
Volume	liter	l, L	$1\text{L} = 1\text{dm}^3$
Mass	ton	t	$1\text{t} = 10^3\text{kg}$
	atomic mass unit	u	$1\text{u} \approx 1.660\,540 \times 10^{-27}\text{kg}$
Rotational speed	revolutions per minute	r/min	$1\text{r/min} = (1/60)\text{s}$
Energy	electronvolt	eV	$1\text{eV} \approx 1.602\,177 \times 10^{-19}\text{J}$
Level difference	decibel	dB	
Line density	tex	tex	$1\text{tex} = 10^{-6}\text{kg/m}$
Area	hectare	ha, hm²	$1\text{ha} = 1\text{hm}^2 = 10^4\text{m}^2$

Appendix 2 Physical and Chemical Constants

Quality	Symbol	Numerical value	Unit	Remarks
speed of light in vacuum	c, c_0	299 792 458	$m \cdot s^{-1}$	exact
magnetic constant in vacuum	μ_0	$4\pi \times 10^{-7}$ $1.256\,637\cdots \times 10^{-6}$	$H \cdot m^{-1}$	exact
electric constant in vacuum $\varepsilon_0 = \dfrac{1}{\mu_0 c_0^2}$	ε_0	$10^7/(4\pi \times 29\,979\,258^2)$ $8.854\,187\,871 \times 10^{-12}$	$F \cdot m^{-1}$	exact
constant of gravitation $F = \dfrac{G m_1 m_2}{r^2}$	G	$(6.674\,08 \pm 0.000\,31) \times 10^{-11}$	$N\,\hbar = \dfrac{h}{2\pi} \cdot m^2 \cdot kg^{-2}$	
Planck constant $\hbar = \dfrac{h}{2\pi}$	h $\hbar$	$(6.626\,070\,040 \pm 0.000\,000\,081) \times 10^{-34}$ $(1.054\,571\,800 \pm 0.000\,000\,013) \times 10^{-34}$	$J \cdot s$ $J \cdot s$	
elementary charge	e	$(1.602\,176\,620\,8 \pm 0.000\,000\,009\,8) \times 10^{-19}$	C	
electron mass	m_e	$(9.109\,383\,56 \pm 0.000\,000\,11) \times 10^{-31}$ $(5.485\,799\,090\,70 \pm 0.000\,000\,000\,16) \times 10^{-4}$	kg u	
proton mass	m_p	$(1.672\,621\,898 \pm 0.000\,000\,021) \times 10^{-27}$ $(1.007\,276\,466\,879 \pm 0.000\,000\,000\,091)$	kg u	
fine-structure constant $\alpha = \dfrac{e^2}{4\pi \varepsilon_0 \hbar c}$	α	$(7.297\,352\,566\,4 \pm 0.000\,000\,001\,7) \times 10^{-3}$	1	
Rydberg constant $R_\infty = \dfrac{e^2}{8\pi \varepsilon_0 a_0 h c}$	R_∞	$(1.097\,373\,156\,850\,8 \pm 0.000\,000\,000\,006\,5) \times 10^7$	m^{-1}	
Avogadro constant $L = \dfrac{N}{n}$	L, N_A	$(6.022\,140\,857 \pm 0.000\,000\,074) \times 10^{23}$	mol^{-1}	
Faraday constant $F = Le$	F	$(9.648\,533\,289 \pm 0.000\,000\,059) \times 10^4$	$C \cdot mol^{-1}$	
molar gas constant $pV_m = RT$	R	$(8.314\,459\,8 \pm 0.000\,004\,8)$	$J \cdot mol^{-1} \cdot K^{-1}$	
Boltzmann constant $k = \dfrac{R}{T}$	k	$(1.380\,648\,52 \pm 0.000\,000\,79) \times 10^{-23}$	$J \cdot K^{-1}$	
Stefan-Boltzmann constant $\sigma = \dfrac{2\pi^5 k^4}{15 h^3 c^2}$	σ	$(5.670\,367 \pm 0.000\,013) \times 10^{-8}$	$W \cdot m^{-2} \cdot K^{-4}$	
atomic mass constant	m_u	$(1.660\,539\,040 \pm 0.000\,000\,020) \times 10^{-27}$	kg	atomic mass unit $1u = (1.660\,539\,040 \pm 0.000\,000\,020) \times 10^{-27} kg$

The values are from Haynes WM. CRC Handbook of Chemistry and Physics. 97th ed. New York: CRC Press, 2016

Appendix 3 Equilibrium Constants

Table 1 Ion Product Constant of Water

Temperature/°C	pK_w	Temperature/°C	pK_w	Temperature/°C	pK_w
0	14.947	35	13.680	70	12.799
5	14.734	40	13.535	75	12.696
10	14.534	45	13.396	80	12.598
15	14.344	50	13.265	85	12.505
20	14.165	55	13.140	90	12.417
25	13.995	60	13.020	95	12.332
30	13.833	65	12.907	100	12.252

The values are from Haynes WM. CRC Handbook of Chemistry and Physics. 97th ed. New York: CRC Press, 2016

Table 2 Dissociation Constants of Weak Electrolytes in Aqueous Solution

Name	Formula	t/°C	Step	K_a^*(or K_b)	pK_a(or pK_b)
Arsenic acid	H_3AsO_4	25	1	5.5×10^{-3}	2.26
			2	1.7×10^{-7}	6.76
			3	5.1×10^{-12}	11.29
Arsenious acid	H_2AsO_3	25	—	5.1×10^{-10}	9.29
Boric acid	H_3BO_3	20	1	5.4×10^{-10}	9.27
			2		>14
Carbonic acid	H_2CO_3	25	1	4.5×10^{-7}	6.35
			2	4.7×10^{-11}	10.33
Chromic acid	H_2CrO_4	25	1	1.8×10^{-1}	0.74
			2	3.2×10^{-7}	6.49
Hydrofluoric acid	HF	25	—	6.3×10^{-4}	3.20
Hydrocyanic acid	HCN	25	—	6.2×10^{-10}	9.21
Hydrogen sulfide	H_2S	25	1	8.9×10^{-8}	7.05
			2	1.0×10^{-19}	19
Hydrogen peroxide	H_2O_2	25	—	2.4×10^{-12}	11.62
Hypobromous acid	HBrO	25	—	2.0×10^{-9}	8.55
Hypochlorous acid	HClO	25	—	3.9×10^{-8}	7.40
Hypoiodous acid	HIO	25	—	3×10^{-11}	10.5
Iodic acid	HIO_3	25	—	1.6×10^{-1}	0.78
Nitrous acid	HNO_2	25	—	5.6×10^{-4}	3.25
Periodic acid	HIO_4	25	—	2.3×10^{-2}	1.64
Phosphoric acid	H_3PO_4	25	1	6.9×10^{-3}	2.16
		25	2	6.1×10^{-8}	7.21
		25	3	4.8×10^{-13}	12.32

Continue

Name	Formula	$t/℃$	Step	K_a^* (or K_b)	pK_a (or pK_b)
Silicic acid	H_4SiO_4	30	1	1.2×10^{-10}	9.9
			2	1.6×10^{-12}	11.8
			3	1×10^{-12}	12
			4	1×10^{-12}	12
Sulfuric acid	H_2SO_4	25	2	1.0×10^{-2}	1.99
Sulfurous acid	H_2SO_3	25	1	1.4×10^{-2}	1.85
			2	6×10^{-8}	7.2
Ammonia	NH_3	25	—	1.8×10^{-5}	4.75
Calcium(II) ion	Ca^{2+}	25	2	4×10^{-2}	1.4
Aluminum(III) ion	Al^{3+}	25	—	1×10^{-9}	9.0
Argentum(II) ion	Ag^+	25	—	1.0×10^{-2}	2.00
Zinc(II) ion	Zn^{2+}	25	—	7.9×10^{-7}	6.10
Formic acid	HCOOH	25	1	1.8×10^{-4}	3.75
Acetic acid	CH_3COOH	25	1	1.75×10^{-5}	4.756
Propanoic acid	C_2H_5COOH	25	1	1.3×10^{-5}	4.87
Chloroacetic acid	$CH_2ClCOOH$	25	1	1.4×10^{-3}	2.85
Oxalic acid	$C_2H_2O_4$	25	1	5.6×10^{-2}	1.25
			2	1.5×10^{-4}	3.81
Citric acid	$C_6H_8O_7$	25	1	7.4×10^{-4}	3.13
			2	1.7×10^{-5}	4.76
			3	4.0×10^{-7}	6.40
Barbituric acid	$C_4H_4N_2O_3$	25	1	9.8×10^{-5}	4.01
Methylammonium chloride	$CH_3NH_2 \cdot HCl$	25	1	2.2×10^{-11}	10.66
Dimethylammonium chloride	$(CH_3)_2NH \cdot HCl$	25	1	1.9×10^{-11}	10.73
Lactic acid	$C_3H_6O_3$	25	1	1.4×10^{-4}	3.86
Ethylamine hydrochloride	$C_2H_5NH_2 \cdot HCl$	20	1	2.2×10^{-11}	10.66
Benzoic acid	C_6H_5COOH	25	1	6.25×10^{-5}	4.204
Phenol	C_6H_5OH	25	1	1.0×10^{-10}	9.99
Phthalic acid	$C_8H_6O_4$	25	1	1.14×10^{-3}	2.943
			2	3.70×10^{-6}	5.432
Tris-HCl	$C_4H_{11}NO_3 \cdot HCl$	37	1	1.4×10^{-8}	7.85
Glycine hydrochloride	$H_2NCH_2COOH \cdot 2HCl$	25	1	4.5×10^{-3}	2.35
			2	1.6×10^{-10}	9.78

The data are mainly from Haynes WM. CRC Handbook of Chemistry and Physics. 97th ed. New York: CRC Press, 2016

Note *: K_a (or K_b) is converted from pK_a (or pK_b) Appendix 1 Legal Units of Measurement

Table 3 Solubility Product Constants (25℃)

Compound	K_{sp}	Compound	K_{sp}	Compound	K_{sp}
AgAc	1.94×10^{-3}	$CdCO_3$	1.0×10^{-12}	$LiCO_3$	8.15×10^{-4}
AgBr	5.35×10^{-13}	CdF_2	6.44×10^{-3}	$MgCO_3$	6.82×10^{-6}
$AgBrO_3$	5.38×10^{-5}	$Cd(IO_3)_2$	2.5×10^{-8}	MgF_2	5.16×10^{-11}
AgCN	5.97×10^{-17}	$Cd(OH)_2$	7.2×10^{-15}	$Mg(OH)_2$	5.61×10^{-12}
AgCl	1.77×10^{-10}	CdS	8.0×10^{-27}	$Mg_3(PO_4)_2$	1.04×10^{-24}
AgI	8.52×10^{-17}	$Cd_3(PO_4)_2$	2.53×10^{-33}	$MnCO_3$	2.24×10^{-11}
$AgIO_3$	3.17×10^{-8}	$Co_3(PO_4)_2$	2.05×10^{-35}	$Mn(IO_3)_2$	4.37×10^{-7}
AgSCN	1.03×10^{-12}	CuBr	6.27×10^{-9}	$Mn(OH)_2$	2.06×10^{-13}
Ag_2CO_3	8.46×10^{-12}	CuC_2O_4	4.43×10^{-10}	MnS	2.5×10^{-13}
$Ag_2C_2O_4$	5.40×10^{-12}	CuCl	1.72×10^{-7}	$NiCO_3$	1.42×10^{-7}
Ag_2CrO_4	1.12×10^{-12}	CuI	1.27×10^{-12}	$Ni(IO_3)_2$	4.71×10^{-5}
Ag_2S	6.3×10^{-50}	CuS	6.3×10^{-36}	$Ni(OH)_2$	5.48×10^{-16}
Ag_2SO_3	1.50×10^{-14}	CuSCN	1.77×10^{-13}	α-NiS	3.2×10^{-19}
Ag_2SO_4	1.20×10^{-5}	Cu_2S	2.5×10^{-48}	$Ni_3(PO_4)_2$	4.74×10^{-32}
Ag_3AsO_4	1.03×10^{-22}	$Cu_3(PO_4)_2$	1.40×10^{-37}	$PbCO_3$	7.40×10^{-14}
Ag_3PO_4	8.89×10^{-17}	$FeCO_3$	3.13×10^{-11}	$PbCl_2$	1.70×10^{-5}
$Al(OH)_3$	1.1×10^{-33}	FeF_2	2.36×10^{-6}	PbF_2	3.3×10^{-8}
$AlPO_4$	9.84×10^{-21}	$Fe(OH)_2$	4.87×10^{-17}	PbI_2	9.8×10^{-9}
$BaCO_3$	2.58×10^{-9}	$Fe(OH)_3$	2.79×10^{-39}	$PbSO_4$	2.53×10^{-8}
$BaCrO_4$	1.17×10^{-10}	FeS	6.3×10^{-18}	PbS	8×10^{-28}
BaF_2	1.84×10^{-7}	HgI_2	2.9×10^{-29}	$Pb(OH)_2$	1.43×10^{-20}
$Ba(IO_3)_2$	4.01×10^{-9}	HgS	4×10^{-53}	$Sn(OH)_2$	5.45×10^{-27}
$BaSO_4$	1.08×10^{-10}	Hg_2Br_2	6.40×10^{-23}	SnS	1.0×10^{-25}
$BiAsO_4$	4.43×10^{-10}	Hg_2CO_3	3.6×10^{-17}	$SrCO_3$	5.60×10^{-10}
CaC_2O_4	2.32×10^{-9}	$Hg_2C_2O_4$	1.75×10^{-13}	SrF_2	4.33×10^{-9}
$CaCO_3$	3.36×10^{-9}	Hg_2Cl_2	1.43×10^{-18}	$Sr(IO_3)_2$	1.14×10^{-7}
CaF_2	3.45×10^{-11}	Hg_2F_2	3.10×10^{-6}	$SrSO_4$	3.44×10^{-7}
$Ca(IO_3)_2$	6.47×10^{-6}	Hg_2I_2	5.2×10^{-29}	$ZnCO_3$	1.46×10^{-10}
$Ca(OH)_2$	5.02×10^{-6}	Hg_2SO_4	6.5×10^{-7}	ZnF_2	3.04×10^{-2}
$CaSO_4$	4.93×10^{-5}	$KClO_4$	1.05×10^{-2}	$Zn(OH)_2$	3×10^{-17}
$Ca_3(PO_4)_2$	2.07×10^{-33}	$K_2[PtCl_6]$	7.48×10^{-6}	α-ZnS	1.6×10^{-24}

The values are mainly from Haynes WM. CRC Handbook of Chemistry and Physics. 97th ed. New York: CRC Press, 2016

The K_{sp} of sulfide is from Lange's Handbook of Chemistry. 16th ed. 2005: 1.331-1.342

Table 4 Cumulative Stability Constants of Complex Ions

Ligand and metal ion	$\lg\beta_1$	$\lg\beta_2$	$\lg\beta_3$	$\lg\beta_4$	$\lg\beta_5$	$\lg\beta_6$
ammonia (NH_3)						
Co^{2+}	2.11	3.74	4.79	5.55	5.73	5.11
Co^{3+}	6.7	14.0	20.1	25.7	30.8	35.2
Cu^{2+}	4.31	7.98	11.02	13.32	12.86	
Hg^{2+}	8.8	17.5	18.5	19.28		
Ni^{2+}	2.80	5.04	6.77	7.96	8.71	8.74
Ag^+	3.24	7.05				
Zn^{2+}	2.37	4.81	7.31	9.46		
Cd^{2+}	2.65	4.75	6.19	7.12	6.80	5.14
chloride (Cl^-)						
Sb^{3+}	2.26	3.49	4.18	4.72		
Bi^{3+}	2.44	4.7	5.0	5.6		
Cu^+		5.5	5.7			
Pt^{2+}		11.5	14.5	16.0		
Hg^{2+}	6.74	13.22	14.07	15.07		
Au^{3+}		9.8				
Ag^+	3.04	5.04				
cyanide (CN^-)						
Au^+		38.3				
Cd^{2+}	5.48	10.60	15.23	18.78		
Cu^+		24.0	28.59	30.30		
Fe^{2+}						35
Fe^{3+}						42
Hg^{2+}				41.4		
Ni^{2+}				31.3		
Ag^+		21.1	21.7	20.6		
Zn^{2+}				16.7		
fluoride (F^-)						
Al^{3+}	6.10	11.15	15.00	17.75	19.37	19.84
Fe^{3+}	5.28	9.30	12.06			
iodide (I^-)						
Bi^{3+}	3.63			14.95	16.80	18.80
Hg^{2+}	12.87	23.82	27.60	29.83		
Ag^+	6.58	11.74	13.68			

Continue

Ligand and metal ion	$\lg\beta_1$	$\lg\beta_2$	$\lg\beta_3$	$\lg\beta_4$	$\lg\beta_5$	$\lg\beta_6$
thiocyanate (SCN$^-$)						
Fe^{3+}	2.95	3.36				
Hg^{2+}		17.47		21.23		
Au^+		23		42		
Ag^+		7.57	9.08	10.08		
thiosulfate ($S_2O_3^{2-}$)						
Ag^+	8.82	13.46				
Hg^{2+}		29.44	31.90	33.24		
Cu^+	10.27	12.22	13.84			
acetate (CH_3COO^-)						
Fe^{3+}	3.2					
Hg^{2+}		8.43				
Pb^{2+}	2.52	4.0	6.4	8.5		
citrate (as L^{3-} ligand)						
Al^{3+}	20.0					
Co^{2+}	12.5					
Cd^{2+}	11.3					
Cu^{2+}	14.2					
Fe^{2+}	15.5					
Fe^{3+}	25.0					
Ni^{2+}	14.3					
Zn^{2+}	11.4					
ethanediamine ($H_2NCH_2CH_2NH_2$)						
Co^{2+}	5.91	10.64	13.94			
Cu^{2+}	10.67	20.00	21.0			
Zn^{2+}	5.77	10.83	14.11			
Ni^{2+}	7.52	13.84	18.33			
oxalate ($C_2O_4^{2-}$)						
Cu^{2+}	6.16	8.5				
Fe^{2+}	2.9	4.52	5.22			
Fe^{3+}	9.4	16.2	20.2			
Hg^{2+}		6.98				
Zn^{2+}	4.89	7.60	8.15			
Ni^{2+}	5.3	7.64	~8.5			

The values are from Lange's Handbook of Chemistry. 16th ed. 2005: 1.358-1.379

Appendix 4 Thermodynamic Data

Table 1 The Standard Enthalpy of Formation, the Standard Free Energy of Formation and the Standard Entropy at 298.15K

Substance	$\dfrac{\Delta_f H_m^\ominus}{kJ \cdot mol^{-1}}$	$\dfrac{\Delta_f H_m^\ominus}{kJ \cdot mol^{-1}}$	$\dfrac{S_m^\theta}{J \cdot K^{-1} mol^{-1}}$
Ag(s)	0	0	42.6
Ag^+(aq)	105.6	77.1	72.7
$AgNO_3$(s)	−124.4	−33.4	140.9
AgCl(s)	−127.0	−109.8	96.3
AgBr(s)	−100.4	−96.9	107.1
AgI(s)	−61.8	−66.2	115.5
Ba(s)	0	0	62.5
Ba^{2+}(aq)	−537.6	−560.8	9.6
$BaCl_2$(s)	−855.0	−806.7	123.7
$BaSO_4$(s)	−1 473.2	−1 362.2	132.2
Br_2(g)	30.9	3.1	245.5
Br_2(l)	0	0	152.2
C(dia)	1.9	2.9	2.4
C(gra)	0	0	5.7
CO(g)	−110.5	−137.2	197.7
CO_2(g)	−393.5	−394.4	213.8
Ca(s)	0	0	41.6
Ca^{2+}(aq)	−542.8	−553.6	−53.1
$CaCl_2$(s)	−795.4	−748.8	108.4
$CaCO_3$(calcite)	−1 207.6	−1 129.1	91.7
$CaCO_3$(aragonile)	−1 207.8	−1 128.2	88.0
CaO(s)	−634.9	−603.3	38.1
$Ca(OH)_2$(s)	−985.2	−897.5	83.4
Cl_2(g)	0	0	223.1
Cl^-(aq)	−167.2	−131.2	56.5
Cu(s)	0	0	33.2
Cu^{2+}(aq)	64.8	65.5	−99.6
F_2(g)	0	0	202.8
F^-(aq)	−332.6	−278.8	−13.8
Fe(s)	0	0	27.3
Fe^{2+}(aq)	−89.1	−78.9	−137.7
Fe^{3+}(aq)	−48.5	−4.7	−315.9

Continue

Substance	$\Delta_f H_m^\ominus$ / kJ·mol^{-1}	$\Delta_f H_m^\ominus$ / kJ·mol^{-1}	$S_m^\ominus$ / J·K^{-1}mol^{-1}
FeO(s)	−272.0	−251	61
Fe$_3$O$_4$(s)	−1 118.4	−1 015.4	146.4
Fe$_2$O$_3$(s)	−824.2	−742.2	87.4
H$_2$(g)	0	0	130.7
H$^+$(aq)	0	0	0
HCl(g)	−92.3	−95.3	186.9
HF(g)	−273.3	−275.4	173.8
HBr(g)	−36.3	−53.4	198.70
HI(g)	26.5	1.7	206.6
H$_2$O(g)	−241.8	−228.6	188.8
H$_2$O(l)	−285.8	−237.1	70.0
H$_2$S(g)	−20.6	−33.4	205.8
I$_2$(g)	62.4	19.3	260.7
I$_2$(s)	0	0	116.1
I$^-$(aq)	−55.2	−51.6	111.3
K(s)	0	0	64.7
K$^+$(aq)	−252.4	−283.3	102.5
KI(s)	−327.9	−324.9	106.3
KCl(s)	−436.5	−408.5	82.6
Mg(s)	0	0	32.7
Mg^{2+}(aq)	−466.9	−454.8	−138.1
MgO(s)	−601.6	−569.3	27.0
MnO$_2$(s)	−520.0	−465.1	53.1
Mn^{2+}(aq)	−220.8	−228.1	−73.6
N$_2$(g)	0	0	191.6
NH$_3$(g)	−45.9	−16.4	192.8
NH$_4$Cl(s)	−314.4	−202.9	94.6
NO(g)	91.3	87.6	210.8
NO$_2$(g)	33.2	51.3	240.1
Na(s)	0	0	51.3
Na$^+$(aq)	−240.1	−261.9	59.0
NaCl(s)	−411.2	−384.1	72.1
O$_2$(g)	0	0	205.2
OH$^-$(aq)	−230.0	−157.2	−10.8
SO$_2$(g)	−296.8	−300.1	248.2

Continue

Substance	$\Delta_f H_m^\ominus$ / kJ·mol^{-1}	$\Delta_f H_m^\ominus$ / kJ·mol^{-1}	$S_m^\ominus$ / J·K^{-1}mol^{-1}
SO$_3$(g)	−395.7	−371.1	256.8
Zn(s)	0	0	41.6
Zn^{2+}(aq)	−153.9	−147.1	−112.1
ZnO(s)	−350.5	−320.5	43.7
CH$_4$(g)	−74.6	−50.5	186.3
C$_2$H$_2$(g)	227.4	209.9	200.9
C$_2$H$_4$(g)	52.4	68.4	219.3
C$_2$H$_6$(g, ethane)	−84.0	−32.0	229.2
C$_6$H$_6$(g, benzene)	82.9	129.7	269.2
C$_6$H$_6$(l)	49.1	124.5	173.4
CH$_3$OH(g)	−201.0	−162.3	239.9
CH$_3$OH(l)	−239.2	−166.6	126.8
HCHO(g)	−108.6	−102.5	218.8
HCOOH(l)	−425.0	−361.4	129.0
C$_2$H$_5$OH(g)	−234.8	−167.9	281.6
C$_2$H$_5$OH(l)	−277.6	−174.8	160.7
CH$_3$CHO(l)	−192.2	−127.6	160.2
CH$_3$COOH(l)	−484.3	−389.9	159.8
H$_2$NCONH$_2$(s)	−333.1	−196.8	104.6
C$_6$H$_{12}$O$_6$(s) (glucose)	−1 273.3	−910.4	212.1
C$_{12}$H$_{22}$O$_{11}$(s) (sucrose)	−2 226.1	−1 544.7	360.2

The data are mainly from Haynes WM. CRC Handbook of Chemistry and Physics. 97th ed. New York: CRC Press, 2016

Table 2 Standard Heat of Combustion

化合物	$\Delta_c H_m^\ominus$ / kJ·mol^{-1}	化合物	$\Delta_c H_m^\ominus$ / kJ·mol^{-1}
CH$_4$(g)	−890.8	HCHO(g)	−570.7
C$_2$H$_2$(g)	−1 301.1	CH$_3$CHO(l)	−1 166.9
C$_2$H$_4$(g)	−1 411.2	CH$_3$COCH$_3$(l)	−1 789.9
C$_2$H$_6$(g)	−1 560.7	HCOOH(l)	-254.6
C$_3$H$_8$(g)	−2 219.2	CH$_3$COOH(l)	−874.2
C$_5$H$_{12}$(l)	−3 509.0	C$_{17}$H$_{35}$COOH stearic acid(s)	−11 281
C$_6$H$_6$(l)	−3 267.6	C$_6$H$_{12}$O$_6$ glucose(s)	−2 803.0
CH$_3$OH(l)	−726.1	C$_{12}$H$_{22}$O$_{11}$ sucrose(s)	−5 640.9
C$_2$H$_5$OH(l)	−1 366.8	CO(NH$_2$)$_2$ uric acid(s)	−632.7

The data are mainly from Haynes WM. CRC Handbook of Chemistry and Physics. 97th ed. New York: CRC Press, 2016

Appendix 5 Standard Potentials in Aqueous Solution at 298.15K

Reaction	$\varphi^{\ominus}/V$	Reaction	$\varphi^{\ominus}/V$
$Sr^+ + e^- \rightleftharpoons Sr$	−4.10	$PbCl_2 + 2e^- \rightleftharpoons Pb + 2Cl^-$	−0.267 5
$Li^+ + e^- \rightleftharpoons Li$	−3.040 1	$Ni^{2+} + 2e^- \rightleftharpoons Ni$	−0.257
$Ca(OH)_2 + 2e^- \rightleftharpoons Ca + 2OH^-$	−3.02	$V^{3+} + e^- \rightleftharpoons V^{2+}$	−0.255
$K^+ + e^- \rightleftharpoons K$	−2.931	$CdSO_4 + 2e^- \rightleftharpoons Cd + SO_4^{2-}$	−0.246
$Ba^{2+} + 2e^- \rightleftharpoons Ba$	−2.912	$Cu(OH)_2 + 2e^- \rightleftharpoons Cu + 2OH^-$	−0.222
$Ca^{2+} + 2e^- \rightleftharpoons Ca$	−2.868	$CO_2 + 2H^+ + 2e^- \rightleftharpoons HCOOH$	−0.199
$Na^+ + e^- \rightleftharpoons Na$	−2.71	$AgI + e^- \rightleftharpoons Ag + I^-$	−0.152 24
$Mg^{2+} + 2e^- \rightleftharpoons Mg$	−2.372	$O_2 + 2H_2O + 2e^- \rightleftharpoons H_2O_2 + 2OH^-$	−0.146
$Mg(OH)_2 + 2e^- \rightleftharpoons Mg + 2OH^-$	−2.690	$Sn^{2+} + 2e^- \rightleftharpoons Sn$	−0.137 5
$Al(OH)_3 + 3e^- \rightleftharpoons Al + 3OH^-$	−2.31	$CrO_4^{2-} + 4H_2O + 3e^- \rightleftharpoons Cr(OH)_3 + 5OH^-$	−0.13
$Be^{2+} + 2e^- \rightleftharpoons Be$	−1.847	$Pb^{2+} + 2e^- \rightleftharpoons Pb$	−0.126 2
$Al^{3+} + 3e^- \rightleftharpoons Al$	−1.662	$O_2 + H_2O + 2e^- \rightleftharpoons HO_2^- + OH^-$	−0.076
$Mn(OH)_2 + 2e^- \rightleftharpoons Mn + 2OH^-$	−1.56	$Fe^{3+} + 3e^- \rightleftharpoons Fe$	−0.037
$ZnO + H_2O + 2e^- \rightleftharpoons Zn + 2OH^-$	−1.260	$Ag_2S + 2H^+ + 2e^- \rightleftharpoons 2Ag + H_2S$	−0.036 6
$H_2BO_3^- + 5H_2O + 8e^- \rightleftharpoons BH_4^- + 8OH^-$	−1.24	$2H^+ + 2e^- \rightleftharpoons H_2$	0
$Mn^{2+} + 2e^- \rightleftharpoons Mn$	−1.185	$Pd(OH)_2 + 2e^- \rightleftharpoons Pd + 2OH^-$	0.07
$2SO_3^{2-} + 2H_2O + 2e^- \rightleftharpoons S_2O_4^{2-} + 4OH^-$	−1.12	$AgBr + e^- \rightleftharpoons Ag + Br^-$	0.071 33
$PO_4^{3-} + 2H_2O + 2e^- \rightleftharpoons HPO_3^{2-} + 3OH^-$	−1.05	$S_4O_6^{2-} + 2e^- \rightleftharpoons 2S_2O_3^{2-}$	0.08
$SO_4^{2-} + H_2O + 2e^- \rightleftharpoons SO_3^{2-} + 2OH^-$	−0.93	$[Co(NH_3)_6]^{3+} + e^- \rightleftharpoons [Co(NH_3)_6]^{2+}$	0.108
$2H_2O + 2e^- \rightleftharpoons H_2 + 2OH^-$	−0.827 7	$S + 2H^+ + 2e^- \rightleftharpoons H_2S(aq)$	0.142
$Zn^{2+} + 2e^- \rightleftharpoons Zn$	−0.761 8	$Sn^{4+} + 2e^- \rightleftharpoons Sn^{2+}$	0.151
$Cr^{3+} + 3e^- \rightleftharpoons Cr$	−0.744	$Cu^{2+} + e^- \rightleftharpoons Cu^+$	0.153
$AsO_4^{3-} + 2H_2O + 2e^- \rightleftharpoons AsO_2^- + 4OH^-$	−0.71	$Fe_2O_3 + 4H^+ + 2e^- \rightleftharpoons 2FeOH^+ + H_2O$	0.16
$AsO_2^- + 2H_2O + 3e^- \rightleftharpoons As + 4OH^-$	−0.68	$SO_4^{2-} + 4H^+ + 2e^- \rightleftharpoons H_2SO_3 + H_2O$	0.172
$SbO_2^- + 2H_2O + 3e^- \rightleftharpoons Sb + 4OH^-$	−0.66	$AgCl + e^- \rightleftharpoons Ag + Cl^-$	0.222 33
$SbO_3^- + H_2O + 2e^- \rightleftharpoons SbO_2^- + 2OH^-$	−0.59	$As_2O_3 + 6H^+ + 6e^- \rightleftharpoons 2As + 3H_2O$	0.234
$Fe(OH)_3 + e^- \rightleftharpoons Fe(OH)_2 + OH^-$	−0.56	$HAsO_2 + 3H^+ + 3e^- \rightleftharpoons As + 2H_2O$	0.248
$2CO_2 + 2H^+ + 2e^- \rightleftharpoons H_2C_2O_4$	−0.49	$Hg_2Cl_2 + 2e^- \rightleftharpoons 2Hg + 2Cl^-$	0.268 08
$B(OH)_3 + 7H^+ + 8e^- \rightleftharpoons BH_4^- + 3H_2O$	−0.481	$Cu^{2+} + 2e^- \rightleftharpoons Cu$	0.341 9
$S + 2e^- \rightleftharpoons S^{2-}$	−0.476 27	$Ag_2O + H_2O + 2e^- \rightleftharpoons 2Ag + 2OH^-$	0.342
$Fe^{2+} + 2e^- \rightleftharpoons Fe$	−0.447	$[Fe(CN)_6]^{3-} + e^- \rightleftharpoons [Fe(CN)_6]^{4-}$	0.358
$Cr^{3+} + e^- \rightleftharpoons Cr^{2+}$	−0.407	$[Ag(NH_3)_2]^+ + e^- \rightleftharpoons Ag + 2NH_3$	0.373
$Cd^{2+} + 2e^- \rightleftharpoons Cd$	−0.403 0	$O_2 + 2H_2O + 4e^- \rightleftharpoons 4OH^-$	0.401
$PbSO_4 + 2e^- \rightleftharpoons Pb + SO_4^{2-}$	−0.358 8	$H_2SO_3 + 4H^+ + 4e^- \rightleftharpoons S + 3H_2O$	0.449
$Tl^+ + e^- \rightleftharpoons Tl$	−0.336	$IO^- + H_2O + 2e^- \rightleftharpoons I^- + 2OH^-$	0.485
$[Ag(CN)_2]^- + e^- \rightleftharpoons Ag + 2CN^-$	−0.31	$Cu^+ + e^- \rightleftharpoons Cu$	0.521
$Co^{2+} + 2e^- \rightleftharpoons Co$	−0.28	$I_2 + 2e^- \rightleftharpoons 2I^-$	0.535 5
$H_3PO_4 + 2H^+ + 2e^- \rightleftharpoons H_3PO_3 + H_2O$	−0.276	$I_3^- + 2e^- \rightleftharpoons 3I^-$	0.536

Continue

Reaction	$\varphi^{\ominus}/V$	Reaction	$\varphi^{\ominus}/V$
$AgBrO_3 + e^- \rightleftharpoons Ag + BrO_3^-$	0.546	$ClO_3^- + 3H^+ + 2e^- \rightleftharpoons HClO_2 + H_2O$	1.214
$MnO_4^- + e^- \rightleftharpoons MnO_4^{2-}$	0.558	$MnO_2 + 4H^+ + 2e^- \rightleftharpoons Mn^{2+} + 2H_2O$	1.224
$AsO_4^{3-} + 2H^+ + 2e^- \rightleftharpoons AsO_3^{2-} + H_2O$	0.559	$O_2 + 4H^+ + 4e^- \rightleftharpoons 2H_2O$	1.229
$H_3AsO_4 + 2H^+ + 2e^- \rightleftharpoons HAsO_2 + 2H_2O$	0.560	$Tl^{3+} + 2e^- \rightleftharpoons Tl^+$	1.252
$MnO_4^- + 2H_2O + 3e^- \rightleftharpoons MnO_2 + 4OH^-$	0.595	$2HNO_2 + 4H^+ + 4e^- \rightleftharpoons N_2O + 3H_2O$	1.297
$Hg_2SO_4 + 2e^- \rightleftharpoons 2Hg + SO_4^{2-}$	0.6125	$HBrO + H^+ + 2e^- \rightleftharpoons Br^- + H_2O$	1.331
$O_2 + 2H^+ + 2e^- \rightleftharpoons H_2O_2$	0.695	$HCrO_4^- + 7H^+ + 3e^- \rightleftharpoons Cr^{3+} + 4H_2O$	1.350
$[PtCl_4]^{2-} + 2e^- \rightleftharpoons Pt + 4Cl^-$	0.755	$Cl_2(g) + 2e^- \rightleftharpoons 2Cl^-$	1.35827
$BrO^- + H_2O + 2e^- \rightleftharpoons Br^- + 2OH^-$	0.761	$Cr_2O_7^{2-} + 14H^+ + 6e^- \rightleftharpoons 2Cr^{3+} + 7H_2O$	1.36
$Fe^{3+} + e^- \rightleftharpoons Fe^{2+}$	0.771	$ClO_4^- + 8H^+ + 8e^- \rightleftharpoons Cl^- + 4H_2O$	1.389
$Hg_2^{2+} + 2e^- \rightleftharpoons 2Hg$	0.7973	$HClO + H^+ + 2e^- \rightleftharpoons Cl^- + H_2O$	1.482
$Ag^+ + e^- \rightleftharpoons Ag$	0.7996	$MnO_4^- + 8H^+ + 5e^- \rightleftharpoons Mn^{2+} + 4H_2O$	1.507
$ClO^- + H_2O + 2e^- \rightleftharpoons Cl^- + 2OH^-$	0.81	$MnO_4^- + 4H^+ + 3e^- \rightleftharpoons MnO_2 + 2H_2O$	1.679
$Hg^{2+} + 2e^- \rightleftharpoons Hg$	0.851	$Au^+ + e^- \rightleftharpoons Au$	1.692
$2Hg^{2+} + 2e^- \rightleftharpoons Hg_2^{2+}$	0.920	$Ce^{4+} + e^- \rightleftharpoons Ce^{3+}$	1.72
$NO_3^- + 3H^+ + 2e^- \rightleftharpoons HNO_2 + H_2O$	0.934	$H_2O_2 + 2H^+ + 2e^- \rightleftharpoons 2H_2O$	1.776
$Pd^{2+} + 2e^- \rightleftharpoons Pd$	0.951	$Co^{3+} + e^- \rightleftharpoons Co^{2+}$	1.92
$Br_2(l) + 2e^- \rightleftharpoons 2Br^-$	1.066	$S_2O_8^{2-} + 2e^- \rightleftharpoons 2SO_4^{2-}$	2.010
$Br_2(aq) + 2e^- \rightleftharpoons 2Br^-$	1.0873	$F_2 + 2e^- \rightleftharpoons 2F^-$	2.866
$2IO_3^- + 12H^+ + 10e^- \rightleftharpoons I_2 + 6H_2O$	1.195		

The data are mainly from Haynes WM. CRC Handbook of Chemistry and Physics. 97th ed. New York: CRC Press, 2016

Appendix 6 Greek Letters

Capital	Lowercase	Name	Pronunciation	Capital	Lowercase	Name	Pronunciation
A	α	alpha	[ˈælfə]	N	ν	nu	[njuː]
B	β	beta	[ˈbiːtə; ˈbeitə]	Ξ	ξ	xi	[ksai; zai; gzai]
Γ	γ	gamma	[ˈgæmə]	O	o	omicron	[ouˈmaikrən]
Δ	δ	delta	[ˈdeltə]	Π	π	pi	[pai]
E	ε	epsilon	[epˈsailnən; ˈepsilnən]	P	ρ	rho	[rou]
Z	ζ	zeta	[ˈziːtə]	Σ	σ,s	sigma	[ˈsigmə]
H	η	eta	[ˈiːtə; ˈeitə]	T	τ	tau	[tɔː]
Θ	θ	theta	[ˈθiːtə]	Y	υ	upsilon	[juːpˈsailən; ˈuːpsilən]
I	ι	iota	[aiˈoutə]	Φ	φ,ϕ	phi	[fai]
K	κ	kappa	[ˈkæpə]	X	χ	chi	[kai]
Λ	λ	lambda	[ˈlæmdə]	Ψ	ψ	psi	[psai]
M	μ	mu	[mjuː]	Ω	ω	omega	[ˈoumigə]

Detailed Solutions to Exercises

Chapter 1

[Exercises]

1. Because the molar mass of biphenyl is $154\text{g}\cdot\text{mol}^{-1}$, 5.50g of biphenyl contains

$$n(\text{biphenyl}) = \frac{5.50\text{g}}{154\text{g}\cdot\text{mol}^{-1}} = 0.035\,7\text{mol}$$

Molality of the biphenyl in the solution is

$$b(\text{biphenyl}) = \frac{0.035\,7\text{mol}}{0.100\,0\text{kg}} = 0.357\,\text{mol}\cdot\text{kg}^{-1}$$

$$K_b = \frac{\Delta T_b}{b(\text{biphenyl})} = \frac{0.903\text{°C}}{0.357\,\text{mol}\cdot\text{kg}^{-1}} = 2.53\text{°C}\cdot\text{kg}\cdot\text{mol}^{-1}$$

2.
$$\Delta T_b = 80.696\text{°C} - 80.099\text{°C}$$
$$= 0.597\text{°C}$$

solving $\Delta T_b = K_b b_B$ for b_B gives

$$b_B = \frac{\Delta T_b}{K_b} = \frac{0.597\text{°C}}{2.53\text{°C}\cdot\text{kg}\cdot\text{mol}^{-1}} = 0.236\,\text{mol}\cdot\text{kg}^{-1}$$

The product of molality of the solution and the mass of solvent is the chemical amount of solute

$$n_B = 0.223\,6\,\text{mol}\cdot\text{kg}^{-1} \times 0.150\text{kg} = 0.035\,4\text{mol},\ MM. = 6.00\text{g} \div 0.035\,4\text{mol} = 178\text{g}\cdot\text{mol}^{-1}$$

Finally, the molar mass of the solute is its mass divided by its chemical amount in the solution

3.
$$M_B = \frac{6.30\text{g}}{0.035\,4\text{mol}} = 178\text{g}\cdot\text{mol}^{-1} \quad \text{by} \quad \Delta T_f = K_f b_B$$

$$0.297\text{°C} = 1.86\text{°C}\cdot\text{kg}\cdot\text{mol}^{-1} \times \frac{\dfrac{m}{62.06\text{g}\cdot\text{mol}^{-1}}}{0.200\,0\text{kg}} \qquad m = 1.98\text{g}$$

4. By $\Delta T_f = K_f b_B$

$$\frac{\dfrac{6.00\text{g}}{M_B}}{0.135\text{kg}} = \frac{0.242\text{°C}}{1.86\text{°C}\cdot\text{kg}\cdot\text{mol}^{-1}} \qquad M_B = 342\text{g}\cdot\text{mol}^{-1}$$

5. $\Delta T_f = K_f b_B$

$$\Delta T_f = 1.86\text{°C}\cdot\text{kg}\cdot\text{mol}^{-1} \times (0.458 + 0.533 + 0.052 + 0.028 + 0.010 + 0.002 + 0.010 + 0.001 + 0.001)\,\text{mol}\cdot\text{kg}^{-1} = 2.04\text{°C}$$

$$T_f = -2.04\text{°C}$$

6.
$$b_B = \frac{36.0\text{g}}{180\text{g}\cdot\text{mol}^{-1} \times 0.400\text{kg}} = 0.500\,\text{mol}\cdot\text{kg}^{-1}$$

7. $\Delta T_f = K_f b_B$

$$\Delta T_f = 1.86\text{°C}\cdot\text{kg}\cdot\text{mol}^{-1} \times \frac{3.60\text{g}}{180.0\text{g}\cdot\text{mol}^{-1} \times 0.050\,0\text{kg}} = 0.744\text{°C}$$

8.
$$T_f = -0.744°C$$
$$\Delta T_f = K_f b_B$$
$$0.930°C = 1.86°C \cdot kg \cdot mol^{-1} \times \frac{\frac{3.25g}{M_B}}{0.125kg}$$
$$M_B = 52.0 g \cdot mol^{-1}$$

9. By
$$\Pi = i b_B RT$$
$$77.4 kPa = 3 \times \frac{\frac{x}{110.9 g \cdot mol^{-1}}}{1L} \times 8.314 kPa \cdot L \cdot mol^{-1} \cdot K^{-1} \times 310K$$
$$x = 1.11g$$

10.
$$\Delta T_f = K_f b_B$$
$$2.5°C = 20.0°C \cdot kg \cdot mol^{-1} \times \frac{\frac{0.2436g}{M_B}}{20.0mL \times 0.779 g \cdot mL^{-1}} \times \frac{1000g}{1kg}$$
$$M_B = 125 g \cdot mol^{-1}$$

11. The concentration in moles per liter is
$$c_B = \frac{\Pi}{RT} = \frac{2.127 kPa}{8.314 kPa \cdot L \cdot mol^{-1} \cdot K^{-1} \times 298K} = 8.585 \times 10^{-4} \, mol \cdot L^{-1}$$

Now, 2.00g of a protein in 0.100L of water gives the same concentration as 20.0g in 1.00L, therefore, 8.585×10^{-4} mol of the protein must weigh 20.0g, and the molar mass is

$$MM_B = \frac{20.0g}{8.585 \times 10^{-4} mol} = 2.33 \times 10^4 g \cdot mol^{-1}$$

[Supplementary Exercises]

1. (a) $c(NaCl) = \dfrac{21.0g/58.5g \cdot mol^{-1}}{0.135L} = 2.659 mol \cdot L^{-1}$

$\Delta T_f = 2 \times 1.86°C \cdot kg \cdot mol^{-1} \times 2.659 mol \cdot L^{-1} = 9.89°C$ $T_f = -9.89°C$

$\Delta T_b = 2 \times 0.512°C \cdot kg \cdot mol^{-1} \times 2.659 mol \cdot L^{-1} = 2.72°C$ $T_b = 102.72°C$

(b) $c(CON_2H_4) = \dfrac{15.4g/60.0g \cdot mol^{-1}}{0.0667L} = 3.848 mol \cdot L^{-1}$

$\Delta T_f = 1.86°C \cdot kg \cdot mol^{-1} \times 3.848 mol \cdot L^{-1} = 7.16°C$ $T_f = -7.16°C$

$\Delta T_b = 0.512°C \cdot kg \cdot mol^{-1} \times 3.848 mol \cdot L^{-1} = 1.97°C$ $T_b = 101.97°C$

2. $\Delta T_f = T_f^\circ - T_f = K_f b_B = 5.5°C - 2.36°C = 3.14°C$

$b_B = \dfrac{3.14°C}{5.10°C \cdot kg \cdot mol^{-1}} = 0.616 mol \cdot kg^{-1}$ $M_r = \dfrac{4.00g / 0.0550 kg}{0.616 mol \cdot kg^{-1}} = 118 g \cdot mol^{-1}$

3. $b_B = \dfrac{1.05°C}{5.10°C \cdot kg \cdot mol^{-1}} = 0.206 mol \cdot kg^{-1}$

$M_r = \dfrac{7.85g}{0.301 kg \times 0.206 mol \cdot kg^{-1}} = 127 g \cdot mol^{-1}$

Since the formula mass of C_5H_4 is $64 g \cdot mol^{-1}$ and the molar mass is found to be $127 g \cdot mol^{-1}$, the molecular formula of the compound is $C_{10}H_8$.

4. $MM(EG) = 62.01 \text{g} \cdot \text{mol}^{-1}$ $b(EG) = \dfrac{651\text{g} / 62.01\text{g} \cdot \text{mol}^{-1}}{2.505\text{kg}} = 4.19 \text{mol} \cdot \text{kg}^{-1}$

$$\Delta T_f = 1.86\text{°C} \cdot \text{kg} \cdot \text{mol}^{-1} \times 4.19 \text{mol} \cdot \text{kg}^{-1} = 7.79\text{°C} \quad T_f = -7.79\text{°C}$$
$$\Delta T_b = 0.512\text{°C} \cdot \text{kg} \cdot \text{mol}^{-1} \times 4.19 \text{mol} \cdot \text{kg}^{-1} = 2.15\text{°C} \quad T_b = 102.15\text{°C}$$

Because the solution will boil at 102.15°C, it would be preferable to leave the antifreeze in your car radiator in summer to prevent the solution from boiling.

5. The concentration of the solution:

$$c(\text{Hb}) = \dfrac{\Pi}{RT} = \dfrac{10.0\text{mmHg} \times \dfrac{101.3\text{kPa}}{760\text{mmHg}}}{8.314\text{kPa} \cdot \text{L} \cdot \text{mol}^{-1} \cdot \text{K}^{-1} \times (273.15 + 25)\text{K}} = 5.38 \times 10^{-4} \text{mol} \cdot \text{L}^{-1}$$

$$M(\text{Hb}) = \dfrac{35.0\text{g}}{5.38 \times 10^{-4} \text{mol} \cdot \text{L}^{-1} \times 1\text{L}} = 6.51 \times 10^4 \text{g} \cdot \text{mol}^{-1}$$

6. $c(\text{NaCl}) = \dfrac{\dfrac{0.86\text{g}}{100\text{g}} \times 1.005 \text{g} \cdot \text{mL}^{-1}}{58.5 \text{g} \cdot \text{mol}^{-1}} \times \dfrac{1\,000\text{mL}}{1\text{L}} = 0.148 \text{mol} \cdot \text{L}^{-1}$

$$\Pi = i c_B RT$$
$$= 2 \times 0.148 \text{mol} \cdot \text{L}^{-1} \times 8.314 \text{kPa} \cdot \text{L} \cdot \text{mol}^{-1} \cdot \text{K}^{-1} \times (273.15 + 37)\text{K}$$
$$= 763 \text{kPa}$$

Chapter 2   The Basis of Chemical Thermodynamics

[Exercises]

2. (1) $\Delta U = Q + W = -2.5\text{kJ} + (-500 \times 10^{-3}\text{kJ}) = -3.0\text{kJ}$
   (2) $\Delta U = -650\text{J} + 350\text{J} = -300\text{J}$

4. (1) $C + D = A + B$ $\Delta_r H_m^{\ominus} = -\Delta_r H_{m,1}^{\ominus} = 40.0 \text{kJ} \cdot \text{mol}^{-1}$
   (2) $2C + 2D = 2A + 2B$ $\Delta_r H_m^{\ominus} = 2 \times (-\Delta_r H_{m,1}^{\ominus}) = 80.0 \text{kJ} \cdot \text{mol}^{-1}$
   (3) $A + B = E$ $\Delta_r H_m^{\ominus} = \Delta_r H_{m,1}^{\ominus} + \Delta_r H_{m,2}^{\ominus} = -40.0 \text{kJ} \cdot \text{mol}^{-1} + 60.0 \text{kJ} \cdot \text{mol}^{-1}$
                              $= 20.0 \text{kJ} \cdot \text{mol}^{-1}$

5. $t = 0 \quad \xi = 0 \quad n(\text{H}_2) = 4.0\text{mol} \quad n(\text{O}_2) = 2.0\text{mol} \quad n(\text{H}_2\text{O}) = 0$
   $t = t \, \xi = t \, n(\text{H}_2) = 3.4\text{mol} \quad n(\text{O}_2) = 1.7\text{mol} \quad n(\text{H}_2\text{O}) = 0.6\text{mol}$

(1) $\qquad\qquad\qquad 2\text{H}_2(\text{g}) + \text{O}_2(\text{g}) = 2\text{H}_2\text{O}(\text{g})$

$$\xi = \dfrac{\Delta n(\text{H}_2\text{O})}{\nu(\text{H}_2\text{O})} = \dfrac{0.60\text{mol} - 0}{2} = 0.30\text{mol}$$

$$\xi = \dfrac{\Delta n(\text{H}_2)}{\nu(\text{H}_2)} = \dfrac{3.4\text{mol} - 4.0\text{mol}}{-2} = 0.30\text{mol}$$

$$\xi = \dfrac{\Delta n(\text{O}_2)}{\nu(\text{O}_2)} = \dfrac{1.7\text{mol} - 2.0\text{mol}}{-1} = 0.30\text{mol}$$

(2) $\qquad\qquad\qquad \text{H}_2(\text{g}) + \dfrac{1}{2}\text{O}_2(\text{g}) = \text{H}_2\text{O}(\text{g})$

$$\xi = \dfrac{\Delta n(\text{H}_2\text{O})}{\nu(\text{H}_2\text{O})} = \dfrac{0.60\text{mol} - 0}{1} = 0.60\text{mol}$$

$$\xi = \dfrac{\Delta n(\text{H}_2)}{\nu(\text{H}_2)} = \dfrac{3.4\text{mol} - 4.0\text{mol}}{-1} = 0.60\text{mol}$$

$$\xi = \frac{\Delta n(O_2)}{\nu(O_2)} = \frac{1.7\text{mol} - 2.0\text{mol}}{-\frac{1}{2}} = 0.60\text{mol}$$

7. $6C(g) + 3H_2(g) = C_6H_6(l)$ from $6\times(2) + 3\times(3) - (1)$

$$\Delta_r H_m^\ominus = 6\Delta_r H_{m,2}^\ominus + 3\Delta_r H_{m,3}^\ominus - \Delta_r H_{m,1}^\ominus$$
$$= 6\times(-393.5)\text{kJ}\cdot\text{mol}^{-1} + 3\times(-285.8)\text{kJ}\cdot\text{mol}^{-1} - (-3\,267.6)\text{kJ}\cdot\text{mol}^{-1}$$
$$= 49.2\text{kJ}\cdot\text{mol}^{-1}$$

8. $2N_2H_4(l) + N_2O_4(g) = 3N_2(g) + 4H_2O(l)$

$$\Delta_r H_m^\ominus = \sum_B \nu_B \Delta_f H_m^\ominus(B)$$
$$= 0 + 4\times(-285.8\text{kJ}\cdot\text{mol}^{-1}) - (2\times 50.63\text{kJ}\cdot\text{mol}^{-1} + 9.16\text{kJ}\cdot\text{mol}^{-1})$$
$$\approx -1\,254\text{kJ}\cdot\text{mol}^{-1}$$

9. the equation $(4) = -\frac{1}{3}\times(3) + \frac{1}{2}\times(1) - \frac{1}{6}\times(2)$

So, $\Delta_r H_{m4}^\ominus = -\frac{1}{3}\times\Delta_r H_{m3}^\ominus + \frac{1}{2}\times\Delta_r H_{m1}^\ominus - \frac{1}{6}\times\Delta_r H_{m2}^\ominus$

$$= -\frac{1}{3}\times 19.4\text{kJ}\cdot\text{mol}^{-1} + \frac{1}{2}\times(-24.8)\text{kJ}\cdot\text{mol}^{-1} - \frac{1}{6}\times(-47.2)\text{kJ}\cdot\text{mol}^{-1}$$
$$= -11.0\text{kJ}\cdot\text{mol}^{-1}$$

$$\Delta_r G_{m4}^\ominus = -\frac{1}{3}\times\Delta_r G_{m3}^\ominus + \frac{1}{2}\times\Delta_r G_{m1}^\ominus - \frac{1}{6}\times\Delta_r G_{m2}^\ominus$$
$$= -\frac{1}{3}\times 5.2\text{kJ}\cdot\text{mol}^{-1} + \frac{1}{2}\times(-29.4)\text{kJ}\cdot\text{mol}^{-1} - \frac{1}{6}\times(-61.4)\text{kJ}\cdot\text{mol}^{-1}$$
$$= -6.20\text{kJ}\cdot\text{mol}^{-1}$$

$$\Delta_r S_{m4}^\ominus = \frac{\Delta_r H_{m4}^\ominus - \Delta_r G_{m4}^\ominus}{298.15\text{K}} = \frac{-11.0\times 1\,000\text{J}\cdot\text{mol}^{-1} - 6.20\times 1\,000\text{J}\cdot\text{mol}^{-1}}{298.15\text{K}}$$
$$= -16.1\text{J}\cdot\text{K}^{-1}\cdot\text{mol}^{-1}$$

10. $\Delta_r H_m^\ominus = -74.6\text{kJ}\cdot\text{mol}^{-1} + \frac{1}{2}\times 0 - (-239.2\text{kJ}\cdot\text{mol}^{-1})$

$$= 164.6\text{kJ}\cdot\text{mol}^{-1}$$

$\Delta_r S_m^\ominus = 186.3\text{J}\cdot\text{K}^{-1}\cdot\text{mol}^{-1} + \frac{1}{2}\times 205.2\text{J}\cdot\text{K}^{-1}\cdot\text{mol}^{-1} - 126.8\text{J}\cdot\text{K}^{-1}\cdot\text{mol}^{-1}$

$$= 162.1\text{J}\cdot\text{K}^{-1}\cdot\text{mol}^{-1}$$

$\Delta_r G_m^\ominus = \Delta_r H_m^\theta - T\Delta_r S_m^\ominus = 164.6\text{kJ}\cdot\text{mol}^{-1} - 298.15\text{K}\times 162.1\text{J}\cdot\text{K}^{-1}\cdot\text{mol}^{-1}$

$$= 116.3\text{kJ}\cdot\text{mol}^{-1} > 0$$

or $\Delta_r G_m^\ominus = \sum_B \nu_B \Delta_f G_m^\theta(B) = -50.5\text{kJ}\cdot\text{mol}^{-1} + 0 - (-166.6\text{kJ}\cdot\text{mol}^{-1})$

$$= 116.3\text{kJ}\cdot\text{mol}^{-1} > 0$$

The reaction do not occur spontaneously at 25℃.

$$T \geq \frac{\Delta_r H_{m,298.15K}^\ominus}{\Delta_r S_{m,298.15K}^\ominus} = \frac{164.6\times 10^3\text{J}\cdot\text{mol}^{-1}}{162.1\text{J}\cdot\text{K}^{-1}\cdot\text{mol}^{-1}} = 1\,015.42\text{K} \quad \text{or } 742.37℃$$

11. $\Delta_r H_m^\ominus = [0 + (-393.5\text{kJ}\cdot\text{mol}^{-1})] + [-241.8\text{kJ}\cdot\text{mol}^{-1} - (-110.5\text{kJ}\cdot\text{mol}^{-1})]$

$$= -41.2\text{kJ}\cdot\text{mol}^{-1}$$

$$\Delta_r G_m^\ominus = [0 + (-394.4 \text{kJ} \cdot \text{mol}^{-1})] + [-228.6 \text{kJ} \cdot \text{mol}^{-1} - (-137.2 \text{kJ} \cdot \text{mol}^{-1})]$$
$$= -28.6 \text{kJ} \cdot \text{mol}^{-1}$$
$$\Delta_r S_m^\ominus = \frac{\Delta_r H_m^\ominus - \Delta_r G_m^\ominus}{298.15 \text{K}} = \frac{-41.2 \times 10^3 \text{J} \cdot \text{mol}^{-1} - (-28.6 \times 10^3 \text{J} \cdot \text{mol}^{-1})}{298.15 \text{K}}$$
$$= -42.26 \text{J} \cdot \text{K}^{-1} \cdot \text{mol}^{-1}$$
$$S_m^\ominus(\text{H}_2\text{O},\text{g}) = S_m^\ominus(\text{H}_2,\text{g}) + S_m^\ominus(\text{CO}_2,\text{g}) - S_m^\ominus(\text{CO},\text{g}) - \Delta_r S_m^\ominus$$
$$= 130.7 \text{J} \cdot \text{K}^{-1} \cdot \text{mol}^{-1} + 213.8 \text{J} \cdot \text{K}^{-1} \cdot \text{mol}^{-1} - 197.77 \text{J} \cdot \text{K}^{-1} \cdot \text{mol}^{-1} - (-42.26 \text{J} \cdot \text{K}^{-1} \cdot \text{mol}^{-1})$$
$$= 189.1 \text{J} \cdot \text{K}^{-1} \cdot \text{mol}^{-1}$$

12. $3.0 \text{kJ} \cdot \text{g}^{-1} \times 250 \text{g} + 12 \text{kJ} \cdot \text{g}^{-1} \times 50 \text{g} + 50 \text{g} \cdot \text{L}^{-1} \times V \times 15.6 \text{kJ} \cdot \text{g}^{-1} = 6\,300 \text{kJ}$

$$V = 6.31 \text{L}$$

13. (1) $\Delta_r H_m^\ominus = -393.5 \text{kJ} \cdot \text{mol}^{-1} \times 12 + (-285.8 \text{kJ} \cdot \text{mol}^{-1}) \times 11 - 0 \times 12 - (-2\,226.1 \text{kJ} \cdot \text{mol}^{-1})$
$$\approx -5\,640 \text{kJ} \cdot \text{mol}^{-1}$$
$$\Delta_r S_m^\ominus = 213.8 \text{J} \cdot \text{K}^{-1} \cdot \text{mol}^{-1} \times 12 + 70.0 \text{J} \cdot \text{K}^{-1} \cdot \text{mol}^{-1} \times 11 - 205.2 \text{J} \cdot \text{K}^{-1} \cdot \text{mol}^{-1} \times 12 - 360.2 \text{J} \cdot \text{K}^{-1} \cdot \text{mol}^{-1}$$
$$= 513 \text{J} \cdot \text{K}^{-1} \cdot \text{mol}^{-1}$$
$$\Delta_r G_m^\ominus = \Delta_r H_m^\theta - T \Delta_r S_m^\ominus$$
$$= -5\,640 \text{kJ} \cdot \text{mol}^{-1} - 298.15 \text{K} \times 513 \times 10^{-3} \text{kJ} \cdot \text{K}^{-1} \cdot \text{mol}^{-1} = -5\,793 \text{kJ} \cdot \text{mol}^{-1}$$
$$\Delta_r G_m^\ominus = \sum_B \nu_B \Delta_f G_m^\ominus(B)$$
$$= -394.4 \text{kJ} \cdot \text{mol}^{-1} \times 12 + (-237.1 \text{kJ} \cdot \text{mol}^{-1}) \times 11 - 0 \times 12 - (-1\,544.6 \text{kJ} \cdot \text{mol}^{-1})$$
$$= -5\,796 \text{kJ} \cdot \text{mol}^{-1}$$

(2) $\Delta_r G_{m,310.5}^\ominus = \Delta_r H_m^\ominus - T \Delta_r S_m^\ominus$
$$= -5\,640 \text{kJ} \cdot \text{mol}^{-1} - 310.5 \text{K} \times 513 \times 10^{-3} \text{kJ} \cdot \text{K}^{-1} \cdot \text{mol}^{-1}$$
$$= -5\,799 \text{kJ} \cdot \text{mol}^{-1}$$
$$W_f = \Delta_r G_{m,310.5}^\ominus \times 30\% = -5\,799.11 \text{kJ} \cdot \text{mol}^{-1} \times 30\% = -1\,740 \text{kJ} \cdot \text{mol}^{-1}$$
$$\Delta_r G_m = \Delta_r G_m^\ominus + RT \ln \frac{\{c(\text{ADP})/c^\ominus\} \times \{c(\text{H}_3\text{PO}_4)/c^\ominus\}}{c(\text{ATP})/c^\ominus} = \Delta_r G_m^\ominus + RT \ln \frac{c(\text{ADP}) \times c(\text{H}_3\text{PO}_4)}{c(\text{ATP}) \times c^\ominus}$$
$$= -31.05 \text{kJ} \cdot \text{mol}^{-1} + 8.314 \text{J} \cdot \text{K}^{-1} \cdot \text{mol}^{-1} \times 310.15 \text{K} \times \frac{1 \text{kJ}}{1\,000 \text{J}}$$
$$\times \ln \frac{3.0 \text{mmol} \cdot \text{L}^{-1} \times 1.0 \text{mmol} \cdot \text{L}^{-1}}{10 \text{mmol} \cdot \text{L}^{-1} \times 1 \text{mol} \cdot \text{L}^{-1} \times \frac{1\,000 \text{mmol} \cdot \text{L}^{-1}}{1 \text{mol} \cdot \text{L}^{-1}}}$$
$$= -51.97 \text{kJ} \cdot \text{mol}^{-1}$$

[Supplementary Exercises]

3. $\text{COCl}_2(\text{g}) + \text{H}_2\text{O}(\text{l}) \rightarrow 2\text{HCl}(\text{g}) + \text{CO}_2(\text{g})$
$$\Delta_r G_m^\ominus = \sum_B \nu_B \Delta_f G_m^\ominus(B)$$
$$= [2 \times (-95.3 \text{kJ} \cdot \text{mol}^{-1}) + (-394.4 \text{kJ} \cdot \text{mol}^{-1})] - [-210 \text{kJ} \cdot \text{mol}^{-1} + (-237.1 \text{kJ} \cdot \text{mol}^{-1})]$$
$$= -137.9 \text{kJ} \cdot \text{mol}^{-1}$$

4. $\Delta_r G_{m,3}^\ominus = 2\Delta_r G_{m,2}^\ominus - \Delta_r G_{m,1}^\ominus = 0.16 \text{kJ} \cdot \text{mol}^{-1}$

5. by $\quad \Delta_r G_m^\ominus = \Delta_r H_m^\ominus - T \Delta_r S_m^\ominus = 0$

$$T = \frac{\Delta_r H_m^\ominus}{\Delta_r S_m^\ominus} = \frac{31.4 \times 1\,000\text{J} \cdot \text{mol}^{-1}}{94.2\text{J} \cdot \text{mol}^{-1} \cdot \text{K}^{-1}} = 333\text{K}$$

6. The necessary data can be found from table in the addenda of the text book:

$\Delta_f H_m^\ominus(CO_2\ g) = -393.5\text{kJ} \cdot \text{mol}^{-1}$

$\Delta_f H_m^\ominus(H_2O,l) = -285.8\text{kJ} \cdot \text{mol}^{-1}$

$\Delta_f H_m^\ominus(O_2,g) = 0$

So $\Delta_r H_m^\ominus = 57\Delta_f H_m^\ominus(CO_2\ g) + 52\Delta_f H_m^\ominus(H_2O,l) - \Delta_f H_m^\ominus(C_{57}H_{104}O_6) - 80\Delta_f H_m^\ominus(O_2\ g)$

$\Delta_f H_m^\ominus(C_{57}H_{104}O_6) = 57 \times (-393.5) + 52 \times (-285.8) - (-3.35 \times 10^{-4}) = -3.79 \times 10^3 \text{kJ} \cdot \text{mol}^{-1}$

7. (a) S > 0;   (b) S < 0;   (c) For no change take place in the number of moles of gas, the entropy change cannot be predicted.

Chapter 3 Chemical Equilibrium

[Exercises]

1.

	standard equilibrium constant expression	experimental equilibrium constant expression
(1)	$K^\ominus = \dfrac{p_{H_2O}/p^\ominus}{(p_{H_2}/p^\ominus)(p_{O_2}/p^\ominus)^{\frac{1}{2}}}$	$K_p = \dfrac{p_{H_2O}}{p_{H_2} p_{O_2}^{1/2}}$
(2)	$K^\ominus = \dfrac{1}{(p_{C_2H_2}/p^\ominus)^3}$	$K_p = \dfrac{1}{p_{C_2H_2}^3}$
(3)	$K^\ominus = (p_{CO_2}/p^\ominus)^2$	$K_p = p_{CO_2}^2$
(4)	$K^\ominus = \dfrac{\{[CN^-]/c^\ominus\}\{[H^+]/c^\ominus\}}{\{[HCN]/c^\ominus\}}$	$K_c = \dfrac{[CN^-][H^+]}{[HCN]}$
(5)	$K^\ominus = \dfrac{\{[Mn^{2+}]/c^\ominus\}(p_{Cl_2}/p^\ominus)}{\{[H^+]/c^\ominus\}^4 \{[Cl^-]/c^\ominus\}^2}$	$K = \dfrac{[Mn^{2+}] \times p_{Cl_2}}{[H^+]^4 [Cl^-]^2}$

2. (a) reaction(2) − reaction (1) = reaction (3),

$$\text{So, } K_3^\ominus = \frac{p_{CO_2}/p^\ominus}{p_{CO}/p^\ominus} = \frac{K_2^\ominus}{K_1^\ominus} = \frac{67}{0.14} = 478.6$$

(b) For reaction (2), under standard conditions,

$\Delta_r G_{m,823K}^\ominus = -RT \ln K_2^\ominus = -8.314\text{J} \cdot \text{K}^{-1} \cdot \text{mol}^{-1} \times 823\text{K} \times \ln 67 = -28.77\text{kJ} \cdot \text{mol}^{-1}$

For reaction (3), under standard conditions,

$\Delta_r G_{m,823K}^\ominus = -RT \ln K_3^\ominus = -8.314\text{J} \cdot \text{K}^{-1} \cdot \text{mol}^{-1} \times 823\text{K} \times \ln 478.6 = -42.22\text{kJ} \cdot \text{mol}^{-1}$

Comparing the value of $\Delta_r G_{m,823}^\ominus$ for the two reactions, the value is more negative for reaction (3) than reaction (2), therefore, the reduced ability of CO(g) is greater than that of H_2(g) under standard conditions.

3. Suppose the change in concentration of SO_3 is x mol·L^{-1} when the equilibrium is reached,

	SO_2(g)	+	NO_2(g)	⇌	SO_3(g)	+	NO(g)
initial concns (mol·L^{-1})	0.003		0.003		0.003		0.003
changes (mol·L^{-1})	−x		−x		+x		+x
equil concns (mol·L^{-1})	0.003−x		0.003−x		0.003+x		0.003+x

At equilibrium, we have

$$K_c = \frac{[NO][SO_3]}{[NO_2][SO_2]} = \frac{(0.003+x)^2}{(0.003-x)^2} = 9.0$$

So $\qquad x = 0.0015 \text{ mol} \cdot L^{-1}$

Therefore $\qquad [SO_3] = 0.003 + 0.0015 = 0.0045 \text{ mol} \cdot L^{-1}$

4. For the reaction $HbO_2(aq) + CO(g) \rightleftharpoons HbCO(aq) + O_2(g)$, when the concentrations of CO and O_2 are at 1×10^{-6} mol·L^{-1} 和 1×10^{-2} mol·L^{-1}, respectively, we can obtain

$$K = \frac{[HbCO][O_2]}{[HbO_2][CO]} = \frac{[HbCO] \times 1 \times 10^{-2} \text{mol} \cdot L^{-1}}{[HbO_2] \times 1 \times 10^{-6} \text{mol} \cdot L^{-1}} = 220$$

So $\dfrac{[HbCO]}{[HbO_2]} = 0.0220$. This means the concentration of HbCO is larger than 2% of the concentration of HbO_2. Therefore, smoking could injure the person's intelligence.

5. Since the conversion rate of PCl_5 is 0.91 at equilibrium, the change in the mole of $PCl_5 = 2 \times 0.91 = 1.82$ mol. According to the reaction, we have

$$PCl_5(g) \rightleftharpoons PCl_3(g) + Cl_2(g)$$

initial amount (mol)	2.00	1.00	0
changes in mol	-1.82	$+1.82$	$+1.82$
equil amount (mol)	0.18	2.82	1.82

Applying the Dalton's Law of Partial Pressures, we can obtain the equilibrium partial pressure for each component:

equilibrium partial pressure of $PCl_5(g) = \dfrac{0.18}{0.18+2.82+1.82} \times 200 \text{kPa} = 7.47 \text{kPa}$

equilibrium partial pressure of $PCl_3(g) = \dfrac{2.82}{0.18+2.82+1.82} \times 200 \text{kPa} = 117.0 \text{kPa}$

equilibrium partial pressure of $Cl_2(g) = \dfrac{1.82}{0.18+2.82+1.82} \times 200 \text{kPa} = 75.52 \text{kPa}$

Therefore, $K^\ominus = \dfrac{(p_{Cl_2}/p^\ominus)(p_{PCl_3}/p^\ominus)}{p_{PCl_5}/p^\ominus} = \dfrac{(75.52 \text{kPa}/100 \text{kPa}) \times (117.0 \text{kpa}/100 \text{kPa})}{7.47 \text{kPa}/100 \text{kPa}} = 11.83$

6. Suppose a change in pressure of N_2O_4 is x kPa and the initial pressure of N_2O_4 is y kPa.

$$N_2O_4(g) \rightleftharpoons 2NO_2(g)$$

initial partial pressure (kPa)	y	0
changes (kPa)	$-x$	$+2x$
equil partial pressure (kPa)	$y-x$	$2x$

At equilibrium, we have

$$\begin{cases} y - x + 2x = 60 \\ K^\ominus = \dfrac{(p_{NO_2}/p^\ominus)^2}{p_{N_2O_4}/p^\ominus} = \dfrac{(2x/100)^2}{(y-x)/100} = 0.15 \end{cases}$$

Then, we obtain $\begin{cases} x = 11.71 \text{kPa} \\ y = 48.29 \text{kPa} \end{cases}$

Therefore, the decomposition rate of $N_2O_4 = \dfrac{11.71}{48.29} \times 100\% = 24\%$, and the initial pressure of N_2O_4 is 48.29kPa.

7. (1) $K^\ominus = \dfrac{\{[Mn^{2+}]/c^\ominus\}(p_{Cl_2}/p^\ominus)}{\{[H^+]/c^\ominus\}^4\{[Cl^-]/c^\ominus\}^2}$

(2) $\Delta_r G_m^\ominus = \Delta_f G_m^\ominus(Mn^{2+}, aq) + \Delta_f G_m^\ominus(Cl_2, g) + 2\Delta_f G_m^\ominus(H_2O, l)$
$\qquad - \Delta_f G_m^\ominus(MnO_2, s) - 4\Delta_f G_m^\ominus(H^+, aq) - 2\Delta_f G_m^\ominus(Cl^-, aq)$
$\quad = -228.1 kJ \cdot mol^{-1} + (-237.1 kJ \cdot mol^{-1}) - (-465.1 kJ \cdot mol^{-1}) - 2\times(-131.2 kJ \cdot mol^{-1})$
$\quad = 25.2 kJ \cdot mol^{-1}$

Now, we substitute the value of $\Delta_r G_m^\ominus$ into Equation (3.1) to find $K^\ominus$

$\ln K^\ominus = -\dfrac{25.2\times 10^3 J \cdot mol^{-1}}{8.314 J \cdot K^{-1} \cdot mol^{-1} \times 298.15K} = -10.17 \qquad K^\ominus = e^{-10.17} = 3.83\times 10^{-5}$

Therefore, the forward reaction cannot proceed spontaneously under standard conditions at 298.15K.

(3) $\Delta_r G_m = \Delta_r G_m^\ominus + RT \ln J$
$\quad = 25.2 kJ \cdot mol^{-1} + 8.314 J \cdot K^{-1} \cdot mol^{-1} \times 298K \times \ln\dfrac{(1/1)(100/100)}{(12/1)^4(12/1)^2} = -11.9 kJ \cdot mol^{-1} < 0$

Therefore, the forward reaction can proceed spontaneously when the concentration of hydrochloric acid is $12.0 mol \cdot L^{-1}$.

8. $\Delta_r H_m^\ominus = 2\times \Delta_f H_m^\ominus(SO_3, g) - 2\times \Delta_f H_m^\ominus(SO_2, g) - \Delta_f H_m^\ominus(O_2, g)$
$\quad = 2\times(-395.7 kJ \cdot mol^{-1}) - 2\times(-296.8 kJ \cdot mol^{-1}) = -197.78 kJ \cdot mol^{-1}$

Now, we substitute the known values of $\Delta_r H_m^\ominus$, $K_1^\ominus$, T_2 and T_1 into Equation (3.7) to find $K^\ominus$ at 900K:

$\ln\dfrac{K_{900}^\ominus}{910} = \dfrac{-197.78\times 10^3 J \cdot mol^{-1}}{8.314 J \cdot mol^{-1} \cdot K^{-1}} \times \left(\dfrac{900K - 800K}{900K \times 800K}\right) = -3.30$

$K_{900}^\ominus = 33.56$

9. $\qquad\qquad\qquad AgCl(s) \rightleftharpoons Ag^+(aq) + Cl^-(aq)$

$\Delta_f G_m^\ominus / (kJ \cdot mol^{-1}) \qquad -109.8 \qquad 77.1 \qquad -131.2$

$\Delta_r G_m^\ominus = \Delta_f G_m^\ominus(Ag^+, aq) + \Delta_f G_m^\ominus(Cl^-, aq) - \Delta_f G_m^\ominus(AgCl, s)$
$\quad = 77.1 kJ \cdot mol^{-1} + (-131.2 kJ \cdot mol^{-1}) - (-109.8 kJ \cdot mol^{-1})$
$\quad = 55.7 kJ \cdot mol^{-1}$

Now, we substitute the value of $\Delta_r G_m^\ominus$ into Equation 3.1 to find $K^\ominus$

$\ln K^\ominus = -\dfrac{55.7\times 10^3 J \cdot mol^{-1}}{8.314 J \cdot K^{-1} \cdot mol^{-1} \times 298.15K} = -22.47$

$K^\ominus = e^{-22.47} = 1.74\times 10^{-10}$

Therefore, $\qquad\qquad K_{sp} = K^\ominus = 1.74\times 10^{-10}$

10. Calculating $\Delta_r H_m^\ominus$ and $\Delta_r S_m^\ominus$ to find $\Delta_r G_m^\ominus$ at 298.15K:

$\Delta_r H_m^\ominus = 2\times \Delta_f H_m^\ominus(Ag^+, aq) + \Delta_f H_m^\ominus(CO_3^{2-}, aq) - \Delta_f H_m^\ominus(Ag_2CO_3, s)$
$\quad = 2\times 105.6 kJ \cdot mol^{-1} + (-667.1 kJ \cdot mol^{-1}) - (-505.8 kJ \cdot mol^{-1})$
$\quad = 49.9 kJ \cdot mol^{-1}$

$$\Delta_r S_m^\ominus = 2 \times S_m^\ominus(Ag^+, aq) + S_m^\ominus(CO_3^{2-}, aq) - S_m^\ominus(Ag_2CO_3, s)$$
$$= 2 \times 72.7 J \cdot K^{-1} \cdot mol^{-1} + (-56.9 J \cdot K^{-1} \cdot mol^{-1}) - 167.4 J \cdot K^{-1} \cdot mol^{-1}$$
$$= 78.9 J \cdot K^{-1} \cdot mol^{-1}$$

$$\Delta_r G_{m,T}^\ominus = \Delta_r H_{m,T}^\ominus - T \Delta_r S_{m,T}^\ominus$$

$$\Delta_r G_{m,298.15}^\ominus = 49.9 kJ \cdot mol^{-1} - 298.15 K \times (-78.9 \times 10^{-3}) kJ \cdot K^{-1} \cdot mol^{-1} = 73.4 kJ \cdot mol^{-1}$$

We substitute the value of $\Delta_r G_m^\ominus$ into Equation (3.1) to find $K^\ominus$ at 298.15K:

$$\ln K^\ominus = -\frac{73.4 \times 10^3 J \cdot mol^{-1}}{8.314 J \cdot K^{-1} \cdot mol^{-1} \times 298.15 K} = -29.61$$

So, at 298.15K $\qquad K_{sp} = K^\ominus = 1.38 \times 10^{-13}$

Since we assume the values of $\Delta_r H_m^\ominus$ do not change with temperature, we substitute the known values of $\Delta_r H_m^\ominus$, $K_1^\ominus$, T_2 and T_1 into Equation 3.7 to find $K^\ominus$ at 373.15K:

$$\ln \frac{K_{373.15}^\ominus}{1.38 \times 10^{-13}} = \frac{49.9 \times 10^3 \ J \cdot mol^{-1}}{8.314 \ J \cdot mol^{-1} \cdot K^{-1}} \times \left(\frac{373.15 K - 298.15 K}{373.15 K \times 298.15 K} \right) = 4.05$$

$$\ln K_{373.15}^\ominus - \ln(1.38 \times 10^{-13}) = 4.05$$

So $\qquad \ln K_{373.15}^\ominus = -25.56$

Therefore, at 373.15K $\qquad K_{sp} = 7.93 \times 10^{-12}$

[Supplementary Exercises]

1. (1) $2N_2O(g) + O_2(g) \rightleftharpoons 4NO(g) \qquad K_1^{\ominus\prime} = 1/(3.54 \times 10^{-25}) = 2.82 \times 10^{24}$

reaction (3) = reaction (1′) + reaction (2) + reaction (2)

Therefore, $\qquad K^\ominus = K_1^{\ominus\prime} \times K_2^\ominus \times K_2^\ominus = 2.82 \times 10^{24} \times (6.26 \times 10^{-13})^2 = 1.10 \times 10^{-2}$

2. Writing the reaction quotient $J = \dfrac{(p_{CO_2}/p^\ominus)^2}{p_{O_2}/p^\ominus}$

(1) The concentration of solid graphite is unchanged as long as some is present, so it does not appear in the reaction quotient J. Increase of the amount of graphite has no effect, so the equilibrium partial pressure of O_2 is unchanged.

(2) Increase of the amount of CO_2 leads to $J > K^\ominus$. The reaction shifts to the left. A portion of O_2 is produced in the reaction and hence the equilibrium partial pressure of O_2 increases.

(3) Increase of the amount of O_2 leads to $J < K^\ominus$. The reaction shifts to the right. Some added O_2 is consumed and the partial pressure of O_2 increases when equilibrium is re-established.

(4) The sign of $\Delta_r H_m^\ominus$ tells us the forward reaction is exothermic. Reduce the reaction temperature favors forward reaction and a shift in the equilibrium condition to the right. Some O_2 is consumed and the equilibrium partial pressure of O_2 decreases.

(5) Adding catalyst to a reaction mixture speeds up both the forward and reverse reactions. Equilibrium is achieved more rapidly, but the equilibrium amounts are unchanged by the catalyst. Thus, the equilibrium partial pressure of O_2 is unchanged.

3. For the reaction $\qquad 2NO_2(g) \rightleftharpoons N_2O_4(g)$

$$K^\ominus = \frac{p_{N_2O_4}/p^\ominus}{(p_{NO_2}/p^\ominus)^2} = \frac{p_{N_2O_4}/100kPa}{(20kPa/100kPa)^2} = 6.67$$

Therefore, $\qquad p_{N_2O_4} = 26.68 kPa$

$$p_{total} = p_{N_2O_4} + p_{NO_2} = 26.68\text{kPa} + 20.0\text{kPa} = 46.68\text{kPa}$$

4. Based on the equation
$$\Delta_r G_m^\ominus = -RT\ln K^\ominus$$

$$\ln K^\ominus = -\frac{-50.5 \times 10^3 \text{J} \cdot \text{mol}^{-1}}{8.314 \text{J} \cdot \text{K}^{-1} \cdot \text{mol}^{-1} \times 298.15\text{K}} = 20.37$$

At 298.15K, $K^\ominus = 7.02 \times 10^5$. The $K^\ominus$ value is is quite large, suggesting that studying this reaction as a mean of methane production is worth pursuing.

5. The reaction is
$$H_2(g) + I_2(g) \rightleftharpoons 2HI(g)$$

$$[H_2] = \frac{4.562 \times 10^{-3} \text{mol}}{1.0\text{L}} = 4.562 \times 10^{-3} \text{mol} \cdot \text{L}^{-1}$$

Similarly, $[I_2] = 7.384 \times 10^{-4} \text{mol} \cdot \text{L}^{-1}$ and $[HI] = 1.355 \times 10^{-2} \text{mol} \cdot \text{L}^{-1}$. At 425.4°C:

$$K_c = \frac{[HI]^2}{[H_2][I_2]} = \frac{(1.355 \times 10^{-2} \text{mol} \cdot \text{L}^{-1})^2}{(4.562 \times 10^{-3} \text{mol} \cdot \text{L}^{-1}) \times (7.384 \times 10^{-4} \text{mol} \cdot \text{L}^{-1})} = 54.5$$

If 1.000×10^{-3} mol of $I_2(g)$ is added, the equilibrium is disturbed and the new initial concentration for $I_2(g)$ changes to $(0.7384 + 1.000)$mmol $\cdot$ L^{-1}. Suppose the change in H_2 concentration is x mmol $\cdot$ L^{-1} after the system has reached equilibrium again:

	$H_2(g)$	+	$I_2(g)$	$\rightleftharpoons$	$2HI(g)$
new initial concns (mmol · L^{-1})	4.562		1.7384		13.55
changes (mmol · L^{-1})	$-x$		$-x$		$+2x$
new equil concns (mmol · L^{-1})	$4.562-x$		$1.7384-x$		$13.55+2x$

$$K_c = \frac{[HI]^2}{[H_2][I_2]} = \frac{[(13.55+2x) \times 10^{-3} \text{mol} \cdot \text{L}^{-1}]^2}{[(4.562-x) \times 10^{-3} \text{mol} \cdot \text{L}^{-1}] \times [(1.7384-x) \times 10^{-3} \text{mol} \cdot \text{L}^{-1}]} = 54.5$$

So
$$x = 0.685 \text{mmol} \cdot \text{L}^{-1}$$

Therefore, the concentrations of $H_2(g)$, $I_2(g)$ and $HI(g)$ after a new equilibrium is established are

$[H_2] = 4.562 - 0.685 = 3.877$mmol · L^{-1} $[I_2] = 1.7384 - 0.685 = 1.0534$mmol · L^{-1}

$[HI] = 13.55 + 2 \times 0.685 = 14.92$mmol · L^{-1}

Chapter 4 Rate of Chemical Reaction

[Exercises]

2. h^{-1}, L · mol^{-1} · h^{-1}, mol · L^{-1} · h^{-1}

3. $\Delta_r H_m = E_{a(fwd)} - E_{a(rev)}$

4. It is reaction (2) by Arrhenius equation

5. $\quad v(SO_2) = 1/2\, v(O_2) = v(SO_3)$

$v(O_2) = (1/2) \times 13.60$ mol · L^{-1} · h^{-1} = 6.8 mol · L^{-1} · h^{-1}

$v(SO_3) = 13.60$ mol · L^{-1} · h^{-1}

6. $\quad t_{1/2} = 0.693/k = 23$d $\qquad k = 3.01 \times 10^{-2}$d^{-1}

7. For the first-order reaction: $\quad t = \dfrac{1}{k}\ln\dfrac{c_0}{c_0 - x}\quad$ and $\quad t_{1/2} = 0.693/k$

$$t = \frac{t_{1/2}}{0.693}\ln\frac{c_0}{c_0-x} = \frac{t_{1/2}}{0.693}\ln\frac{1}{0.001} = 9.97 t_{1/2} \approx 10 t_{1/2}$$

8. (1) zero-order reaction $\quad v = k \quad k = 0.014$ mol · L^{-1} · s^{-1}

(2) first-order reaction $\quad v=kc \quad k=v/c=0.014 \text{mol} \cdot \text{L}^{-1} \cdot \text{s}^{-1} / 0.50 \text{mol} \cdot \text{L}^{-1}=0.028 \text{s}^{-1}$

(3) second-order reaction $\quad v=kc^2 \quad k=v/c^2=0.014 \text{mol} \cdot \text{L}^{-1} \cdot \text{s}^{-1} / (0.50 \text{mol} \cdot \text{L}^{-1})^2$
$$=0.056 \text{L} \cdot \text{mol}^{-1} \cdot \text{s}^{-1}$$

9. For the first-order reaction:

$$k=\frac{1}{t}\ln\frac{c_0}{c}=\frac{1}{12}\ln\frac{c_0}{(1-0.9)c_0}=0.19\text{s}^{-1} \qquad k=\frac{1}{12\times 60}\ln\frac{c_0}{(1-0.9)c_0}=3.2\times 10^{-3}\text{s}^{-1}$$

$$\ln\frac{3.2\times 10^{-3}\text{s}^{-1}}{3.5\times 10^{-2}\text{s}^{-1}}=\frac{1.440\times 10^4 \text{J}\cdot\text{mol}^{-1}}{8.314 \text{J}\cdot\text{mol}^{-1}\cdot\text{K}^{-1}}\left(\frac{T_2-553\text{K}}{T_2\times 553\text{K}}\right) \qquad T_2=313\text{K}$$

10. For the second-order reaction:

$$\frac{1}{c}-\frac{1}{c_0}=kt$$

$$\frac{1}{0.005\times(1-27.8\%)\text{mol}\cdot\text{L}^{-1}}-\frac{1}{0.005\text{mol}\cdot\text{L}^{-1}}=k_1\times 300\text{s}$$

at 500℃: $k_1=0.257 \text{L}\cdot\text{mol}^{-1}\cdot\text{s}^{-1}$

at 510℃:

$$\frac{1}{0.005\times(1-36.2\%)\text{mol}\cdot\text{L}^{-1}}-\frac{1}{0.005\text{mol}\cdot\text{L}^{-1}}=k_2\times 300\text{s}$$

$$k_2=0.378 \text{L}\cdot\text{mol}^{-1}\cdot\text{s}^{-1}$$

$$E_a=R\frac{T_2 T_1}{(T_1-T_2)}\ln\frac{k_2}{k_1}=\frac{8.314\text{J}\cdot\text{mol}^{-1}\cdot\text{K}^{-1}\times 773\text{K}\times 783\text{K}}{(783-773)\text{K}}\ln\frac{0.378}{0.257}$$

$$=194.2 \text{kJ}\cdot\text{mol}^{-1}$$

At 673K (400℃), rate constant is k_3.

$$\ln\frac{0.378}{k_3}=\frac{194.2\times 10^3 \text{J}\cdot\text{mol}^{-1}\times(783\text{K}-673\text{K})}{8.314\text{J}\cdot\text{mol}^{-1}\cdot\text{K}^{-1}\times 783\text{K}\times 673\text{K}}$$

$$k_3=2.88\times 10^{-3}\text{L}\cdot\text{mol}^{-1}\cdot\text{s}^{-1}$$

11. This is first-order reaction by problem

At 3℃: $\quad \ln\dfrac{c_0}{(1-30\%)c_0}=k_1\times 365\times 2\text{d} \qquad k_1=4.9\times 10^{-4}\text{d}^{-1}$

Rate constant is k_2 at 25℃:

$$\ln\frac{k_2}{4.9\times 10^{-4}}=\frac{135.0\times 10^3 \text{J}\cdot\text{mol}^{-1}\times(298\text{K}-276\text{K})}{8.314\text{J}\cdot\text{mol}^{-1}\cdot\text{K}^{-1}\times 298\text{K}\times 276\text{K}}$$

$$k_2=3.8\times 10^{-2}\text{d}^{-1}$$

x is decomposition ratio after keeping two weeks, so

$$\ln\frac{c_0}{(1-x)c_0}=3.8\times 10^{-2}\text{d}^{-1}\times 14\text{d} \qquad x=41.2\%$$

Decomposition of drug is over 30%, It is invalid.

12. $v(O_2)=v(HBO_2)=kc(O_2)c(HB)$

(1) $v(O_2)=v(HBO_2)$

$=1.98\times 10^6 \text{L}\cdot\text{mol}^{-1}\cdot\text{s}^{-1}\times 8.0\times 10^{-6}\text{mol}\cdot\text{L}^{-1}\times 1.6\times 10^{-6}\text{mol}\cdot\text{L}^{-1}$

$=2.53\times 10^{-5}\text{mol}\cdot\text{L}^{-1}\cdot\text{s}^{-1}$

(2) $c(O_2) = \dfrac{v(HBO_2)}{kc(HB)} = \dfrac{1.3 \times 10^{-4} \text{mol} \cdot L^{-1} \cdot s^{-1}}{1.98 \times 10^6 L \cdot \text{mol}^{-1} \cdot s^{-1} \times 8.0 \times 10^{-6} \text{mol} \cdot L^{-1}}$

$= 8.2 \times 10^{-6} \text{mol} \cdot L^{-1}$

13.
$$\ln \dfrac{k_2}{k_1} = \dfrac{E_a}{R}(\dfrac{T_2 - T_1}{T_1 T_2})$$

$E_a = R\dfrac{T_2 T_1}{(T_1 - T_2)} \ln \dfrac{k_2}{k_1} = \dfrac{8.314 \text{J} \cdot \text{mol}^{-1} \cdot K^{-1} \times 310K \times 316K}{(316 - 310)K} \ln \dfrac{4.05 \times 10^{-2}}{2.16 \times 10^{-2}}$

$= 85.5 \text{kJ} \cdot \text{mol}^{-1}$

Or calculate the average activation energy, it is $84.7 \text{kJ} \cdot \text{mol}^{-1}$

and $\qquad \ln k = -\dfrac{E_a}{RT} + \ln A$

$\ln A = \ln 2.16 \times 10^{-2} + \dfrac{85.5 \times 10^3 \text{J} \cdot \text{mol}^{-1}}{8.314 \text{J} \cdot \text{mol}^{-1} \cdot K^{-1} \times 310K}$

$A = 5.47 \times 10^{12}$

Or calculate the average A, it is 4.05×10^{12}

14. By the question: $\qquad v \propto k$, or $v \propto 1/t$, so $k \propto 1/t$

$E_a = R\dfrac{T_2 T_1}{(T_1 - T_2)} \ln \dfrac{k_2}{k_1} = R(\dfrac{T_2 - T_1}{T_1 T_2}) \ln \dfrac{t_1}{t_2}$

$= \dfrac{8.314 \text{J} \cdot \text{mol}^{-1} \cdot K^{-1} \times 301K \times 278K}{(301 - 278)K} \ln \dfrac{48}{4} = 75.0 \text{kJ} \cdot \text{mol}^{-1}$

15. Rate constant of the reaction without catalyst is k_1, Rate constant of it by catalyst Au and Pt is k_2 and k_3 respectively

$$\ln k = -\dfrac{E_a}{RT} + \ln A$$

$\ln \dfrac{k_2}{k_1} = \dfrac{E_{a1} - E_{a2}}{RT} = \dfrac{(184 - 105) \times 10^3 \text{J} \cdot \text{mol}^{-1}}{8.314 \text{J} \cdot \text{mol}^{-1} \cdot K^{-1} \times 298K} = 31.89$

$\dfrac{k_2}{k_1} = 7.1 \times 10^{13}$ it is 7.1×10^{13} times when Au acts as catalyst.

$\ln \dfrac{k_3}{k_1} = \dfrac{E_{a1} - E_{a3}}{RT} = \dfrac{(184 - 42) \times 10^3 \text{J} \cdot \text{mol}^{-1}}{8.314 \text{J} \cdot \text{mol}^{-1} \cdot K^{-1} \times 298K} = 57.31$

$\dfrac{k_3}{k_1} = 7.8 \times 10^{24}$ it is 7.8×10^{24} times when Pt acts as catalyst.

16. $\ln \dfrac{k_2}{k_1} = \dfrac{E_a}{R}(\dfrac{T_2 - T_1}{T_1 T_2}) = \dfrac{50.0 \times 10^3 \text{J} \cdot \text{mol}^{-1}}{8.314 \text{J} \cdot \text{mol}^{-1} \cdot K^{-1}} \times \dfrac{(313 - 310)K}{313K \times 310K} = 0.19$

$\dfrac{k_2}{k_1} = 1.2$

Catalytic reaction rate by enzyme at 40℃ is 1.2 times as fast as normal body at 37℃

[Supplementary Exercises]

1. (a)

2. (a) This is a first-order reaction, by $\ln\dfrac{c_0}{c} = kt$

$$\ln\dfrac{2.00\,\text{mol}\cdot\text{L}^{-1}}{c} = 87\,\text{s}^{-1} \times 0.001\,0\,\text{s} \qquad c = 1.83\,\text{mol}\cdot\text{L}^{-1}$$

(b) Fraction decomposed $= (c_0 - c)/c_0$
$$= (2.00\,\text{mol}\cdot\text{L}^{-1} - 1.83\,\text{mol}\cdot\text{L}^{-1})/2.00\,\text{mol}\cdot\text{L}^{-1} = 0.085$$

3. $\qquad v = kc(NO_2)^m c(CO)^n$

$$\text{rate2/rate1} = kc_2(NO_2)^m c_2(CO)^n / kc_1(NO_2)^m c_1(CO)^n$$
$$= c_2(NO_2)^m / c_1(NO_2)^m = (0.40\,\text{mol}\cdot\text{L}^{-1}/0.10\,\text{mol}\cdot\text{L}^{-1})^m$$
$$16 = 4.0^m \qquad m = 2$$

$$\text{rate3/rate1} = kc_3(NO_2)^2 c_3(CO)^n / kc_1(NO_2)^2 c_1(CO)^n$$
$$= c_3(CO)^n / c_1(CO)^n$$
$$= (0.20\,\text{mol}\cdot\text{L}^{-1}/0.10\,\text{mol}\cdot\text{L}^{-1})^n$$
$$1.0 = 2^n \qquad n = 0$$
$$v = kc(NO_2)^2 c(CO)^0 = kc(NO_2)^2$$

This reaction is zero-order in CO, second-order in NO_2 and second-order overall respectively.

4. According to question, $\qquad t_{1/2} = 6\,\text{h},\ t = 1\,\text{h}$

For first-order reaction, $\quad k = 0.693/t_{1/2} = 0.693/6\,\text{h} = 0.116\,\text{h}^{-1}$

$$\ln\dfrac{c_0}{c} = kt = 0.116\,\text{h}^{-1} \times 1\,\text{h} = 0.116$$

$\dfrac{c_0}{c} = 1.12$, the artificial red blood cells will be left $\dfrac{c}{c_0} \times 100\% = 89.3\%$

5. We suppose that tentative rate law for this reaction is $v = kc^\alpha(A)\,c^\beta(B)\,c^\gamma(E)$

(a) In comparing experiment 1 with 2, If
$$v_1 = kc^\alpha(A)\,c^\beta(B)\,c^\gamma(E) = R_1$$
then, $\qquad v_2 = kc^\alpha(A)\,c^\beta(B)\,c^\gamma(E) = (1/2)R_1$

we can get the ratio, $\qquad v_2/v_1 = (1/2)^\alpha = 1/2,\quad \alpha = 1$

Similarly, in comparing experiment 2 with 3 and 1 with 4 respectively, we have the values $\beta = 2$ and $\gamma = -1$. In summary, the reaction is first-order in A ($\alpha = 1$), second-order in B ($\beta = 2$) and the order of -1 for E ($\gamma = -1$). The overall order of reaction is 2.

(b) $\qquad v_5 = \dfrac{k\dfrac{c(A)}{2}[\dfrac{c(B)}{2}]^2}{\dfrac{c(E)}{2}} = \dfrac{1}{4}v_1$

6. (a) the reaction equation is
$$2NO + 2H_2 \rightleftharpoons N_2 + 2H_2O$$

(b) In this complex reaction, the step (1) is fast, $K_1 = \dfrac{c(N_2O_2)}{c^2(NO)}$

The step (2) is the rate-determining step. So the rate law is
$$v = k_2 c(N_2O_2) c(H_2) = K_1 k_2 c^2(NO)\,c(H_2)$$

(c) 3rd-order

Chapter 5 Electrolyte Solutions

[Exercises]

1.

Acid	H_2O	H_3O^+	H_2CO_3	HCO_3^-	NH_4^+	$NH_3^+CH_2COO^-$	H_2S	HS^-
Conjugate base	OH^-	H_2O	HCO_3^-	CO_3^{2-}	NH_3	$NH_2CH_2COO^-$	HS^-	S^{2-}

2.

Base	H_2O	NH_3	HPO_4^{2-}	NH_2^-	$[Al(H_2O)_5OH]^{2+}$	CO_3^{2-}	$NH_3^+CH_2COO^-$
Conjugate acid	H_3O^+	NH_4^+	$H_2PO_4^-$	NH_3	$[Al(H_2O)_6]^{3+}$	HCO_3^-	$NH_3^+CH_2COOH$

3. (1) There exist following proton transfer equilibriums in H_3PO_4 Solution

$$H_3PO_4 (aq) + H_2O(l) \rightleftharpoons H_2PO_4^-(aq) + H_3O^+(aq)$$
$$H_2PO_4^-(aq) + H_2O(l) \rightleftharpoons HPO_4^{2-}(aq) + H_3O^+(aq)$$
$$HPO_4^{2-}(aq) + H_2O(l) \rightleftharpoons PO_4^{3-}(aq) + H_3O^+(aq)$$
$$H_2O(l) + H_2O(l) \rightleftharpoons H_3O^+(aq) + OH^-(aq)$$

Known H_3PO_4: $K_{a1} = 6.9 \times 10^{-3}$, $K_{a2} = 6.1 \times 10^{-8}$, $K_{a3} = 4.8 \times 10^{-13}$, $K_w = 1.0 \times 10^{-14}$.

Due to $K_{a1} \gg K_{a2} \gg K_{a3}$, therefore, $[H_3O^+] \approx [H_2PO_4^-] > [HPO_4^{2-}] \approx K_{a2}$

And due to $\dfrac{[PO_4^{3-}][H_3O^+]}{[HPO_4^{2-}]} = K_{a3}$ get $\dfrac{[PO_4^{3-}]K_w}{[OH^-]} = K_{a2} \cdot K_{a3}$

i.e. $[PO_4^{3-}]K_w = [OH^-]K_{a2} \cdot K_{a3}$ ∵ $K_{a2} \cdot K_{a3} < K_w$, ∴ $[OH^-] > [PO_4^{3-}]$

Therefore, the ionic concentration order is: $[H_3O^+] \approx [H_2PO_4^-] > [HPO_4^{2-}] > [OH^-] > [PO_4^{3-}]$
where the concentration of H_3O^+ is not as three times that of PO_4^{3-}.

(2) There exist following proton transfer equilibriums in $NaHCO_3$ Solution

$$HCO_3^-(aq) + H_2O(l) \rightleftharpoons CO_3^{2-}(aq) + H_3O^+(aq)$$
$$K_a(HCO_3^-) = K_{a2}(H_2CO_3) = 4.7 \times 10^{-11}$$
$$HCO_3^-(aq) + H_2O(l) \rightleftharpoons H_2CO_3(aq) + OH^-(aq)$$
$$K_b(HCO_3^-) = K_w / K_{a1}(H_2CO_3) = 1.0 \times 10^{-14} / (4.5 \times 10^{-7}) = 2.2 \times 10^{-8}$$

There exist following proton transfer equilibriums in NaH_2PO_4 Solution

$$H_2PO_4^-(aq) + H_2O(l) \rightleftharpoons HPO_4^{2-}(aq) + H_3O^+(aq)$$
$$K_a(H_2PO_4^-) = K_{a2}(H_3PO_4) = 6.1 \times 10^{-8}$$
$$H_2PO_4^-(aq) + H_2O(l) \rightleftharpoons H_3PO_4(aq) + OH^-(aq)$$
$$K_b(H_2PO_4^-) = K_w / K_{a1}(H_3PO_4) = 1.0 \times 10^{-14} / (6.9 \times 10^{-3}) = 1.4 \times 10^{-12}$$

Because $K_b(HCO_3^-) > K_b(H_2PO_4^-)$, the aqueous $NaHCO_3$ solution is more basic than the aqueous NaH_2PO_4 solution.

4. (1) $K = \dfrac{[NO_2^-][HCN]}{[HNO_2][CN^-]} = \dfrac{K_a(HNO_2)}{K_a(HCN)} = \dfrac{5.6 \times 10^{-4}}{6.2 \times 10^{-10}} = 9.0 \times 10^5$, the reaction proceed forwardly.

(2) $K = \dfrac{[SO_4^{2-}][HNO_2]}{[HSO_4^-][NO_2^-]} = \dfrac{K_{a2}(H_2SO_4)}{K_a(HNO_2)} = \dfrac{1.0 \times 10^{-2}}{5.6 \times 10^{-4}} = 18$, the reaction proceed forwardly.

(3) $K = \dfrac{[NH_3][HAc]}{[NH_4^+][Ac^-]} = \dfrac{K_a(NH_4^+)}{K_a(HAc)} = \dfrac{\dfrac{K_w}{K_b(NH_3)}}{K_a(HAc)} = \dfrac{1.0 \times 10^{-14}}{1.8 \times 10^{-5} \times 1.75 \times 10^{-5}} = 3.2 \times 10^{-5}$, the reaction proceed reversely.

(4) $K = \dfrac{[HSO_4^-][OH^-][H_3O^+]}{[SO_4^{2-}][H_3O^+]} = \dfrac{K_w}{K_{a2}(H_2SO_4)} = \dfrac{1.0 \times 10^{-14}}{1.0 \times 10^{-2}} = 1.0 \times 10^{-12}$, the reaction proceed reversely.

5. The pH of gastric juice of normal adult is 1.4, so $[H_3O^+]_{adult} = 0.040 \text{mol} \cdot L^{-1}$

pH of gastric juice of infants is 5.0, so $[H_3O^+]_{infant} = 1.0 \times 10^{-5} \text{mol} \cdot L^{-1}$

$[H_3O^+]_{adult} / [H_3O^+]_{infant} = 0.040 \text{mol} \cdot L^{-1} / (1.0 \times 10^{-5} \text{mol} \cdot L^{-1}) = 4\,000$

6. $\qquad\qquad\qquad H_2S(aq) + H_2O(l) \rightleftharpoons HS^-(aq) + H_3O^+(aq)$

Initial/mol·L^{-1} $\qquad\qquad$ 0.10

At equilibrium/mol·L^{-1} $\qquad$ $0.10 - x \approx 0.10$ $\qquad\qquad x \qquad\qquad x$

$$K_{a1} = \dfrac{[H_3O^+][HS^-]}{[H_2S]} = 8.9 \times 10^{-8} \qquad \dfrac{x^2}{0.10} = 8.9 \times 10^{-8}$$

$$x = \sqrt{8.9 \times 10^{-8} \times 0.10} \text{ mol} \cdot L^{-1} = 9.4 \times 10^{-5} \text{mol} \cdot L^{-1}$$

$$[HS^-] \approx [H_3O^+] = 9.4 \times 10^{-5} \text{mol} \cdot L^{-1}$$

$$HS^-(aq) + H_2O(l) \rightleftharpoons S^{2-}(aq) + H_3O^+(aq)$$

$$K_{a2} = \dfrac{[H_3O^+][S^{2-}]}{[HS^-]} = 1.2 \times 10^{-13}$$

Due to $[HS^-] \approx [H_3O^+]$ so $[S^{2-}] \approx K_{a2}$ $\quad [S^{2-}] = 1.2 \times 10^{-13} \text{mol} \cdot L^{-1}$

7. $\because c / K_b = 0.015 / 7.9 \times 10^{-7} > 500$, and $c \cdot K_b > 20K_w$

$\therefore [OH^-] = \sqrt{K_b \cdot c} = \sqrt{7.9 \times 10^{-7} \times 0.015} \text{mol} \cdot L^{-1} = 1.1 \times 10^{-4} \text{ mol} \cdot L^{-1}$

$\qquad\qquad$ pOH = 3.96 $\quad$ pH = 10.04

8. set $[OH^-] = x \text{ mol} \cdot L^{-1}$

$\qquad\qquad\qquad N_3^-(aq) + H_2O(l) \rightleftharpoons HN_3(aq) + OH^-(aq)$

Initial /mol·L^{-1} $\qquad\qquad$ 0.010

At equilibrium /mol·L^{-1} $\qquad 0.010 - x \qquad\qquad x \qquad\qquad x$

$$K_b = \dfrac{[HN_3][OH^-]}{[N_3^-]} = \dfrac{x^2}{0.010 - x} = \dfrac{1.00 \times 10^{-14}}{1.9 \times 10^{-5}} = 5.3 \times 10^{-10}$$

$$x = 2.3 \times 10^{-6}$$

$$[OH^-] = [HN_3] = 2.3 \times 10^{-6} \text{ mol} \cdot L^{-1}$$

$$[N_3^-] = [Na^+] = 0.010 \text{mol} \cdot L^{-1}$$

$[H_3O^+] = K_w/[OH^-] = 1.00 \times 10^{-14} / (2.3 \times 10^{-6}) = 4.3 \times 10^{-9} \text{mol} \cdot L^{-1}$

9. $\because K_{a2}$ is very small so omit the H_3O^+ from the secondary dissociation, but $c/K_{a1} < 500$

$\qquad\qquad C_7H_6O_3(aq) + H_2O(l) \rightleftharpoons C_7H_5O_3^-(aq) + H_3O^+(aq)$

Initial /mol·L^{-1} $\qquad\qquad$ 0.065

At equilibrium /mol·L^{-1} $\qquad 0.065 - x \qquad\qquad x \qquad\qquad x$

$$K_{a1} = \dfrac{[C_7H_5O_3^-][H_3O^+]}{[C_7H_6O_3]} = \dfrac{x^2}{0.065 - x} = 1.06 \times 10^{-3}$$

$$x = 7.8 \times 10^{-3}, [C_7H_5O_3^-] \approx [H_3O^+] = 7.8 \times 10^{-3} \text{ mol} \cdot L^{-1}$$

$$pH = 2.11$$
$$[C_7H_4O_3^{2-}] \approx K_{a2}, \text{ i.e, } [C_7H_4O_3^{2-}] = 3.6 \times 10^{-14} \text{ mol} \cdot L^{-1}$$

10. (1) $n(H_3PO_4) : n(NaOH) = (100mL \times 0.10 mol \cdot L^{-1}) : (100mL \times 0.20 mol \cdot L^{-1})$
$$= 1 : 2$$

the reaction is $\quad H_3PO_4(aq) + 2NaOH(aq) = Na_2HPO_4(aq) + 2H_2O(l)$

Formed $[Na_2HPO_4] = 0.10 mol \cdot L^{-1} \times 100mL / (100mL + 100mL) = 0.050 mol \cdot L^{-1}$

$$pH = \frac{1}{2}(pK_{a2} + pK_{a3}) = \frac{1}{2}(7.21 + 12.32) = 9.76$$

(2) $n(Na_3PO_4) : n(HCl) = (100mL \times 0.10 mol \cdot L^{-1}) : (100mL \times 0.20 mol \cdot L^{-1}) = 1 : 2$,

the reaction is $\quad Na_3PO_4 + 2HCl = NaH_2PO_4 + 2NaCl$

Formed $[NaH_2PO_4] = 0.10 mol \cdot L^{-1} \times 100mL / (100mL + 100mL) = 0.050 mol \cdot L^{-1}$

$$pH = \frac{1}{2}(pK_{a1} + pK_{a2}) = \frac{1}{2}(2.16 + 7.21) = 4.68$$

11. $\quad CH_3COOH(aq) + NH_3(l) \rightarrow CH_3COO^-(aq) + NH_4^+(aq)$

Because the basicity of liquid ammonia is stronger than water, in other words, the ability of accepting proton for liquid ammonia is stronger than that for water; therefore, the acidity of acetate acid in liquid ammonia is stronger than water.

12. $\quad pH = 2.45, [H_3O^+] = 3.5 \times 10^{-3} mol \cdot L^{-1}$

Set the concentration of lactic acid is $c(HC_3H_5O_3)$.

$$HC_3H_5O_3(aq) + H_2O(l) \rightleftharpoons C_3H_5O_3^-(aq) + H_3O^+(aq)$$

Due to: $\quad K_a = \dfrac{[C_3H_5O_3^-][H_3O^+]}{[HC_3H_5O_3]} = \dfrac{(3.5 \times 10^{-3})^2}{c(HC_3H_5O_3) - 3.5 \times 10^{-3}} = 1.4 \times 10^{-4}$

$$1.4 \times 10^{-4} \times c(HC_3H_5O_3) = 1.4 \times 10^{-4} \times 3.5 \times 10^{-3} + (3.5 \times 10^{-3})^2$$

$$c(HC_3H_5O_3) = 0.091 mol \cdot L^{-1}$$

13. (1) Basic NaAc should be added.

(2) After adding NaAc of the equal volume, because NaAc is excessive, the concentration of formed HAc is:

$$[HAc] = 0.20 mol \cdot L^{-1} / 2 = 0.10 mol \cdot L^{-1}$$

The concentration of NaAc in solution is:

$$[NaAc] = \frac{2.0 mol \cdot L^{-1} \times V(NaAc) - 0.20 mol \cdot L^{-1} \times V(HCl)}{V(HCl) + V(NaAc)} = \frac{1.8 mol \cdot L^{-1} \times V}{2V} = 0.90 mol \cdot L^{-1}$$

$$[H_3O^+] = \frac{K_a[HAc]}{[Ac^-]} = \frac{1.76 \times 10^{-5} \times 0.10}{0.90} mol \cdot L^{-1} = 2.0 \times 10^{-6} mol \cdot L^{-1} \qquad pH = 5.70$$

(3) NaOH is excessive, so

$$[OH^-] = \frac{2.0 mol \cdot L^{-1} \times V(NaOH) - 0.20 mol \cdot L^{-1} \times V(HCl)}{V(HCl) + V(NaOH)} = \frac{1.8 mol \cdot L^{-1} \times V}{2V} = 0.90 mol \cdot L^{-1}$$

$$pOH = 0.046 \qquad pH = 13.95$$

14. $\quad c_b = 1.0 g/(1.90 L \times 324.4 g \cdot mol^{-1}) = 1.6 \times 10^{-3} mol \cdot L^{-1}$

$$pK_{b1} = 5.1 \qquad K_{b1} = 7.9 \times 10^{-6}$$

$\because c/K_{b1} < 500$, $\qquad$ setting $[OH^-] = x \text{ mol} \cdot L^{-1}$

$$\therefore K_{b1} = \frac{[C_{20}H_{24}N_2O_2H^+][OH^-]}{[C_{20}H_{24}N_2O_2]} = \frac{x^2}{1.6 \times 10^{-3} - x} = 7.9 \times 10^{-6}$$

$x = 1.1 \times 10^{-4} \text{mol} \cdot \text{L}^{-1}$ pOH = 3.9   pH = 10.1

15. (1)   As $n(\text{HCl}) = n(\text{NH}_3)$

forming $0.10 \text{mol} \cdot \text{L}^{-1} / 2 = 0.050 \text{mol} \cdot \text{L}^{-1}$ of NH$_4$Cl in solution

$$[H_3O^+] = \sqrt{K_a \times c(\text{NH}_4^+)} = \sqrt{\frac{1.0 \times 10^{-14}}{1.8 \times 10^{-5}} \times 0.050} \text{ mol} \cdot \text{L}^{-1} = 5.3 \times 10^{-6} \text{mol} \cdot \text{L}^{-1}$$

pH = 5.28

(2) As $n(\text{HAc}) = n(\text{NH}_3)$

forming $0.10 \text{mol} \cdot \text{L}^{-1} / 2 = 0.050 \text{mol} \cdot \text{L}^{-1}$ of NH$_4$Ac

$$[H_3O^+] = \sqrt{K_a(\text{NH}_4^+) \cdot K_a(\text{HAc})} = \sqrt{\frac{1.0 \times 10^{-14}}{1.8 \times 10^{-5}} \times 1.75 \times 10^{-5}} \text{ mol} \cdot \text{L}^{-1} = 9.8 \times 10^{-8} \text{ mol} \cdot \text{L}^{-1}$$

pH = 7.01

(3) As $n(\text{HCl}) = n(\text{Na}_2\text{CO}_3)$

forming $0.10 \text{mol} \cdot \text{L}^{-1}/2 = 0.050 \text{mol} \cdot \text{L}^{-1}$ of NaHCO$_3$

$$[H_3O^+] = \sqrt{K_{a1}(\text{H}_2\text{CO}_3) \cdot K_{a2}(\text{H}_2\text{CO}_3)} = \sqrt{4.5 \times 10^{-7} \times 4.7 \times 10^{-11}} \text{ mol} \cdot \text{L}^{-1} = 4.6 \times 10^{-9} \text{ mol} \cdot \text{L}^{-1}$$

pH = 8.34

16. (1) H$_3$PO$_4$(aq) + Na$_3$PO$_4$(aq) = NaH$_2$PO$_4$(aq) + Na$_2$HPO$_4$(aq)

Initial: $V \times 0.20 \text{mol} \cdot \text{L}^{-1}$ $V \times 0.20 \text{mol} \cdot \text{L}^{-1}$

After reaction: $V \times 0.20 \text{mol} \cdot \text{L}^{-1}$ $V \times 0.20 \text{mol} \cdot \text{L}^{-1}$

In solution, $[\text{Na}_2\text{HPO}_4] = [\text{NaH}_2\text{PO}_4] = (V/2V) \times 0.20 \text{mol} \cdot \text{L}^{-1} = 0.10 \text{mol} \cdot \text{L}^{-1}$

$$K_{a2} = \frac{[H_3O^+][HPO_4^{2-}]}{[H_2PO_4^-]} \qquad [H_3O^+] = 6.1 \times 10^{-8} \text{mol} \cdot \text{L}^{-1}$$

pH = 7.21

(2) HCl(aq) + Na$_2$CO$_3$(aq) = NaHCO$_3$(aq) + NaCl(aq)

Initial: $V \times 0.10 \text{mol} \cdot \text{L}^{-1}$ $V \times 0.20 \text{mol} \cdot \text{L}^{-1}$

after reaction: $V \times 0.10 \text{mol} \cdot \text{L}^{-1}$ $V \times 0.10 \text{mol} \cdot \text{L}^{-1}$

In solution, $[\text{Na}_2\text{CO}_3] = [\text{NaHCO}_3] = (V/2V) \times 0.10 \text{mol} \cdot \text{L}^{-1} = 0.050 \text{mol} \cdot \text{L}^{-1}$

$$K_{a2} = \frac{[H_3O^+][CO_3^{2-}]}{[HCO_3^-]} \qquad [H_3O^+] = 4.7 \times 10^{-11} \text{mol} \cdot \text{L}^{-1}$$

pH = 10.32

17. (1) before reaction,

$n(\text{H}_3\text{PO}_4) = 0.10 \text{mol} \cdot \text{L}^{-1} \times 1.0 \text{L} = 0.10 \text{mol}$

$n(\text{NaOH}) = 6.0 \text{g} / (40 \text{g} \cdot \text{mol}^{-1}) = 0.15 \text{mol}$

$n(\text{NaOH}) - n(\text{H}_3\text{PO}_4) = 0.15 \text{mol} - 0.10 \text{mol} = 0.05 \text{mol}$

Reaction	H$_3$PO$_4$(aq) + NaOH(aq) = NaH$_2$PO$_4$(aq) + H$_2$O(l)	
Initial /mol	0.10	0.15
At equilibrium /mol		0.15 − 0.10
	= 0.05	0.10

Reaction continued $NaH_2PO_4(aq) + NaOH(aq) = Na_2HPO_4(aq) + H_2O(l)$

Initial/mol   0.10   0.05

At equilibrium/mol   0.10 − 0.05
                     = 0.05   0.05

So, at equilibrium $[Na_2HPO_4] = [NaH_2PO_4] = 0.050 \text{mol} \cdot L^{-1}$

From $K_{a2} = \dfrac{[H_3O^+][HPO_4^{2-}]}{[H_2PO_4^-]}$ get $[H_3O^+] = 6.1 \times 10^{-8} \text{mol} \cdot L^{-1}$

$$pH = 7.21$$

(2) $\Pi = \sum icRT = [c(HPO_4^{2-}) + c(H_2PO_4^-) + c(Na^+)]RT$

$= (0.050 + 0.050 + 3 \times 0.050) \text{mol} \cdot L^{-1} \times 8.314 J \cdot mol^{-1} \cdot K^{-1} \times (273 + 37) K \times \dfrac{1 kPa \cdot L}{1J}$

$= 644 kPa$

(3) The osmolarity of the solution is

$18 g \cdot L^{-1} / (180.2 g \cdot mol^{-1}) + (0.050 + 0.050 + 3 \times 0.050) \text{mol} \cdot L^{-1} = 0.35 \text{mol} \cdot L^{-1}$

It is a hypertonic solution compared with normal blood.

[Supplementary Exercises]

1. $c(HPr) = \dfrac{0.40 \text{mol} \cdot L^{-1} \times 125.0 mL}{500.0 mL} = 0.10 \text{mol} \cdot L^{-1}$

    $[H_3O^+] = \sqrt{K_a c} = \sqrt{1.3 \times 10^{-5} \times 0.10} \text{mol} \cdot L^{-1} = 1.1 \times 10^{-3} \text{mol} \cdot L^{-1}$

    $pH = 2.94$

2. $pH = 11.86 \quad pOH = 14.00 - 11.86 = 2.14 \quad [OH^-] = 7.2 \times 10^{-3} \text{mol} \cdot L^{-1}$

    $K_b(CH_3CH_2NH_2) = \dfrac{[CH_3CH_2NH_3^+][OH^-]}{[CH_3CH_2NH_2]} = \dfrac{(7.2 \times 10^{-3})^2}{0.10} = 5.2 \times 10^{-4}$

    $K_a(CH_3CH_2NH_3^+) = \dfrac{1.0 \times 10^{-14}}{5.2 \times 10^{-4}} = 1.9 \times 10^{-11}$

3. $K_a(HPi) = \dfrac{[H_3O^+][Pi^-]}{[HPi]} = \dfrac{(1.0 \times 10^{-3})^2}{0.100} = 1.0 \times 10^{-5}$

    $K_b(NaPi) = K_w / K_a(HPi) = 1.0 \times 10^{-14} / (1.0 \times 10^{-5}) = 1.0 \times 10^{-9}$

    $[OH^-] = \sqrt{K_b \cdot c_b} = \sqrt{1.0 \times 10^{-9} \times 0.100} \text{mol} \cdot L^{-1} = 1.0 \times 10^{-5} \text{mol} \cdot L^{-1}$,

    $pH = 9.00$

4. (1) $K_a(HA) = c\alpha^2 = 0.086 \times (3.2 \times 10^{-2})^2 = 8.8 \times 10^{-5}$

    (2) $pH = 2.48 \quad [H_3O^+] = 3.3 \times 10^{-3} \text{mol} \cdot L^{-1}$

    $[HA] = \dfrac{[H_3O^+]^2}{K_a} = \dfrac{(3.3 \times 10^{-3})^2}{8.8 \times 10^{-5}} \text{mol} \cdot L^{-1} = 0.12 \text{mol} \cdot L^{-1}$

Chapter 6   Buffer Solution

[Exercises]

3. (1), (2), (4) and (5)

4. C_5H_5N reacts with HCl to produce $C_5H_5NH^+Cl^-$. The reaction is

$$C_5H_5N + HCl \rightleftharpoons C_5H_5NH^+Cl^-$$

The residual C_5H_5N and $C_5H_5NH^+$ make up a buffer pair.

$$c(C_5H_5N) = (0.30 - 0.10) \text{ mol} \cdot L^{-1} / 2 = 0.10 \text{mol} \cdot L^{-1}$$

$$c(C_5H_5NH^+) = 0.10 \text{mol} \cdot L^{-1} / 2 = 0.050 \text{mol} \cdot L^{-1}$$

For $C_5H_5NH^+$: $pK_a = 14.00 - 8.77 = 5.23$

$$pH = pK_a + \lg \frac{c(C_5H_5N)}{c(C_5H_5NH^+)} = 5.23 + \lg \frac{0.10}{0.050} = 5.53$$

5. They are calculated as followings:

$$n(HCO_3^-) = \frac{10.0 \text{g}}{84.0 \text{g} \cdot \text{mol}^{-1}} = 0.119 \text{mol}$$

$$n(CO_3^{2-}) = \frac{10.0 \text{g}}{106 \text{g} \cdot \text{mol}^{-1}} = 0.094 \text{mol}$$

$$pH = pK_a + \lg \frac{n(CO_3^{2-})}{n(HCO_3^-)} = 10.33 + \lg \frac{0.094 \text{mol}}{0.119 \text{mol}} = 10.23$$

6. Assume $c(HCOONa) = x \text{ mol} \cdot L^{-1}$, $c(HCOOH) = 0.400 \text{mol} \cdot L^{-1} - x \text{ mol} \cdot L^{-1}$

$$pH = 3.75 + \lg \frac{x \text{ mol} \cdot L^{-1}}{(0.400 - x) \text{mol} \cdot L^{-1}} = 3.90$$

$$c(HCOO^-) = x \text{ mol} \cdot L^{-1} = 0.234 \text{mol} \cdot L^{-1}$$

$$c(HCOOH) = (0.400 - 0.234) \text{ mol} \cdot L^{-1} = 0.166 \text{mol} \cdot L^{-1}$$

7.
$$c(NaOH) = \frac{0.20 \text{g} / 40 \text{g} \cdot \text{mol}^{-1}}{100 \text{mL}} \times \frac{1\,000 \text{mL}}{1 \text{L}} = 0.050 \text{mol} \cdot L^{-1}$$

After adding NaOH,

$$pH = 5.30 + \lg \frac{c(B^-) + 0.050 \text{mol} \cdot L^{-1}}{(0.25 - 0.050) \text{mol} \cdot L^{-1}} = 5.60$$

$$c(B^-) = 0.35 \text{mol} \cdot L^{-1}$$

In initial buffer solution,

$$pH = 5.30 + \lg \frac{0.35 \text{mol} \cdot L^{-1}}{0.25 \text{mol} \cdot L^{-1}} = 5.45$$

8. $$pH = pK_a + \lg \frac{n(Asp^-)}{n(HAsp)} = 3.48 + \lg \frac{n(Asp^-)}{n(HAsp)} = 2.95$$

$$\frac{n(Asp^-)}{n(HAsp)} = 0.295 \qquad n(Asp^-) + n(HAsp) = \frac{0.65 \text{g}}{180.2 \text{g} \cdot \text{mol}^{-1}} = 0.003\,6 \text{mol}$$

$n(HAsp) = 0.002\,8 \text{mol} \qquad m(HAsp) = 0.002\,8 \text{mol} \times 180.2 \text{g} \cdot \text{mol}^{-1} = 0.50 \text{g}$

9. Assume $V(HAc) = 3V$, $V(NaOH) = V$

$$c(HAc) = (0.10 \times 3V - 0.10 \times V) \text{ mol} \cdot L^{-1} / (3V + V) = 0.050 \text{mol} \cdot L^{-1}$$

$$c(Ac^-) = 0.10 \text{mol} \cdot L^{-1} \times V / (3V + V) = 0.025 \text{mol} \cdot L^{-1}$$

$$pH = 4.756 + \lg \frac{0.025 \text{mol} \cdot L^{-1}}{0.050 \text{mol} \cdot L^{-1}} = 4.45$$

10. H_2Bar reacts with NaOH to produce NaHBar. The reaction as following:

$$H_2Bar(aq) + NaOH(aq) \rightleftharpoons NaHBar(aq) + H_2O(l)$$

The residual H_2Bar and NaHBar make up a buffer pair.

$$n(NaHBar) = c(NaOH)V(NaOH) = 6.00 \text{mol} \cdot L^{-1} \times 4.17 \text{mL} = 25 \text{mmol}$$

$$n(H_2Bar) = n(H_2Bar) - n(NaOH) = \frac{18.4g}{184g \cdot mol^{-1}} \times 1\,000 - 25mmol = 75mmol$$

$$pH = pK_a + \lg\frac{n(HBar^-)}{n(H_2Bar)} = 7.43 + \lg\frac{25mmol}{75mmol} = 6.95$$

11.

Solution	Buffer system	Anti-acid component.	Anti-base component	Effective buffer range	Volume ratio at β_{max}
$Na_2HCit + HCl$	$H_2Cit^- - HCit^{2-}$	$HCit^{2-}$	H_2Cit^-	3.76~5.76	2:1
$Na_2HCit + HCl$	$H_3Cit - H_2Cit^-$	H_2Cit^-	H_3Cit	2.13~4.13	2:3
$Na_2HCit + NaOH$	$HCit^{2-} - Cit^{3-}$	Cit^{3-}	$HCit^{2-}$	5.40~7.40	2:1

12. (1) $V(HCl) = 50mL$

(2) $HCl(aq) + NH_3 \cdot H_2O(aq) \rightleftharpoons NH_4Cl(aq) + H_2O(l)$

For NH_4^+, $pK_a = 14.00 - 4.75 = 9.25$

$$7.00 = 9.25 + \lg\frac{0.10mol \cdot L^{-1} \times 50mL - 0.10mol \cdot L^{-1} \times V(HCl)}{0.10mol \cdot L^{-1} \times V(HCl)}$$

$$V(HCl) = 49.7mL$$

(3) $HCl(aq) + Na_2HPO_4(aq) \rightleftharpoons NaH_2PO_4(aq) + NaCl(aq)$

For H_3PO_4, $pK_{a2} = 7.21$

$$7.00 = 7.21 + \lg\frac{0.10mol \cdot L^{-1} \times 50mL - 0.10mol \cdot L^{-1} \times V(HCl)}{0.10mol \cdot L^{-1} \times V(HCl)}$$

$$V(HCl) = 31mL$$

13. Assume $m(NH_4Cl) = x$ g

For NH_4^+, $pK_a = 14.00 - 4.75 = 9.25$

$$\frac{x\,g / 53.5g \cdot mol^{-1}}{1L} = 0.125mol \cdot L^{-1}$$

$$x = m(NH_4Cl) = 6.69g$$

The residual NH_4^+ and NH_3 make up a buffer pair.

$$9.00 = 9.25 + \lg\frac{1.00mol \cdot L^{-1} \times V(NaOH)}{6.69/53.5g \cdot mol^{-1} - 1.00mol \cdot L^{-1} \times V(NaOH)}$$

$$V(NaOH) = 0.045L = 45mL$$

14. According to preparing rule, $H_2PO_4^- - HPO_4^{2-}$ buffer system is the suitable conjugate acid-base pair. The reactions that occur when H_3PO_4 and NaOH solution are mixed include two steps:

(1) $H_3PO_4(aq) + NaOH(aq) = NaH_2PO_4(aq) + H_2O(l)$

In this step, H_3PO_4 is neutralized completely to produce NaH_2PO_4, which need x mL NaOH solution and x mL H_3PO_4 solution

The amount of NaH_2PO_4 produced is as follows:

$$n(NaH_2PO_4) = 0.020mol \cdot L^{-1} \times x\,mL = 0.020x\,mmol$$

(2) $NaH_2PO_4(aq) + NaOH(aq) = Na_2HPO_4(aq) + H_2O(l)$

In this step, NaH_2PO_4 is neutralized partly to form Na_2HPO_4, which need y mL NaOH solution.

The amount of Na_2HPO_4 produced: $n(Na_2HPO_4) = 0.020y\,mmol$

The amount of surplus NaH_2PO_4: $n(NaH_2PO_4) = 0.020x - 0.020y$ mmol $= 0.020(x-y)$ mmol

$$7.40 = 7.21 + \lg \frac{0.020y \text{ mmol}}{0.020(x-y)\text{mmol}}$$

$$\frac{y}{x-y} = 1.55$$

Because
$$2x + y = 100\text{mL}$$
$$x = 38.4 \quad y = 23.2$$

That is, $V(H_3PO_4) = 38.4$mL, $V(NaOH)_{total} = (38.4 + 23.2)$ mL $= 61.6$mL.

15. (1)

$$7.40 = 8.3 + \lg \frac{n(\text{Tris})}{n(\text{Tris} \cdot \text{HCl})}$$

$$7.40 = 8.3 + \lg \frac{0.050 \text{mol} \cdot \text{L}^{-1} \times 100\text{mL} - 0.050\text{mol} \cdot \text{L}^{-1} \times V(\text{HCl})}{0.050\text{mol} \cdot \text{L}^{-1} \times 100\text{mL} + 0.050\text{mol} \cdot \text{L}^{-1} \times V(\text{HCl})}$$

$$0.13 = \frac{100 - V(\text{HCl})}{100 + V(\text{HCl})}$$

$$V(\text{HCl}) = 77\text{mL}$$

(2) Assume after adding x g NaCl, c_{os} of blood is 300mmol $\cdot$ L^{-1}.

$$c(\text{Tris}) = \frac{0.050\text{mol} \cdot \text{L}^{-1} \times 100\text{mL} - 0.050\text{mol} \cdot \text{L}^{-1} \times 77\text{mL}}{177\text{mL}} = 0.006\,5\text{mol} \cdot \text{L}^{-1}$$

$$c(\text{Tris} \cdot \text{HCl}) = \frac{0.050\text{mol} \cdot \text{L}^{-1} \times 100\text{mL} + 0.050\text{mol} \cdot \text{L}^{-1} \times 77\text{mL}}{177\text{mL}} = 0.050\text{mol} \cdot \text{L}^{-1}$$

$$(0.006\,5 + 2 \times 0.050)\text{mol} \cdot \text{L}^{-1} + \frac{2x \text{ g} \times 1\,000\text{mL} \cdot \text{L}^{-1}}{58.5\text{g} \cdot \text{mol}^{-1} \times 177\text{mL}} = 0.300\text{mol} \cdot \text{L}^{-1}$$

$$x = 1.0\text{g}$$

16. $$\text{pH} = \text{p}K_{a1} + \lg \frac{[\text{HCO}_3^-]}{[\text{CO}_2(\text{aq})]} = 6.10 + \lg \frac{24.0\text{mmol} \cdot \text{L}^{-1} \times 0.90}{1.20\text{mmol} \cdot \text{L}^{-1}} = 7.36$$

It does not result in acidosis.

[Supplementary Exercises]

3. (a) HAc and Ac$^-$ (b) $H_2PO_4^-$ and HPO_4^{2-} (c) NH_4^+ and NH_3

4. Choose (d) and (e). Buffer capacity depends on both the concentration of the reservoirs and the buffer-component ratio. The more concentrated the components of a buffer, the greater the buffer capacity. When the component ratio is close to one, a buffer is most effective.

5. $$c(\text{CH}_3\text{COO}^-) + c(\text{CH}_3\text{COOH}) = 0.150\text{mol} \cdot \text{L}^{-1}$$

$$n(\text{CH}_3\text{COO}^-) + n(\text{CH}_3\text{COOH}) = 0.150\text{mol} \cdot \text{L}^{-1} \times 500\text{mL} \times \frac{1\text{L}}{1\,000\text{mL}} = 0.075\,0\text{mol}$$

$$5.00 = 4.756 + \lg \frac{n(\text{CH}_3\text{COO}^-)}{n(\text{CH}_3\text{COOH})} \quad \frac{n(\text{CH}_3\text{COO}^-)}{n(\text{CH}_3\text{COOH})} = 1.75$$

$$n(\text{CH}_3\text{COOH}) = 0.027\,0\text{mol}, \; n(\text{CH}_3\text{COO}^-) = 0.048\,0\text{mol}$$

Mass of sodium acetate $= 0.048\,0\text{mol} \times 136.1\text{g} \cdot \text{mol}^{-1} = 6.53\text{g}$

$$V(\text{CH}_3\text{COOH}) = \frac{0.027\,0\text{mol} \times 10^3 \text{mL} \cdot \text{L}^{-1}}{17.45\text{mol} \cdot \text{L}^{-1}} = 1.55\text{mL}$$

6.

$$7.40 = 6.80 + \lg\frac{[HPO_4^{2-}]}{[H_2PO_4^-]} \quad \frac{[HPO_4^{2-}]}{[H_2PO_4^-]} = 3.98 \approx 4.00$$

Chapter 7 Equilibria of Slightly Soluble Ionic Compounds

[Exercises]

1. The solubility product constant K_{sp} is the equilibrium constant for insoluble strong electrolyte called precipitation generally in aqueous solution. The product of the ionic concentrations to the power of the stoichiometric coefficients is called ion product I_p.

The Relationship between K_{sp} and I_p is as follows:

(1) If $I_p = K_{sp}$, the solution is saturated, then the solution is at equilibrium.

(2) If $I_p < K_{sp}$, the solution is undersaturated and the solute will continue to dissolve.

(3) If $I_p > K_{sp}$, the solution is supersaturated and precipitation will occur.

2. The solubility of $BaSO_4$ was increased in normal saline by salt effect, while the solubility of AgCl was decreased in normal saline because of common ion effect.

3.
$$NH_3 \cdot H_2O \rightleftharpoons NH_4^+ + OH^-$$
$$Mg^{2+} + 2OH^- \rightleftharpoons Mg(OH)_2 \downarrow$$

overall reaction $\quad Mg^{2+} + 2NH_3 \cdot H_2O \rightleftharpoons Mg(OH)_2 \downarrow + 2NH_4^+$

When NH_4Cl solution was added to the solution, the dissociation of ammonia was inhibited and the concentration of OH^- in the solution decreased. $I_p\{Mg(OH)_2\} < K_{sp}\{Mg(OH)_2\}$, so the precipitation dissolved.

4. (1) HCl and (2) $AgNO_3$, due to the common ion effect, will lower the solubility of AgCl; But if the concentration of HCl is high enough, Ag^+ and Cl^- can form complex ion $[AgCl_2]^-$ when the solubility will be decreased instead.

(3) KNO_3, due to salt effect, will decrease the solubility of AgCl slightly.

(4) $NH_3 \cdot H_2O$ will increase the solubility of AgCl greatly because NH_3 and Ag^+ can form complex ion which will greatly decrease the concentration of Ag^+ ion.

5. The reaction is as follows: $\quad Zn^{2+} + H_2S \rightleftharpoons ZnS \downarrow + 2H^+$

With the reaction, the concentration of H^+ in the solution increased gradually, the reverse reaction increased gradually, which make ZnS precipitation cannot completely. When NaAc was added, the weak electrolyte HAc was formed, and the H^+ concentration in the solution was decreased, reaction shifted the right, the precipitation is complete.

6.
$$Mn(OH)_2(s) \rightleftharpoons Mn^{2+}(aq) + 2OH^-(aq)$$
$$[Mn^{2+}][OH^-]^2 = K_{sp}$$

(1) $S(2S)^2 = K_{sp}, \quad S = \sqrt[3]{\dfrac{K_{sp}}{4}} = \sqrt[3]{\dfrac{2.06 \times 10^{-13}}{4}} \text{ mol} \cdot L^{-1} = 3.72 \times 10^{-5} \text{ mol} \cdot L^{-1}$

(2) $[OH^-] = 0.10 \text{mol} \cdot L^{-1}$

$$S = [Mn^{2+}] = \frac{K_{sp}}{[OH^-]^2} = \frac{2.06 \times 10^{-13}}{(0.10)^2} \text{ mol} \cdot L^{-1} = 2.1 \times 10^{-11} \text{ mol} \cdot L^{-1}$$

(3) $[Mn^{2+}] = 0.20 \text{mol} \cdot L^{-1}$

$$S = \sqrt{\frac{K_{sp}}{4[Mn^{2+}]}} = \sqrt{\frac{2.06 \times 10^{-13}}{4 \times 0.20}} \text{mol} \cdot L^{-1} = 5.1 \times 10^{-7} \text{mol} \cdot L^{-1}$$

7. When AgCl begins to precipitate:

$$[Ag^+]_{AgCl} = \frac{K_{sp}(AgCl)}{[Cl^-]} = \frac{1.77 \times 10^{-10}}{0.010} \text{mol} \cdot L^{-1} = 1.77 \times 10^{-8} \text{mol} \cdot L^{-1}$$

When Ag_2CrO_4 begins to precipitate:

$$[Ag^+]_{Ag_2CrO_4} = \sqrt{\frac{K_{sp}(Ag_2CrO_4)}{[CrO_4^{2-}]}} = \sqrt{\frac{1.12 \times 10^{-12}}{0.010}} \text{mol} \cdot L^{-1} = 1.06 \times 10^{-5} \text{mol} \cdot L^{-1}$$

Precipitating Cl^- needs less $[Ag^+]$, so AgCl precipitates firstly. When $[Ag^+] = 1.1 \times 10^{-5}$ mol·L^{-1}, Ag_2CrO_4 begins to precipitate, when the concentration of Cl^- left in solution is:

$$[Cl^-] = \frac{K_{sp}(AgCl)}{[Ag^+]} = \frac{1.77 \times 10^{-10}}{1.1 \times 10^{-5}} \text{mol} \cdot L^{-1} = 1.67 \times 10^{-5} \text{mol} \cdot L^{-1}$$

8.

$$[Ca^{2+}] = \frac{0.10}{1.4 \times 40} = 1.8 \times 10^{-3} \text{mol} \cdot L^{-1}$$

$$[PO_4^{3-}] \leq \sqrt{\frac{2.07 \times 10^{-33}}{(1.8 \times 10^{-3})^3}} = 5.9 \times 10^{-13} \text{mol} \cdot L^{-1}$$

9. The required S^{2-} concentration for precipitation of PbS is

$$[S^{2-}] = \frac{K_{sp}(PbS)}{[Pb^{2+}]} = \frac{8.0 \times 10^{-28}}{0.10} \text{mol} \cdot L^{-1} = 8.0 \times 10^{-27} \text{mol} \cdot L^{-1}$$

When FeS does not produce precipitation

$$[S^{2-}] = \frac{K_{sp}(FeS)}{[Fe^{2+}]} = \frac{6.3 \times 10^{-18}}{0.10} \text{mol} \cdot L^{-1} = 6.3 \times 10^{-17} \text{mol} \cdot L^{-1}$$

Therefore, in order to make Pb^{2+} form PbS precipitation, the concentration of S^{2-} in solution should be controlled between $8.0 \times 10^{-27} \sim 6.3 \times 10^{-17}$ mol·L^{-1}. When FeS begins to precipitate, the residual concentration of Pb^{2+} in the solution is:

$$[Pb^{2+}] = \frac{K_{sp}(PbS)}{[S^{2-}]} = \frac{8.0 \times 10^{-28}}{6.3 \times 10^{-17}} \text{mol} \cdot L^{-1} = 1.3 \times 10^{-11} \text{mol} \cdot L^{-1} \ll 1.0 \times 10^{-5} \text{mol} \cdot L^{-1}$$

At this time, the PbS has been completely precipitated. Pb^{2+} and Fe^{2+} can be separated completely by means of fractional precipitation.

10. (1) $[F^-] = \sqrt{\frac{K_{sp}(CaF_2)}{[Ca^{2+}]}} = \sqrt{\frac{3.45 \times 10^{-11}}{0.0020}} = 1.3 \times 10^{-4} \text{mol} \cdot L^{-1}$

(2) The concentration of F^- mentioned above is

$$1.3 \times 10^{-4} \text{mol} \cdot L^{-1} \times 19 \text{g} \cdot \text{mol}^{-1} = 2.47 \times 10^{-3} \text{g} \cdot L^{-1} = 2.47 \text{mg} \cdot L^{-1}$$

It is more than 1mg·L^{-1}, so the concentration of F^- in fluorinated water is exceeding the standard.

11. If the precipitation of $Fe(OH)_3$ began to occur,

$$[OH^-] = \sqrt[3]{\frac{K_{sp}\{Fe(OH)_3\}}{[Fe^{3+}]}} = \sqrt[3]{\frac{2.79 \times 10^{-39}}{0.010}} \text{mol} \cdot L^{-1} = 6.54 \times 10^{-13} \text{mol} \cdot L^{-1}$$

$$pH = 1.82$$

When Fe(OH)$_3$ completely, $[Fe^{3+}] = 1.0 \times 10^{-5}$ mol·L^{-1}

$$[OH^-] = \sqrt[3]{\frac{K_{sp}\{Fe(OH)_3\}}{[Fe^{3+}]}} = \sqrt[3]{\frac{2.79 \times 10^{-39}}{1 \times 10^{-5}}} \text{ mol·L}^{-1} = 6.54 \times 10^{-12} \text{ mol·L}^{-1}$$

$$pH = 2.82$$

If Mg(OH)$_2$ will not occur, $Ip\{Mg(OH)_2\} \leqslant K_{sp}\{Mg(OH)_2\}$

$$[OH^-] = \sqrt{\frac{K_{sp}\{Mg(OH)_2\}}{[Mg^{2+}]}} = \sqrt{\frac{5.61 \times 10^{-12}}{0.010}} \text{ mol·L}^{-1} = 2.37 \times 10^{-5} \text{ mol·L}^{-1}$$

$$pH = 9.37$$

So, in order to separate Fe^{3+} ion and Mg^{2+} ion, control the pH in the range of 2.82~9.37.

[Supplementary Exercises]

1. (1) $K_{sp}(PbCrO_4) = [Pb^{2+}][CrO_4^{2-}] = S^2$

$$S = \sqrt{K_{sp}} = \sqrt{2.8 \times 10^{-13}} \text{ mol·L}^{-1} = 5.3 \times 10^{-7} \text{ mol·L}^{-1}$$

(2) $PbCrO_4(s) \rightleftharpoons Pb^{2+}(aq) + CrO_4^{2-}(aq)$

$$S \qquad S + 0.001 \approx 0.001$$

$$K_{sp}(PbCrO_4) = [Pb^{2+}][CrO_4^{2-}] = S \times 0.001$$

$$S = \frac{K_{sp}(PbCrO_4)}{0.001} \text{ mol·L}^{-1} = \frac{2.8 \times 10^{-13}}{0.001} \text{ mol·L}^{-1} = 2.8 \times 10^{-10} \text{ mol·L}^{-1}$$

3. $AgCl(s) \rightleftharpoons Ag^+(aq) + Cl^-(aq)$

$$K_{sp}(AgCl) = [Ag^+][Cl^-] = S^2$$

$$S = \sqrt{K_{sp}} = \sqrt{1.77 \times 10^{-10}} \text{ mol·L}^{-1} = 1.33 \times 10^{-5} \text{ mol·L}^{-1}$$

Solubility (g·L^{-1}) = 1.33×10^{-5} mol·L^{-1} × 143.4 g·mol^{-1} = 1.91×10^{-3} g·L^{-1}

$$Ag_2CrO_4(s) \rightleftharpoons 2Ag^+(aq) + CrO_4^{2-}(aq)$$

$$K_{sp}(Ag_2CrO_4) = [Ag^+]^2[CrO_4^{2-}] = 4S^3$$

$$S = \sqrt[3]{\frac{K_{sp}}{4}} = \sqrt[3]{\frac{1.12 \times 10^{-12}}{4}} \text{ mol·L}^{-1} = 6.54 \times 10^{-5} \text{ mol·L}^{-1}$$

Solubility (g·L^{-1}) = 6.54×10^{-5} mol·L^{-1} × 331.6 g·mol^{-1} = 2.17×10^{-2} g·L^{-1}, So, Ag$_2$CrO$_4$ has both the higher molar and gram solubility.

4. $c(M^+) = c(X^-) = \dfrac{\Pi}{iRT} = \dfrac{100 \text{kPa} \times 75.4 \text{mmHg}/760 \text{mmHg}}{2 \times 8.314 \text{kPa·L·K}^{-1}\text{·mol}^{-1} \times 298 \text{K}} = 2.00 \times 10^{-3}$ mol·L^{-1}

$$K_{sp}(MX) = [M^+][X^-] = c(M^+)c(X^-) = (2.00 \times 10^{-3})^2 = 4.00 \times 10^{-6}$$

5. The dissolution of Ca(OH)$_2$ is described by the equation

$$Ca(OH)_2(s) \rightleftharpoons Ca^{2+}(aq) + 2OH^-(aq)$$

$$K_{sp} = [Ca^{2+}][OH^-]^2 = 4S^3 = 5.02 \times 10^{-6}$$

$$S = 1.08 \times 10^{-2} \text{ mol·L}^{-1}, \quad [OH^-] = 2S = 2.16 \times 10^{-2} \text{ mol·L}^{-1}$$

$$pOH = -\lg[OH^-] = 1.67 \qquad pH = 14.00 - pOH = 12.33$$

6. $\quad CaC_2O_4(s) \rightleftharpoons Ca^{2+}(aq) + C_2O_4^{2-}(aq)$

$$3.00 \times 10^{-8} \qquad x$$

$$K_{sp}(CaC_2O_4) = [Ca^{2+}][C_2O_4^{2-}] = 3.00 \times 10^{-8} \times x = 2.32 \times 10^{-9}$$

$$x = 0.0773 \text{ mol·L}^{-1}$$

7.
$$BaSO_4(s) \rightleftharpoons Ba^{2+}(aq) + SO_4^{2-}(aq)$$
$$K_{sp}(BaSO_4) = [Ba^{2+}][SO_4^{2-}] = (1.04 \times 10^{-5})^2 = 1.08 \times 10^{-10}$$

8.
$$Mn(OH)_2(s) \rightleftharpoons Mn^{2+}(aq) + 2OH^-(aq)$$
$$K_{sp} = [Mn^{2+}][OH^-]^2$$
$$2.06 \times 10^{-13} = (1.8 \times 10^{-6})[OH^-]^2$$

$[OH^-] = 3.38 \times 10^{-4}\ mol \cdot L^{-1}$ $pOH = -lg[OH^-] = 3.47$ $pH = 14.00 - pOH = 10.53$

Chapter 8 Oxidation-Reduction Reaction and Electrode Potential

[Exercises]

1. $+6; +2; +4; +4; +5; -1; -1; +6$

2. (1) $2MnO_4^-(aq) + 5H_2O_2(aq) + 6H^+(aq) = 2Mn^{2+}(aq) + 5O_2(g) + 8H_2O(l)$

   (2) $Cr_2O_7^{2-}(aq) + 3SO_3^{2-}(aq) + 8H^+(aq) = 2Cr^{3+}(aq) + 3SO_4^{2-}(aq) + 4H_2O(l)$

   (3) $As_2S_3(s) + 5ClO_3^-(aq) + 5H_2O(l) = 5Cl^-(aq) + 2H_3AsO_4(sln) + 3SO_4^{2-}(aq) + 10H^+(aq)$

3. According to the standard electrode potentials, $\varphi^{\ominus}(Cl_2/Cl^-) = 1.358V$, $\varphi^{\ominus}(H_2O_2/H_2O) = 1.776V$, chlorine and hydrogen peroxide are all strong oxidizing agents, and they can easily oxidize many reducing substances as sanitizers.

4. (1) oxidizing ability: $Zn^{2+} < Fe^{3+} < MnO_2 < Cr_2O_7^{2-} < Cl_2 < MnO_4^-$

   (2) reducing ability: $Cl^- < Cr^{3+} < Fe^{2+} < H_2 < Li$

5. (1) $(-)\ Zn(s)|Zn^{2+}(aq)\|Ag^+(aq)|Ag(s)\ (+)$

   The cell reaction will proceed spontaneously as written.

   (2) $(-)\ Pt|Cr^{3+}(aq), Cr_2O_7^{2-}(aq), H^+(aq)\|Cl^-(aq)|Cl_2(g),Pt\ (+)$

   The cell reaction will proceed spontaneously as written.

   (3) $(-)\ Pt|Fe^{2+}(aq), Fe^{3+}(aq)\|IO_3^-(aq),H^+(aq)|I_2(s),Pt\ (+)$

   The cell reaction will proceed spontaneously in the opposite direction.

6. The redox electric couple, whose electrode potential value is between the electrode potential values of the two given redox electric couples, is in accord with the conditions.

7.
Cathode: $H_2O_2(aq) + 2H^+(aq) + 2e^- \rightarrow 2H_2O(l)$
Anode: $H_2O_2(aq) - 2e^- \rightarrow O_2(g) + 2H^+(aq)$
Cell reaction: $2H_2O_2(aq) \rightarrow O_2(g) + 2H_2O(l)$
$\therefore E^{\ominus} = 1.776V - 0.695V > 0$

$\therefore H_2O_2$ will spontaneously decompose to H_2O and O_2 under standard conditions.

8. $\because \varphi^{\ominus}(Cr_2O_7^{2-}/Cr^{3+}) = 1.36V$, $\varphi^{\ominus}(Br_2/Br^-) = 1.066V$

(1) $\varphi = \varphi^{\ominus} + \dfrac{0.05916V}{2} \lg \dfrac{c(H^+)^2}{p_{H_2}/p^{\ominus}} = 0.00000 + \dfrac{0.05916}{2} \lg \dfrac{0.1^2}{200/100} = -0.068V$

(2) $\varphi = \varphi^{\ominus} + \dfrac{0.05916V}{6} \lg \dfrac{c(H^+)^{14} c(Cr_2O_7^{2-})}{c(Cr^{3+})^2} = 1.36 + \dfrac{0.05916}{6} \lg \dfrac{0.001^{14} \times 1.0}{1.0^2} = 0.946V$

(3) $\varphi = \varphi^{\ominus} + \dfrac{0.05916V}{2} \lg \dfrac{1}{c(Br^-)^2} = 1.066 + \dfrac{0.05916}{2} \lg \dfrac{1}{0.2^2} = 1.107V$

9. (1) $\because \varphi^{\ominus}(MnO_4^-/Mn^{2+}) = 1.507V$, $\varphi^{\ominus}(Br_2/Br^-) = 1.066V$, $\varphi^{\ominus}(I_2/I^-) = 0.535V$

∴ When pH = 0.0, the system is at the standard state

∴ MnO_4^- ion can oxidize I^- ion and Br^- ion.

(2) When pH = 5.5,

$$\varphi(MnO_4^-/Mn^{2+}) = \varphi^{\ominus}(MnO_4^-/Mn^{2+}) + \frac{0.059\,16V}{5} lg\frac{c(MnO_4^-)c(H^+)^8}{c(Mn^{2+})}$$

$$= 1.507V - 0.059\,16V \times 5.5 \times 8/5 = 0.986V$$

∴ MnO_4^- ion can oxidize I^- ion only but cannot oxidize Br^- ion.

10. (1) $E^{\ominus} = 1.358V - 0.954V = 0.404V$

$\Delta_r G_m^{\ominus} = -2 \times 96\,500C \cdot mol^{-1} \times 0.404V = -77\,972J \cdot mol^{-1}$

$lg\,K^{\ominus} = 2 \times 0.404V/0.059\,16V$ $K^{\ominus} = 4.5 \times 10^{13}$

(2) $6ClO_2(g) + 3H_2O \rightleftharpoons 5ClO_3^-(aq) + Cl^-(aq) + 6H^+$

11. (1) Under the standard conditions, when Co metal dissolves in $1.0 mol \cdot L^{-1}$ HNO_3 solution, it will be oxidized to Co^{2+} ion.

(2) The changes of concentration of HNO_3 solution cannot change above conclusion for $\Delta\varphi > 0.3V$.

12. $\varphi(H^+/H_2) = \varphi^{\ominus}(H^+/H_2) + \frac{0.059\,16V}{2} lg\frac{c(H^+)^2}{p_{H_2}/100} = -0.059\,16V(-pH)$

∵ $E = \varphi(SCE) - \varphi(H^+/H_2) = 0.241\,2V - 0.059\,16V(-pH) = 0.420V$

∴ pH = 3.02

13. The oxidizing abilities of $Cr_2O_7^{2-}$, MnO_4^- and H_2O_2 will be strengthened, and the oxidizing abilities of Hg_2^{2+}, Cl_2 and Cu^{2+} will be maintained.

14. $\varphi_{right} = \varphi^{\ominus}(Cu^{2+}/Cu) + \frac{0.059\,16V}{2} lg(1.0 \times 10^{-1}) = \varphi^{\ominus}(Cu^{2+}/Cu) - 0.029\,6V$

$\varphi_{left} = \varphi^{\ominus}(Cu^{2+}/Cu) + \frac{0.059\,16V}{2} lg(1.0 \times 10^{-4}) = \varphi^{\ominus}(Cu^{2+}/Cu) - 0.118\,4V$

∴ $Cu^{2+}(1.0 \times 10^{-4} mol \cdot L^{-1})/Cu$ is the cathode of the cell and $Cu^{2+}(1.0 \times 10^{-1} mol \cdot L^{-1})/Cu$ is the anode of the cell.

∴ $E = 0.118\,4V - 0.029\,6V = 0.088\,8V$

15.   ∵ $\varphi^{\ominus}(Cd^{2+}/Cd) = -0.403V$; $\varphi^{\ominus}(Zn^{2+}/Zn) = -0.762V$

∴ $E^{\ominus} = \varphi^{\ominus}(Cd^{2+}/Cd) - \varphi^{\ominus}(Zn^{2+}/Zn) = -0.403V - (-0.762V) = 0.359V$

∴ $0.388\,4V = 0.359V - \frac{0.059\,16V}{2} lg\frac{0.2}{c(Zn^{2+})}$

∴ $c(Zn^{2+}) = 0.02 mol \cdot L^{-1}$

16. Design a primary cell with above two electrodes and its cell reaction is

$$Hg_2^{2+} + SO_4^{2-} = Hg_2SO_4(s)$$

∴ $E^{\ominus} = 0.797V - 0.612V = 0.185V$

∵ $lg\,K^{\ominus} = 2 \times E^{\ominus}/0.059\,16$ ∴ $K^{\ominus} = 1.8 \times 10^6$

∴ $K_{sp} = \frac{1}{K^{\ominus}} = 5.6 \times 10^{-7}$

17. $\varphi(Hg_2Cl_2/Hg) = \varphi^{\ominus}(Hg_2Cl_2/Hg) - \frac{0.059\,16V}{2} lg\frac{1}{c(Cl^-)^2}$

$$0.327V = 0.268V - \frac{0.05916V}{2} \lg \frac{1}{c(Cl^-)^2}$$

$$\therefore c(Cl^-) = 0.1 \text{mol} \cdot L^{-1}$$

18.
$$pH = pH_s + \frac{(E-E_s)F}{2.303RT}$$

$$pH = 6.0 + \frac{(0.231V - 0.350V)}{0.05916V} = 4.0$$

$$\therefore [H^+] = 1.0 \times 10^{-4} \text{mol} \cdot L^{-1} \qquad \because c = 0.01 \text{mol} \cdot L^{-1}$$

$$\therefore K_a = \frac{[H^+]^2}{c} = 1.0 \times 10^{-6}$$

[Supplementary Exercises]

1.
$$\varphi^{\ominus}(I_2/I^-) = 0.5355V; \quad \varphi^{\ominus}(Br_2/Br^-) = 1.066V$$

$$\lg K^{\ominus} = nFE^{\ominus}/RT = \frac{2 \times (0.5355V - 1.066V)}{0.05916V} = -17.94$$

$$K^{\ominus} = 1.15 \times 10^{-18}$$

2. $\Delta_r G_m^{\ominus} = -RT\ln K^{\ominus} = -8.314 J \cdot K^{-1} \cdot mol^{-1} \times 298K \times \ln(5 \times 10^3) = -21100 J \cdot mol^{-1}$

$$\because \Delta_r G_m^{\ominus} = -nFE^{\ominus}, n=1 \qquad \therefore E^{\ominus} = -\frac{-21100 J \cdot mol^{-1}}{96500 C \cdot mol^{-1}} = 0.219V$$

3. (1) $3Fe(OH)_2(s) + MnO_4^-(aq) + 2H_2O \rightleftharpoons MnO_2(s) + 3Fe(OH)_3(s) + OH^-(aq)$

$MnO_4^-(aq)$ is the oxidizing agent

(2) $5Zn(s) + 2NO_3^-(aq) + 12H^+ \rightleftharpoons 5Zn^{2+}(aq) + N_2(g) + 6H_2O$

$NO_3^-(aq)$ is the oxidizing agent

4. (1) $\quad (-) Al(s) \mid Al^{3+}(c_1) \parallel Cr^{3+}(c_2) \mid Cr(s) \quad (+)$

(2) $\quad (-) Pt(s) \mid SO_2(g) \mid SO_4^{2-}(c_1), H^+(c_2) \parallel Cu^{2+}(c_3) \mid Cu(s) (+)$

5.
$$\varphi^{\ominus}(Cu^{2+}/Cu) = 0.3419V$$

$$\varphi(Cu^{2+}/Cu) = \varphi^{\ominus}(Cu^{2+}/Cu) + \frac{0.05916V}{2} \lg c(Cu^{2+})$$

$$E = \varphi(Cu^{2+}/Cu) - \varphi(SHE)$$

$$0.25V = 0.3419V + \frac{0.05916V}{2} \lg c(Cu^{2+}) - 0.000V$$

$$c(Cu^{2+}) = 7.8 \times 10^{-4} \text{mol} \cdot L^{-1}$$

6. (1) $\quad \varphi^{\ominus}(Co^{2+}/Co) = -0.28V, \varphi^{\ominus}(Ni^{2+}/Ni) = -0.257V$

$$\varphi(Co^{2+}/Co) = \varphi^{\ominus}(Co^{2+}/Co) + \frac{0.05916V}{2} \lg 0.2 = -0.300V$$

$$\varphi(Ni^{2+}/Ni) = \varphi^{\ominus}(Ni^{2+}/Ni) + \frac{0.05916V}{2} \lg 0.8 = -0.260V$$

The initial $\quad E = -0.260V - (-0.300V) = 0.040V$

Cell reaction: $\quad Ni^{2+}(aq) + Co \rightarrow Co^{2+}(aq) + Ni, \quad n=2$

(2) When $\quad [Co^{2+}] = 0.4 \text{mol} \cdot L^{-1}, [Ni^{2+}] = 0.6 \text{mol} \cdot L^{-1}$

$$\varphi(Co^{2+}/Co) = \varphi^{\ominus}(Co^{2+}/Co) + \frac{0.05916V}{2} \lg 0.4 = -0.292V$$

$$\varphi(Ni^{2+}/Ni) = \varphi^{\ominus}(Ni^{2+}/Ni) + \frac{0.05916V}{2} \lg 0.6 = -0.264V$$

(3)
$$E = -0.264V - (-0.292V) = 0.028V$$
$$E^\ominus = -0.257V - (-0.28V) = 0.023V, \text{ and } n = 2$$
$$\lg K^\ominus = nE^\ominus / 0.059\,16V = \frac{2 \times 0.023V}{0.059\,16V} = 0.778, \quad K^\ominus = 6.0$$

(4)
$$E = E^\ominus - \frac{0.059\,16V}{n} \lg Q = E^\ominus + \frac{0.059\,16V}{2} \lg \frac{c(\text{Ni}^{2+})}{c(\text{Co}^{2+})}$$

$$0.025V = 0.023V + \frac{0.059\,16V}{2} \lg \frac{c(\text{Ni}^{2+})}{c(\text{Co}^{2+})}$$

$$\frac{c(\text{Ni}^{2+})}{c(\text{Co}^{2+})} = 1.17$$

7. hydrogen electrode half-reaction $2\text{H}^+(\text{aq}) + 2\text{e}^- \rightleftharpoons \text{H}_2(\text{g})$, $n = 2$

Electrode A: $\quad \varphi_A = \varphi_A^\ominus + \frac{0.059\,16V}{2} \lg \frac{0.1^2}{0.9} = -0.058V$

Electrode B: $\quad \varphi_B = \varphi_B^\ominus + \frac{0.059\,16V}{2} \lg \frac{2.0^2}{0.5} = 0.026\,7V$

Electrode A is the anode.

$$E = 0.026\,7V - (-0.058V) = 0.084\,7V$$
$$\because E^\ominus = 0.0V, \quad \lg K^\ominus = nE^\ominus/0.059\,16V \quad \therefore K^\ominus = 1.$$

8.
$$\varphi^\ominus(\text{Zn}^{2+}/\text{Zn}) = -0.761\,8V$$
$$\varphi(\text{Zn}^{2+}/\text{Zn}) = \varphi^\ominus(\text{Zn}^{2+}/\text{Zn}) + \frac{0.059\,16V}{2} \lg 0.01 = -0.821V$$
$$\varphi(\text{H}^+/\text{H}_2) = \varphi^\ominus(\text{H}^+/\text{H}_2) + \frac{0.059\,16V}{2} \lg \frac{2.5^2}{0.3} = 0.039V$$
$$E = 0.039V - (-0.821V) = 0.860V$$

Chapter 9 Atomic Structure and Periodic Law

[Exercises]

1. Since an electron with high velocity does not have definite position and momentum simultaneously, it is impossible to depict its motion trail like that of a macroscopic object. The wave-like motion of electron is statistic and can be expressed through probability. In modern quantum mechanics, the squared modulus of the wave function, $|\psi|^2$ that is probability density, is used to represent the intensity of electron wave. Therefore, the electron wave is a probability wave. The electromagnetic waves are synchronized oscillations of electric and magnetic fields that propagate at the speed of light through a vacuum. They are not probability waves, but energy waves.

2. It is false. Since there is not classical, planetary-like circule orbital in an atom, the motion of an electron cannot be described through a trace, but by wave function. The "1s" here means the wave function whose shape is spherical. The 1s electron can be found everywhere in the space outside the nucleus. But the probability of appearing in a spherical shell at different distance from the nucleus is diverse.

3. According to de Broglie relation,

$$\lambda = \frac{h}{mv} = \frac{6.626\times10^{-34}\,\text{kg}\cdot\text{m}^2\cdot\text{s}^{-1}}{9.1\times10^{-31}\,\text{kg}\times7\times10^5\,\text{m}\cdot\text{s}^{-1}} = 10.4\times10^{-10}\,\text{m} = 1\,040\,\text{pm}$$

4. According to de Broglie relation,

$$\lambda = \frac{h}{mv} = \frac{6.626\times10^{-34}\,\text{kg}\cdot\text{m}^2\cdot\text{s}^{-1}}{10\times10^{-3}\,\text{kg}\times1\,000\,\text{m}\cdot\text{s}^{-1}} = 6.626\times10^{-35}\,\text{m}$$

The de Broglie wavelength of the bullet is so small that its wave properties can be neglected.

According to uncertainty principle,

$$\Delta x \geqslant \frac{h}{4\pi m\Delta v_x} = \frac{6.626\times10^{-34}\,\text{kg}\cdot\text{m}^2\cdot\text{s}^{-1}}{4\pi\times10\times10^{-3}\,\text{kg}\times10^{-3}\,\text{m}\cdot\text{s}^{-1}} = 5.3\times10^{-30}\,\text{m}$$

The position uncertainty of the bullet is so small that it can be neglected. Therefore, the bullet can fly precisely along the trajectory.

5. According to Pauli Exclusion Principle, no two electrons could exist in the same quantum state, identified by four quantum numbers n, l, m, and s. If more than two electrons are in one orbital, at least two electrons have the same quantum state, meaning the break of Pauli Exclusion Principle. Therefore, one atomic orbital only can be occupied by up to 2 electrons in antiparallel-spin state.

6. (1) 2p energy level, (2) 3d energy level, (3) 5f energy level, (4) $2p_x$ or $2p_y$ orbital, (5) 4s orbital

7. (2, 0, 0, +1/2), (2, 0, 0, −1/2), (2, 1, −1, +1/2), (2, 1, 0, +1/2), (2, 1, 1, +1/2)

8. (1) reasonable, and it only has one orbital. (2) unreasonable, since $n=3$, l must be smaller than 3. (3) reasonable, and it has three orbitals. (4) reasonable, and it has five orbitals.

9.

Atomic number	Electron configuration	Valence electronic configuration	Period	Group
49	$[Kr]4d^{10}5s^25p^1$	$5s^25p^1$	5	ⅢA
10	$1s^22s^22p^6$	$2s^22p^6$	2	0
24	$[Ar]3d^54s^1$	$3d^54s^1$	4	ⅥB
80	$[Xe]4f^{14}5d^{10}6s^2$	$5d^{10}6s^2$	6	ⅡB

10. (1) $[Ar]3d^54s^2$ and five unpaired electrons, (2) $[Ar]3d^{10}4s^24p^6$ and no unpaired electrons, (3) $[Kr]5s^2$ for the atom, $[Kr]5s^0$ for its most stable ion and no unpaired electron, (4) $[Ar]3d^{10}4s^24p^3$ and three unpaired electrons.

11. Ag^+: $[Kr]4d^{10}$, Zn^{2+}: $[Ar]\,3d^{10}$, Fe^{3+}: $[Ar]\,3d^5$, Cu^+: $[Ar]\,3d^{10}$

12. The electronic configuration of the atom in the ground state is $[Ar]3d^54s^1$. The element is in Period 4, Group ⅥB, d-block and belongs to the transition elements. Therefore, the difference of atomic radius between the element and either of its two neighbors in periodic table is approximately 5 pm.

13. For the elements from left to right along a period, the atomic radius decreases, and the effective nuclear charge increases, therefore, to ionize the outermost electrons needs more energy. According to the supplement of Hund's rule, the valence electrons in 3p orbitals of $_{15}P$ are more stable due to half-filled, if compared to those of $_{16}S$. As a result, the I_1 of $_{15}P$ is higher than that of $_{16}S$.

14. The positions of the five elements in periodic table are as follows. Since the electronegativity of elements increases on passing from left to right along a period and decreases down a group, the arrangement of elements in the order of electronegativity decreasing is F, S, As, Zn, Ca.

Period \ Group	IIA	IIB	VA	VIA	VIIA
2					F
3				S	
4	Ca	Zn	As		

15. (1) The valence electron configuration of the element is ns^2np^2. The element is in Group IVA. (2) The valence electron configuration is $3d^64s^2$. The element is Fe which is in Period 4, Group VIIIB. (3) The valence electron configuration is $3d^{10}4s^1$. It is element Cu which is in Period 4, Group IB.

16. The electronic configuration of Fe^{2+} is $[Ar]3d^6$. After losing one electron, Fe^{2+} is changed into Fe^{3+} whose 3d orbitals are half-filled. According to the supplement of Hund's rule, Fe^{3+} is more stable. This means Fe^{2+} is easily to be oxidized into Fe^{3+}.

17. Element Se is in Period 4, Group VIA, p-block of periodic table. Its valence electron configuration is $4s^24p^4$. It can lose up to 6 electrons. Since the oxidation number of oxygen is generally -2, the oxide of Se with the highest oxidation state is SeO_3.

[Supplementary Exercises]

1. According to Bohr's theory, the energy level can be expressed as

$$E = -2.18 \times 10^{-18} \frac{1}{n^2} (J)$$

The frequency of radiation emitted or absorbed by an atom as a result of a transition between two energy levels is determined by the frequency rule, $hv = |E_2 - E_1|$.

$$v = \frac{E}{h} = \frac{2.18 \times 10^{-18} J}{6.626 \times 10^{-34} J \cdot s} \times \left| -\frac{1}{3^2} + \frac{1}{5^2} \right| = 2.34 \times 10^{14} s^{-1}$$

2. The mass of a neutron (m_n) is 1.67×10^{-27} kg, therefore the wavelength of a neutron traveling at a speed of $3.90 \times 10^3 m \cdot s^{-1}$ is obtained by using de Broglie relation.

$$\lambda = \frac{h}{m_n v} = \frac{6.626 \times 10^{-34} kg \cdot m^2 \cdot s^{-1}}{1.67 \times 10^{-27} kg \times 3.90 \times 10^3 m \cdot s^{-1}} = 1.02 \times 10^{-10} m = 102 pm$$

3. $n = 3$ for M shell, so where there are 3 subshells. As $l = 3$ stands for f subshell, the number of orbitals is $2l + 1 = 7$.

4. All the electron configurations are possible except the one in answer (3), which allows 8 electrons being filled in three 2p orbitals.

5. Thallium is a representative group element in Period 6, Group IIIA of periodic table.

6. (1) Cs, Ba, Sr   (2) Ca, Ga, Ge   (3) As, P, S

Chapter 10 Covalent Bond and Intermolecular Forces

[Exercises]

3.

molecules or ions	Molecular geometry	Valence electron-pair geometry
CO_3^{2-}	Trigonal planar	Trigonal planar
SO_2	Bent	Trigonal planar

molecules or ions	Molecular geometry	Valence electron-pair geometry
NH_4^+	Tetrahedral	Tetrahedral
H_2S	Bent	Tetrahedral
PCl_5	Trigonal bipyramidal	Trigonal bipyramidal
SF_4	Distortion tetrahedral	Trigonal bipyramidal
SF_6	Octahedral	Octahedral
BrF_5	Square pyramidal	Octahedral

4. C_2H_6: sp^3 C_2H_4: sp^2 C_2H_2: sp

5. The type of hybrid orbitals of central atom B is sp^2 equivalent hybridization and that of N is sp^3 nonequivalent hybridization.

6. (1) sp^2 equivalent hybridization → sp^3 equivalent hybridization, trigonal planar → tetrahedral

(2) no changes in the hybrid types of center atom, V shape → triangular pyramid

(3) sp^3 nonequivalent hybridization → sp^3 equivalent hybridization, triangular pyramid → tetrahedral

8. (1) sp^3 nonequivalent hybridization   triangular pyramid

(2) sp hybridization   linear

(3) sp^3 equivalent hybridization   tetrahedral

(4) sp^3 nonequivalent hybridization   V shape

(5) sp^3 nonequivalent hybridization   triangular pyramid

9. (1) HNO_2 (2) $:\ddot{O}\overset{\sigma}{\frown}\overset{\ddot{N}}{}\overset{\sigma}{\frown}\ddot{O}\overset{\sigma}{-}H$ (3) Π_3^4

10.(1) $[(\sigma_{1s})^2(\sigma_{1s}^*)^2(\sigma_{2s})^2(\sigma_{2s}^*)^2(\pi_{2p_y})^1(\pi_{2p_z})^1]$, two single-electron π bonds, bond order is 1, paramagnetic.

(2) $[KK(\sigma_{2s})^2(\sigma_{2s}^*)^2(\sigma_{2p_x})^2(\pi_{2p_y})^2(\pi_{2p_z})^2(\pi_{2p_y}^*)^2(\pi_{2p_z}^*)^2]$, one σ bond, bond order is 1, diamagnetic.

(3) $[KK(\sigma_{2s})^2(\sigma_{2s}^*)^2(\sigma_{2p_x})^2(\pi_{2p_y})^2(\pi_{2p_z})^2(\pi_{2p_y}^*)^2(\pi_{2p_z}^*)^1]$, one σ bond and one three-electron π bond, bond order is 1.5, paramagnetic.

(4) $[(\sigma_{1s})^2(\sigma_{1s}^*)^1]$, one three-electron σ bond, bond order is 0.5, paramagnetic.

F_2^+ is the most stable, He_2^+ is the most unstable.

11. The electron configuration of oxygen molecular is

$O_2[KK(\sigma_{2s})^2(\sigma_{2s}^*)^2(\sigma_{2p_x})^2(\pi_{2p_y})^2=(\pi_{2p_z})^2(\pi_{2p_y}^*)^1=(\pi_{2p_z}^*)^1]$, bond order is 2. There is one electron in two three-electron π bonds respectively, therefore oxygen is paramagnetic.

The electron configuration of O_2^- is

$(\sigma_{1s})^2(\sigma_{1s}^*)^2(\sigma_{2s})^2(\sigma_{2s}^*)^2(\sigma_{2p_x})^2(\pi_{2p_y})^2(\pi_{2p_z})^2(\pi_{2p_y}^*)^2(\pi_{2p_z}^*)^1$, bond order is 1.5. There is one electron in the oxygen molecular ion O_2^-, therefore it is paramagnetic.

The magnetic property of O_2^- is weaker than that of O_2 and the stability: $O_2 > O_2^-$.

12. The electron configuration of hydrogen atom is $1s^1$ and that of He is $1s^2$. According to VB, there is one single-electron in hydrogen atom, after two single-electrons with opposite spin are paired, they

can form a stable covalent bond. However, there is no single-electron in helium atom, it cannot form covalent bond, which means that He₂ cannot exist.

According to MO, the electron configuration of hydrogen molecule is $(\sigma_{1s})^2$. The bond order is 1, therefore the H₂ can exist stable. While the electron configuration of helium molecule is $(\sigma_{1s})^2(\sigma_{1s}^*)^2$. The bond order is zero, which means that He₂ cannot exist.

13. (1) Π_3^3 (2) two Π_3^4 (3) Π_4^6 (4) Π_4^4 (5) Π_4^6

15.

Molecule	Molecular geometry	Electric dipole moment	Molecular polarity
SiF₄	Tetrahedral	=0	Non-polar molecule
NF₃	Trigonal pyramidal	≠0	Polar molecule
BCl₃	Trigonal planar	=0	Non-polar molecule
H₂S	Bent	≠0	Polar molecule
CHCl₃	Distortion tetrahedral	≠0	Polar molecule

16. (1) HCl   (2) H₂O   (3) NH₃   (4) nonpolar molecule   (5) CHCl₃   (6) NF₃

18. Rank the following in order of increasing boiling points, and explain your ranking.

(1) H₂ < Ne < CO < HF   (2) CF₄ < CCl₄ < CBr₄ < Cl₄

21. Determine the intermolecular forces of the following groups.

(1) Dispersion force

(2) Orientation force, induction force, dispersion force and hydrogen bond

(3) Dispersion force and induction force

(4) Orientation force, induction force, dispersion force and hydrogen bond

22. (1) > (2) > (3)

23. (1) sp³ equivalent hybridization

(2) A-B is polar covalent bond, AB₄ is nonpolar molecule.

(3) Dispersion force

(4) The melting point and boiling point of SiCl₄ is higher.

[Supplementary Exercises]

1. (a) sp hybridization, linear

(b) sp² hybridization, trigonal planar

(c) sp³ hybridization, trigonal pyramidal

2. (a) trigonal pyramidal   (b) bent   (c) tetrahedral   (d) linear

3. (a) $(\sigma_{1s})^2(\sigma_{1s}^*)^2(\sigma_{2s})^2(\sigma_{2s}^*)^2(\pi_{2p_y})^2(\pi_{2p_z})^2(\sigma_{2p_x})^1$

(b) 2.5

(c) Paramagnetic

(d) Longer bond than N₂

4. It is delocalized.

5. F⁻ and HCOOH

Chapter 11 Coordination Compounds

[Exercises]

4. (1) × (2) × (3) × (4) × (5) √ (6) × (7) √

5. $[PdCl_4]^{2-}$, dsp^2 hybridization.

$\mu \approx \sqrt{n(n+2)}\mu_B = \sqrt{0 \times (0+2)}\mu_B = 0\mu_B$, $[PdCl_4]^{2-}$ is diamagnetism.

$[Cd(CN)_4]^{2+}$, sp^3 hybridization

$\mu \approx \sqrt{n(n+2)}\mu_B = \sqrt{0 \times (0+2)}\mu_B = 0\mu_B$, $[Cd(CN)_4]^{2-}$ is diamagnetism.

6. Octahedron, outer orbital complex

(1) octahedron, outer orbital complex

(2) octahedron, inner orbital complex

7. Magnetic moment is $5.10\mu_B$, $\mu \approx \sqrt{n(n+2)}\mu_B$ $n=4$。

When the oxidation value of ferric ion is +2, $d_\varepsilon^4 d_\gamma^2$, $n=4$; high-spin

8. According to the Valence Bond Theory, the low spin $[Co(CN)_6]^{4-}$ is easy to lose one electron to form low spin $[Co(CN)_6]^{3-}$.

According to the Crystal Field Theory, one electron of $[Co(CN)_6]^{4-}$ at d_r orbital is easy to lose and form lower energy-level $[Co(CN)_6]^{3-}$.

9.

$[Co(NH_3)_6]^{2+}$	$d_\varepsilon^5 d_\gamma^2$	$\mu=3.87$	high-spin
$[Fe(H_2O)_6]^{2+}$	$d_\varepsilon^4 d_\gamma^2$	$\mu=4.90$	high-spin
$[Co(NH_3)_6]^{3+}$	$d_\varepsilon^6 d_\gamma^0$	$\mu=0$	low-spin

10. The shorter the wavelength is, the higher the splitting energy is. So the splitting energy of $[Mn(H_2O)_6]^{2+}$ is larger than that of $[Cr(H_2O)_6]^{2+}$.

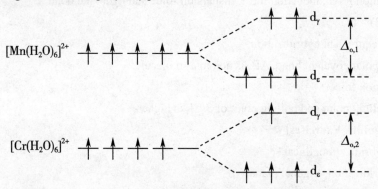

11. Given $[Fe(H_2O)_6]^{2+}$ is high spin,

$$CFSE = 4 \times (-0.4\Delta_o) + 2 \times 0.6\Delta_o$$
$$= -0.4 \times 124.38 \, kJ \cdot mol^{-1}$$
$$= -49.75 \, kJ \cdot mol^{-1}$$

Given $[Fe(CN)_6]^{4-}$ is low spin,

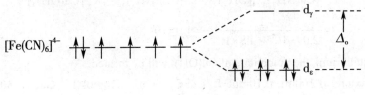

$$\begin{aligned}
CFSE &= 6 \times (-0.4\Delta_o) + 0 \times 0.6\Delta_o + (3-1)P \\
&= -2.4\Delta_o + 2P \\
&= -2.4 \times 394.68 \text{kJ} \cdot \text{mol}^{-1} + 2 \times 179.40 \text{kJ} \cdot \text{mol}^{-1} \\
&= -588.43 \text{kJ} \cdot \text{mol}^{-1}
\end{aligned}$$

12. The equilibrium would shift to the complex ion with higer K_s.

$K_s\{[Hg(NH_3)_4]^{2+}\} = 1.90 \times 10^{19}$, $K_s\{[HgY]^{2-}\} = 6.3 \times 10^{21}$
$K_s\{[Cu(NH_3)_4]^{2+}\} = 2.1 \times 10^{13}$, $K_s\{[Zn(NH_3)_4]^{2+}\} = 2.9 \times 10^9$
$K_s\{[Fe(C_2O_4)_3]^{3-}\} = 1.6 \times 10^{20}$, $K_s\{[Fe(CN)_6]^{3-}\} = 1.0 \times 10^{42}$

(1) $$K = \frac{[HgY^{2-}][NH_3]^4}{[Hg(NH_3)_4^{2+}][Y^{4-}]} = \frac{[HgY^{2-}][NH_3]^4[Hg^{2+}]}{[Hg(NH_3)_4^{2+}][Y^{4-}][Hg^{2+}]}$$

$$= \frac{K_s\{[HgY]^{2-}\}}{K_s\{[Hg(NH_3)_4]^{2+}\}} = \frac{6.3 \times 10^{21}}{1.90 \times 10^{19}}$$

$$= 3.3 \times 10^2$$

Direction is forward. $[Hg(NH_3)_4]^{2+} + Y^{4-} \rightarrow [HgY]^{2-} + 4NH_3$.

(2) $$K = \frac{[Zn(NH_3)_4^{2+}][Cu^{2+}]}{[Cu(NH_3)_4^{2+}][Zn^{2+}]} = \frac{[Zn(NH_3)_4^{2+}][Cu^{2+}]}{[Cu(NH_3)_4^{2+}][Zn^{2+}]} \times \frac{[NH_3]^4}{[NH_3]^4}$$

$$= \frac{K_s\{[Zn(NH_3)_4]^{2+}\}}{K_s\{[Cu(NH_3)_4]^{2+}\}} = \frac{2.9 \times 10^9}{2.1 \times 10^{13}}$$

$$= 1.4 \times 10^{-4}$$

Direction is backward. $[Zn(NH_3)_4]^{2+} + Cu^{2+} \rightarrow [Cu(NH_3)_4]^{2+} + Zn^{2+}$

(3) $$K = \frac{[Fe(CN)_6^{3-}][C_2O_4^{2-}]^3}{[Fe(C_2O_4)_3^{3-}][CN^-]^6} = \frac{[Fe(CN)_6^{3-}][C_2O_4^{2-}]^3[Fe^{3+}]}{[Fe(C_2O_4)_3^{3-}][CN^-]^6[Fe^{3+}]}$$

$$= \frac{K_s\{[Fe(CN)_6]^{3-}\}}{K_s\{[Fe(C_2O_4)_3]^{3-}\}} = \frac{1.0 \times 10^{42}}{1.6 \times 10^{20}}$$

$$= 2.6 \times 10^{21}$$

Direction is forward. $[Fe(C_2O_4)_3]^{3-} + 6CN^- \rightarrow [Fe(CN)_6]^{3-} + 3C_2O_4^{2-}$

13. (1) $K_{sp} = [Cu(OH)_2] = 4.8 \times 10^{-20}$, in the equal volume mixture of $0.10 \text{mol} \cdot \text{L}^{-1} \text{CuSO}_4$ and $0.10 \text{mol} \cdot \text{L}^{-1}$ NaOH, $c(Cu^{2+}) = 0.05 \text{mol} \cdot \text{L}^{-1}$, $c(OH^-) = 0.05 \text{mol} \cdot \text{L}^{-1}$, so

$$I_p = c(Cu^{2+})c^2(OH^-) = 0.05 \times (0.05)^2 = 1.25 \times 10^{-4} > K_{sp}[Cu(OH)_2] = 4.8 \times 10^{-20}$$

So the precipitate of $Cu(OH)_2$ will be formed.

(2) In the equal volume mixture of $0.10 \text{mol} \cdot \text{L}^{-1}$ $[Cu(NH_3)_4]SO_4$ and $0.10 \text{mol} \cdot \text{L}^{-1}$ NaOH

$$[Cu(NH_3)_4]^{2+}(aq) + 2OH^-(aq) \rightleftharpoons Cu(OH)_2 \downarrow + 4NH_3(aq)$$
$\quad 0.05 \text{mol} \cdot \text{L}^{-1} \qquad 0.05 \text{mol} \cdot \text{L}^{-1} \qquad\quad 0 \qquad\qquad\quad 0$

$$K = \frac{[NH_3]^4}{[Cu(NH_3)_4^{2+}][OH^-]^2} = \frac{1}{K_s\{[Cu(NH_3)_4]^{2+}\}K_{sp}[Cu(OH)_2]}$$

$$= \frac{1}{2.09 \times 10^{13} \times 4.8 \times 10^{-20}} = 1.0 \times 10^6$$

$Q \ll K$, at equilibrium, the precipitate of $Cu(OH)_2$ will be formed.

(3) In the mixture of 100mL $0.10\,mol\cdot L^{-1}$ $CuSO_4$, 50mL $0.10\,mol\cdot L^{-1}$ NaOH, 50mL $0.10\,mol\cdot L^{-1}$ NH_3.

$$I_p(CuOH_2) = c(Cu^{2+})c^2(OH^-) = 0.050 \times (0.025)^2 \gg K_{sp}(CuOH_2) = 4.8 \times 10^{-20}$$

So the precipitate of $Cu(OH)_2$ will be formed.

14. At 25℃, $K_s\{[Ni(NH_3)_6]^{2+}\} = 5.5 \times 10^8$, $K_s\{[Ni(en)_3]^{2+}\} = 2.1 \times 10^{18}$, if en is added in $[Ni(NH_3)_6]^{2+}$ solution, the reaction is

$$[Ni(NH_3)_6]^{2+}(aq) + 3en(aq) \rightleftharpoons [Ni(en)_3]^{2+}(aq) + 6NH_3(aq)$$

0.10	2.30	0	1.00
x	$2.30 - 3 \times 0.10 + 3x$	$0.10 - x$	$1.00 + 6 \times 0.10 - 6x$
	≈ 2.00	≈ 0.10	≈ 1.6

$$K = \frac{[Ni(en)_3^{2+}][NH_3]^6}{[Ni(NH_3)_6^{2+}][en]^3} = \frac{K_s\{[Ni(en)_3]^{2+}\}}{K_s\{[Ni(NH_3)_4]^{2+}\}} = \frac{2.1 \times 10^{18}}{5.49 \times 10^8} = 3.8 \times 10^9$$

$$\frac{0.10 \times (1.6)^6}{x \times (2.00)^3} = 3.8 \times 10^9$$

$$[Ni(NH_3)_6^{2+}] = x\,mol \cdot L^{-1} = \frac{0.10(1.6)^6}{(2.00)^3 \times 3.8 \times 10^9}\,mol \cdot L^{-1} = 5.5 \times 10^{-11}\,mol \cdot L^{-1}$$

$$[NH_3] = (1.00 + 6 \times 0.10 - 6x)\,mol \cdot L^{-1} = 1.60\,mol \cdot L^{-1}$$

$$[Ni(en)_3^{2+}] = (0.10 - x)\,mol \cdot L^{-1} \approx 0.10\,mol \cdot L^{-1}$$

$$[en] = (2.30 - 3 \times 0.10 + 3x)\,mol \cdot L^{-1} = 2.00\,mol \cdot L^{-1}$$

15. Before the formation of $[Ag(NH_3)_2]^+$, $c(Ag^+) = \frac{0.10 \times 50.0}{100}\,mol \cdot L^{-1} = 0.050\,mol \cdot L^{-1}$

$$c(NH_3) = \frac{0.929\,kg \cdot L^{-1} \times (1\,000g/1kg) \times 18.3\% \times 30.0mL \times (1L/1\,000mL)}{17.03g \cdot mol^{-1} \times 100mL \times (1L/1\,000mL)} = 3.00\,mol \cdot L^{-1}$$

$Ag^+(aq)$	+ $2NH_3(aq)$	$\rightleftharpoons$	$[Ag(NH_3)_2]^{2+}(aq)$
0.05	3.00		0
x	$3.00 - 0.05 \times 2 + 2x$		$0.05 - x$
	≈ 2.90		≈ 0.05

$$K_s\{[Ag(NH_3)_2]^+ = \frac{[Ag(NH_3)_2^+]}{[Ag^+][NH_3]^2} = 1.1 \times 10^7$$

$$[Ag^+] = \frac{[Ag(NH_3)_2^+]}{K_s\{[Ag(NH_3)_2]^+\} \cdot [NH_3]^2} = \frac{0.05}{1.1 \times 10^7 \times (2.90)^2}\,mol \cdot L^{-1} = 5.4 \times 10^{-10}\,mol \cdot L^{-1}$$

In solution,

$$[Ag^+] = 5.4 \times 10^{-10}\,mol \cdot L^{-1}, [Ag(NH_3)_2^+] = 0.05\,mol \cdot L^{-1}, [NH_3] = 2.90\,mol \cdot L^{-1}$$

after the addition of 10.0mL KCl, the total volume of solution is 110mL,

$$c\{[Ag(NH_3)_2]^+\} = (0.05 \times 100/110) \text{mol} \cdot \text{L}^{-1} = 0.045 \text{mol} \cdot \text{L}^{-1}$$

$$c(NH_3) = (2.90 \times 100/110) \text{mol} \cdot \text{L}^{-1} = 2.64 \text{mol} \cdot \text{L}^{-1}$$

$$c(Cl^-) = (0.100 \times 10.0/110) \text{mol} \cdot \text{L}^{-1} = 9.1 \times 10^{-3} \text{mol} \cdot \text{L}^{-1}$$

$$[Ag(NH_3)]^+(aq) + Cl^-(aq) = AgCl(s) + 2NH_3(aq)$$

$$K = \frac{[NH_3]^2}{[Ag(NH_3)_2^+][Cl^-]} = \frac{1}{K_s\{[Ag(NH_3)_2]^+\} \cdot K_{sp}(AgCl)} = \frac{1}{1.1 \times 10^7 \times 1.77 \times 10^{-10}} = 514$$

$$Q = \frac{c^2(NH_3)}{c\{[Ag(NH_3)_2]^+\} \cdot c(Cl^-)} = \frac{(2.64)^2}{0.0455 \times 9.1 \times 10^{-3}} = 1.7 \times 10^4$$

Because of $Q > K$, this forward reaction cannot occur and no AgCl precipitate.
$Q \geqslant K$ can prevent the precipitation of AgCl.

$$\frac{c^2(NH_3)}{c\{[Ag(NH_3)_2]^+\}c(Cl^-)} \geqslant \frac{1}{K_s\{[Ag(NH_3)_2]^+\} \cdot K_{sp}(AgCl)}$$

$$c(NH_3) \geqslant \sqrt{\frac{c\{[Ag(NH_3)_2]^+\} \cdot c(Cl^-)}{K_s\{[Ag(NH_3)_2]^+\} \cdot K_{sp}(AgCl)}} \text{ mol} \cdot \text{L}^{-1}$$

$$= \sqrt{\frac{0.0455 \times 9.1 \times 10^{-3}}{1.1 \times 10^7 \times 1.77 \times 10^{-10}}} \text{ mol} \cdot \text{L}^{-1} = 0.46 \text{mol} \cdot \text{L}^{-1}$$

16.
$$c(CN^-) = \frac{0.250 \text{mol} \cdot \text{L}^{-1} \times 35.0 \text{mL}}{30.0 \text{mL} + 35.0 \text{mL}} = 0.135 \text{mol} \cdot \text{L}^{-1}$$

$$c(Ag^+) = \frac{0.10 \text{mol} \cdot \text{L}^{-1} \times 30.0 \text{mL}}{30.0 \text{mL} + 35.0 \text{mL}} = 0.046 \text{mol} \cdot \text{L}^{-1}$$

$$\begin{array}{cccc} Ag^+(aq) & + & 2CN^-(aq) & \rightleftharpoons & [Ag(CN)_2]^-(aq) \\ x & & 0.135 - 2 \cdot 0.046 + 2x & & 0.046 \end{array}$$

$$\approx 0.043 \text{mol} \cdot \text{L}^{-1}$$

$$K_s = \frac{[Ag(CN)_2^-]}{[Ag^+][CN^-]^2} = 1.3 \times 10^{21}$$

$$[Ag^+] = \frac{[Ag(CN)_2^-]}{[CN^-]K_s} = \frac{0.046}{(0.043)^2 \times 1.3 \times 10^{21}} \text{mol} \cdot \text{L}^{-1} = 1.91 \times 10^{-20} \text{mol} \cdot \text{L}^{-1}$$

$$[CN^-] = (0.135 - 2 \times 0.046 + 2x) \text{mol} \cdot \text{L}^{-1} = 0.043 \text{mol} \cdot \text{L}^{-1}$$

$$[Ag(CN)_2^-] = (0.046 - x) \text{mol} \cdot \text{L}^{-1} = 0.046 \text{mol} \cdot \text{L}^{-1}$$

17. $\varphi^{\ominus}(Zn^{2+}/Zn) = -0.7618 \text{V}$, $K_{sp}[Zn(OH)_2] = 3 \times 10^{-17}$

$$Zn(OH)_2(s) + 2OH^-(aq) \rightleftharpoons [Zn(OH)_4]^{2-}(aq)$$

$$K = \frac{[Zn(OH)_4^{2-}]}{[OH^-]^2}$$

$$= \frac{[Zn(OH)_4^{2-}]}{[OH^-]^2} \times \frac{[Zn^{2+}]}{[Zn^{2+}]} \times \frac{[OH^-]^2}{[OH^-]^2}$$

$$= K_s\{[Zn(OH)_4]^{2-}\} \cdot K_{sp}(Zn(OH)_2)$$

$$K_s\{[Zn(OH)_4]^{2-}\} = \frac{K}{K_{sp}(Zn(OH)_2)} = \frac{4.786}{3 \times 10^{-17}} = 2 \times 10^{17}$$

$$\varphi^{\ominus}\{[Zn(OH)_4]^{2-}/Zn\} = \varphi^{\ominus}(Zn^{2+}/Zn) - \frac{0.05916}{n}\lg K_s\{[Zn(OH)_4]^{2-}\}$$

$$= [-0.763 - \frac{0.05916}{2} \times \lg(3.99 \times 10^{17})]V$$

$$= -1.28V$$

18.(1) to prevent the precipitation of AgCl, the concentration of Ag^+ in solution should be:

$$c(Ag^+) < \frac{K_{sp}(AgCl)}{c(Cl^-)} = \frac{1.77 \times 10^{-10}}{9 \times 10^{-3}} mol \cdot L^{-1} = 1.97 \times 10^{-8} mol \cdot L^{-1}$$

In the mixture of $AgNO_3$ and NH_3, suppose the lowest concentration of ammonia x mol·L^{-1},

$$Ag^+(aq) + 2NH_3(aq) \rightleftharpoons [Ag(NH_3)_2]^+(aq)$$

$$1.97 \times 10^{-8} \quad x - 2 \times 0.05 + 2 \times 1.97 \times 10^{-8} \quad 0.05 - 1.97 \times 10^{-8}$$

$$\approx x - 0.1 \quad\quad\quad \approx 0.05$$

$$K_s\{[Ag(NH_3)_4]^+\} = \frac{[Ag(NH_3)_4^+]}{[Ag^+][NH_3]^2} = \frac{0.05}{(1.97 \times 10^{-8}) \times (x-0.1)^2} = 1.1 \times 10^7$$

$$K_s([Ag(NH_3)_2]^+) = \frac{[Ag(NH_3)_2^+]}{[Ag^+][NH_3]^2} = \frac{0.05}{(1.97 \times 10^{-8}) \times (x-0.1)^2} = 1.1 \times 10^7$$

$$x - 0.1 = 0.48, \quad c(NH_3) = x \, mol \cdot L^{-1} = (0.48 + 0.1) mol \cdot L^{-1} = 0.58 mol \cdot L^{-1}$$

(2) at equilibrium, $[Ag(NH_3)_2^+] = 0.05 mol \cdot L^{-1}$, $[Ag^+] = 1.97 \times 10^{-8} mol \cdot L^{-1}$

$$[NH_3] = c(NH_3) - 0.1 mol \cdot L^{-1} = (0.58 - 0.1) mol \cdot L^{-1} = 0.48 mol \cdot L^{-1}$$

$$[Cl^-] = [K^+] = 9 \times 10^{-3} mol \cdot L^{-1}$$

(3) $$\varphi^{\ominus}\{[Ag(NH_3)_2]^+/Ag\} = \varphi^{\ominus}(Ag^+/Ag) - 0.05916 \lg K_s\{[Ag(NH_3)_2]^+/Ag\}$$

$$= 0.7996V - 0.05916V \times \lg(1.1 \times 10^7)$$

$$= 0.3830V$$

$$[Ag(NH_3)_2]^+(aq) + e^- \rightleftharpoons Ag(s) + 2NH_3(aq)$$

$$\varphi\{[Ag(NH_3)_2]^+/Ag\} = \varphi^{\ominus}\{[Ag(NH_3)_2]^+/Ag\} + 0.05916 \lg\{\frac{[Ag(NH_3)_2^+]}{[NH_3]^2}\}$$

$$= 0.3830V + 0.05916V \times \lg(\frac{0.05}{(0.48)^2})$$

$$= 0.3437V$$

19. (1) $$\varphi(Ag^+/Ag) = \varphi^{\ominus}(Ag^+/Ag) + 0.05916 \lg[Ag^+]$$

$$= 0.7996V + 0.05916V \times \lg[Ag^+]$$

$$= -0.505V$$

$$\lg[Ag^+] = \frac{-0.0505 - 0.7996}{0.05916} = \frac{-0.8501}{0.05916} = -14.3695$$

$$[Ag^+] = 4.27 \times 10^{-15} mol \cdot L^{-1}$$

(2) $$K_s\{[Ag(S_2O_3)_2]^{3-}\} = \frac{[Ag(S_2O_3)_2^{3-}]}{[Ag^+][S_2O_3^{2-}]^2}$$

$$= \frac{0.50}{4.27 \times 10^{-15} \times 2^2} = 2.93 \times 10^{13}$$

(3)
$$[Ag(S_2O_3)_2]^{3-}(aq) + 2CN^-(aq) \rightleftharpoons [Ag(CN)_2]^-(aq) + 2S_2O_3^{2-}$$

$$\begin{array}{cccc} x & 2-2\times0.5+2x & 0.5-x & 3.0-2x \\ & =1+2x\approx1 & \approx0.5 & \approx3.0 \end{array}$$

$$K = \frac{[Ag(CN)_2^-][S_2O_3^{2-}]^2}{[Ag(S_2O_3)_2^{3-}][CN^-]^2} = \frac{K_s\{[Ag(CN)_2]^-\}}{K_s\{[Ag(S_2O_3)_2]^{3-}\}}$$

$$= \frac{1.3\times10^{21}}{2.93\times10^{13}} = 4.44\times10^7$$

$$K = \frac{0.5\times(3.0)^2}{x\times1^2} \quad 4.44\times10^7$$

$$[Ag(S_2O_3)_2^{3-}] = x\,\text{mol}\cdot\text{L}^{-1} = \frac{0.5\times(3.0)^2}{4.44\times10^7\times1^2}\,\text{mol}\cdot\text{L}^{-1} = 1.01\times10^{-7}\,\text{mol}\cdot\text{L}^{-1}$$

$$[CN^-] = (2-0.5\times2+2x)\,\text{mol}\cdot\text{L}^{-1} \approx 1\,\text{mol}\cdot\text{L}^{-1}$$

$$[Ag(CN)_2^-] = (0.5-x)\,\text{mol}\cdot\text{L}^{-1} \approx 0.5\,\text{mol}\cdot\text{L}^{-1}$$

$$[S_2O_3^{2-}] = (3.0-2x)\,\text{mol}\cdot\text{L}^{-1} \approx 3.0\,\text{mol}\cdot\text{L}^{-1}$$

20. $K_{sp} = [Cu(OH)_2] = 4.8\times10^{-20}$

$$Cu(OH)_2(s) + 2OH^-(aq) \rightleftharpoons [Cu(OH)_4]^{2-}(aq)$$

$$K^{\ominus} = \frac{[Cu(OH)_4^{2-}]}{[OH^-]^2} \times \frac{[Cu^{2+}]}{[Cu^{2+}]} \times \frac{[OH^-]^2}{[OH^-]^2} = K_s K_{sp}$$

$$K_s = K^{\ominus}/K_{sp} = (1.66\times10^{-3})/(4.8\times10^{-20}) = 3.5\times10^{16}$$

Suppose $[OH^-] = x\,\text{mol}\cdot\text{L}^{-1}$

$$Cu(OH)_2(s) + 2OH^-(aq) \rightleftharpoons [Cu(OH)_4]^{2-}(aq)$$

$$\begin{array}{cc} x\,\text{mol}\cdot\text{L}^{-1} & 0.1\,\text{mol}\cdot\text{L}^{-1} \end{array}$$

$$K^{\ominus} = \frac{[Cu(OH)_4^{2-}]}{[OH^-]^2} = \frac{0.1}{x^2} = 1.66\times10^{-3}$$

$$[OH^-] = x\,\text{mol}\cdot\text{L}^{-1} = 7.1\,\text{mol}\cdot\text{L}^{-1}$$

The total concentration of OH^- ions is

$$c(OH^-) = (7.1+0.1\times2)\,\text{mol}\cdot\text{L}^{-1} = 7.3\,\text{mol}\cdot\text{L}^{-1}$$

21. $Fe^{3+}(aq) + e^- \rightarrow Fe^{2+}(aq)$, $\varphi^{\ominus}(Fe^{3+}/Fe^{2+})$ (1)

$[Fe(bipy)_3]^{3+}(aq) + e^- \rightarrow [Fe(bipy)_3]^{2+}(aq)$, $\varphi^{\ominus}\{[Fe(bipy)_3]^{3+}/[Fe(bipy)_3]^{2+}\}$ (2)

(1) − (2):

$$[Fe(bipy)_3]^{2+}(aq) + Fe^{3+}(aq) \rightarrow [Fe(bipy)_3]^{3+}(aq) + Fe^{2+}(aq)$$

$$K = \frac{[Fe^{2+}][Fe(bipy)_3^{3+}]}{[Fe^{3+}][Fe(bipy)_3^{2+}]} = \frac{K_{s,2}}{K_{s,1}}$$

$$\lg K = \lg\frac{K_{s,2}}{K_{s,1}} = \frac{\varphi^{\ominus}(Fe^{3+}/Fe^{2+}) - \varphi^{\ominus}\{[Fe(bipy)_3]^{3+}/[Fe(bipy)_3]^{2+}\}}{0.059\,16}$$

$$\varphi^{\ominus}\{[Fe(bipy)_3]^{3+}/[Fe(bipy)_3]^{2+}\} = \varphi^{\ominus}(Fe^{3+}/Fe^{2+}) - 0.059\,16\lg\frac{K_{s,2}}{K_{s,1}}$$

22. $Fe^{3+}(aq) + e^- \rightarrow Fe^{2+}(aq)$, $\varphi^{\ominus}(Fe^{3+}/Fe^{2+}) = 0.77\,V$

$$\varphi^{\ominus}\{[\text{Fe(bipy)}_3]^{3+}/[\text{Fe(bipy)}_3]^{2+}\} = \varphi^{\ominus}(\text{Fe}^{3+}/\text{Fe}^{2+}) - 0.059\,16\lg\frac{K_{s,2}}{K_{s,1}}$$

$$= (0.77 - 0.059\,16\lg\frac{K_{s,2}}{1.818\times10^{17}})\text{V}$$

$$= 0.96\text{V}$$

$K_{s,2} = 1.74\times10^{14}$. $[\text{Fe(bipy)}_3]^{2+}$ is more stable because it has bigger K_s.

23.

$$[\text{Ag(NH}_3)_2]^+(\text{aq}) + 2\text{S}_2\text{O}_3^{2-}(\text{aq}) \rightleftharpoons [\text{Ag(S}_2\text{O}_3)_2]^{3-}(\text{aq}) + 2\text{NH}_3(\text{aq})$$

$$\begin{array}{cccc} 0.10 & a & 0 & 0 \\ 10^{-5} & a - 0.1\times2 + 10^{-5}\times2 & 0.10 - 10^{-5} & 0.10\times2 - 10^{-5}\times2 \\ & \approx a - 0.20 & \approx 0.10 & \approx 0.20 \end{array}$$

$$K = \frac{[\text{Ag(S}_2\text{O}_3)_2^{3-}][\text{NH}_3]^2}{[\text{Ag(NH}_3)_2^+][\text{S}_2\text{O}_3^{2-}]^2}\times\frac{[\text{Ag}^+]}{[\text{Ag}^+]} = \frac{K_s\{[\text{Ag(S}_2\text{O}_3)_2]^{3-}\}}{K_s\{[\text{Ag(NH}_3)_2]^+\}} = \frac{3.2\times10^{13}}{1.6\times10^7} = 2.0\times10^6$$

$$K = \frac{[\text{Ag(S}_2\text{O}_3)_2^{3-}]\cdot[\text{NH}_3]^2}{[\text{Ag(NH}_3)_2^+]\cdot[\text{S}_2\text{O}_3^{2-}]^2} = \frac{0.10\,\text{mol}\cdot\text{L}^{-1}\times(0.20\,\text{mol}\cdot\text{L}^{-1})^2}{10^{-5}\,\text{mol}\cdot\text{L}^{-1}\times[(a-0.20)\,\text{mol}\cdot\text{L}^{-1}]^2} = 2.0\times10^6$$

$$[\text{S}_2\text{O}_3^{2-}] = (a-0.20)\,\text{mol}\cdot\text{L}^{-1} = \sqrt{\frac{0.10\times(0.20)^2}{10^{-5}\times2.0\times10^6}}\,\text{mol}\cdot\text{L}^{-1} = 1.4\times10^{-2}\,\text{mol}\cdot\text{L}^{-1}$$

At equilibrium, $[\text{Ag(S}_2\text{O}_3)_2^{3-}] \approx 0.10\,\text{mol}\cdot\text{L}^{-1}$,

$[\text{Ag(S}_2\text{O}_3)_2^{3-}] \approx 0.10\,\text{mol}\cdot\text{L}^{-1}$ $[\text{NH}_3] = 0.20\,\text{mol}\cdot\text{L}^{-1}$,

$[\text{Ag(NH}_3)_2^+] \approx 1.0\times10^{-5}\,\text{mol}\cdot\text{L}^{-1}$

the mole of $\text{Na}_2\text{S}_2\text{O}_3$ has been added is

$$n(\text{Na}_2\text{S}_2\text{O}_3) = aV = (0.20 + 1.4\times10^{-2})\,\text{mol}\cdot\text{L}^{-1}\times1\text{L} = 0.214\,\text{mol}$$

[Supplementary Exercises]

1. According to $\Delta T_f = iK_f b_B$, then

$$i = \frac{\Delta T_f}{K_f b_B} \approx \frac{[273.15 - (-0.56)]\text{K} - 273.15\text{K}}{1.86\,\text{K}\cdot\text{kg}\cdot\text{mol}^{-1}\times0.1\,\text{mol}\times\text{kg}^{-1}} = 3$$

Thus, the complex in the problem is $[\text{Co(NH}_3)_4(\text{H}_2\text{O})\text{Cl}]\text{Cl}_2$.

2. For $[\text{CuY}]^{2-}$, the number of ligand is 1, for $[\text{Cu(en)}_2]^{2+}$, the number of ligand is 2, their stability are different. For $[\text{CuY}]^{2-}$

$$\text{Cu}^{2+}(\text{aq}) + \text{Y}^{4-}(\text{aq}) \rightleftharpoons [\text{CuY}]^{2-}(\text{aq})$$

$$\begin{array}{ccc} x_1 & x_1 & 0.10 - x_1 \approx 0.10\,\text{mol}\cdot\text{L}^{-1} \end{array}$$

$$K_s\{[\text{CuY}]^{2-}\} = \frac{[\text{CuY}^{2-}]}{[\text{Cu}^{2+}][\text{Y}^{4-}]} = \frac{0.1}{x_1^2} = 5.0\times10^{18}$$

$$x_1 = 1.4\times10^{-10}\,\text{mol}\cdot\text{L}^{-1}$$

For $[\text{Cu(en)}_2]^{2+}$,

$$\text{Cu}^{2+}(\text{aq}) + 2\text{en}(\text{aq}) \rightleftharpoons [\text{Cu(en)}_2]^{2+}(\text{aq})$$

$$\begin{array}{ccc} x_2 & 2x_2 & 0.10 - x_2 \approx 0.1\,\text{mol}\cdot\text{L}^{-1} \end{array}$$

$$K_s\{[\text{Cu(en)}]^{2+}\} = \frac{[\text{Cu(en)}_2^{2+}]}{[\text{Cu}^{2+}][\text{en}]^2} = \frac{0.1}{x_2\times(2x_2)^2} = 1.0\times10^{20}$$

$$x_2 = 6.3\times10^{-8}\,\text{mol}\cdot\text{L}^{-1}$$

$[\text{CuY}]^{2-}$ is the more stable than $[\text{Cu(en)}_2]^{2+}$ because of $x_2 > x_1$.

3. $K_{sp}(AgCl) = 1.77 \times 10^{-10}$, $K_s\{[Ag(NH_3)_2]^+\} = 1.1 \times 10^7$

$$[Ag(NH_3)_2]^+(aq) + Cl^-(aq) \rightleftharpoons AgCl(s) + 2NH_3(aq)$$

Initiation of   0.05   0.05 a
the reaction

$$K = \frac{[NH_3]^2}{[Ag(NH_3)_2^+][Cl^-]} = \frac{1}{K_s\{[Ag(NH_3)_2]^+\} \cdot K_{sp}(AgCl)}$$

$$= \frac{1}{1.1 \times 10^7 \times 1.77 \times 10^{-10}} = 5.1 \times 10^2$$

To no precipitation of AgCl

$$Q = \frac{a^2}{0.05 \times 0.05} \geqslant K$$

Then, $a \geqslant \sqrt{(0.05)^2 K}$ mol·L^{-1} = $\sqrt{(0.05)^2 \times 5.1 \times 10^2}$ mol·L^{-1} = 1.13 mol·L^{-1}

4. $K_{sp}(AgI) = 8.52 \times 10^{-17}$, $K_s\{[Ag(CN)_2]^-\} = 1.3 \times 10^{21}$

$$[Ag(CN)_2]^-(aq) + I^-(aq) \rightleftharpoons AgI(s) + 2CN^-(aq)$$

Initiation of   0.1   0.1   0.1
the reaction

$$K = \frac{[CN^-]^2}{[Ag(CN)_2^-][I^-]} = \frac{1}{K_s\{[Ag(CN)_2]^-\} \cdot K_{sp}(AgI)}$$

$$= \frac{1}{1.3 \times 10^{21} \times 8.52 \times 10^{-17}} = 9.03 \times 10^{-6}$$

$$Q = \frac{(0.1)^2}{0.1 \times 0.1} = 1 \geqslant K = 9.03 \times 10^{-6}$$

Not AgI precipitate forms in the solution.

5.
$$K = \frac{[Ag(NH_3)_2^+][CN^-]^2}{[Ag(CN)_2^-][NH_3]^2} = \frac{[Ag(NH_3)_2^+][CN^-]^2}{[Ag(CN)_2^-][NH_3]^2} \times \frac{[Ag^+]}{[Ag^+]}$$

$$= \frac{K_s\{[Ag(NH_3)_2]^+\}}{K_s\{[Ag(CN)_2]^-\}} = \frac{1.1 \times 10^7}{1.3 \times 10^{21}} = 8.5 \times 10^{-15}$$

The reaction is in reverse direction.

$$[Ag(NH_3)_2]^+(aq) + 2CN^-(aq) \rightleftharpoons [Ag(CN)_2]^-(aq) + 2NH_3(aq)$$

6. The cell in the problem is as follows:

$$(-) | [Cu(NH_3)_4]^{2+}(1.00 \text{mol} \cdot L^{-1}), NH_3(1.00 \text{mol} \cdot L^{-1}) \| SHE(+)$$

$$E_{cell} = \varphi_{SHE} - \varphi\{[Cu(NH_3)_4]^{2+}/Cu\} = -\varphi\{[Cu(NH_3)_4]^{2+}/Cu\} = 0.052V$$

$$\varphi\{[Cu(NH_3)_4]^{2+}/Cu\} = -0.052V$$

$$\varphi\{[Cu(NH_3)_4]^{2+}/Cu\} = \varphi^{\ominus}\{[Cu(NH_3)_4]^{2+}/Cu\} + \frac{0.05916}{2} \lg \frac{c\{[Cu(NH_3)_4]^{2+}\}}{c^4(NH_3)}$$

$$= \varphi^{\ominus}\{[Cu(NH_3)_4]^{2+}/Cu\} + \frac{0.05916}{2} \lg \frac{1.00}{(1.00)^4}$$

$$= \varphi^{\ominus}\{[Cu(NH_3)_4]^{2+}/Cu\}$$

$$= -0.052V$$

$$\varphi^{\ominus}\{[Cu(NH_3)_4]^{2+}/Cu\} = \varphi^{\ominus}(Cu^{2+}/Cu) - \frac{0.05916}{2}\lg K_s\{[Cu(NH_3)_4]^{2+}\}$$

$$= 0.3419\,V - \frac{0.05916\,V}{2}\lg K_f\{[Cu(NH_3)_4]^{2+}\}$$

$$= -0.052\,V$$

$$\lg K_s\{[Cu(NH_3)_4]^{2+}\} = \frac{2\times[0.3419\,V-(-0.052\,V)]}{0.05916\,V} = 13.32,\ K_s\{[Cu(NH_3)_4]^{2+}\} = 2.09\times10^{13}$$

7. Cathode $\varphi(Ag^+/Ag) = \varphi^{\ominus}(Ag^+/Ag) + 0.05916\lg c(Ag^+)$

Anode $\varphi\{[Ag(CN)_2]^-/Ag\} = \varphi^{\ominus}\{[Ag(CN)_2]^-/Ag\} + 0.05916\lg\dfrac{c\{[Ag(CN)_2]^-\}}{c^2(CN^-)}$

$$= \varphi^{\ominus}(Ag^+/Ag) - 0.05916\lg K_s\{[Ag(CN)_2]^-\}$$

$$+ 0.05916\lg\dfrac{c\{[Ag(CN)_2]^-\}}{c^2(CN^-)}$$

$$E_{cell} = \varphi(Ag^+/Ag) - \varphi\{[Ag(CN)_2]^-/Ag\}$$

$$= -0.05916\,V \times 2 + 0.05916\,V \times \lg(1.26\times10^{21})$$

$$= 1.130\,V$$

Chapter 12 Colloids

[Exercises]

1. (3) causes the greatest risk of mercury poisoning. That is because the surface area increases when dispersing liquid mercury into tiny droplets of mercury. Thus more mercury atoms are at the surface. They tend to evaporate due to their higher energy level, causing more opportunities for contact with the human organs and more chances for mercury poisoning.

2. Suppose the radii of water droplets before and after dispersion are $r_1 = 1.00\,mm$ and $r_2 = 1.00\times10^{-3}\,mm$, respectively. For a droplet number N, the surface areas of water before and after dispersion are $A_1 = 4\pi r_1^2$ and $A_2 = N\times(4\pi r_2^2)$, respectively. Since the total volume of water before dispersion are the same, hence $\dfrac{4}{3}\pi r_1^3 = N\times\dfrac{4}{3}\pi r_2^3$ and $N = \left(\dfrac{r_1}{r_2}\right)^3$. Thus the work in the dispersion process is

$$W = \sigma(A_2 - A_1) = \sigma\left[\left(\dfrac{r_1}{r_2}\right)^3 \times(4\pi r_2^2) - 4\pi r_1^2\right] = 4\pi r_1^2 \sigma\left(\dfrac{r_1}{r_2} - 1\right)$$

$$= 4\times3.14\times(1.00\times10^{-3}\,m)^2 \times 0.0728\,N\cdot m^{-1}\times\left(\dfrac{1.00\times10^{-3}\,m}{1.00\times10^{-6}\,m} - 1\right)$$

$$= 9.13\times10^{-4}\,J$$

3. For molecules that stay near the surface of the liquid, the uneven attractive forces by their surrounding molecules result in the attraction to the inside of the liquid. To increase the surface area, work must be done to overcome the attraction by molecules in the liquid phase. Subsequently, the work is stored in surface molecules as potential energy. Thus surface molecules have more energy than the internal molecules. The increased energy is termed as surface energy. All molecules at the liquid

surface are subject to an inward force. At constant temperature and pressure, the force on unit length along the surface is defined as surface tension. Solution surface tends to adsorb solutes to reduce the surface tension. Thus surface energy and surface tension are two different concepts for the same physical concept.

4. Sols are thermodynamically unstable because the dispersed particles in sols are much larger than that in solutions and have coagulation instabilities in gravitational field. Hence when stabilization factors are destroyed, colloidal particles in sols will collide, stick to each other and precipitate from the dispersion medium. However, the electrostatic repulsion between two colloidal particles with the same charges prevents them from getting close, merging and becoming larger. Furthermore, hydration membranes on colloidal particles and Brownian motion also help to make sols stable.

5. River water contains sediment particles, while seawater contains large amount of NaCl and other electrolytes. In the intersection between Yangtze River, Pearl River and other rivers to the sea, charges on sediment particles in river water will be neutralization by ions of opposite charge in seawater, causing the reduced repulsion between sediment particles and the subsequent coagulation and sedimentary. After long term accumulation, deltas are formed at the intersection.

6. Sols are dispersions formed by colloidal particles with a diameter of 1 to 100nm and dispersion mediums. Tyndall effect can be attributed to the scattered light or opalescence formed by the scattering of light around the colloidal particles whose diameter is comparable to or shorter than the wavelength of the light.

7.
$$n(KCl) = 0.02 \text{mol} \cdot L^{-1} \times 0.012L = 2.4 \times 10^{-4} \text{mol}$$
$$n(AgNO_3) = 0.05 \text{mol} \cdot L^{-1} \times 0.100L = 5.0 \times 10^{-3} \text{mol}$$

Because of excess $AgNO_3$, the micellar formula of AgCl is
$$[(AgCl)_m \cdot nAg^+ \cdot (n-x)NO_3^-]^{x+} \cdot x(NO_3^-)$$

8. Since colloidal particles are positively charged with excess $AgNO_3$, anions are the key factor causing coagulation. Hence the sequence of the electrolytic coagulation abilities is
$$K_3[Fe(CN)_6] > MgSO_4 > AlCl_3$$

Similarly, colloidal particles are negatively charged with excess KI. Thus cations are the key factor causing coagulation. The sequence of the electrolytic coagulation abilities is
$$AlCl_3 > MgSO_4 > K_3[Fe(CN)_6]$$

9. Suppose χ mL $0.005 \text{mol} \cdot L^{-1}$ $AgNO_3$ solution is added to prepare negatively charged AgI sol. Since
$$25\text{ml} \times 0.016 \text{mol} \cdot L^{-1} > \chi \text{ mL} \times 0.005 \text{mol} \cdot L^{-1}$$
$x < 80$, meaning 80mL $0.005 \text{mol} \cdot L^{-1}$ $AgNO_3$ solution should be added at the most.

10. According to Schulze-Hardy rule, Ba^{2+} and Na^+ coagulate sol A, while SO_4^{2-} and Cl^- coagulate sol B. Thus sol A is negatively charged, while sol B is positively charged.

11. Sols are relatively stable because of the following facts. (1) The surface of colloidal particles are charged, hence the electrostatic repulsion between two colloidal particles of the same charges prevents them from coagulation. (2) The hydration membranes on colloidal particles prevent colloidal particles from colliding with each other and coagulating. (3) Brownian motion of the highly dispersed colloidal particles is also one important reason for keeping a sol stable in gravitational field. Coagulation of a sol

can be caused by heating, radiation and electrolytes that can destroy the sol's stability. Coagulation also occurs when oppositely charged sols are mixed.

On the contrary, macromolecular solutions are stable because of the charged colloidal particles and their hydration, with the latter is the key factor for stability. Hydration is the key factor to stabilize a protein solution. If a large amount of inorganic salt were added to a protein solution, the hydration degree of protein molecules would decrease dramatically due to the intense hydration of inorganic ions, often resulting in the precipitation of protein. Furthermore, organic solvents that have strong affinity to water can also decrease the hydration degree of proteins, resulting in the precipitation of proteins from water.

12. Both of them are uniform solutions. They are both thermodynamically stable. The dispersed phase can both pass through filter paper. However, particle diameters in macromolecular solutions are in the range of 1 to 100nm. Thus dispersed phase in macromolecular solutions cannot pass through semipermeable membranes, diffuse slowly, and has increasing viscosity with decreased temperature.

13. In an applied electrical field, both the solution pH and the protein isoelectric point determine whether proteins will move in electrophoresis. If the pH value does not equal to pI, electrophoresis would be observed since protein molecules are ionized. The protein molecules will be negatively charged in a solution of pH 8.6, hence would move towards the anode.

14. Under certain circumstances, the viscosity of many macromolecular solutions would become larger and larger. Finally, these macromolecular solutions would lose their mobility and form a mesh-structured semi-solid material called gel.

Main properties of gels are: (1) If a dried elastic gel is put into a proper solvent, it will absorb the solvent and its volume will expand, called *swelling*. (2) Gels absorb water on swelling. Water molecules tightly bonding to gels are called combined water. The permittivity, the vapor pressure, the freezing point and the boiling point of combined water are lower than those of pure water. (3) When an elastic gel is exposed to air for a period of time, part of water will automatically separate from the gel, causing its volume diminish gradually. This phenomenon is called syneresis.

Asymmetric chain-shaped polymer that can form mesh structure is the prerequisite for effective gelation. Macromolecular concentration must reach a certain concentration, so that all liquid can be combined into the mesh structure. Furthermore, circumstances such as temperature drops or solubility decreases must be satisfied.

15. Substance that can dramatically decrease water surface tension is called surfactant. There are generally two kinds of groups in the molecules of surfactants. One is the nonpolar lipophilic group, the other is the polar *hydrophilic* group. When dissolved in water, the hydrophilic groups get into water, while the long lipophilic hydrocarbon chains tend to leave the water phase. Thus the surface tension is reduced.

16. The lowest concentration of a surfactant needed to form micelles is called critical micelle concentration (CMC). Before reaching CMC, surfactant molecules directionally concentrate on the water surface to reduce the surface tension. On the contrary, when the surfactant concentration is close to CMC, micelles are in globular shape and display similar association degree. With the concentration increasing, the micelles varied to circular cylinder shapes or even layered structures.

17. Emulsions can be classified into two types, the O/W emulsions and the W/O emulsions. The O/W emulsions are emulsions formed by dispersing oil (insoluble organics) in water, while the W/O emulsions are emulsions formed by dispersing water droplets in oil.

[Supplementary Exercises]

1. The diameter of dispersed phase particles in true solution is smaller than 1nm. Usually they were consisted of low molecular and ion. It is homogeneous and has faster rate of dispersion, and can pass through the filter paper and semi-permeable membrane. The Tyndall effect is weak. On the other hand, the diameter of colloidal particles ranges from 1nm 10 to 100nm, they are consisted of colloidal particles, high molecular and micelle. Its dispersion is slow and It has obvious Tyndall effect.

2. The preparation methods of sol are dispersion method and coagulation. The dispersion method has four means (mechanical grind, ultrasonic dispersion, electro dispersion and chemical methods). Coagulation, which is produced by various chemical reactions, makes the size of insoluble product in the ranges of the size of colloidal particles when insoluble product is dialyzed.

3. The dispersion particles in true solution are smaller than 1nm. They scatter the visible light weakly. The Tyndall effect is almost cannot be observed. Otherwise the diameter of colloidal dispersion particles ranges from 1nm to 100nm. They scatter the visible light strongly and the Tyndall effect is obvious. Therefore, we can distinguish colloidal solution from true solution by means of the Tyndall effect.

4. It keeps the particles in a specific colloid dispersed that the particles coated a solvent molecular membrane, the identically charged particles repelling each other, and the Brownian movement making them diffusing.

5. Suppose the number of smaller droplets is N, then

$$\frac{4}{3}\pi r_1^3 = \frac{4}{3}\pi r_1^3 N$$

$$N = \left(\frac{r_1}{r_2}\right)^3 = 10^{18}$$

$$\frac{A_2}{A_1} = \frac{4\pi r_2^2 N}{4\pi r_1^2} = N\left(\frac{r_2}{r_1}\right)^2 = 10^6$$

6. Sols are thermodynamically unstable dispersions formed by dispersion medium and colloidal particles with a diameter of 1 to 100nm. PM2.5, particulate matter whose diameter is under less than or equals to 2.5μm, generally possesses properties of sols since main component owns a diameter of less than 100nm. The basic properties of sols are heterogeneity, highly distribution and aggregation instability. The optical, kinetic and electrical properties of sols are all based on these basic properties. Hence main methods that can be used for the detection of PM2.5 include:

(1) Optical methods. Content of PM2.5 can be determined based on **scattering** since the intensity of the scattered light increases with the number of particles in unit volume. Particle sizes can even be determined since the larger the particle volume is, the greater is the intensity of scattered light.

(2) Kinetic methods. Based on the diffusion and sedimentation equilibrium, size of PM2.5 can be determined with ultracentrifugation since sedimentation equilibrium of PM2.5 particles can be rapidly

achieved.

(3) Electrical properties of sols. Electrophoresis and electro-osmosis **may** also be used for the determination of PM2.5, since both particle sizes and charges carried affect the migration rate.

The stability of a colloidal dispersion is attributed to several factors include surface charges, hydration membranes, and Brownian motion. Generally, when stabilization factors are destroyed, coagulation happens. There are many factors that may cause PM2.5 to coagulate, such as heating, radiation and the addition of electrolytes. Thus main methods that can be used for the removal of PM2.5 include spraying water or saline solution.

Chapter 13 Titrimetric Analysis
[Exercises]

1. The significant figures are: (1) 5, (2) 3, (3) 2, (4) 1, and (5) 2

2. $E_r = \dfrac{E}{T} \times 100\%$ $0.1 = \dfrac{0.000\,2}{T} \times 100$ $T = 0.2\,\text{g}$

0.2g of the sample is required at least to guaranty the relative weighing error less than 0.1%.

3. Mean of measured values for student A is

$$\overline{X}(A) = (0.361\,0 + 0.361\,2 + 0.360\,3)/3 = 0.360\,8$$

Mean of measured values for student B is

$$\overline{X}(B) = (0.364\,1 + 0.364\,2 + 0.364\,3)/3 = 0.364\,2$$

$$d = x - \overline{x}$$

The absolute deviations $d(A)$ for student A are respectively.

$d(A)$: $0.361\,0 - 0.360\,8 = 0.000\,2$; $0.361\,2 - 0.360\,8 = 0.000\,4$; $0.360\,3 - 0.360\,8 = -0.000\,5$

The relative deviation $d_r(A)$ for student A is respectively

$$d_r(A) = \dfrac{\overline{d}}{\overline{x}} \times 100\% = \dfrac{(|0.000\,2| + |0.000\,4| + |0.000\,5|)/3}{0.360\,8} \times 100\% = 0.037\%$$

The absolute deviations $d(B)$ for student B are respectively

$d(B)$: $0.364\,1 - 0.364\,2 = -0.000\,1$; $0.364\,2 - 0.364\,2 = 0.000\,0$; $0.364\,3 - 0.364\,2 = 0.000\,1$

The relative deviation $d_r(B)$ for student B is:

$$d_r(B) = \dfrac{\overline{d}}{\overline{x}} \times 100\% = \dfrac{(|0.000\,1| + |0.000\,0| + |0.000\,1|)/3}{0.364\,2} \times 100\% = 0.033\%$$

According to above calculation about means, absolute deviation and relative deviation, if we compare means between measured value of sample $\overline{X}(A)$ is 0.360 8g, $\overline{X}(B)$ is 0.364 2g, and 0.360 6g true value of sample, we can conclude that accuracy of measured values for student A is higher than student B, but the measured values for student B have higher precision than student A as $d_r(A) > d_r(B)$, that is, measured values for student B has higher repeatability than A.

4. The concentration of NaOH solution can be calculated as following,

$$c(\text{NaOH}) = \dfrac{m(\text{KHC}_8\text{H}_4\text{O}_4)}{V(\text{NaOH}) \times Mr(\text{KHC}_8\text{H}_4\text{O}_4)}$$

(1) If 1.00mL of the initial reading of NaOH volume in buret was recorded as 0.10mL by mistake, the result of $V(\text{NaOH})$ titrated is increased by the mistake, according to above formula, and the

concentration of NaOH is lower than true value.

(2) If the mass of 0.351 8g potassium hydrogen phthalate was recorded as 0.357 8g by mistake, the high $m(KHC_8H_4O_4)$ is caused and the concentration of NaOH is higher than true value according above equation.

5. $pK_{In^-} = -lg\ K_{In^-} = -lg(1.3 \times 10^{-5}) = 4.89$ $pK_{HIn} = 14 - 4.89 = 9.11$

The color transition pH range of this indicator is between pH 8.11 ~ 10.11.

6. The mass of primary standard substance in 25.00mL Na_2CO_3 solution is got by following

$$m(Na_2CO_3) = \frac{1.335\ 0g \times 25.00mL}{250.0mL} = 0.133\ 5g$$

The concentration of HCl solution is given by the following stoichiometric relationship of reaction between Na_2CO_3 and HCl:

$$Na_2CO_3 + 2HCl = 2NaCl + CO_2 + H_2O$$

$$c(HCl)V(HCl) = 2n(Na_2CO_3) = 2 \times \frac{m(Na_2CO_3)}{M.(Na_2CO_3)}$$

$$c(HCl) = \frac{2 \times m(Na_2CO_3)}{M.(Na_2CO_3) \times V(HCl)} = \frac{2 \times 0.133\ 5g}{106.0g \cdot mol^{-1} \times 24.50 \times 10^{-3}L} = 0.102\ 8 mol \cdot L^{-1}$$

7. The mass of oxalic acid ($H_2C_2O_4 \cdot 2H_2O$) primary standard substance required to standardize about 25mL of $0.1 mol \cdot L^{-1}$ NaOH solution is calculated by following as,

$$2n(H_2C_2O_4 \cdot 2H_2O) = c(NaOH)V(NaOH)$$

$$\frac{2m(H_2C_2O_4 \cdot 2H_2O)}{Mr(H_2C_2O_4 \cdot 2H_2O)} = c(NaOH)V(NaOH)$$

$$m(H_2C_2O_4 \cdot 2H_2O) = \frac{c(NaOH)V(NaOH) \times M.(H_2C_2O_4 \cdot 2H_2O)}{2}$$

$$= \frac{0.1 mol \cdot L^{-1} \times 25 \times 10^{-3}L \times 126g \cdot mol^{-1}}{2}$$

$$= 0.15g$$

If oxalic acid is weighed by method of weighing by difference, the absolute weighing error caused by analytical balance would be $0.000\ 1g \times 2$, and then the relative weighing error is

$$d_r = \frac{0.000\ 2}{0.15} \times 100\% = 0.133\%$$

The relative weighing error is over 0.05%.

If the primary standard substance is replaced by potassium hydrogen phthalate ($KHC_8H_4O_4$), as molecular weight $Mr(KHC_8H_4O_4) = 204g \cdot mol^{-1}$, the mass of $KHC_8H_4O_4$ required to standardize about 25mL of $0.1 mol \cdot L^{-1}$ NaOH solution is got as follow

$$n(KHC_8H_4O_4) = c(NaOH)V(NaOH)$$

$$m(KHC_8H_4O_4) = c(NaOH)V(NaOH) \times MM.(KHC_8H_4O_4)$$

$$= 0.1 mol \cdot L^{-1} \times 25 \times 10^{-3}\ L \times 204g \cdot mol^{-1}$$

$$= 0.51g$$

Then, $$d_r = \frac{0.000\ 2}{0.51} \times 100\% = 0.039\%$$

The relative weighing error is controlled within 0.05% when the primary standard substance is replaced by potassium hydrogen phthalate, hence the mass of 0.51g $KHC_8H_4O_4$ is allowed to standardize about 25mL of $0.1 \text{mol} \cdot \text{L}^{-1}$ NaOH solution.

8. Refer to the concepts in textbook.

9. Assume HA is a monoprotic weak acid,

$$\text{pH} = \text{p}K_a + \lg\frac{n(\text{NaA})}{n(\text{HA})} = \text{p}K_a + \lg\frac{16.00\text{mL} \times c(\text{NaOH})}{(40.00-16.00)\text{mL} \times c(\text{NaOH})}$$

$$6.20 = \text{p}K_a + \lg\frac{16.00\text{mL}}{24.00\text{mL}}$$

$$\text{p}K_a = 6.38$$

10. As long as the condition of weak acid titrated by strong base, $cK_a \geqslant 10^{-8}$ is meted, the monoprotic weak acid can be titrated by NaOH standard solution directly. For polyprotic acid, which each hydrogen is titrated by NaOH standard solution directly, in addition to $cK_a \geqslant 10^{-8}$, the successive acid-dissociation constants K_a differ by factor of 10^4 should be satisfied.

(1) Formic acid can be titrated by NaOH standard solution directly because $\text{p}K_a = 3.75$, $K_a = 1.7 \times 10^{-4}$, $cK_a \geqslant 10^{-8}$. The phenolphthalein can be use as indicator to signal the endpoint as inflection point of formic acid is between $6.75 \sim 9.70$.

(2) Succinic acid as a monoprotic weak acid can be titrated by NaOH standard solution directly as succinic acid only has one inflection point, although $cK_{a1} \geqslant 10^{-8}$ and $cK_{a2} \geqslant 10^{-8}$, $K_{a1}/K_{a2} < 10^4$. The thymolphthalein can be use as indicator to indicate the endpoint as equivalence point of succinic acid is pH 9.6.

(3) Each hydrogen in maleic acid can be titrated by NaOH standard solution stepwise as maleic acid has two inflection points: $cK_{a1} \geqslant 10^{-8}$, $cK_{a2} \geqslant 10^{-8}$ and $K_{a1}/K_{a2} > 10^4$. First equivalence point pH 3.95 is indicated by indicator of methyl orange and second equivalence point pH 9.30 is indicated by indicator of thymolphthalein.

(4) Phthalic acid as a monoprotic weak acid can be titrated by NaOH standard solution directly as phthalic acid only has one inflection point, although $cK_{a1} \geqslant 10^{-8}$ and $cK_{a2} \geqslant 10^{-8}$, $K_{a1}/K_{a2} < 10^4$. The thymolphthalein or phenolphthalein can be use as indicator to indicate the endpoint because equivalence point of phthalic acid is pH 9.0.

(5) $\because c \times K_{a1} = 0.1000 \times 10^{-3.14} = 7.2 \times 10^{-5} \geqslant 10^{-8}$

$$c \times K_{a2} = 0.1000 \times 10^{-4.77} = 1.7 \times 10^{-6} \geqslant 10^{-8}$$

$$c \times K_{a3} = 0.1000 \times 10^{-6.39} = 4.1 \times 10^{-8} \geqslant 10^{-8}$$

but $\quad \dfrac{K_{a1}}{K_{a2}} = \dfrac{10^{-3.14}}{10^{-4.77}} = 42.7 < 10^4$

$$\frac{K_{a2}}{K_{a3}} = \frac{10^{-4.77}}{10^{-6.39}} = 41.7 < 10^4$$

$\therefore$ Citric acid can be considered as a monoprotic acid which has only one inflection point and is titrated by $0.1000 \text{mol} \cdot \text{L}^{-1}$ NaOH standard solution.

12.

$$\omega(\text{Na}_2\text{CO}_3) = \frac{c(\text{HCl}) \times V_1(\text{HCl}) \times MM.(\text{HCl})}{m(\text{sample})}$$

$$= \frac{0.2000 \text{mol} \cdot \text{L}^{-1} \times 23.10 \times 10^{-3} \text{L} \times 106.0 \text{g} \cdot \text{mol}^{-1}}{0.6839 \text{g}} = 0.7160$$

$$\omega(\text{NaHCO}_3) = \frac{c(\text{HCl}) \times [V_2(\text{HCl}) - V_1(\text{HCl})] \times MM.(\text{HCl})}{m(\text{sample})}$$

$$= \frac{0.200\,0\,\text{mol}\cdot\text{L}^{-1} \times (26.81 - 23.10) \times 10^{-3}\,\text{L} \times 84.0\,\text{g}\cdot\text{mol}^{-1}}{0.683\,9\,\text{g}} = 0.091\,14$$

13. According to the following reactions

$$\text{NH}_4^+ + \text{OH}^- = \text{NH}_3 + \text{H}_2\text{O}$$

$$2\text{NH}_3 + \text{H}_2\text{SO}_4 = (\text{NH}_4)_2\text{SO}_4 + 2\text{H}_2\text{O}$$

$$\text{H}_2\text{SO}_4 + 2\text{NaOH} = \text{Na}_2\text{SO}_4 + 2\text{H}_2\text{O}$$

the stoichiometric equation is given between reactants,

$$\frac{1}{2}n(\text{NH}_3) = n(\text{H}_2\text{SO}_4) = \frac{1}{2}n(\text{NaOH})$$

$$n(\text{NH}_3) = 2n(\text{H}_2\text{SO}_4)$$
$$= 2 \times (0.250\,0\,\text{mol}\cdot\text{L}^{-1} \times 50.00 \times 10^{-3}\,\text{L} - 0.500\,0\,\text{mol}\cdot\text{L}^{-1} \times 1.56 \times 10^{-3}\,\text{L})$$
$$= 2.422 \times 10^{-2}\,\text{mol}$$

$$\omega(\text{NH}_3) = \frac{n(\text{NH}_3) \times MM.(\text{NH}_3)}{m(\text{sample})} = \frac{2.422 \times 10^{-2}\,\text{mol} \times 17.03\,\text{g}\cdot\text{mol}^{-1}}{1.000\,0\,\text{g}} = 0.412\,5$$

14. Base on following reaction

$$5\text{H}_2\text{O}_2 + 2\text{KMnO}_4 + 3\text{H}_2\text{SO}_4 = \text{K}_2\text{SO}_4 + 2\text{MnSO}_4 + 5\text{O}_2 + 8\text{H}_2\text{O}$$

The stoichiometric relationship between H_2O_2 and KMnO_4 is got

$$\frac{1}{5}n(\text{H}_2\text{O}_2) = \frac{1}{2}n(\text{KMnO}_4)$$

$$\rho(\text{H}_2\text{O}_2) = \frac{m(\text{H}_2\text{O}_2)}{V(\text{sample})} = \frac{n(\text{H}_2\text{O}_2)MM.(\text{H}_2\text{O}_2)}{V(\text{sample})} = \frac{\frac{5}{2}c(\text{KMnO}_4)V(\text{KMnO}_4)MM.(\text{H}_2\text{O}_2)}{V(\text{sample})}$$

$$= \frac{\frac{5}{2} \times 0.027\,32\,\text{mol}\cdot\text{L}^{-1} \times 35.86 \times 10^{-3}\,\text{L} \times 34.01\,\text{g}\cdot\text{mol}^{-1}}{0.025\,00\,\text{L} \times \frac{25.00\,\text{mL}}{250.0\,\text{mL}}}$$

$$= 33.30\,\text{g}\cdot\text{L}^{-1}$$

15. Base on the reactions of titration

$$\text{Ca}^{2+} + \text{C}_2\text{O}_4^{2-} = \text{CaC}_2\text{O}_4 \downarrow$$

$$5\text{C}_2\text{O}_4^{2-} + 2\text{MnO}_4^- + 16\text{H}^+ = 2\text{Mn}^{2+} + 10\text{CO}_2 \uparrow + 8\text{H}_2\text{O}$$

the stoichiometric relationship between Ca^{2+}, $\text{C}_2\text{O}_4^{2-}$ and MnO_4^- is got

$$n(\text{Ca}^{2+}) = n(\text{CaC}_2\text{O}_4) = \frac{5}{2}n(\text{MnO}_4^-)$$

$$n(\text{Ca}^{2+}) = \frac{5}{2} \times \frac{c(\text{MnO}_4^-)V(\text{MnO}_4^-)}{V(\text{blood})} = \frac{5}{2} \times \frac{0.002\,00\,\text{mol}\cdot\text{L}^{-1} \times 1.15 \times 10^{-3}\,\text{L}}{10.00 \times 10^{-3}\,\text{L} \times \frac{5.00\,\text{mL}}{50.00\,\text{mL}}} = 0.005\,75\,\text{mol}\cdot\text{L}^{-1}$$

The mass of Ca^{2+} in 100mL blood is got:

$$m(\text{Ca}^{2+}) = c(\text{Ca}^{2+}) \times V(\text{blood}) \times MM.(\text{Ca}^{2+})$$
$$= 0.005\,75\,\text{mol}\cdot\text{L}^{-1} \times 100.00 \times 10^{-3}\,\text{L} \times 40\,\text{g}\cdot\text{mol}^{-1}$$
$$= 23.0 \times 10^{-3}\,\text{g} = 23.0\,\text{mg}$$

16. Base on the following reactions

$$Ca(OCl)_2 + 4HCl + 4KI = CaCl_2 + 2I_2 + 4KCl + 2H_2O$$
$$I_2 + 2S_2O_3^{2-} = 2I^- + S_4O_6^{2-}$$

The stoichiometric relationship between reactants is got

$$n[Ca(ClO)_2] = \frac{1}{2}n(I_2) = \frac{1}{4}n(Na_2S_2O_3) = \frac{1}{4}c(Na_2S_2O_3)V(Na_2S_2O_3)$$

The mass fraction of available chlorine $Ca(OCl)_2$ in the sample of bleach is

$$\omega[Ca(ClO)_2] = \frac{m[Ca(ClO)_2]}{m(sample)} = \frac{n[Ca(ClO)_2] \times MM.[Ca(ClO)_2]}{m(sample)}$$

$$= \frac{\frac{1}{4}c(Na_2S_2O_3)V(Na_2S_2O_3)MM.[Ca(ClO)_2]}{m(sample)}$$

$$= \frac{\frac{1}{4} \times 0.1109 mol \cdot L^{-1} \times 35.58 \times 10^{-3} L^{-1} \times 143.0 g \cdot mol^{-1}}{2.6220 g} = 0.05380$$

17. Base on the stoichiometric relationship between reactants following

$$n(C_6H_8O_6) = n(I_2) = c(I_2)V(I_2)$$

the mass fraction of ascorbic acid in the sample of vitamin C is got

$$\omega(C_6H_8O_6) = \frac{m(C_6H_8O_6)}{m(sample)} = \frac{c(I_2)V(I_2)MM.(C_6H_8O_6)}{m(sample)}$$

$$\omega(C_6H_8O_6) = \frac{0.05000 mol \cdot L^{-1} \times 22.14 \times 10^{-3} L \times 176.12 g \cdot mol^{-1}}{0.1988 g} = 0.9807$$

18. According to the reaction between ZnO and HCl

$$ZnO + 2HCl = ZnCl_2 + H_2O$$

the stoichiometric relationship between reactants following is got

$$n(ZnO) = \frac{1}{2}(HCl)$$

then

$$n(ZnO) = \frac{0.1000 mol \cdot L^{-1} \times 20.00 \times 10^{-3} L}{2} = 1.000 \times 10^{-3} mol$$

$$\because n(ZnO) = n(EDTA)$$

$$\therefore V(EDTA) = \frac{n(ZnO)}{c(EDTA)} = \frac{1.000 \times 10^{-3} mol}{0.05000 mol \cdot L^{-1}} = 20.00 mL$$

19.
$$n(CaCO_3) = n(EDTA) = c(EDTA)V(EDTA)$$

$$c(EDTA) = \frac{n(CaCO_3)}{V(EDTA)} = \frac{m(CaCO_3)}{M(CaCO_3) \times V(EDTA)}$$

$$= \frac{0.2100 g \times \frac{25.00 mL}{250.00 mL}}{100 g \cdot mol^{-1} \times 20.15 \times 10^{-3} L} = 0.01042 mol \cdot L^{-1}$$

20. The molar concentration of Ca^{2+} and Mg^{2+} respectively in the sample of water is

$$c(Ca^{2+}) = \frac{c(EDTA)V_2(EDTA)}{V(water\ sample)}$$

$$= \frac{0.01048 mol \cdot L^{-1} \times 10.54 mL}{100.00 mL} = 0.001105 mol \cdot L^{-1}$$

$$c(Mg^{2+}) = \frac{c(EDTA)V_1(EDTA)}{V(\text{water sample})}$$

$$= \frac{0.010\,48\,\text{mol}\cdot L^{-1} \times (14.20-10.54)\text{mL}}{100.00\text{mL}} = 3.836\times 10^{-4}\,\text{mol}\cdot L^{-1}$$

21. $\because$ $Ag^+(aq) + SCN^-(aq) = AgSCN(s)$

$\therefore$ $n(Ag^+) = n(SCN^-) = c(NH_4SCN)\,V(NH_4SCN)$

$$\rho(Ag^+) = \frac{n(Ag^+)MM.(Ag^+)}{V(\text{sample})} = \frac{c(NH_4SCN)V(NH_4SCN)MM.(Ag^+)}{V(\text{sample})}$$

$$= \frac{0.043\,82\,\text{mol}\cdot L^{-1} \times 23.48\text{mL} \times 107.87\,\text{g}\cdot\text{mol}^{-1}}{10.00\text{mL}} = 11.09\,\text{g}\cdot L^{-1}$$

[Supplementary Exercises]

1. The significant figure would be 0.019.

2. (a) HX is weaker. The higher the pH at the stoichiometric point, the stronger the conjugate base(X^-) and the weaker the conjugate acid(HX).

(b) Phenolphthalein, which changes color in the pH 8.2~10.0 range, is perfect for HX and probably appropriate for HY.

3.
(a) $HA + NaOH = NaA + H_2O$

$n(HA) = n(NaOH) = c(NaOH)\,V(NaOH)$

$$\frac{m(HA)}{Mr(HA)} = c(NaOH)\,V(NaOH)$$

$$\frac{0.127\,6\,\text{g}}{M(HA)} = 0.063\,3\,\text{mol}\cdot L^{-1} \times 18.40\,\text{mL}$$

$$Mr(HA) = \frac{0.127\,6}{0.063\,3 \times 18.40 \times 25} = 365\,\text{g}\cdot\text{mol}^{-1}$$

(b) $pH = pKa + \lg\frac{[NaA]}{[HA]} = pKa + \lg\frac{V(NaA)}{V(HA)}$

$$pKa = 5.87 - \lg\frac{10.00\text{mL}}{(18.40-10.00)\text{mL}} = 5.87 - 0.076 = 5.79$$

$$K_a = 1.6 \times 10^{-6}$$

4. $\dfrac{x}{126\,\text{g}\cdot\text{mol}^{-1}} = 250.0 \times 10^{-3}\,L \times 0.100\,0\,\text{mol}\cdot L^{-1}$

$x = 3.150\,\text{g}$ of $H_2C_2O_4\cdot 2H_2O$ are needed to prepare 250.0 mL of a $0.1\,000\,\text{mol}\cdot L^{-1}$ $C_2O_4^{2-}$ standard solution.

$$\because \frac{K_{a1}}{K_{a2}} = \frac{5.9\times 10^{-2}}{6.4\times 10^{-5}} \geq 10^2$$

Almost all the H_3O^+ in the solution comes from the first ionization reaction. The concentration in H_3O^+ can be determined by considering only K_{a1}.

$$H_2C_2O_4 = CH_2O_4^- + H_3O^+$$
$$Y - 0.1 \times 0.25 \quad 0.1\times 0.25 \quad 0.1\times 0.25$$

$$K_{a1} = \frac{[HC_2O_4^-][H_3O^+]}{[H_2C_2O_4]} = \frac{(0.100\,0\,\text{mol}\cdot L^{-1} \times 0.250\,0\,\text{mL})^2}{(y - 0.100\,0\,\text{mol}\cdot L^{-1} \times 0.250\,0\,\text{mL})} = 5.9 \times 10^{-2}$$

$$y = 0.036\,\text{mol}$$

$$m(H_2C_2O_4 \cdot 2H_2O) = y \times MM.(H_2C_2O_4 \cdot 2H_2O) = 0.036 \text{mol} \times 126 \text{g} \cdot \text{mol}^{-1} = 4.5\text{g}$$

4.5g of $H_2C_2O_4 \cdot 2H_2O$ are needed to prepare 250.0mL of a $0.1000 \text{mol} \cdot \text{L}^{-1}$ H_3O^+ standard solution.

5.
$$2NaOH + H_2SO_4 = Na_2SO_4 + 2H_2O$$

$$\frac{1}{2}n(NaOH) = n(H_2SO_4)$$

$$\frac{1}{2}c(NaOH)\,V(NaOH) = c(H_2SO_4)\,V(H_2SO_4)$$

$$c(NaOH) = \frac{2 \times c(H_2SO_4)\,V(H_2SO_4)}{V(NaOH)} = \frac{2 \times 0.05550 \text{mol} \cdot \text{L}^{-1} \times 13.40 \text{mL}}{25.00 \text{mL}} = 0.05950 \text{mol} \cdot \text{L}^{-1}$$

6. (a)
$$\frac{1}{2}c(NaOH)\,V(NaOH) = c(H_2SO_4)\,V(H_2SO_4)$$

$$c(NaOH) = \frac{0.1024 \text{mol} \cdot \text{L}^{-1} \times 21.28 \text{mL}}{19.46 \text{mL}} \times 2 = 0.2240 \text{mol} \cdot \text{L}^{-1}$$

(b)
$$m(NaOH) = c(NaOH)\,V(NaOH)\,MM.(NaOH)$$
$$= 0.2240 \text{mol} \cdot \text{L}^{-1} \times 1\text{L} \times 40\text{g} \cdot \text{mol}^{-1} = 8.958\text{g}$$

7.
$$Na_2CO_3 + HCl = NaCl + NaHCO_3 \quad \text{(Indicator: phenolphthalein)}$$

$$n(Na_2CO_3) = n(HCl) = c(HCl)V_1(HCl)$$

$$\omega = \frac{m(Na_2CO_3)}{m(\text{sample})} = \frac{n(Na_2CO_3)MM.(Na_2CO_3)}{m(\text{sample})} \times 100\%$$

$$= \frac{c(HCl)V_1(HCl)MM.(Na_2CO_3)}{m(\text{sample})} \times 100\%$$

$$= \frac{0.1090 \text{mol} \cdot \text{L}^{-1} \times 15.70 \text{mL} \times 106\text{g} \cdot \text{L}^{-1}}{0.5720} \times \frac{1\text{L}}{1000\text{mL}} \times 100\% = 31.71\%$$

$$NaHCO_3 + HCl = NaCl + CO_2 + H_2O \quad \text{(Indicator: methyl orange)}$$

$$n(NaHCO_3) = n(HCl) = c(HCl)[V_2(HCl) - V_1(HCl)]$$

$$\omega = \frac{m(NaHCO_3)}{m(\text{sample})} = \frac{n(NaHCO_3)MM.(NaHCO_3)}{m(\text{sample})} \times 100\%$$

$$= \frac{c(HCl)[V_2(HCl) - V_1(HCl)]MM.(NaHCO_3)}{m(\text{sample})} \times 100\%$$

$$= \frac{0.1090 \text{mol} \cdot \text{L}^{-1} \times (43.80 - 15.70)\text{mL} \times 84.0\text{g} \cdot \text{L}^{-1}}{0.5720} \times \frac{1\text{L}}{1000\text{mL}} \times 100\% = 44.98\%$$

$$\omega(NaHCO_3) = \frac{m(NaHCO_3)}{m(\text{sample})} = \frac{n(NaHCO_3)MM.(NaHCO_3)}{m(\text{sample})} \times 100\%$$

$$= \frac{c(HCl)[V_2(HCl) - V_1(HCl)]MM.(NaHCO_3)}{m(\text{sample})} \times 100\%$$

$$= \frac{0.1090 \text{mol} \cdot \text{L}^{-1} \times (43.80 - 15.70) \times 10^{-3} \text{L} \times 84.0\text{g} \cdot \text{L}^{-1}}{0.5720\text{g}} \times 100\% = 44.98\%$$

Chapter 14 Ultraviolet-Visible Spectrophotometry

[Exercises]

1. Spectrophotometry is sensitive, accurate, rapid, and convenient.

2. In spectrophotometry, when concentration is expressed in mass concentration (ρ) with unit of $g \cdot L^{-1}$, the coefficient is called as mass absorptivity, a (with unit of $L \cdot g^{-1} \cdot cm^{-1}$); when concentration is expressed in molarity of solution with unit of $mol \cdot L^{-1}$, ε is the analyte's molar absorptivity with unit of $L \cdot mol^{-1} \cdot cm^{-1}$.

The mass absorptivity a and molar absorptivity ε can be converted to each other as: $\varepsilon = aM$, where M is molar mass of analyte with unit of $g \cdot mol^{-1}$.

Beer-Lambert law can be applied only for monochromatic light. The determination will deviate more from Beer-Lambert law when the wavelength coverage gets wider. λ_{max} is chosen as detection wavelength because at which the maximum absorptivity is obtained and the highest sensitivity can be obtained.

3. An ultraviolet-visible absorption spectrum or absorption curve is essentially a graph of absorbance represented by the symbol A versus the wavelength represented by the symbol λ in a range of ultraviolet or visible regions.

Different compound has different absorption spectrum, so it can be applied to the qualitative analysis. Additionally, the absorbance increases with the concentration around the absorption peak, so we can determine the content of a given substance according to its absorbance at specific wavelength.

The general procedure is as following, a series of standard solutions are prepared by varying concentrations, selection of wavelength (the wavelength with maximum absorbance is ordinarily selected as detection wavelength), absorbance determination of standard solutions using cuvettes with the same path length, plotting the absorbance value versus concentration in x/y system (x, concentration of standard solution; y, absorbance), then obtaining a straight calibration line passing through origin which is called a calibration curve or analytical curve.

Calibration curve is the frequently used in quantitative analysis.

4. A spectrophotometer consists of light source, monochromator (wavelength selector), sample containers, light detector, signal processor and readout.

The light source of incident light is used to supply complex light of high intensity and stability. Monochromator (wavelength selector) is used to provide monochromatic light. Sample must be placed into sample containers to measure its absorbance. Light detector can detect the transmitted light so as to certain the absorbance of the sample. Signal processor and readout help us obtain A or T of the sample.

5. When concentration is changed to $0.5c_1$ or $2c_1$ keeping the same path length, because the solution is consistent with Beer-Lambert law, so $T_2 = T_1^{1/2}$, $T_3 = T_1^2$, T_2 is the largest one.

6.
$$A = ab\rho$$
$$0.198 = a \times 2.00cm \times 0.500\mu g \cdot mL^{-1}, a = 0.198 mL \cdot \mu g^{-1} \cdot cm^{-1}$$
$$\varepsilon = aM = 0.198 \times 1\,000 L \cdot g^{-1} \cdot cm^{-1} \times 55.8 g \cdot mol^{-1} = 1.10 \times 10^4 L \cdot mol^{-1} \cdot cm^{-1}$$

7.
$$A = \varepsilon bc$$
$$0.120 = \varepsilon_1 \times 3.00cm \times 2.0 \times 10^{-4} mol \cdot L^{-1}$$
$$\varepsilon_1 = 200 L \cdot mol^{-1} \cdot cm^{-1}$$
$$0.200 = \varepsilon_2 \times 5.00cm \times 1.0 \times 10^{-4} mol \cdot L^{-1}$$

$$\varepsilon_2 = 400 \text{L} \cdot \text{mol}^{-1} \cdot \text{cm}^{-1}$$

$\varepsilon_1 \neq \varepsilon_2$, the solution is not consistent with Beer-Lambert law.

8.
$$A = \varepsilon bc$$
$$0.687 = 703 \text{L} \cdot \text{mol}^{-1} \cdot \text{cm}^{-1} \times 1.00 \text{cm} \times c, \ c = 9.77 \times 10^{-4} \text{mol} \cdot \text{L}^{-1}$$

Mass of dobutamine in this tablet $= 9.77 \times 10^{-4} \text{mol} \cdot \text{L}^{-1} \times 2.00 \text{L} \times 270 \text{g} \cdot \text{mol}^{-1}$
$$= 0.528 \text{g}$$

9.
$$A = \varepsilon bc$$
$$0.600 = 2.5 \times 10^5 \text{L} \cdot \text{mol}^{-1} \cdot \text{cm}^{-1} \times 1.00 \text{cm} \times c, \ c = 2.4 \times 10^{-6} \text{mol} \cdot \text{L}^{-1}$$

Mass of this compound $= 2.4 \times 10^{-6} \text{mol} \cdot \text{L}^{-1} \times 200 \text{L} \times 125 \text{g} \cdot \text{mol}^{-1}$
$$= 0.060 \text{g}$$

10.
$$A = \varepsilon bc = -\lg T$$

When path length is 2.00cm,
$$A_1 = 2.00 \times \varepsilon bc = 2A, \quad T_1 = 10^{-2\varepsilon bc} = T^2$$

11.
$$A = \varepsilon bc$$
$$0.380 = \varepsilon \times 1.00 \text{cm} \times 1/V \times 0.53 \text{mL/L}$$
$$0.200 = \varepsilon \times 1.00 \text{cm} \times 0.2 \text{mL/L}$$
$$V = 4.97 \text{L}$$

12.
$$A = \varepsilon bc = E_{1\text{cm}}^{1\%} M bc / 10$$
$$0.551 = 560 \times 176 \times 1.00 \times c / 10, \ c = 5.59 \times 10^{-5} \text{mol} \cdot \text{L}^{-1}$$

Before dilution, $c = 5.59 \times 10^{-5} \text{mol} \cdot \text{L}^{-1} \times 50 = 2.80 \times 10^{-3} \text{mol} \cdot \text{L}^{-1}$
$$m = 2.80 \times 10^{-3} \times 176 \times 0.1 = 4.93 \times 10^{-2} \text{g}$$
$$\omega = 0.0493 / 0.0500 = 0.986$$

[Supplementary Exercises]

1.
$$A = \varepsilon bc$$
$$0.48 = \varepsilon \times 1.00 \text{cm} \times 4.12 \times 10^{-5} \text{mol} \cdot \text{L}^{-1}$$
$$\varepsilon = 1.17 \times 10^4 \text{mol} \cdot \text{L}^{-1}$$
$$\varepsilon = aM$$
$$a = \varepsilon / M = 1.17 \times 10^4 \text{mol} \cdot \text{L}^{-1} / 55.85 \text{g} \cdot \text{mol}^{-1}$$
$$= 2.09 \times 10^2 \text{L} \cdot \text{g}^{-1} \cdot \text{cm}^{-1} = 0.209 \text{L} \cdot \text{mg}^{-1} \cdot \text{cm}^{-1}$$

2. $A = \varepsilon bc = -\lg T = -\lg(I_t/I_0), \ A_1 = -\lg 0.25 = 0.602$

When path length is 2.00cm, $A_2 = 2.00 \times \varepsilon bc = 2A_1 = 1.20$
$$T_2 = 10^{-2\varepsilon bc} = T_1^2 = 0.25^2 = 0.062$$

3.
$$A = \varepsilon bc = -\lg T, \ A = -\lg 0.70 = 0.155$$

After concentrated four times, $c_2 = 4c, \ A_2 = 4A = -\lg T_2$
$$T_2 = 10^{-4\varepsilon bc} = T^4 = 0.24$$

Chapter 15 Brief Introduction to the Modern Instrumental Analysis

[Exercises]

2. From the experimental data of the standard additions method, a standard working curve of

absorption vs. concentration of cadmium solution is plotted. Extend the working curve obtained by linking the plots. The concentration (c_x) of cadmium solution is the distance between the origin and the intersecting point of the working curve on the abscissa and is $0.232 \mu g \cdot mL^{-1}$.

The amount of cadmium = $0.232 \times 5 \times 5/0.256\,6 = 22.6 \mu g \cdot g^{-1}$

3. $\Delta c = 0.3mg \times 50mg \cdot L^{-1}/(50mL + 0.3mL) = 0.298 mg \cdot L^{-1}$

$$A = k'c \qquad \frac{A_x}{A_{x+s}} = \frac{c_x}{c_x + \Delta c}$$

$$c_x = \frac{A_x}{(A_{x+s} - A_x)} \Delta c = \frac{0.325}{(0.670 - 0.325)} 0.298 mg \cdot L^{-1} = 0.280 mg \cdot L^{-1}$$

7. From the experimental data of calibration curve method, a standard working curve of the fluorescence intensity vs. concentration of salicylic acid solution is plotted. The unknown concentration is determined by use of the bracketing method from the plot, so the fluorescence of the he diluting solution is 8.69, the concentration is $4.56 mg \cdot L^{-1}$.

w/w%(acetylsalicylic acid) = $1 - 4.56 mg \cdot L^{-1} \times 10 \times 1L / 0.101\,3g \times 100\% = 65.0\%$

[Supplementary Exercises]

1. Excitation 367nm, emission 483nm, pH = 9.35

2.
$$F = kc \qquad \frac{F_s}{F_x} = \frac{c_s}{c_x}$$

$$c_x = \frac{F_x \times c_s}{F_s} = \frac{4.85 \times 5.00 \times 10^{-5} mol \cdot L^{-1}}{3.74} = 3.85 \times 10^{-5} mol \cdot L^{-1}$$

3. (1) $n_A = 16(\frac{t_{RA}}{W})^2 = 16(\frac{8.04}{0.15})^2 = 46\,000$; $n_B = 16(\frac{t_{RB}}{W})^2 = 16(\frac{8.26}{0.15})^2 = 48\,500$

$n_C = 16(\frac{t_{RC}}{W})^2 = 16(\frac{8.43}{0.16})^2 = 44\,400$; $n_{average} = 46\,300$

(2) $H = \frac{L}{n_{average}} = \frac{20m}{46\,300} = 0.43mm$

(3) $R_{AB} = \frac{2(t_{RB} - t_{RA})}{(W_B + W_A)} = \frac{2(8.26 - 8.04)}{(0.15 + 0.15)} = 1.5$

$R_{BC} = \frac{2(t_{RC} - t_{RB})}{(W_C + W_B)} = \frac{2(8.43 - 8.26)}{(0.15 + 0.16)} = 1.1$

Chapter 16 Nuclear Chemistry and Its Applications

[Exercises]

3. (1) $^{210}_{83}Bi \rightarrow ^{210}_{84}Po + ^{0}_{-1}e$

(2) $^{232}_{90}Th \rightarrow ^{228}_{88}Ra + ^{4}_{2}He$

(3) $^{84}_{39}Y \rightarrow ^{84}_{38}Sr + ^{0}_{1}e$

(4) $^{44}_{22}Ti + ^{0}_{-1}e \rightarrow ^{44}_{21}Sc$

[Supplementary Exercises]

1. (1) $^{6}_{3}Li + ^{1}_{0}n \rightarrow ^{4}_{2}\alpha + ^{3}_{1}H$

(2) $^{238}_{92}U + ^{1}_{0}n \rightarrow ^{239}_{92}U$ $^{239}_{92}U \rightarrow ^{0}_{-1}e + ^{239}_{93}Np$ $^{239}_{93}Np \rightarrow ^{0}_{-1}e + ^{239}_{94}Pu$

(3) $^{14}_{7}N + ^{1}_{0}n \rightarrow ^{14}_{6}C + ^{1}_{1}H$

(4) $^{1}_{1}H + ^{1}_{1}H \rightarrow ^{3}_{2}He + ^{1}_{0}n$

2. $\because$ 5.3 a are 3 half-lifes, $\therefore m = \dfrac{1}{8} \times 1.000\,\text{mg} = 0.125\,\text{mg}$

References

[1] Raymond C., Jason O.. General Chemistry. 6th ed. New York, McGraw-Hill, 2011.

[2] Silberberg M.. Chemistry: The Molecular Nature of Matter and Change. 6th ed. New York: McGraw-Hill Higher Education, 2011.

[3] Petrucci P.H., Harwood H.S., Herring F.G.. General Chemistry. 8th ed. Higher Education Press, 2015.

[4] Levine I.N.. Quantum Chemistry. 7th ed. London: Pearson, 2013.

[5] Ballentine L.E.. Quantum Mechanics: A Modern Development. 2nd ed. London: World Scientific Publishing Company, 2014.

[6] Susskind L., Friedman A.. Quantum Mechanics: The Theoretical Minimum. 2nd ed. New York: Basic Books, 2015.

[7] Cook D.B.. Quantum Chemistry: A Unified Approach. 2nd ed. London: Imperial College Press, 2012.

[8] Hammers, Gordon G., Sharon.Physical Chemistry for the Biological Science. Methods of Biochemical Analysis, 2007.

[9] Umland J.B., Bellaman J.M.. General chemistry. 3rd ed. Brooks/cole Publishing Company, 1999.

[10] Oxtoby D.W., Gillis H.P., Macntrieb N.H.. Principles of Modern Chemistry. 5th ed. Thomson Learning Inc. 2002.

[11] Brady J.E., Russell J.W., Holum J.R.. Chemistry Matter and Its Changes. New York: John Wiley & Sons, Inc. 2000.

[12] Umbland J.B., Bellama J.M.. General Chemistry. 3rd ed. Brooks/cole publishing company, 1999.

[13] Loveland W., Morrissey D., Seaborg G.. Modern Nuclear Chemistry. New Jersey: John Wiley & Sons, 2006.

[14] Ebbing D., Gammon S.. General Chemistry. 9th ed. Boston: Cengage Learning, 2009.

[15] Vertes A., Nagy S., Klencar Z., et. al. Handbook of Nuclear Chemistry. 2nd ed. New York Springer-Verlag, 2011.

[16] Silberberg M.S.. Chemistry. 7th ed. New York, McGraw-Hill, 2014.

[17] Silberberg M.. Chemistry: The Molecular Nature of Matter and Change. 8th ed. New York: McGraw-Hill Higher Education, 2018.

Name, Atomic Number and Relative Atomic Weight of Elements

English name	Symbol	Atomic number	Relative atomic weight	English name	Symbol	Atomic number	Relative atomic weight
Hydrogen	H	1	1.007 94(7)	Bromine	Br	35	79.904(1)
Helium	He	2	4.002 602(2)	Krypton	Kr	36	83.798(2)
Lithium	Li	3	[6.941(2)]	Rubidium	Rb	37	85.467 8(3)
Beryllium	Be	4	9.012 182(3)	Strontium	Sr	38	87.62(1)
Boron	B	5	10.811(7)	Yttrium	Y	39	88.905 85(2)
Carbon	C	6	12.010 7(8)	Zirconium	Zr	40	91.224(2)
Nitrogen	N	7	14.006 7(2)	Niobium	Nb	41	92.906 38(2)
Oxygen	O	8	15.999 4(3)	Molybdenum	Mo	42	95.96(2)
Fluorine	F	9	18.998 403 2(5)	Technetium	Tc	43	[98]
Neon	Ne	10	20.179 7(6)	Ruthenium	Ru	44	101.07(2)
Sodium	Na	11	22.989 769 28(2)	Rhodium	Rh	45	102.905 50(2)
Magnesium	Mg	12	24.305 0(6)	Palladium	Pd	46	106.42(1)
Aluminium	Al	13	26.981 538 6(8)	Silver	Ag	47	107.868 2(2)
Silicon	Si	14	28.085 5(3)	Cadmium	Cd	48	112.411(8)
Phosphorus	P	15	30.973 762(2)	Indium	In	49	114.818(3)
Sulfur	S	16	32.065(5)	Tin	Sn	50	118.710(7)
Chlorine	Cl	17	35.453(2)	Antimony	Sb	51	121.760(1)
Argon	Ar	18	39.948(1)	Tellurium	Te	52	127.60(3)
Potassium	K	19	39.098 3(1)	Iodine	I	53	126.904 47(3)
Calcium	Ca	20	40.078(4)	Xenon	Xe	54	131.293(6)
Scandium	Sc	21	44.955 912(6)	Caesium	Cs	55	132.905 451 9(2)
Titanium	Ti	22	47.867(1)	Barium	Ba	56	137.327(7)
Vanadium	V	23	50.941 5(1)	Lanthanum	La	57	138.905 47(7)
Chromium	Cr	24	51.996 1(6)	Cerium	Ce	58	140.116(1)
Manganese	Mn	25	54.938 045(5)	Praseodymium	Pr	59	140.907 65(2)
Iron	Fe	26	55.845(2)	Neodymium	Nd	60	144.242(3)
Cobalt	Co	27	58.933 195(5)	Promethium	Pm	61	[145]
Nickel	Ni	28	58.693 4(4)	Samarium	Sm	62	150.36(2)
Copper	Cu	29	63.546(3)	Europium	Eu	63	151.964(1)
Zinc	Zn	30	65.38(2)	Gadolinium	Gd	64	157.25(3)
Gallium	Ga	31	69.723(1)	Terbium	Tb	65	158.925 35(2)
Germanium	Ge	32	72.64(1)	Dysprosium	Dy	66	162.500(1)
Arsenic	As	33	74.921 60(2)	Holmium	Ho	67	164.930 32(2)
Selenium	Se	34	78.96(3)	Erbium	Er	68	167.259(3)

Name, Atomic Number and Relative Atomic Weight of Elements

续表

English name	Symbol	Atomic number	Relative atomic weight	English name	Symbol	Atomic number	Relative atomic weight
Thulium	Tm	69	168.934 21(2)	Plutonium	Pu	94	[244]
Ytterbium	Yb	70	173.054(5)	Americium	Am	95	[243]
Lutetium	Lu	71	174.966 8(1)	Curium	Cm	96	[247]
Hafnium	Hf	72	178.49(2)	Berkelium	Bk	97	[247]
Tantalum	Ta	73	180.947 88(2)	Californium	Cf	98	[251]
Tungsten	W	74	183.84(1)	Einsteinium	Es	99	[252]
Rhenium	Re	75	186.207(1)	Fermium	Fm	100	[257]
Osmium	Os	76	190.23(3)	Mendelevium	Md	101	[258]
Iridium	Ir	77	192.217(3)	Nobelium	No	102	[259]
Platinum	Pt	78	195.084(9)	Lawrencium	Lr	103	[262]
Gold	Au	79	196.966 569(4)	Rutherfordium	Rf	104	[267]
Mercury	Hg	80	200.59(2)	Dubnium	Db	105	[268]
Thallium	Tl	81	204.383 3(2)	Seaborgium	Sg	106	[271]
Lead	Pb	82	207.2(1)	Bohrium	Bh	107	[272]
Bismuth	Bi	83	208.980 40(1)	Hassium	Hs	108	[270]
Polonium	Po	84	[209]	Meitnerium	Mt	109	[276]
Astatine	At	85	[210]	Darmstadtium	Ds	110	[281]
Radon	Rn	86	[222]	Roentgenium	Rg	111	[280]
Francium	Fr	87	[223]	Copernicium	Cn	112	[285]
Radium	Ra	88	[226]	Nihonium	Nh	113	[284]
Actinium	Ac	89	[227]	Flerovium	Fl	114	[289]
Thorium	Th	90	232.038 06(2)	Moscovium	Mc	115	[288]
Protactinium	Pa	91	231.035 88(2)	Livermorium	Lv	116	[293]
Uranium	U	92	238.028 91(3)	Tennessine	Ts	117	[294]
Neptunium	Np	93	[237]	Oganesson	Og	118	[294]

From the IUPAC data in 2007, the error of the relative atomic weight is in the parenthesis, and the data in square brackets are the mass numbers of the most stable isotopes

Periodic Table of the Elements

1	2											13	14	15	16	17	18
1 **H** hydrogen 1.008 [1.0078, 1.0082]																	2 **He** helium 4.0026
3 **Li** lithium 6.94 [6.938, 6.997]	4 **Be** beryllium 9.0122											5 **B** boron 10.81 [10.806, 10.821]	6 **C** carbon 12.011 [12.009, 12.012]	7 **N** nitrogen 14.007 [14.006, 14.008]	8 **O** oxygen 15.999 [15.999, 16.000]	9 **F** fluorine 18.998	10 **Ne** neon 20.180
11 **Na** sodium 22.990	12 **Mg** magnesium 24.305 [24.304, 24.307]	3	4	5	6	7	8	9	10	11	12	13 **Al** aluminium 26.982	14 **Si** silicon 28.085 [28.084, 28.086]	15 **P** phosphorus 30.974	16 **S** sulfur 32.06 [32.059, 32.076]	17 **Cl** chlorine 35.45 [35.446, 35.457]	18 **Ar** argon 39.95 [39.792, 39.963]
19 **K** potassium 39.098	20 **Ca** calcium 40.078(4)	21 **Sc** scandium 44.956	22 **Ti** titanium 47.867	23 **V** vanadium 50.942	24 **Cr** chromium 51.996	25 **Mn** manganese 54.938	26 **Fe** iron 55.845(2)	27 **Co** cobalt 58.933	28 **Ni** nickel 58.693	29 **Cu** copper 63.546(3)	30 **Zn** zinc 65.38(2)	31 **Ga** gallium 69.723	32 **Ge** germanium 72.630(8)	33 **As** arsenic 74.922	34 **Se** selenium 78.971(8)	35 **Br** bromine 79.904 [79.901, 79.907]	36 **Kr** krypton 83.798(2)
37 **Rb** rubidium 85.468	38 **Sr** strontium 87.62	39 **Y** yttrium 88.906	40 **Zr** zirconium 91.224(2)	41 **Nb** niobium 92.906	42 **Mo** molybdenum 95.95	43 **Tc** technetium	44 **Ru** ruthenium 101.07(2)	45 **Rh** rhodium 102.91	46 **Pd** palladium 106.42	47 **Ag** silver 107.87	48 **Cd** cadmium 112.41	49 **In** indium 114.82	50 **Sn** tin 118.71	51 **Sb** antimony 121.76	52 **Te** tellurium 127.60(3)	53 **I** iodine 126.90	54 **Xe** xenon 131.29
55 **Cs** caesium 132.91	56 **Ba** barium 137.33	57–71 lanthanoids	72 **Hf** hafnium 178.49(2)	73 **Ta** tantalum 180.95	74 **W** tungsten 183.84	75 **Re** rhenium 186.21	76 **Os** osmium 190.23(3)	77 **Ir** iridium 192.22	78 **Pt** platinum 195.08	79 **Au** gold 196.97	80 **Hg** mercury 200.59	81 **Tl** thallium 204.38 [204.38, 204.39]	82 **Pb** lead 207.2	83 **Bi** bismuth 208.98	84 **Po** polonium	85 **At** astatine	86 **Rn** radon
87 **Fr** francium	88 **Ra** radium	89–103 actinoids	104 **Rf** rutherfordium	105 **Db** dubnium	106 **Sg** seaborgium	107 **Bh** bohrium	108 **Hs** hassium	109 **Mt** meitnerium	110 **Ds** darmstadtium	111 **Rg** roentgenium	112 **Cn** copernicium	113 **Nh** nihonium	114 **Fl** flerovium	115 **Mc** moscovium	116 **Lv** livermorium	117 **Ts** tennessine	118 **Og** oganesson

Key:
atomic number
Symbol
name
conventional atomic weight
standard atomic weight

57 **La** lanthanum 138.91	58 **Ce** cerium 140.12	59 **Pr** praseodymium 140.91	60 **Nd** neodymium 144.24	61 **Pm** promethium	62 **Sm** samarium 150.36(2)	63 **Eu** europium 151.96	64 **Gd** gadolinium 157.25(3)	65 **Tb** terbium 158.93	66 **Dy** dysprosium 162.50	67 **Ho** holmium 164.93	68 **Er** erbium 167.26	69 **Tm** thulium 168.93	70 **Yb** ytterbium 173.05	71 **Lu** lutetium 174.97
89 **Ac** actinium	90 **Th** thorium 232.04	91 **Pa** protactinium 231.04	92 **U** uranium 238.03	93 **Np** neptunium	94 **Pu** plutonium	95 **Am** americium	96 **Cm** curium	97 **Bk** berkelium	98 **Cf** californium	99 **Es** einsteinium	100 **Fm** fermium	101 **Md** mendelevium	102 **No** nobelium	103 **Lr** lawrencium